T0338180

Plate and Shell Structures

Plate and Shell Structures

Selected Analytical and Finite Element Solutions

Maria Radwańska
Anna Stankiewicz
Adam Wosatko
Jerzy Pamin
Cracow University of Technology, Poland

WILEY

This edition first published 2017
© 2017 John Wiley & Sons Ltd

Registered office
John Wiley & Sons Ltd, The Atrium, Southern Gate, Chichester, West Sussex, PO19 8SQ, United Kingdom

For details of our global editorial offices, for customer services and for information about how to apply for permission to reuse the copyright material in this book please see our website at www.wiley.com.

The right of Maria Radwańska, Anna Stankiewicz, Adam Wosatko and Jerzy Pamin to be identified as the authors of this work has been asserted in accordance with the Copyright, Designs and Patents Act 1988.

All rights reserved. No part of this publication may be reproduced, stored in a retrieval system, or transmitted, in any form or by any means, electronic, mechanical, photocopying, recording or otherwise, except as permitted by the UK Copyright, Designs and Patents Act 1988, without the prior permission of the publisher.

Wiley also publishes its books in a variety of electronic formats. Some content that appears in print may not be available in electronic books.

Designations used by companies to distinguish their products are often claimed as trademarks. All brand names and product names used in this book are trade names, service marks, trademarks or registered trademarks of their respective owners. The publisher is not associated with any product or vendor mentioned in this book.

Limit of Liability/Disclaimer of Warranty: While the publisher and author have used their best efforts in preparing this book, they make no representations or warranties with respect to the accuracy or completeness of the contents of this book and specifically disclaim any implied warranties of merchantability or fitness for a particular purpose. It is sold on the understanding that the publisher is not engaged in rendering professional services and neither the publisher nor the author shall be liable for damages arising herefrom. If professional advice or other expert assistance is required, the services of a competent professional should be sought

Library of Congress Cataloging-in-Publication data applied for

ISBN: 9781118934548

A catalogue record for this book is available from the British Library.

Cover image: Jasmin Kämmerer/EyeEm/Gettyimages

Typeset in 10/12pt Warnock Pro by SPi Global, Chennai, India
Printed and bound in Malaysia by Vivar Printing Sdn Bhd

10 9 8 7 6 5 4 3 2 1

To our families

Contents

Preface

This book deals with the mechanics and numerical simulations of plates and shells, which are flat and curved thin-walled structures, respectively (called shell structures for short in this book). They have very important applications as complete structures or structural elements in many branches of engineering. Examples of shell structures in civil and mechanical engineering include slabs, vaults, roofs, domes, chimneys, cooling towers, pipes, tanks, containers and pressure vessels; in shipbuilding – ship and submarine hulls, in the vehicle and aerospace industries – automobile bodies and tyres and the wings and fuselages of aeroplanes.

The scope of the book is limited to the presentation of the theory of elastic plates and shells undergoing small deformation (thus assuming linear constitutive and kinematic equations).

The book is aimed at the large international community of engineering students, university teachers, professional engineers and researchers interested in the mechanics of shell structures, as well as developers testing new simulation software. The book can be the basis of an intermediate-level course on (computational) mechanics of shell structures at the level of doctoral, graduate and undergraduate studies. The reader should have the basic knowledge of the strength of materials, theory of elasticity, structural mechanics and FEM technology; basic information in these areas is not repeated in the book.

The strength of the book results from the fact that it not only provides the theoretical formulation of fundamental problems of mechanics of plates and shells, but also several examples of analytical and numerical solutions for different types of shell structures. The book also contains some advanced aspects related to the stability analysis and a brief description of classical and modern finite element formulations for plates and shells, including the discussion of mixed/hybrid models and so-called locking phenomena.

The book contains a comprehensive presentation of the theory of elastic plates and shells, formulations and solutions of fundamental mechanical problems (statics, stability, free vibrations) for these structures using exact approaches and computational (approximate) methods, with emphasis on modern capabilities of the finite element (FE) technology. In the book we introduce a large number of examples that illustrate various physical phenomena associated with the behaviour of shell structures under external actions. Comparisons of analytical and numerical solutions are given for several benchmark tests. The book includes plenty of boxes and tables that contain sets of formulae or data and check values describing the examples. They help the reader to find and integrate the information provided and draw conclusions.

The authors are researchers and teachers from the Institute for Computational Civil Engineering of Cracow University of Technology. They have done research on structural mechanics for years, in particular on the theories and advanced computational methods for shell structures, and they also have a long history of teaching the subject to students and practitioners. The selection of the contents of the book is based on this experience. The motivation to write the present book has also come from the fact that there are no books that contain, in one volume, the foundation of the theory and solutions of selected problems using simultaneously analytical and numerical methods.

Following a sequence of subjects: mathematics, theoretical mechanics, strength of materials, structural mechanics, computer science, numerical methods and the finite element method – we have developed a comprehensive course on the mechanics of shell structures. This course contains: (i) discussion of the assumptions and limits of applicability of selected theories on which mathematical models are based, (ii) choice of a method to solve the problem efficiently, (iii) analytical and/or numerical calculations simulating physical phenomena or processes, (iv) confrontation of the results of theoretical and numerical analysis and (v) evaluation of the calculation methodology and results.

Maria Radwańska and Jerzy Pamin were members of Professor Zenon Waszczyszyn's research team, who implemented the finite element code ANKA for buckling and nonlinear analysis of structures at the end of the twentieth century. This resulted in the 1994 Elsevier book: Waszczyszyn, Z. and Cichoń, Cz. and Radwańska, M., *Stability of Structures by Finite Element Method*.

Next, we briefly describe the contents of the book, which is divided into five parts. Part 1 is the introductory part that gives a compact encyclopedic overview of the fundamentals of the theory and modelling of plates and shells in the linear elastic range. A description of static analysis of (plane) plates is contained in Part 2 and of (curved) shells in Part 3. Part 4 includes information on the selected problems of buckling and free vibrations of shell structures. In Part 5, the authors discuss the general aspects of finite element analysis, including the modelling process, evaluation of the quality of finite elements and accuracy of solutions, Part 5 also contains a brief presentation of advanced formulations of finite elements for plates and shells.

While working on the book, we felt special gratitude to two of our teachers: Professors Zenon Waszczyszyn and Michał Życzkowski, who we always thought of as scientific authorities in the field of structural mechanics. In particular, we are deeply indebted to Professor Zenon Waszczyszyn for his invaluable contribution to our knowledge, motivation to do research and to participate in high-level university education. Under his guidance we got to know the theory of plates and shells, computational mechanics applied in civil engineering and modern numerical methods; in particular, the finite element method.

The authors wish to express their appreciation to several colleagues from the Institute for Computational Civil Engineering for discussions and help during the preparation of the book, in particular to A. Matuszak, E. Pabisek, P. Pluciński, R. Putanowicz and T. Żebro. We also record our gratitude to our students who cooperated with us in the computation of numerous examples: M. Abramowicz, M. Bera, I. Bugaj, M. Florek, S. Janowiak, A. Kornaś and K. Kwinta.

Notation

Detailed notation for theoretical analysis

Indices

$\alpha, \beta = 1, 2$	Greek indices (for curvature lines and surface coordinates)
$i, j, k = 1, 2, 3$	Latin indices (for 3D space)
n, m, t	indices for membrane, bending, transverse shear states
$j = 0, 1, 2, \ldots$	number of a components of trigonometric series or number of circumferential wave (half-wave)
(i, j)	indices describing number of waves of deformation in two directions

Coefficients and variables

α_T	coefficient of thermal expansion of a material
$A_\alpha : A_1, A_2$	Lame coefficients
$\hat{\mathbf{b}} = [\hat{b}_x, \hat{b}_y, \hat{b}_z]^{\mathrm{T}}$	prescribed body forces
$b_{\alpha\beta}$	components of II (second) metric tensor
$\beta = \sqrt{\frac{1}{R\,h}}\,\sqrt[4]{3(1-\nu^2)}$	coefficient in equation of local bending state in cylindrical shell
C^0, C^*	initial and current configuration of a body (shell)
$\mathbf{C}, \mathbf{E} = \mathbf{C}^{-1}$	matrices of local flexibility and stiffness in constitutive equations
$D^n = \frac{Eh}{1-\nu^2}$	cross-sectional stiffness in membrane state
$D^m = \frac{Eh^3}{12(1-\nu^2)}$	cross-sectional stiffness in bending state
$D^t = \frac{k\,Eh}{2(1+\nu)}, k = \frac{5}{6}$	cross-sectional stiffness in transverse shear state
$(\mathbf{e}_\alpha, \mathbf{n}), (\mathbf{e}_\alpha^{(z)}, \mathbf{n})$	local base versors on middle surface and on equidistant surface from the middle surface in initial configuration
$(\mathbf{e}_\alpha^*, \mathbf{n}^*)$	local base versors on middle surface in current configuration
$\mathbf{e}^{\mathrm{T}} = [\mathbf{e}^n, \mathbf{e}^m, \mathbf{e}^t]$	generalized strain vector (membrane, bending and transverse shear components)

$\mathbf{e}^n = [\varepsilon_{11}, \varepsilon_{22}, \gamma_{12}]^T$	membrane strain vector
$\varepsilon_{11}, \varepsilon_{22}, \gamma_{12} = \varepsilon_{12} + \varepsilon_{21}$	membrane strains: normal and shear in middle surface
$\varepsilon_x, \varepsilon_\theta, \gamma_{x\theta}$	membrane strains in cylindrical system
$\varepsilon_\varphi, \varepsilon_\theta, \gamma_{\varphi\theta}$	membrane strains in spherical system
$\mathbf{e}^m = [\kappa_{11}, \kappa_{22}, \chi_{12}]^T$	bending strain vector
$\kappa_{11}, \kappa_{22}, \chi_{12} = \kappa_{12} + \kappa_{21}$	bending strains: changes of curvature and warping of middle surface
$\kappa_x, \kappa_\theta, \chi_{x\theta}$	bending strains in cylindrical system
$\kappa_\varphi, \kappa_\theta, \chi_{\varphi\theta}$	bending strains in spherical system
$\mathbf{e}^t = [\gamma_{1z}, \gamma_{2z}]^T$	transverse shear strain vector
$\gamma_{xz}, \gamma_{\theta z}$	transverse shear strains in cylindrical system
$\gamma_{\varphi z}, \gamma_{\theta z}$	transverse shear strains in spherical system
$E, \nu, G = E/(2 + 2\,\nu)$	material constants: Young's modulus, Poisson's ratio, Kirchhoff's modulus
f	rise of shallow shell
F	Airy's stress function
$\mathbf{i}_k$	base versors related to Cartesian coordinates x^k
$g_{\alpha\beta}$	components of I (first) metric tensor
h	thickness of shell
K	Gaussian curvature of surface
$\lambda = \pi/\beta$	length of half-wave for exponential-trigonometric function in local membrane-bending state of cylindrical shell
$m_{11}, m_{22}, m_{12} = m_{21}$	moments: bending and twisting in middle surface
$m_x, m_\theta, m_{x\theta}$	moments in cylindrical system
$m_\varphi, m_\theta, m_{\varphi\theta}$	moments in spherical system
$n_{11}, n_{22}, n_{12} = n_{21}$	membrane forces: normal and tangential in middle surface
$n_x, n_\theta, n_{x\theta}$	membrane forces in cylindrical system
$n_\varphi, n_\theta, n_{\varphi\theta}$	membrane forces in spherical system
n_I, n_{II}, m_I, m_{II}	principal membrane forces and bending moments
$\tilde{n}_{\nu s}, \tilde{t}_\nu$	effective boundary forces (tangential membrane and transverse shear)
$n_\nu, \tilde{n}_{\nu s}, \tilde{t}_\nu, m_\nu$	generalized boundary forces
$\hat{n}_\nu, \hat{n}_{\nu s}, \hat{t}_\nu, \hat{m}_\nu$	prescribed generalized boundary loads
ν, s, n	directions of boundary base vectors
$\hat{\mathbf{p}} = [\hat{p}_1, \hat{p}_2, \hat{p}_n]^T$	vector of prescribed surface loads
$\hat{\mathbf{p}}_b, \hat{\mathbf{u}}_b$	vectors of prescribed generalized boundary loads and displacements
$\hat{P}_i$	prescribed concentrated force in corner i
$\Pi, \Pi^{(z)}$	middle and equidistant surfaces in initial configuration
$\Pi^*, \Pi^{*\,(z)}$	middle and equidistant surfaces in current configuration

Π_ν, Π_s	cross-sectional planes: normal and tangent to middle surface
$\Pi_\alpha: \Pi_1, \Pi_2$	two transverse cross-sectional planes normal to middle surface
Π, U, W	total potential energy, internal energy, external load work
$r = f(x), r = f(\varphi)$	meridian equation for axisymmetric shell
$R_\alpha: R_1, R_2$	principal curvature radii for middle surface of a shell
s	arch coordinate for a line on surface
$\mathbf{s}^\mathrm{T} = [\mathbf{s}^n, \mathbf{s}^m, \mathbf{s}^t]$	vector of generalized resultant forces for membrane, bending and transverse shear states
$\mathbf{s}_b = [n_\nu, \tilde{n}_{\nu s}, \tilde{t}_\nu, m_\nu]^\mathrm{T}$	vector of generalized boundary forces
$\hat{\mathbf{s}}_b = [\hat{n}_\nu, \hat{n}_{\nu s}, \hat{t}_\nu, \hat{m}_\nu]^\mathrm{T}$	vector of presribed boundary forces
$\mathbf{s}^n = \mathbf{n} = [n_{11}, n_{22}, n_{12}]^\mathrm{T}$	vector of membrane forces
$\mathbf{s}^m = \mathbf{m} = [m_{11}, m_{22}, m_{12}]^\mathrm{T}$	vector of bending and twisting moments
$\mathbf{s}^t = \mathbf{t} = [t_1, t_2]^\mathrm{T}$	vector of transverse shear forces
$\varsigma = \sqrt{\frac{R}{h}}\sqrt[4]{3(1-\nu^2)}$	coefficient in equation of local bending state in spherical shell
T_i	effective force in a corner used in static boundary conditions
$\vartheta = [\vartheta_1, \vartheta_2, \vartheta_n]^\mathrm{T}$	vector of rotations
$\vartheta_\alpha: \vartheta_1, \vartheta_2$	two rotations of normal to middle surface
ϑ_n	rotation around normal to middle surface
$\vartheta_x = \varphi_y, \vartheta_y = -\varphi_x$	two rotations of normal to middle plane of plate under bending in Cartesian system (two alternative notations)
$\sigma_{nn}, \sigma_{ns}, \sigma_{nz}$	stresses: in-plane normal, in-plane tangential, transverse shear
$u = u_x, v = u_y, w$	translations with respect to local system (x, y, z)
U, V, W	translations with respect to global system (X, Y, Z)
$\mathbf{u} = [u_1, u_2, w, \vartheta_1, \vartheta_2, \vartheta_n]^\mathrm{T}$	generalized displacement vector
$\mathbf{u} = [u_1, u_2, w]^\mathrm{T}$	translation vector in three-parameter thin shell theory
$\mathbf{u} = [w, \vartheta_1, \vartheta_2]^\mathrm{T}$	generalized displacement vector in three-parameter moderately thick plate theory
$\mathbf{u} = [u_1, u_2, w, \vartheta_1, \vartheta_2]^\mathrm{T}$	generalized displacement vector in five-parameter moderately thick shell theory
$\mathbf{u}_b = [u_\nu, u_s, w, \vartheta_\nu]$	vector of generalized boundary displacements
$\hat{\mathbf{u}}_b = [\hat{u}_\nu, \hat{u}_s, \hat{w}, \hat{\vartheta}_\nu]$	vector of prescribed generalized boundary displacements
U^n, U^m, U^t	strain energy in membrane, bending and transverse shear states
$\xi_\alpha: \xi_1, \xi_2$	curvilinear surface coordinates on middle surface $z = 0$
$\xi_\alpha = \text{const.}$	coordinate lines on middle surface
$\xi_1 = x, \xi_2 = y$	Cartesian coordinates
$\xi_1 = \varphi, \xi_2 = \theta$	spherical coordinates

$\xi_1 = r, \xi_2 = \theta$	polar coordinates
$\xi_1 = x, \xi_2 = \theta$	cylindrical coordinates
z	coordinate in direction normal to the middle surface Π (distance of equidistant surface $\Pi^{(z)}$ from middle surface Π, $z = 0$ corresponds to the middle surface Π)
$x^k: x^1, x^2, x^3$	Cartesian coordinates with respect to versors $\mathbf{i}_k$
(X, Y, Z)	Cartesian coordinate system
Ω	problem domain
$\partial\Omega$	boundary of domain
$\partial\Omega_\sigma, \partial\Omega_u$	boundary with prescribed loads and displacements, respectively
T_r	reference temperature
T_0	temperature on middle surface
$\Delta T_0 = T_0 - T_\mathrm{r}$	temperature change (independent of z) with respect to reference temperature
$\Delta T_h = \Delta T(h/2) - \Delta T(-h/2)$	temperature difference between limiting shell surfaces $z = \pm h/2$

Detailed notation for numerical analysis

Indices

e	index of finite element (FE)
(ef)	index of interelement boundary
n, *node*	index of FE node

Abbreviations

NNDOF, NEDOF, NSDOF	number of degrees of freedom (dofs) for node, element and structure
NSE	number of FEs in a structure
NEN, NSN	number of FE nodes and of structure nodes
NGP	number of Gauss points

Coefficients and variables

$\boldsymbol{\alpha}_u, \boldsymbol{\alpha}_\sigma, \boldsymbol{\alpha}_\varepsilon$	mathematical dofs for interpolation of displacement, stress, strain fields
$\mathbf{B}^n, \mathbf{B}^m, \mathbf{B}^t$	matrices in kinematic relations for membrane, bending and transverse shear states
$\mathbf{D}^n, \mathbf{D}^m, \mathbf{D}^t$	matrices in constitutive equations for membrane, bending and transverse shear states

$\mathbf{f}^e, \mathbf{f}_b^e$	vector of substitute nodal forces which represent loads in FE and on FE boundary
$\mathbf{F}$	global vector of substitute nodal forces after assembly process
$I_p[\mathbf{u}], I_c[\boldsymbol{\sigma}]$	potential and complementary energy functionals
$I_{p,m}, I_{c,m}$	modified potential and complementary energy functionals
$I_{H\text{-}R}[\mathbf{u}, \boldsymbol{\sigma}]$	two-field Hellinger–Reissner functional
$I_{H\text{-}W}[\mathbf{u}, \boldsymbol{\sigma}, \boldsymbol{\epsilon}]$	three-field Hu–Washizu functional
$G^{(ef)}[\boldsymbol{\sigma}, \mathbf{u}^{(ef)}]$	component added to functional and associated with the equilibrium of tractions on interelement boundary
$H^{(ef)}[\mathbf{u}, \mathbf{t}^{(ef)}]$	component added to functional and associated with the continuity of displacements on interelement boundary
J, $\det \mathbf{J}$	jacobian, determinant of Jacobi matrix
$\mathbf{k}^{en}, \mathbf{k}^{em}, \mathbf{k}^{et}$	element stiffness matrix for membrane, bending and transverse shear states
$\mathbf{k}_\sigma^e$	stress stiffness matrix in initial and linearized buckling analysis
$\mathbf{k}_u^e$	displacement stiffness matrix for FE in linearized buckling analysis
$\mathbf{L}$	matrix of differential operators in kinematic strain-displacement equations $\varepsilon = \mathbf{L}\,\mathbf{u}$
$\mathbf{N}$	matrix of shape functions used for displacement field approximation
$\mathbf{N}_u, \mathbf{N}_\sigma, \mathbf{N}_\varepsilon$	matrices for approximation of displacement, stress, strain fields in two- or three-field formulation in mixed FEs
$\mathbf{P}^*, \mathbf{Q}^*$	vectors of reference loads and displacements for one-parameter loading process
$\mathbf{q}^e = \mathbf{q}_u^e$	element generalized displacement vector for displacement-based FE model
$\mathbf{q}_{node}$	nodal generalized displacement (dof) vector for displacement-based FE model
$\mathbf{q}_u^e, \mathbf{q}_\sigma^e, \mathbf{q}_\varepsilon^e$	vectors of element generalized displacement, stress and strain dofs, respectively, for different FE models
$\mathbf{q}_u^{(ef)}, \mathbf{q}_t^{(ef)}$	vectors of generalized displacement or, respectively, traction dofs on interelement boundary
$\mathbf{Q}$	vector of generalized displacements for structure
$\mathbf{u}(\mathbf{x}), \boldsymbol{\sigma}(\mathbf{x}), \boldsymbol{\varepsilon}(\mathbf{x})$	displacement, stress, strain fields approximated within FE domain
$\mathbf{u}^{(ef)}(s), \mathbf{t}^{(ef)}(s)$	displacement and traction function approximated along interelement boundary
$\mathbf{R}_{supp}$	support reaction vector for structure
ξ, η, ζ	natural normalized dimensionless coordinates
$\Omega^e, \partial\Omega^e$	area and boundary of FE
$\partial\Omega^{(ef)}$	interelement boundary

Conversions between imperial and metric system units

Quantity	Imperial units	International System of Units (SI)	
length	1 in.	= 2.54 cm	= 0.0254 m
	1 ft.	= 30.48 cm	= 0.3048 m
area	1 in.^{2}	= 6.45 cm^2	= 0.000645 m^2
	1 ft^2	= 929 cm^2	= 0.0929 m^2
force	1 lb-f = 1 lbf	= 4.45 N	= 0.00445 kN
moment	1 lbf-in.	= 11.31 Ncm	= 0.0001131 kNm
intensity of membrane force	1 lbf/in.	= 1.751 N/cm	= 0.175 kN/m
intensity of moment	1 lbf-in./in.	= 4.45 N/cm/cm	= 0.00445 kNm/m
pressure	1 psi = 1 lbf/in.^{2}	= 0.690 N/cm^2	= 6.90 kN/m^2

Part 1

Fundamentals: Theory and Modelling

1

General Information

1.1 Introduction

In the classification of mechanical structures, somewhere between one-dimensional (1D) bar structures and three-dimensional (3D) solid structures, a class of two-dimensional (2D) plates and shells (thin-walled flat and curved structures) can be distinguished. The attention is focused on a deformable solid body, which is limited by two surfaces (top and bottom) and lateral surfaces, see Figure 1.1. The distance between the top and bottom surfaces, identified as the thickness, is small compared to the other dimensions of the body (e.g. radius of curvature or span), measured referring to the so-called primary surface (2D physical model), most often taken as the middle surface defined as equidistant from the top and bottom surfaces.

The following, generally accepted nomenclature is going to be used throughout the book:

- shells = thin-walled curved shells
- curved membranes = special shells that have no bending rigidity
- plates = thin plane structures that have some subclasses:
  - flat membranes = plates with load in the middle plane, sometimes also called panels
  - plates under bending = plates with transverse load (normal to the middle plane), sometimes also called slabs

In the general description, for all these classes we will use the name 'shell structures' or, in brief, 'shells'. In other words, we understand that shell structures can be flat.

Scientists, teachers, students, engineers and even the authors of software are interested in the mechanics of plates and shells. Due to the variety of potential users, the following variants of the mechanical theory have evolved:

- general advanced tensorial shell theory
- technical (engineering) shell theory

The scope of this book is limited to the case of linear constitutive and kinematic equations.

The theory is the basis for the construction of appropriate mathematical models (sets of differential and algebraic equations) and is associated with the calculation method that can be used to solve general or particular mechanical problems.

Plate and Shell Structures: Selected Analytical and Finite Element Solutions, First Edition.
Maria Radwańska, Anna Stankiewicz, Adam Wosatko and Jerzy Pamin.
© 2017 John Wiley & Sons Ltd. Published 2017 by John Wiley & Sons Ltd.

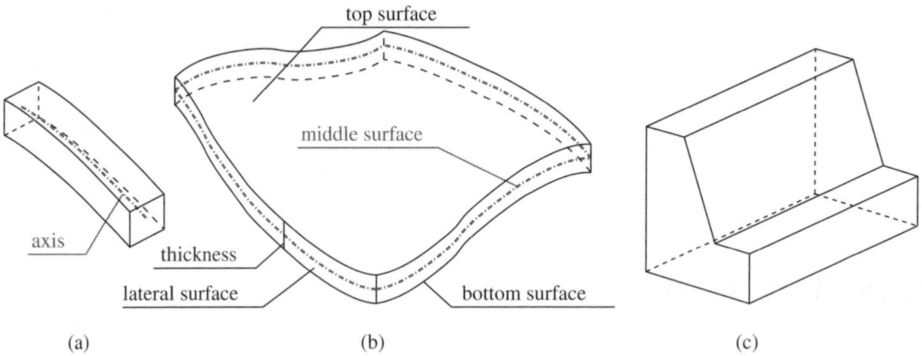

Figure 1.1 Structures: (a) bar (1D), (b) surface (2D) and (c) solid (3D)

In Section 1.2, encyclopedic information on the development of theories describing elastic plates and shells is included.

The description of shell structures, which makes them different from bar (1D) and solid (3D) bodies, must contain the following aspects:

- information on the coordinate systems and geometry of representative surfaces
- specification of kinematic constraints related to the mode of deformation
- definitions of so-called generalized strains with respect to the middle surface
- definitions of resultant forces and moments on the middle surface
- characteristics of fundamental stress and strain states

Detailed discussion is given in Sections 1.3 and 1.4.

A classification of plates and shells can be performed taking into account the slenderness (thickness to span ratio), the shape of the middle surface, the definitions and assumptions presented further in Section 1.4 and the character of stress distribution along the thickness, related to the stress state. In Section 1.5 and in Box 1.1, we present the classification of surface structures according to these aspects.

Thin-walled shell structures of various types are very important structural elements. Examples of shell structures can be encountered in civil and mechanical engineering (slabs, vaults, roofs, domes, chimneys, cooling towers, pipes, tanks, containers, pressure vessels), shipbuilding (ship hulls, submarine hulls) and in the vehicle and aerospace industries (car bodies and tyres, wings and fuselages of aeroplanes).

From an engineering point of view, it is necessary to predict different modes of behaviour of plates and shells under applied loading. In the case of a (flat) plate subjected to a transverse load, static equilibrium is preserved by the action of bending and twisting moments and transverse shear forces. On the other hand, the (curved) shell structure is able to carry the load inducing membrane tension or compression, distributed uniformly throughout the thickness (it is an optimal case from the viewpoint the material strength). This feature of shell structures makes them more economical and stiffer in comparison to plates.

Familiarity with the technical shell theory is necessary for engineers who are responsible for the safety of structures and are supposed to take into account various safety factors using computer-aided design.

Box 1.1 Summary of classification of shell structures

Thin plates for $\dfrac{h}{L_{min}} < \dfrac{1}{10}$

Moderately thick plates for $\dfrac{1}{10} \leq \dfrac{h}{L_{min}}$

Thin shells for $\dfrac{h}{R_{min}} \leq \dfrac{1}{20}$

Moderately thick shells for $\dfrac{1}{20} < \dfrac{h}{R_{min}} \leq \dfrac{1}{6}$

Thick shells for $\dfrac{1}{6} < \dfrac{h}{R_{min}}$

Shallow shell for $\dfrac{f}{L} < \dfrac{1}{5}$

h – thickness of plate or shell

L – characteristic dimension of plate or shell

L_{min} – the smallest dimension in the middle plane of a plate

R_{min} – smaller of two principal radii of curvature

f – distance of shell from the horizontal plane, of its projection, that is rise

w – representative deflection

Geometrically linear theory of plates with small deflections for $|w| < \dfrac{h}{5}$

von Kármán theory of plates with moderately large deflections for $|w| \approx h$

Geometrically nonlinear theory of plates with large deflections for $|w| > 5h$

As emphasized in Ramm and Wall (2004), shell structures exhibit the strong influence of initial geometry, slenderness, type of loading and boundary conditions on the deformation and load carrying capacity. Small variations or even imperfections of these parameters can change the structural response significantly and, in particular, cause loss of stability.

Shells are characterized by an advantageous ratio of stiffness to weight, which makes them suitable for lightweight and long-spanned structures. Moreover, optimal shells are designed to carry predominantly membrane forces with minimum bending effects. It is therefore extremely important to understand the principal mechanical features of plates and shells before using computer-aided design involving numerical simulation.

1.2 Review of Theories Describing Elastic Plates and Shells

The general description of the historical development of plate and shell theory, as well as details of specific theories are referred to in a lot of books and monographs. Here, the authors do not try to present the developmental trends of this branch of mechanics, even limiting interest to the theory of elastic plates and shells undergoing small deformations.

The beginnings of the linear theory of plates and shells date back to the nineteenth century, however, the vibration problem of bells was considered by Leonhard Euler in 1764. The name of Sophie Germain is associated with the theory of plates: in 1811 she submitted work on plates for a contest announced by the French Academy of Sciences.

Following the two encyclopaedic elaborations:

- Mechanics of Elastic Plates and Shells, vol. 8 in *Technical Mechanics* (Borkowski et al. 2001)
- Models and Finite Elements for Thin-walled Structures, Chapter 3 by Bischoff et al. in vol. 2 of *The Encyclopedia of Computational Mechanics* (Stein et al. 2004)

the authors of this book would like to mention the names of researchers associated with the theories of plates and shells from three different, consecutive periods (listing names in alphabetical order):

- nineteenth century:
  A. Cauchy, S. Germain, A.E. Green, G. Kirchhoff, A.H. Love, S.D. Poisson and L. Rayleigh
- first half of the twentieth century:
  E. Cosserat and F. Cosserat, A.L. Gol'denveizer, Th. von Kármán, S. Lévy, A.I. Lur'e and E. Reissner
- second half of the twentieth century:
  Y. Başar, B. Budiansky, L.H. Donnell, J.L. Ericksen, W. Flügge, J.M. Gere, K. Girkmann, K.Z. Golimov, R. Harte, Z. Kączkowski, W.T. Koiter, W.B. Krätzig, H. Kraus, R.D. Mindlin, K.M. Mushtari, P.M. Naghdi, F.I. Niordson, W. Nowacki, V.V. Novozhilov, W. Pietraszkiewicz, E. Ramm, J.L. Sanders, J.G. Simmonds, I. Szabó, S.P. Timoshenko, C. Truesdell, V.Z. Vlasov, W. Wunderlich, S. Woinowski-Krieger, Cz. Woźniak and W. Zerna

We also mention some previous works relevant to the subject of this book, dividing them into:

- books dealing with the basis of mechanics: Timoshenko and Goodier (1951), Fung (1965), Washizu (1975), Reddy (1986), Borkowski et al. (2001) and Stein et al. (2004)
- monographs related to the theories of plates and shells: Girkmann (1956), Timoshenko and Woinowsky-Krieger (1959), Kolkunov (1972), Nowacki (1980), Niordson (1985), Noor et al. (1989), Waszczyszyn and Radwańska (1995), Reddy (1999), Başar and Krätzig (2001), Borkowski et al. (2001), Reddy (2007), Radwańska (2009), Wiśniewski (2010) and Oñate (2013)

The general formulation of the theory of thin-walled structures is determined by their specific geometry with one dimension (thickness) much smaller in comparison to the other two dimensions. There are two essential concepts that can be used to formulate the mathematical description of the problem.

One possibility is to start from the equations of three-dimensional continuum, describing a body with a specified geometry. Applying a power series representation of certain quantities as a function of coordinate z (measured in the direction of a thickness) the reduction to a two-dimensional theory is performed. Using a specified number of terms of this representation a 2D problem with varying accuracy of approximation is obtained.

Alternatively, one can adopt suitable kinematic assumptions and treat a thin-walled structure as a two-dimensional continuum representation of a substitute problem, (see Borkowski et al. 2001). This option is associated with direct methods of formulating two-dimensional models of plates and shells, based on appropriate static and kinematic hypotheses. The approximation in this theory is that the deformed state of the shell is determined entirely by the configuration of its middle surface.

Beside the two approaches based on three-dimensional continuum mechanics or two-dimensional surface-based theories we mention a so-called Cosserat surface concept, see for instance Chapter 3 in vol. 2 of Stein et al. (2004). This approach is an extension of classical continuum formulation by adding information about the orientation of a material point equipped with rotational degrees of freedom.

Among the developed theories for shells a few specific approaches can be distinguished:

- general theory applying any parametrization of the curved middle surface
- theory that uses the orthogonal parameterization of the middle surface based on principal curvature coordinates
- general membrane-bending shell theory with or without the consideration of transverse shear deformation
- theories for particular cases of shells (e.g. for cylindrical or spherical shells of revolution)
- theory of plates
- theory of flat membranes

The full set of equations of the linear theory of shells, which contains Kirchhoff plate equations as a special case, are given in pages 173–174 of Love (1944). This theory is called the Kirchhoff–Love (K–L) theory of first approximation or order. In theory based on assumptions of K–L the effects of transverse shear and normal strains in the thickness direction are neglected. The weakening of these assumptions leads to enhanced variants of the equations, the so-called second and third approximations. This involves more complex forms of measurement of deformation and construction of constitutive equations. In fact, the first approximation theory is mathematically and physically incorrect. When the kinematic equations and constitutive equations, used in this approach, are substituted into the sixth equilibrium equation (expressing equilibrium of moments around the normal to the middle surface), the equation is not satisfied. The sixth equilibrium equation guarantees that all strains vanish for small rigid-body rotations of the shell.

The inconsistency of Love's first order theory was removed in the improved theory for thin shells by Sanders (1959), formulating the equations in principal curvature coordinates. For this new improved theory modified equilibrium equations, strain-displacement relations and boundary conditions were derived using the principle of virtual work. The detailed information about the basics of theory of Sanders is presented in Chapter 3.

Koiter checked and corrected Love's theory (see Koiter 1960). An assessment of the order of magnitude of the terms in Love's strain-energy expression was carried out. Appropriate consistent stress-strain relations for stress resultants and equilibrium equations in tensorial form were presented. In the theory the sixth equation of equilibrium is satisfied identically.

In work of Budiansky and Sanders (1963) the equations of the 'best' first-order linear elastic shell theory were formulated for shells of arbitrary shape in a coordinate system related to the middle surface using general tensor notation.

In the broad literature a variety of kinematic and constitutive equations can be found, because different simplifications were used in their derivation. The summary of various descriptions of the strain state and kinematic relations (even for linear analysis) is also presented in the work by Lewiński (1980). The following four essential features of the improved first approximation shell theory are cited here from this paper:

- matrices of generalized strains (membrane strains and changes of curvature) and stress resultants (forces and moments) are symmetric
- constitutive equations are decoupled
- the sixth equation of equilibrium is identically satisfied
- a rigid motion of the shell does not cause strains or stresses

Now, a little information about the classical three-parameter Sanders thin shell theory is given, because this formulation is applied in our book. The equations are considered to be the most suitable with respect to both theoretical and numerical applications. In the geometry description the orthogonality of the coordinate lines implies that the first metric tensor is diagonal, and the surface is described by only two Lame parameters and two radii of curvature (or curvatures themselves), see Subsection 1.3.2. The following fields are used in the shell problem description: translation(s), rotation(s), generalized strain(s) and stress resultants, all defined with respect to the two-dimensional middle surface. In this three-parameter thin shell theory three translations u_1, u_2, w are adopted as independent variables in the description of the deformation (see Subsection 1.4.1).

The five-parameter theory is used to describe moderately thick shells with five independent generalized displacements: three translations u_1, u_2, w and two rotations ϑ_1, ϑ_2 (see Subsection 1.4.2).

At this point the assumptions adopted in this book are specified:

- translations, rotations and strains are assumed to be small enough for nonlinear components in the kinematic and equilibrium equations to be omitted (thus taking into account only the first order terms)
- the initial undeformed configuration of the structure is the reference configuration
- the material is treated as isotropic linearly elastic, described by Hooke's constitutive equations, that is to define the material only two parameters are used: Young's modulus and Poisson's ratio

A more advanced tensorial formulation of the theory of shell structures can be found, for instance, in Başar and Krätzig (2001). The theoretical foundations there are coupled with:

- local formulation using differential and algebraic equations
- global formulations employing energy theorems and variational principles for plates and shells

In Chapter 3 of Vol. 2 of Stein et al. (2004), entitled 'Models and Finite Elements for Thin-walled Structures', both the mathematical and mechanical foundations of the theory of plates and shells and the description of FE formulations are presented. The chapter includes an extensive derivation of kinematic equations and strains, constitutive equations and stresses as well as the parametrization of displacements and rotations, both in linear and nonlinear range. The long list of references contains 211 items from 1833 to 2003.

Most recent efforts of scientists are aimed at the analysis of:

- anisotropic, composite (in particular layered) shells
- shells undergoing large deformations (with varying magnitude of displacements, rotations and strains)
- shell in inelastic (in particular plastic) states

However, these issues are beyond the scope of this book. The reader is referred to the following works on nonlinear theories of plates and shells: Woźniak (1966), Pietraszkiewicz (1977, 1979, 2001), Crisfield (1982), Hinton et al. (1982), Kleiber (1985), Borkowski et al. (2001), Wiśniewski (2010), de Borst et al. (2012).

1.3 Description of Geometry for 2D Formulation

The description of the geometry of 2D surfaces is based on the works by Waszczyszyn and Radwańska (1995) and Radwańska (2009).

1.3.1 Coordinate Systems, Middle Surface, Cross Section, Principal Coordinate Lines

The analysis of thin and moderately thick shell structures is most often performed with respect to the middle surface, that is to a geometrically two-dimensional object; only thick shells are treated as three-dimensional bodies.

The geometry of a shell structure is defined when the shape of the middle surface, the boundary contour and the thickness distribution have been specified. In the theoretical consideration we assume for simplicity that the thickness is constant.

Two families of curves are introduced. They are parametrized with so-called curvilinear coordinates ξ_1, ξ_2, see Figure 1.2a, used for an explicit definition of the position of a point on the surface, In most cases a general curvilinear coordinate system will be employed, and further a discussion of particular cases will be provided, for instance the cylindrical (x, θ), spherical (φ, θ) or Cartesian (x, y) coordinate systems will be applied. In the two-dimensional description of shells analogous pairs of variables (e.g. R_α) or pairs of formulae (e.g. $\mathrm{d}s_\alpha = A_\alpha \mathrm{d}\xi_\alpha$) will be used, where the Greek index α represents numbers 1 or 2.

On the middle surface the so-called principal curvature lines related to principal curvature radii are specified. Many equations formulated for particular shells refer to these principal (extreme) curvature lines.

At any point P on the middle surface a cross section can be defined. We consider two normal section planes Π_1 and Π_2, see Figure 1.2b. These planes are perpendicular to each other and their intersections with the middle surface generate arc segments of unit length $ds_\alpha = 1$. We emphasize that the intersection of the planes Π_α is a

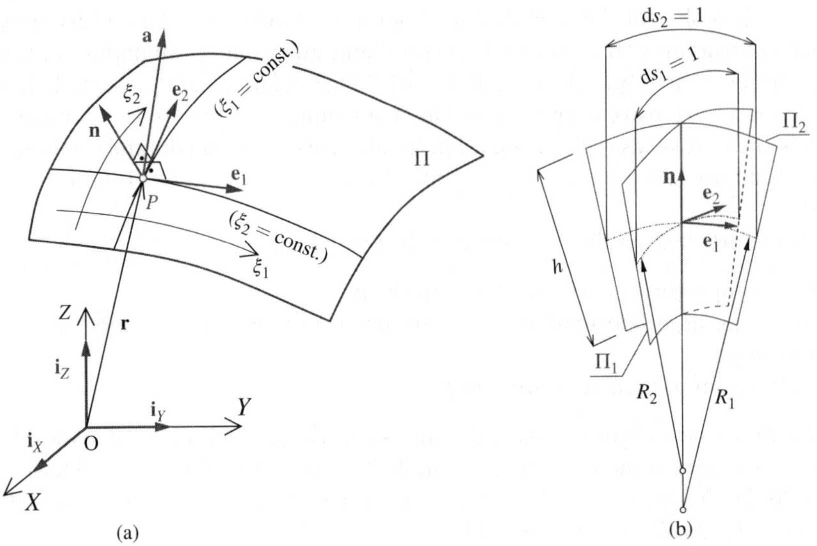

Figure 1.2 (a) A middle surface with curvilinear coordinates ξ_α and local base vectors $\mathbf{e}_\alpha$, $\mathbf{n}$ at point P, (b) straight fibre – intersection of planes Π_α, $\alpha = 1, 2$. Source: Waszczyszyn and Radwańska (1995). Reproduced with permission of Waszczyszyn.

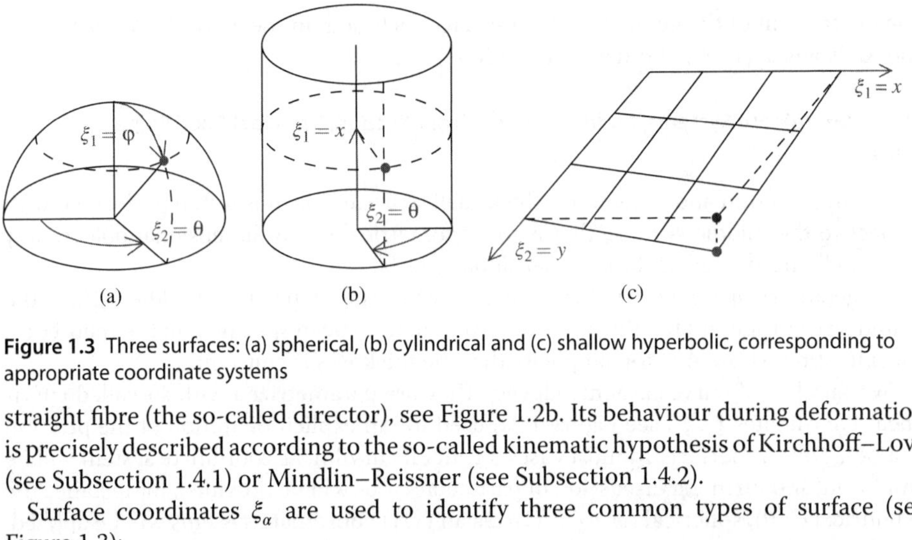

Figure 1.3 Three surfaces: (a) spherical, (b) cylindrical and (c) shallow hyperbolic, corresponding to appropriate coordinate systems

straight fibre (the so-called director), see Figure 1.2b. Its behaviour during deformation is precisely described according to the so-called kinematic hypothesis of Kirchhoff–Love (see Subsection 1.4.1) or Mindlin–Reissner (see Subsection 1.4.2).

Surface coordinates ξ_α are used to identify three common types of surface (see Figure 1.3):

(a) spherical surface described in a spherical coordinate system (φ, θ)
(b) cylindrical surface in a cylindrical system (x, θ)
(c) shallow ruled hyperbolic surface in a Cartesian system (x, y)

1.3.2 Geometry of Middle Surface

For point P on the middle surface Π the connection between global Cartesian coordinates X, Y, Z and local curvilinear coordinates ξ_1, ξ_2 is expressed by the relation

$$\mathbf{r} = X\,\mathbf{i}_X + Y\,\mathbf{i}_Y + Z\,\mathbf{i}_Z \tag{1.1}$$

where:

$$X = f_1(\xi_1, \xi_2), \qquad Y = f_2(\xi_1, \xi_2), \qquad Z = f_3(\xi_1, \xi_2) \tag{1.2}$$

On the middle surface, a two-dimensional segment $P - P_1 - M - P_2$ is identified, resulting from the intersection of four lines $\xi_1 = \text{const.}$, $\xi_1 + \mathrm{d}\xi_1 = \text{const.}$, $\xi_2 = \text{const}$, $\xi_2 + \mathrm{d}\xi_2 = \text{const.}$ (see Figure 1.4a). Next, curve l is considered. The curve, parametrized by coordinate λ, passes through points P and M that are located on the elementary surface subdomain, with lengths of sides $\mathrm{d}s_\alpha$, $\alpha = 1, 2$ measured by so-called Lame parameters A_α, which are magnitudes of the tangential vectors $\mathbf{r}_{,\alpha}$:

$$\mathrm{d}s_\alpha = A_\alpha \mathrm{d}\xi_\alpha, \qquad A_\alpha = |\mathbf{r}_{,\alpha}| = |\mathbf{g}_\alpha|, \qquad (\ )_{,\alpha} = \frac{\partial(\ )}{\partial \xi_\alpha}, \qquad \alpha = 1, 2 \tag{1.3}$$

$$\mathbf{r} = \mathbf{r}[\xi_1(\lambda), \xi_2(\lambda)], \qquad \mathrm{d}\mathbf{r} = \left(\frac{\partial \mathbf{r}}{\partial \xi_1} \frac{\mathrm{d}\xi_1}{\mathrm{d}\lambda} + \frac{\partial \mathbf{r}}{\partial \xi_2} \frac{\mathrm{d}\xi_2}{\mathrm{d}\lambda} \right) \mathrm{d}\lambda = \mathbf{r}_{,1}\, \mathrm{d}\xi_1 + \mathbf{r}_{,2}\, \mathrm{d}\xi_2 \tag{1.4}$$

The length of the arch between points P and M on line l is calculated using the formula

$$\begin{aligned}(\mathrm{d}s)^2 &= \mathbf{r}_{,1} \cdot \mathbf{r}_{,1}\, (\mathrm{d}\xi_1)^2 + 2\mathbf{r}_{,1} \cdot \mathbf{r}_{,2}\, \mathrm{d}\xi_1 \mathrm{d}\xi_2 + \mathbf{r}_{,2} \cdot \mathbf{r}_{,2}\, (\mathrm{d}\xi_2)^2 \\ &= (A_1)^2\, (\mathrm{d}\xi_1)^2 + 2A_1 A_2 \cos(\mathbf{g}_1, \mathbf{g}_2)\, \mathrm{d}\xi_1 \mathrm{d}\xi_2 + (A_2)^2\, (\mathrm{d}\xi_2)^2\end{aligned} \tag{1.5}$$

The product of tangential vectors $\mathbf{g}_\alpha$ defines the first (I) metric tensor $g_{\alpha\beta}$

$$g_{\alpha\beta} = \mathbf{g}_\alpha \cdot \mathbf{g}_\beta = \mathbf{r}_{,\alpha} \cdot \mathbf{r}_{,\beta} \tag{1.6}$$

Moreover, the I fundamental quadratic form of the surface is derived

$$(\mathrm{d}s)^2 = g_{11}\, (\mathrm{d}\xi_1)^2 + 2g_{12}\, \mathrm{d}\xi_1 \mathrm{d}\xi_2 + g_{22}\, (\mathrm{d}\xi_2)^2 \tag{1.7}$$

Next, base vectors with unit length (versors) $\mathbf{e}_\alpha$, $\mathbf{n}$ are obtained:

$$\mathbf{e}_\alpha = \frac{\mathbf{r}_{,\alpha}}{A_\alpha} = \frac{\mathbf{g}_\alpha}{A_\alpha}, \qquad \mathbf{n} = \mathbf{e}_3 = \mathbf{e}_1 \times \mathbf{e}_2 \tag{1.8}$$

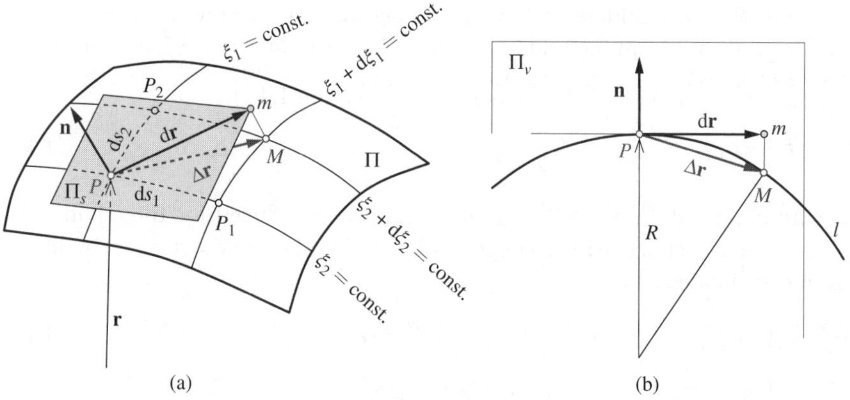

Figure 1.4 Description of objects: (a) on middle surface Π and in plane Π_s, (b) in plane Π_v. Source: Waszczyszyn and Radwańska (1995). Reproduced with permission of Waszczyszyn.

where × denotes the vector product of two vectors. The components of load $\hat{\mathbf{p}}$ and displacement vectors $\mathbf{u}$ can be defined using the local base versors $(\mathbf{e}_\alpha, \mathbf{n})$:

$$\hat{\mathbf{p}} = \hat{p}_1\mathbf{e}_1 + \hat{p}_2\mathbf{e}_2 + \hat{p}_n\mathbf{n} \tag{1.9}$$

$$\mathbf{u} = u_1\mathbf{e}_1 + u_2\mathbf{e}_2 + w\mathbf{n} \tag{1.10}$$

The measure of the middle surface curvature for a shell, denoted by m, can be calculated as the length of projection of vector $\Delta\mathbf{r}$ on direction $\mathbf{n}$, see Figure 1.4b

$$m = \mathbf{n}\cdot\Delta\mathbf{r} = \mathbf{n}\cdot\left(\mathrm{d}\mathbf{r} + \frac{1}{2}\mathrm{d}^2\mathbf{r} + \ldots\right) = \frac{1}{2}\mathbf{n}\cdot\mathrm{d}^2\mathbf{r} + \ldots . \tag{1.11}$$

To this end the second (II) metric tensor $b_{\alpha\beta}$

$$b_{\alpha\beta} = \mathbf{r}_{,\alpha\beta}\cdot\mathbf{n} = -\mathbf{r}_{,\alpha}\cdot\mathbf{n}_{,\beta} \tag{1.12}$$

and the II fundamental form of the surface

$$2m = b_{11}\,(\mathrm{d}\xi_1)^2 + 2b_{12}\,\mathrm{d}\xi_1\mathrm{d}\xi_2 + b_{22}\,(\mathrm{d}\xi_2)^2 \tag{1.13}$$

are defined. For line l its curvature radius R and curvature k are calculated as

$$\frac{1}{R} \equiv k = \lim_{|\Delta\mathbf{r}|\to 0}\frac{2m}{|\Delta\mathbf{r}|^2} = \mathbf{n}\cdot\frac{\mathrm{d}^2\mathbf{r}}{\mathrm{d}s^2} \tag{1.14}$$

In relation to the so-called principal coordinate lines, for which $g_{12} = b_{12} = 0$, two extreme principal curvature radii $R_{\alpha\alpha}$, as well as two characteristic parameters, mean curvature H and so-called Gaussian curvature K, are calculated using the formulae:

$$k_{\alpha\alpha} = -\frac{1}{R_{\alpha\alpha}} = \frac{b_{\alpha\alpha}}{g_{\alpha\alpha}} = \frac{b_{\alpha\alpha}}{(A_\alpha)^2} \tag{1.15}$$

$$k^2 - 2Hk + K = 0, \qquad H = \frac{1}{2}(k_1 + k_2), \qquad K = k_1k_2 \tag{1.16}$$

1.3.3 Geometry of Surface Equidistant from Middle Surface

Similar to point P on the middle surface Π (see Figure 1.5), we consider point $P^{(z)}$ on surface $\Pi^{(z)}$, equidistant from the middle surface. The position vector $\mathbf{r}^{(z)}$ of point $P^{(z)}$ is the sum of position vector $\mathbf{r}$ of point P and vector $z\mathbf{n}$:

$$\mathbf{r}^{(z)} = \mathbf{r} + z\mathbf{n}, \qquad -\frac{h}{2} \le z \le \frac{h}{2} \tag{1.17}$$

The following objects $\mathrm{d}s_\alpha^{(z)}$, $\mathbf{e}_\alpha^{(z)}$, $A_\alpha^{(z)}$, $R_\alpha^{(z)}$, $\alpha = 1, 2$, are introduced for the equidistant surface. They are associated with analogous objects for the middle surface by linear functions of coordinate z:

$$\mathrm{d}s_\alpha^{(z)} = A_\alpha^{(z)}\mathrm{d}\xi_\alpha, \qquad \mathbf{e}_\alpha^{(z)} = \frac{1}{A_\alpha^{(z)}}\mathbf{r}_{,\alpha}^{(z)}, \qquad \mathbf{n}^{(z)} \equiv \mathbf{n} \tag{1.18}$$

$$A_\alpha^{(z)} = A_\alpha\left(1 + \frac{z}{R_\alpha}\right), \qquad R_\alpha^{(z)} = R_\alpha\left(1 + \frac{z}{R_\alpha}\right) \tag{1.19}$$

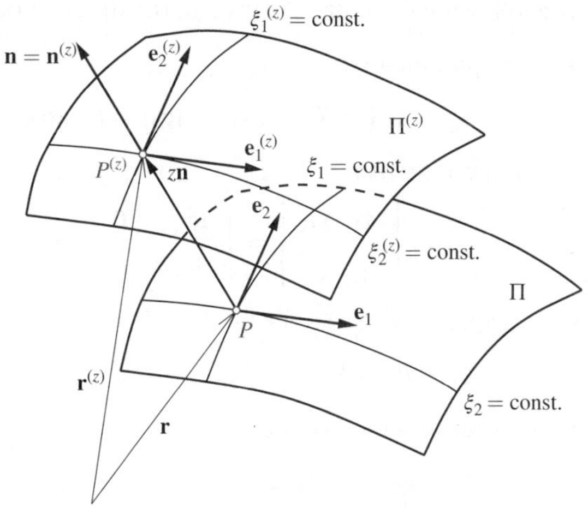

Figure 1.5 Middle surface Π and equidistant surface $\Pi^{(z)}$. Source: Waszczyszyn and Radwańska (1995). Reproduced with permission of Waszczyszyn.

1.3.4 Geometry of Selected Surfaces

We will now present three typical coordinate systems and three selected surfaces as well as scalar, vector and tensor quantities, useful in the description of a surface identified with the middle surface of a shell structure. We will specify base vectors and the first metric tensor. Omitting detailed derivations, we will provide formulae used for the description of geometry of these surface. For more information on the subject, the reader is referred, for instance, to Başar and Krätzig (2001).

1.3.4.1 Spherical Surface

A spherical surface is located in a 3D space with a Cartesian coordinate system ($x^1 = x, x^2 = y, x^3 = z$). On this surface point P is considered, whose position is defined using two spherical surface coordinates $\xi_1 = \varphi, \xi_2 = \theta$ and radius $R_1 = R_2 = R$ (see Figure 1.6).

In the global system of axes x^i, $i = 1, 2, 3$, the position vector $\mathbf{r}$ of point P is written first with Cartesian coordinates x^i, and next using two spherical coordinates ξ_α

$$\mathbf{r} = x^i\,\mathbf{i}_i = R\,\sin\varphi\sin\theta\,\mathbf{i}_1 + R\,\cos\varphi\,\mathbf{i}_2 + R\,\sin\varphi\cos\theta\,\mathbf{i}_3 \tag{1.20}$$

Base vectors $(\mathbf{e}_\alpha, \mathbf{n})$ are derived from the formulae:

$$\begin{bmatrix} \mathbf{g}_1 \\ \mathbf{g}_2 \end{bmatrix} = R \begin{bmatrix} \cos\varphi\sin\theta\,\mathbf{i}_1 - \sin\varphi\,\mathbf{i}_2 + \cos\varphi\cos\theta\,\mathbf{i}_3 \\ \sin\varphi\cos\theta\,\mathbf{i}_1 + 0\,\mathbf{i}_2 - \sin\varphi\sin\theta\,\mathbf{i}_3 \end{bmatrix}$$
$$\mathbf{e}_\alpha = \frac{1}{R}\,\mathbf{g}_\alpha, \qquad \mathbf{n} = \sin\varphi\sin\theta\,\mathbf{i}_1 + \cos\varphi\,\mathbf{i}_2 + \sin\varphi\cos\theta\,\mathbf{i}_3 \tag{1.21}$$

Figure 1.6 Spherical surface

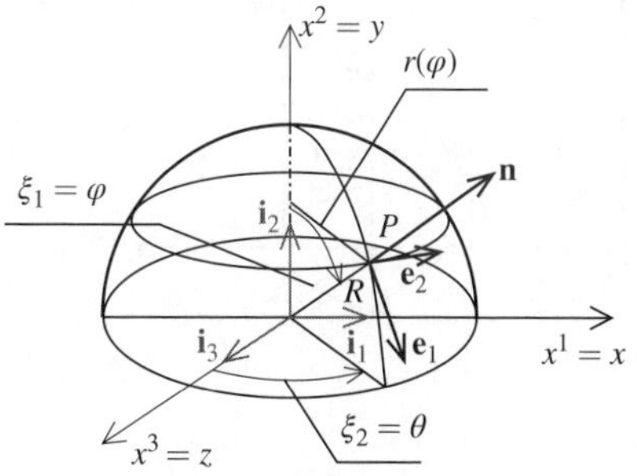

The following formulae are used in the description of a sphere:

- Lame parameters:

$$A_1 = |\mathbf{g}_1| = R, \qquad A_2 = |\mathbf{g}_2| = R\,\sin\varphi \tag{1.22}$$

- first metric tensor

$$g_{\alpha\beta} = \begin{bmatrix} g_{11} & g_{12} \\ g_{21} & g_{22} \end{bmatrix} = \begin{bmatrix} R^2 & 0 \\ 0 & R^2\sin^2\varphi \end{bmatrix} \tag{1.23}$$

- principal curvature radii

$$R_1 = R_2 = R \tag{1.24}$$

- Gaussian and mean curvatures:

$$K = \frac{1}{R^2}, \qquad H = \frac{1}{R} \tag{1.25}$$

1.3.4.2 Cylindrical Surface

A cylindrical surface, for which the symmetry axis is identical to axis $x^1 = x$ of the Cartesian coordinate system (x, y, z), is shown in Figure 1.7. The position of point P from the cylindrical surface is specified using three Cartesian coordinates x^i, which are related to two cylindrical surface coordinates $\xi_1 = x$ and $\xi_2 = \theta$ and radius R.

The main formulae for the calculation of characteristic parameters, vectors and tensor are (the names are as for the previous surface):

$$\mathbf{r} = x^i\,\mathbf{i}_i = x\,\mathbf{i}_1 + R\,\sin\theta\,\mathbf{i}_2 + R\,\cos\theta\,\mathbf{i}_3 \tag{1.26}$$

$$\begin{aligned} \begin{bmatrix} \mathbf{g}_1 \\ \mathbf{g}_2 \end{bmatrix} &= \begin{bmatrix} 1\,\mathbf{i}_1 \\ R\cos\theta\,\mathbf{i}_2 - R\,\sin\theta\,\mathbf{i}_3 \end{bmatrix} \\ \begin{bmatrix} \mathbf{e}_1 \\ \mathbf{e}_2 \end{bmatrix} &= \begin{bmatrix} 1\,\mathbf{i}_1 \\ \cos\theta\,\mathbf{i}_2 - \sin\theta\,\mathbf{i}_3 \end{bmatrix} \\ \mathbf{n} &= \sin\theta\,\mathbf{i}_2 + \cos\theta\,\mathbf{i}_3 \end{aligned} \tag{1.27}$$

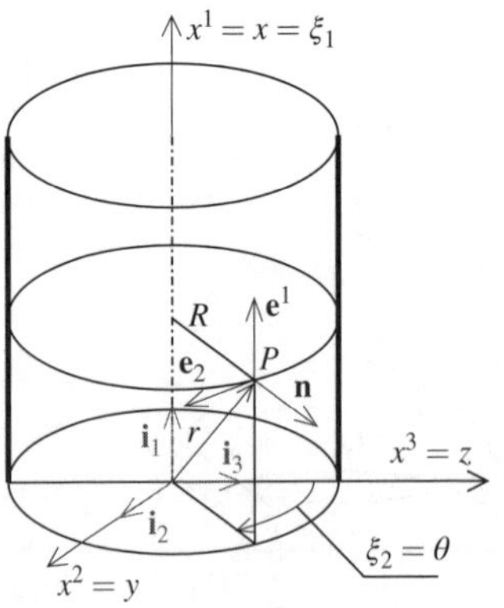

Figure 1.7 Cylindrical surface

$$A_1 = |\mathbf{g}_1| = 1, \qquad A_2 = |\mathbf{g}_2| = R \tag{1.28}$$

$$g_{\alpha\beta} = \begin{bmatrix} g_{11} & g_{12} \\ g_{21} & g_{22} \end{bmatrix} = \begin{bmatrix} 1 & 0 \\ 0 & R^2 \end{bmatrix} \tag{1.29}$$

$$R_1 = \infty, \qquad R_2 = R \tag{1.30}$$

$$K = 0, \qquad H = \frac{1}{2R} \tag{1.31}$$

1.3.4.3 Hyperbolic Paraboloid

The surface called the hyperbolic paraboloid is defined over a rectangle with dimensions $2a \times 2b$ on plane $x^3 = z = 0$ with two Cartesian coordinates $\xi_1 = x$, $\xi_2 = y$. The surface (see Figure 1.8) is defined by the equation:

$$z(x,y) = kxy, \qquad k = \frac{f}{ab}, \quad m = z_{,y} = kx, \quad n = z_{,x} = ky \tag{1.32}$$

The characteristic formulae used to describe the surface in question are:

$$\mathbf{r} = x^i\,\mathbf{i}_i = x\,\mathbf{i}_1 + y\,\mathbf{i}_2 + kxy\,\mathbf{i}_3 \tag{1.33}$$

$$\begin{gathered} \begin{bmatrix} \mathbf{g}_1 \\ \mathbf{g}_2 \end{bmatrix} = \begin{bmatrix} 1\,\mathbf{i}_1 + n\,\mathbf{i}_3 \\ 1\,\mathbf{i}_2 + m\,\mathbf{i}_3 \end{bmatrix} \\ \begin{bmatrix} \mathbf{e}_1 \\ \mathbf{e}_2 \end{bmatrix} = \frac{1}{\sqrt{1+m^2+n^2}} \begin{bmatrix} \mathbf{g}_1 \\ \mathbf{g}_2 \end{bmatrix} \\ \mathbf{n} = \frac{1}{\sqrt{1+m^2+n^2}}(-n\,\mathbf{i}_1 - m\,\mathbf{i}_2 + \mathbf{i}_3) \end{gathered} \tag{1.34}$$

$$A_1 = |\mathbf{g}_1| \approx 1, \qquad A_2 = |\mathbf{g}_2| \approx 1 \tag{1.35}$$

$$g_{\alpha\beta} = \begin{bmatrix} g_{11} & g_{12} \\ g_{21} & g_{22} \end{bmatrix} = \begin{bmatrix} 1+n^2 & mn \\ mn & 1+m^2 \end{bmatrix} \tag{1.36}$$

Figure 1.8 Hyperbolic paraboloid

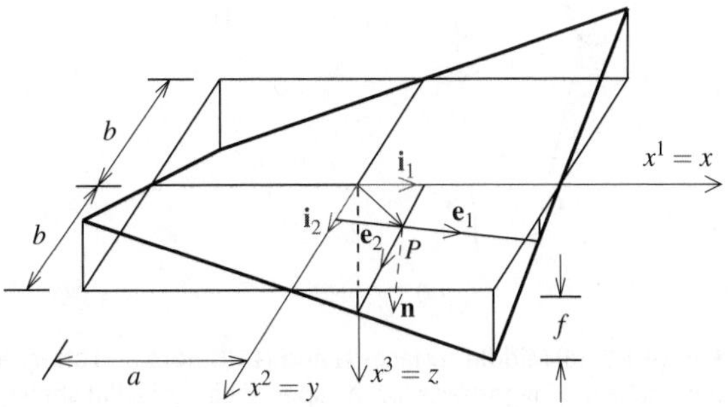

1.4 Definitions and Assumptions for 2D Formulation

1.4.1 Generalized Displacements and Strains Consistent with the Kinematic Hypothesis of Three-Parameter Kirchhoff–Love Shell Theory

The Kirchhoff–Love (K–L) kinematic hypothesis, adopted for thin shell structures, can be formulated in the following manner:

> *A straight fibre, located at the intersection of two cross-sectional planes, normal to the undeformed (initial) middle surface of a shell, after application of external actions remains straight and normal to the deformed (current) middle surface and has an unchanged length.*

To describe the fields of generalized displacements and strains it is necessary to use two surfaces Π, $\Pi^{(z)}$ in the initial configuration, as well as two analogous surfaces Π^*, $\Pi^{*(z)}$, marked by $*$ and related to the current configuration (after deformation).

In the description of kinematics two middle surfaces Π and Π^* (in initial and current configurations, respectively) are used (see Figure 1.9).

In the analysis of the current configuration the following vectors are distinguished: position vector $\mathbf{r}$, displacement (translation) vector $\mathbf{u}$ and rotation vector $\boldsymbol{\vartheta}$, whose components are related to the local base $(\mathbf{e}_\alpha, \mathbf{n})$ from the initial middle surface Π:

$$\mathbf{r}^* = \mathbf{r} + \mathbf{u} \tag{1.37}$$

$$\mathbf{u} = u_1\mathbf{e}_1 + u_2\mathbf{e}_2 + w\mathbf{n} \tag{1.38}$$

$$\boldsymbol{\vartheta} = -\vartheta_2\mathbf{e}_1 + \vartheta_1\mathbf{e}_2 + \vartheta_n\mathbf{n} = \varphi_1\mathbf{e}_1 + \varphi_2\mathbf{e}_2 + \varphi_n\mathbf{n} \tag{1.39}$$

For the rotation vector we can use two types of components: ϑ_α, ϑ_n or φ_α, φ_n (see Figure 1.10).

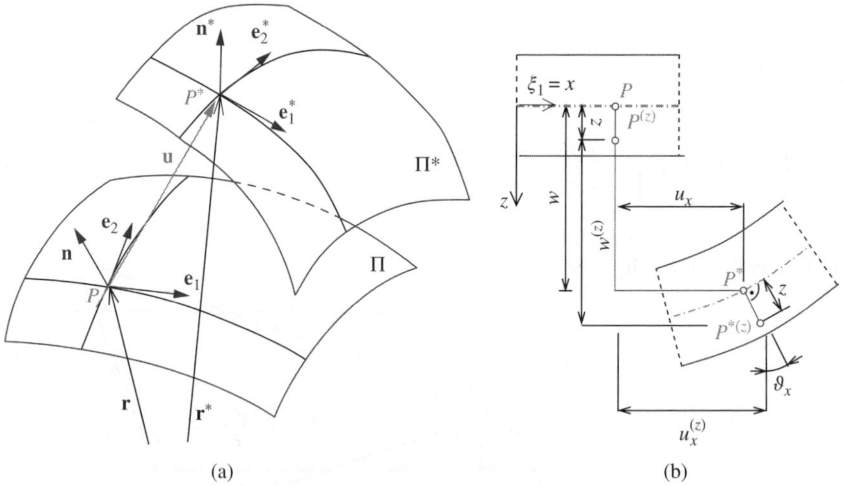

Figure 1.9 (a) Middle surfaces Π and Π^* (before and after deformation), (b) graphical interpretation of kinematic K–L hypothesis for the special case of a flat shell (plate) on plane (ξ_1, z); analogical section can be shown for plane (ξ_2, z). Source: Waszczyszyn and Radwańska (1995). Reproduced with permission of Waszczyszyn.

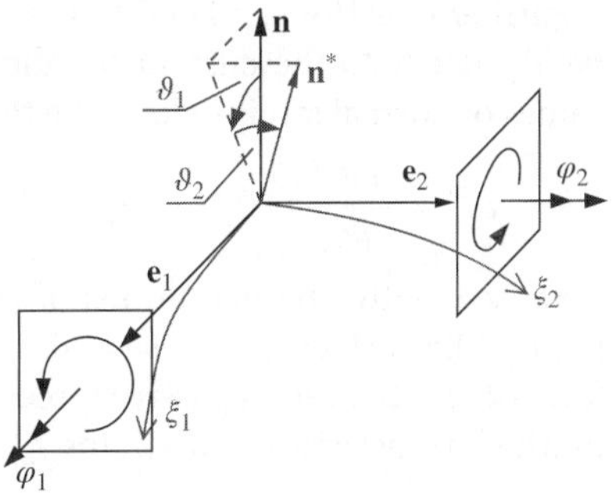

Figure 1.10 Description of rotations of vector **n** normal to middle surface. Source: Waszczyszyn and Radwańska (1995). Reproduced with permission of Waszczyszyn.

During deformation of the middle surface the orthogonal unit base $(\mathbf{e}_\alpha, \mathbf{n})$ changes into a different base $(\mathbf{e}^*_\alpha, \mathbf{n}^*)$, in general nonorthogonal:

$$\begin{aligned}\mathbf{e}^*_\alpha &= \frac{1}{A^*_\alpha}\mathbf{r}^*_{,\alpha} \approx \mathbf{e}_\alpha + \delta\mathbf{e}_\alpha \\ &= \mathbf{e}_\alpha + \frac{1}{A_\alpha}\left(u_{\beta,\alpha} - \frac{A_{\alpha,\beta}}{A_\beta}u_\alpha\right)\mathbf{e}_\beta + \frac{1}{A_\alpha}\left(w_{,\alpha} - \frac{A_\alpha}{R_\alpha}u_\alpha\right)\mathbf{n}\end{aligned} \tag{1.40}$$

$$\mathbf{n}^* = \mathbf{e}^*_1 \times \mathbf{e}^*_2 = \mathbf{n} + \delta\mathbf{n} = \mathbf{n} - \vartheta_2\mathbf{e}_1 + \vartheta_1\mathbf{e}_2 \tag{1.41}$$

The formula (1.41) expresses the change of normal vector **n** into new vector $\mathbf{n}^*$ by means of two rotations ϑ_α (see Figure 1.10).

The displacements at point $P^{(z)}$ on surface $\Pi^{(z)}$, equidistant from the middle surface Π, are calculated on the basis of translations u_α, w and rotations ϑ_α, defined at point P on the middle surface Π:

$$u_1^{(z)} = u_1 + z\,\vartheta_1, \qquad u_2^{(z)} = u_2 + z\,\vartheta_2, \qquad w^{(z)} = w \tag{1.42}$$

The K–L kinematic constraints imply the following relations between two rotations ϑ_α and three translations u_α, w:

$$\begin{aligned}&\vartheta_1 = -\frac{1}{A_1}\frac{\partial w}{\partial \xi_1} + \frac{u_1}{R_1}, \qquad \vartheta_2 = -\frac{1}{A_2}\frac{\partial w}{\partial \xi_2} + \frac{u_2}{R_2} \\ &\vartheta_\alpha = -\frac{1}{A_\alpha}\frac{\partial w}{\partial \xi_\alpha} + \frac{u_\alpha}{R_\alpha}, \qquad \alpha = 1, 2\end{aligned} \tag{1.43}$$

Equations (1.43) show the possibility of using a shortened notation of two analogous formulae to describe two-dimensional shell structures.

The third rotation ϑ_n, around the normal, is related to the translations by the following equation

$$\vartheta_n = \frac{1}{2}\left[\left(\frac{1}{A_2}\frac{\partial u_1}{\partial \xi_2} - \frac{1}{A_1}\frac{\partial u_2}{\partial \xi_1}\right) - \left(\frac{A_{1,2}u_1}{A_1A_2} - \frac{A_{2,1}u_2}{A_1A_2}\right)\right] \tag{1.44}$$

The name three-parameter theory of shell structures originates from the fact that only three translations of points from the middle surface, written in a vector

$$\mathbf{u} = [u_1, u_2, w]^{\mathrm{T}} \tag{1.45}$$

and treated as independent components, suffice to describe generalized displacements of the shell (three translations and three rotations).

Further equations, related to the unchanging length of the straight fibre and its perpendicularity to the current middle surface, express the information concerning zero values of normal strains along the thickness and transverse shear strains:

$$\varepsilon_{zz}^{(z)} = \frac{\partial w^{(z)}}{\partial z} = 0, \qquad \gamma_{\alpha z} = \vartheta_\alpha + \left[\frac{1}{A_\alpha}\frac{\partial w}{\partial \xi_\alpha} - \frac{u_\alpha}{R_\alpha}\right] = 0, \qquad \alpha = 1, 2 \tag{1.46}$$

Subsequently, the definitions of generalized strains and resultant forces referred to the middle surface are presented. It should be emphasized here that transverse shear forces t_α are treated as passive forces as a consequence of the constraints in K–L theory, in which transverse shear strains $\gamma_{\alpha z}$ equal zero.

1.4.2 Generalized Displacements and Strains Consistent with the Kinematic Hypothesis of Five-Parameter Mindlin–Reissner Shell Theory

Unlike the three-parameter K–L theory, in five-parameter Mindlin–Reissner (M–R) theory orthogonality of the straight fibre to the deformed middle surface is not imposed (see Figure 1.11) and a vector of five independent generalized displacements, three translations u_1, u_2, w and two rotations ϑ_1, ϑ_2, is introduced

$$\mathbf{u} = [u_1, u_2, w, \vartheta_1, \vartheta_2]^{\mathrm{T}} \tag{1.47}$$

In the kinematic equations for moderately thick plates and shells, three translations and two rotations appear as independent variables. In this approach, next to membrane and bending strains, nonzero averaged transverse shear strains are taken into account. In the M–R theory the transverse shear forces are included, they appear in constitutive relations that are written in terms of appropriate strains, forces and the local stiffness operator. The equations of the M–R theory for plates will be given in Section 8.5.

1.4.3 Force and Moment Resultants Related to Middle Surface

In the K–L theory, beside the kinematic hypothesis, a static hypothesis is introduced, which simplifies the continuum equations:

> *In comparison with the other stress tensor components stress σ_z is so small that for all points of a thin shell it can be ignored in the constitutive relations, so that*

$$\sigma_z(\xi_1, \xi_2, z) \equiv 0 \tag{1.48}$$

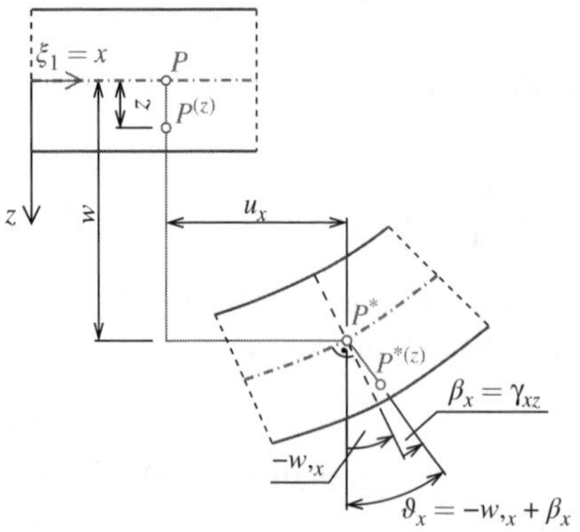

Figure 1.11 Position of a straight fibre before and after deformation of a plate according to the kinematic M–R hypothesis; analogical section can be shown for plane (ξ_2, z)

As a consequence of the static K–L hypothesis further simplifications are made in the description of thin shell structures.

At any point on the middle surface Π, one of two sections obtained using plane Π_α with the normal versor $\mathbf{e}_\alpha$ is shown in Figure 1.12, and the following stress vector is revealed at the intersection

$$\boldsymbol{\sigma}_\alpha(\xi_1,\xi_2,z) = \sigma_{\alpha 1}\mathbf{e}_1 + \sigma_{\alpha 2}\mathbf{e}_2 + \sigma_{\alpha n}\mathbf{n}, \qquad \alpha = 1,2 \tag{1.49}$$

Next, integrating along the thickness, the intensities of resultant forces $\mathbf{f}_\alpha$ [kN/m] and moments $\mathbf{m}_\alpha$ [kNm/m] with adequate components, visible in section with normal $\mathbf{e}_\alpha$, are first calculated for $\alpha = 1$:

$$\mathbf{f}_1 = f_{11}\mathbf{e}_1 + f_{12}\mathbf{e}_2 + f_{1n}\mathbf{n}, \qquad \mathbf{m}_1 = m_{11}\mathbf{e}_2 - m_{12}\mathbf{e}_1 \tag{1.50}$$

where:

$$f_{11} = n_{11} = \int_{-h/2}^{+h/2} \sigma_{11}\left(1+\frac{z}{R_2}\right)\mathrm{d}z, \quad f_{12} = n_{12} = \int_{-h/2}^{+h/2} \sigma_{12}\left(1+\frac{z}{R_2}\right)\mathrm{d}z$$

$$f_{1n} = t_1 = \int_{-h/2}^{+h/2} \sigma_{1z}\left(1+\frac{z}{R_2}\right)\mathrm{d}z$$

$$m_{11} = \int_{-h/2}^{+h/2} \sigma_{11}z\left(1+\frac{z}{R_2}\right)\mathrm{d}z, \qquad m_{12} = \int_{-h/2}^{+h/2} \sigma_{12}z\left(1+\frac{z}{R_2}\right)\mathrm{d}z \tag{1.51}$$

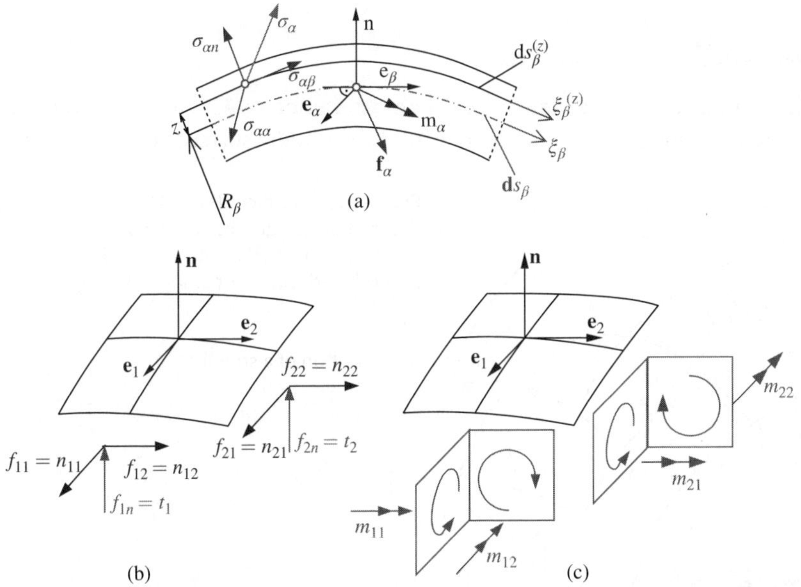

Figure 1.12 (a) Section of a shell with characteristic stresses and vectors of: force $\mathbf{f}_\alpha$ and moment $\mathbf{m}_\alpha$, (b) and (c) elementary surface segments with all resultant forces revealed on section lines in the membrane-bending state. Source: Waszczyszyn and Radwańska (1995). Reproduced with permission of Waszczyszyn.

The derivations are repeated for $\alpha = 2$:

$$\mathbf{f}_2 = f_{21}\mathbf{e}_1 + f_{22}\mathbf{e}_2 + f_{2n}\mathbf{n}, \qquad \mathbf{m}_2 = -m_{22}\mathbf{e}_1 + m_{21}\mathbf{e}_2 \tag{1.52}$$

where:

$$f_{22} = n_{22} = \int_{-h/2}^{+h/2} \sigma_{22}\left(1+\frac{z}{R_1}\right)\mathrm{d}z, \quad f_{21} = n_{21} = \int_{-h/2}^{+h/2} \sigma_{21}\left(1+\frac{z}{R_1}\right)\mathrm{d}z$$
$$f_{2n} = t_2 = \int_{-h/2}^{+h/2} \sigma_{2z}\left(1+\frac{z}{R_1}\right)\mathrm{d}z$$
$$m_{22} = \int_{-h/2}^{+h/2} \sigma_{22}z\left(1+\frac{z}{R_1}\right)\mathrm{d}z, \qquad m_{21} = \int_{-h/2}^{+h/2} \sigma_{21}z\left(1+\frac{z}{R_1}\right)\mathrm{d}z \tag{1.53}$$

In these equations a set of quantities are defined for shells. They are membrane forces: normal n_{11}, n_{22} and tangential n_{12}, n_{21}, moments: bending m_{11}, m_{22} and twisting m_{12}, m_{21}, as well as transverse shear forces t_1, t_2. In the definitions a factor $(1 + z/R_\alpha)$ appears, resulting from the relation between the lengths of arc segments from the middle and equidistant surfaces

$$\mathrm{d}s_\alpha^{(z)} = \mathrm{d}s_\alpha(1 + z/R_\alpha) \tag{1.54}$$

In the case of thin shell structures or weakly curved shells ($R_\alpha \to \infty$), when $z/R_\alpha \ll 1$, these definitions can be reduced, omitting the factor $(1 + z/R_\alpha) \approx 1$ and then the following equalities are valid $n_{12} = n_{21}$, $m_{12} = m_{21}$.

1.4.4 Generalized Strains in Middle Surface

In the current configuration (see Figure 1.13) Lame parameters $A_\alpha^{*(z)}$ and base versors $\mathbf{e}_\alpha^{*(z)}$ are calculated for surface $\Pi^{*(z)}$ equidistant from the middle surface Π^*. The normal

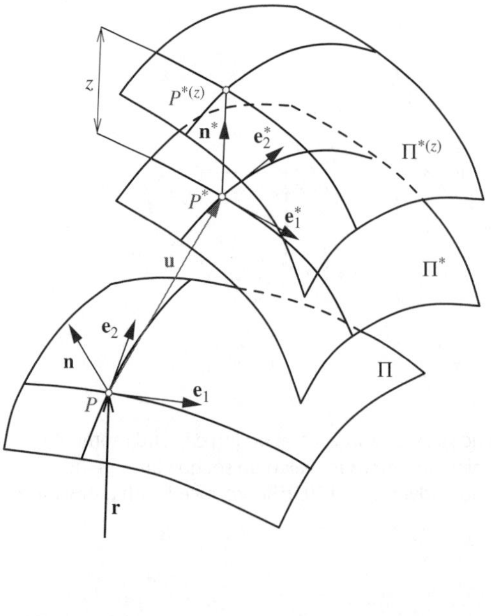

Figure 1.13 Surfaces $\Pi, \Pi^*, \Pi^{*(z)}$ defined for description of deformation. Source: Waszczyszyn and Radwańska (1995). Reproduced with permission of Waszczyszyn.

and shear strains are expressed in terms of the generalized strains from the middle surface using the linear function of coordinate z:

$$\varepsilon_{\alpha\alpha}^{(z)} = \frac{A_\alpha^{*(z)}}{A_\alpha^{(z)}} - 1 \approx \varepsilon_{\alpha\alpha} + \kappa_{\alpha\alpha}\, z, \qquad \gamma_{12}^{(z)} = \mathbf{e}_1^{*(z)} \cdot \mathbf{e}_1^{*(z)} \approx \gamma_{12} + \chi_{12}\, z \tag{1.55}$$

The symbol $\approx$ means that necessary simplifications were made when the relations were formulated, resulting from the assumption $(1 + z/R_\alpha)^{-1} \approx 1 - z/R_\alpha$ and the omission of higher order terms.

Further, we quote, without derivation, the kinematic equations proposed by Sanders for the three-parameter thin shell theory of Sanders (1959):

$$\begin{aligned}
\varepsilon_{11} &= \frac{1}{A_1}\frac{\partial u_1}{\partial \xi_1} + \frac{1}{A_1A_2}\frac{\partial A_1}{\partial \xi_2}u_2 + \frac{w}{R_1}\\
\varepsilon_{22} &= \frac{1}{A_2}\frac{\partial u_2}{\partial \xi_2} + \frac{1}{A_1A_2}\frac{\partial A_2}{\partial \xi_1}u_1 + \frac{w}{R_2}\\
\gamma_{12} &= \frac{A_2}{A_1}\frac{\partial}{\partial \xi_1}\left(\frac{u_2}{A_2}\right) + \frac{A_1}{A_2}\frac{\partial}{\partial \xi_2}\left(\frac{u_1}{A_1}\right)\\
\kappa_{11} &= -\frac{1}{A_1}\frac{\partial}{\partial \xi_1}\left(\frac{1}{A_1}\frac{\partial w}{\partial \xi_1} - \frac{u_1}{R_1}\right) - \frac{1}{A_1A_2}\left(\frac{1}{A_2}\frac{\partial w}{\partial \xi_2} - \frac{u_2}{R_2}\right)\frac{\partial A_1}{\partial \xi_2}\\
\kappa_{22} &= -\frac{1}{A_2}\frac{\partial}{\partial \xi_2}\left(\frac{1}{A_2}\frac{\partial w}{\partial \xi_2} - \frac{u_2}{R_2}\right) - \frac{1}{A_1A_2}\left(\frac{1}{A_1}\frac{\partial w}{\partial \xi_1} - \frac{u_1}{R_1}\right)\frac{\partial A_2}{\partial \xi_1}\\
\chi_{12} &= -\frac{A_2}{A_1}\frac{\partial}{\partial \xi_1}\left[\frac{1}{A_2}\left(\frac{1}{A_2}\frac{\partial w}{\partial \xi_2} - \frac{u_2}{R_2}\right)\right] - \frac{A_1}{A_2}\frac{\partial}{\partial \xi_2}\left[\frac{1}{A_1}\left(\frac{1}{A_1}\frac{\partial w}{\partial \xi_1} - \frac{u_1}{R_1}\right)\right]\\
&\quad + \frac{1}{2A_1A_2}\left(\frac{1}{R_2} - \frac{1}{R_1}\right)\left[\frac{\partial(A_2u_2)}{\partial \xi_1} - \frac{\partial(A_1u_1)}{\partial \xi_2}\right]
\end{aligned} \tag{1.56}$$

They are general equations, which after the introduction of specified coordinates ξ_α, Lame parameters A_α and principal curvature radii R_α $(\alpha = 1, 2)$, can be suitably modified in order to describe selected types of shells.

1.5 Classification of Shell Structures

To propose a classification of shells the two geometrical parameters are used:

- h – thickness of plate or shell
- L – characteristic dimension of plate or shell $L = \min(L_{\min}, R_{\min})$, where $L_{\min}$ – the smallest dimension in the middle plane of a plate, $R_{\min}$ – smaller of two principal radii of curvature of the middle surface in a shell

In next subsections the following properties of shell structures are discussed:

- geometry (shape)
- measure of slenderness h/L
- types of stress state
- magnitude of considered displacements

The classification is summarized in Box 1.1.

1.5.1 Curved, Shallow and Flat Shell Structures

Shell structures, represented by two-dimensional middle surfaces, are divided into curved, shallow and flat, see Figure 1.14.

When the middle surface is curved, we deal with thin-walled curved shell structures and the following cases are possible:

- shells curved in one direction, for example cylindrical or conical shells, for which Gaussian curvature $K = 1/(R_1R_2) = 0$
- doubly curved shells, for example a spherical shell with $K > 0$ or rotational hyperbolic shell with $K < 0$
- shells with a ruled surface, for example a hyperbolic paraboloid shell
- shells of completely arbitrary shapes

Flat structures represented by a two-dimensional middle plane have two subclasses with respect to load type:

- flat membranes, that is plates with load (and deformation) in the middle plane, with no transverse displacements
- plates with transverse load, that is load normal to the middle plane, exhibiting transverse displacements (deflections)

Shallow shells are characterized by rise f, in other words by the surface deviation from its projection on the horizontal plane. The shell is considered shallow when $\frac{f}{L} < \frac{1}{5}$ and then the Cartesian set of coordinates can be used for the approximate description of geometry, but in the corresponding equations curvature radii R_x and R_y, resulting from the surface definition $z(x, y)$, are taken into account.

1.5.2 Thin, Moderately Thick, Thick Structures

The structure can be treated as thin-walled if $\frac{h}{L} \ll 1$. As regards the thickness the following limitations are adopted:

- thin flat structures (plates and membranes) $\frac{h}{L_{min}} < \frac{1}{10}$
- moderately thick flat structures $\frac{1}{10} \leq \frac{h}{L_{min}}$
- thin curved structures (shells and membranes) $\frac{h}{R_{min}} \leq \frac{1}{20}$
- moderately thick curved structures $\frac{1}{20} < \frac{h}{R_{min}} \leq \frac{1}{6}$
- thick shells $\frac{1}{6} < \frac{h}{R_{min}}$

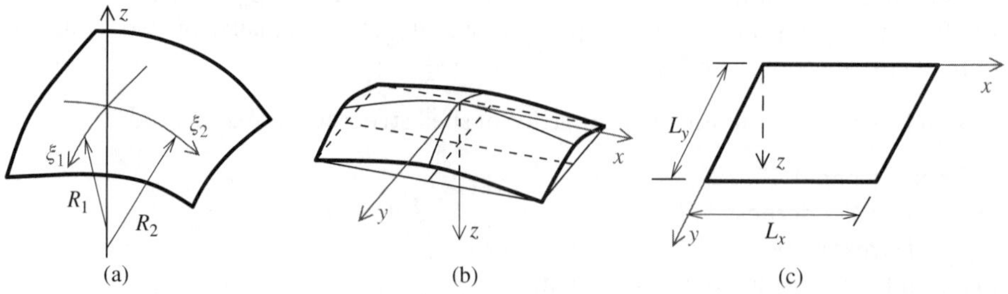

Figure 1.14 Middle surfaces of structures with different shapes: (a) curved, (b) shallow and (c) flat

1.5.3 Plates and Shells with Different Stress Distributions Along Thickness

If normal and shear stress distributions are uniform in the thickness direction, it is a state without bending and shear effects, understood as the membrane state in flat membranes and curved membrane shells.

If normal and in-plane shear stresses change linearly along the thickness, then we deal with a plate or shell in the bending state.

In thin plates and shells the effects related to transverse shear are insignificant and disregarded. Non-deformability of the thin structures in the thickness direction is additionally assumed.

In the case of moderately thick plates and shells, the effects related to transverse shear stresses (varying quadratically along the thickness) are taken into account next to the bending effects.

1.5.4 Range of Validity of Geometrically Linear and Nonlinear Theories for Plates and Shells

While defining the generalized strains in the middle surface and formulating the kinematic equations some simplifications are made, resulting from the approximation

$$(1 + z/R_\alpha)^{-1} \approx 1 - z/R_\alpha \tag{1.57}$$

The reader looking for a deeper knowledge of the theories of plates and shells can find a classification of geometrically nonlinear theories for elastic thin shell structures, for example in Pietraszkiewicz (1979). Taking as a starting point for analysis general nonlinear equilibrium and kinematic equations, many steps of simplification are proposed depending on the expected magnitudes of strains, translations and rotations. In a geometrically nonlinear theory of shells (or plates) the following components are present in the kinematic equations: (i) linear components related to three displacements $u_\alpha (\alpha = 1, 2)$ and w, (ii) nonlinear components dependent only on normal displacement w and (iii) nonlinear components related to three displacements u_α, w.

The geometrically linear theory for bending plates (with linear kinematic equations) is applied when deflections (translations normal to the middle plane) are expected of the order $|w| < h/5$. In the case of deflections of the order of the thickness $|w| \approx h$, the previously mentioned kinematic equations must be extended by components nonlinear only with respect to w. The behaviour of a plate in membrane-bending state is described by the equations from the von Kármán theory of plates with moderately large deflections. Expecting deflections of the order $|w| > 5h$, the kinematic equations are augmented by additional nonlinear terms dependent on u_α next to deflection w. These ranges of applicability of respective plate theories are listed in the second part of Box 1.1.

In the works devoted to geometrically nonlinear theories of shells, among others Pietraszkiewicz (1979), the classification of particular cases is presented taking into account the estimation of rotation angles, which provides the theories of: (i) small rotations, (ii) moderate rotations and (iii) large rotations, (iv) finite rotations and strains.

The considerations and examples connected with geometrically as well as physically nonlinear theories are beyond the scope of this book. Summarizing, the information contained in Subsections 1.5.1–1.5.4 allows one to apply an appropriate theory for a considered problem and to choose a suitable solution method.

References

Başar Y and Krätzig WB 2001 *Theory of Shell Structures* Fortschritt Berichte 18, Nr. 258 2nd edn. Düsseldorf: VDI Verlag Gmbh, Ruhr–Universität Bochum.

Bischoff M, Wall WA, Bletzinger KU and Ramm E 2004 Models and finite elements for thin-walled structures. In Stein E, de Borst R and Hughes TJR (eds), *Encyclopedia of Computational Mechanics: Solids and Structures*, vol. **2**. John Wiley & Sons, Ltd, Chichester UK. chapter 3, pp. 59–137.

Borkowski A, Jemielita G, Michalak B, Nagórski R, Pietraszkiewicz W, Rudnicki M, Woźniak M and Woźniak C 2001 *Mechanics of Elastic Plates and Shells*, vol. **8** of *Technical Mechanics*. PWN, Warsaw (in Polish).

Budiansky B and Sanders JL 1963 On the 'best' first-order shell theory *Progress in Applied Mechanics, The Prager Anniversary Volume*, pp. 129–140. MacMillan, New York.

Crisfield MA 1982 Solution procedures for non-linear structural problems. In Hinton E, Owen DR and Taylor C (eds), *Recent Advances in Non-linear Computational Mechanics*. Pineridge Press, Swansea. pp. 59–137.

de Borst R, Crisfield MA, Remmers JJC and Verhoosel CV 2012 *Non-linear Finite Element Analysis of Solids and Structures* 2nd edn. John Wiley & Sons, Ltd, Chichester, UK.

Fung YC 1965 *Foundations of Solid Mechanics*. Prentice-Hall, Englewood Cliffs.

Girkmann K 1956 *Flächentragwerke*. Springer-Verlag, Wien.

Hinton E, Owen DR and Taylor C (eds) 1982 *Recent Advances in Non-linear Computational Mechanics*. Pineridge Press, Swansea.

Kleiber M 1985 *Finite Element Method in Nonlinear Continuum Mechanics*. PWN, Warsaw-Poznan (in Polish).

Koiter WT 1960 A consistent first approximation in the general theory of thin elastic shells, In *Proc. Symp. on Theory of Thin Elastic Shells* Ashwell D and Koiter W (eds), pp. 12–33 North-Holland, Amsterdam.

Kolkunov NV 1972 *Foundations of the Analysis of Elastic Shells*. Izdatel'stvo Vishaya Shkola, Moskow (in Russian).

Lewiński T 1980 An analysis of various description of state of strain in the linear Kirchhoff–Love type shell theory. *Engineering Transactions* **28**(4), 635–652.

Love AEH (ed.) 1944 *A Treatise on the Mathematical Theory of Elasticity*, 4th edn. Dover Publications.

Niordson FI 1985 *Shell Theory*. North-Holland, Amsterdam-New York-Oxford.

Noor AK, Belytschko T and Simo JC (ed.) 1989 *Analytical and Computational Models of Shells*. ASME.

Nowacki W 1980 *Plates and Shells*. PWN, Warsaw (in Polish).

Oñate E 2013 *Structural Analysis with the Finite Element Method. Linear Statics. Volume 2: Beams, Plates and Shells*, Lecture Notes on Numerical Methods in Engineering and Sciences. CIMNE Springer.

Pietraszkiewicz W 1977 *Introduction to the Non-Linear Theory of Shells* Nr. 10. Mitteilungen aus dem Institut für Mechanik, Ruhr–Universität Bochum.

Pietraszkiewicz W 1979 *Finite Rotations and Lagrangean Description in the Non-linear Theory of Shells*. PWN, Warsaw-Poznan (in Polish).

Pietraszkiewicz W 2001 Nonlinear shell theories. In Borkowski A, Jemielita G, Michalak B, Nagòrski R, Pietraszkiewicz W, Rudnicki M, Wozniak M and Wozniak C, *Mechanics of Elastic Plates and Shells*, vol. **8** of Technical Mechanics. PWN, Warsaw, pp. 424–497 (in Polish).

Radwańska M 2009 *Shell Structures*. Cracow University of Technology, Cracow (in Polish).
Ramm E and Wall WA 2004 Shell structures – a sensitive interrelation between physics and numerics. *International Journal for Numerical Methods in Engineering* **60**(1), 381–427.
Reddy JN 1986 *Applied Functional Analysis and Variational Methods in Engineering*. McGraw-Hill, New York.
Reddy JN 1999 *Theory and Analysis of Elastic Plates*. Taylor & Francis.
Reddy JN 2007 *Theory and Analysis of Elastic Plates and Shells* 2nd edn. CRC Press/Taylor & Francis, Boca Raton-London-New York.
Sanders JL 1959 An improved first-approximation theory of thin shells. Technical Report TR-R24, NASA.
Stein E, de Borst R and Hughes TJR (ed.) 2004 *Encyclopedia of Computational Mechanics: Solids and Structures*, vol. **2**. John Wiley & Sons, Ltd, Chichester, UK.
Timoshenko S and Goodier JN 1951 *Theory of Elasticity* 2nd edn. McGraw-Hill, New York-Toronto-London.
Timoshenko S and Woinowsky-Krieger S 1959 *Theory of Plates and Shells*. McGraw-Hill, New York-Auckland.
Washizu K 1975 *Variational Methods in Elasticity and Plasticity* 2nd edn. Pergamon Press.
Waszczyszyn Z and Radwańska M 1995 Basic equations and calculations methods for elastic shell structures. In Borkowski A, Cichoń C, Radwańska M, Sawczuk A and Waszczyszyn Z, *Structural Mechanics: Computer Approach*, vol. **3**. Arkady, Warsaw, chapter 9, pp. 11–190 (in Polish).
Wiśniewski K 2010 *Finite Rotation Shells. Basic Equations and Finite Elements for Reissner Kinematics*, Lecture Notes on Numerical Methods in Engineering and Sciences. CIMNE Springer.
Woźniak C 1966 *Nonlinear Shell Theory*. PWN, Warsaw (in Polish).

2

Equations for Theory of Elasticity for 3D Problems

The subject matter of *strength of materials* and *theory of elasticity* (Timoshenko and Goodier 1951) includes the problem of a deformable three-dimensional (3D) body, described in a 3D space with, for example, a Cartesian coordinate system (x, y, z) or $(x^i),\ i = 1, 2, 3$. The volume of the body is denoted by Ω and its surface by $\partial\Omega$. We are interested in the behaviour of the body with specified supports (kinematic constraints), under a given load. The state of the body is defined when stresses, strains and displacements are identified as a solution of a boundary value problem (BVP), which is a stationary 3D mechanical problem. Hence, there is no necessity of introducing a time variable.

The local formulation, that is the set of equations describing the BVP, can be given in index, engineering or matrix notation.

In the matrix notation the quantities of the same type are grouped together in respective vectors, defined at a point in the volume of the body or on its surface (with outer normal vector $\boldsymbol{\nu} = [\nu_1, \nu_2, \nu_3]^{\mathrm{T}}$ introduced at a boundary point).

At the initial stage of analysis the following external actions (loads and kinematic constraints) are assumed to be known:

- vector of body forces $\hat{\mathbf{b}}_{(3\times1)} = [\hat{b}_x, \hat{b}_y, \hat{b}_z]^{\mathrm{T}}$ [N/m^3] at a point in volume Ω
- vector of distributed boundary (surface) loads $\hat{\boldsymbol{\sigma}}_{b(3\times1)} = [\hat{\sigma}_{\nu1}, \hat{\sigma}_{\nu2}, \hat{\sigma}_{\nu3}]^{\mathrm{T}}$ [N/m^2] at a point on surface part $\partial\Omega_\sigma$
- vector of boundary displacements $\hat{\mathbf{u}}_{b(3\times1)} = [\hat{u}_{\nu1}, \hat{u}_{\nu2}, \hat{u}_{\nu3}]^{\mathrm{T}}$ [m] at a point on surface part $\partial\Omega_u$

The fields to be found, that is the unknown quantities at a point in the domain Ω are:

- stress vector $\boldsymbol{\sigma}_{(6\times1)}(x, y, z) = [\sigma_{xx}, \sigma_{yy}, \sigma_{zz}, \tau_{xy}, \tau_{xz}, \tau_{yz}]^{\mathrm{T}}$ [N/m^2]
- strain vector $\boldsymbol{\varepsilon}_{(6\times1)}(x, y, z) = [\varepsilon_{xx}, \varepsilon_{yy}, \varepsilon_{zz}, \gamma_{xy}, \gamma_{xz}, \gamma_{yz}]^{\mathrm{T}}$ [–]
- displacement vector $\mathbf{u}_{(3\times1)}(x, y, z) = [u_x, u_y, u_z]^{\mathrm{T}} = [u, v, w]^{\mathrm{T}}$ [m]

The following equality conditions are taken into account for the tangent stresses $\tau_{xy} = \tau_{yx}, \tau_{xz} = \tau_{zx}, \tau_{yz} = \tau_{zy}$ and the shear strains $\gamma_{xy} = \gamma_{yx}, \gamma_{xz} = \gamma_{zx}, \gamma_{yz} = \gamma_{zy}$.

The formulation of a BVP in the *theory of elasticity* is reduced to writing a set of 15 differential and algebraic equations. The set of equations contains: six kinematic relations (I), three equilibrium equations (II) and six constitutive equations (III). These equations must be completed with the conditions describing boundary constraints and surface loads, called kinematic and static boundary conditions (IV), respectively.

Plate and Shell Structures: Selected Analytical and Finite Element Solutions, First Edition.
Maria Radwańska, Anna Stankiewicz, Adam Wosatko and Jerzy Pamin.
© 2017 John Wiley & Sons Ltd. Published 2017 by John Wiley & Sons Ltd.

Next, the set of equations for a deformable body in a 3D space with a Cartesian coordinate system is presented in engineering and matrix notations:

(I) kinematic equations, that is differential equations describing the relations between displacements $\mathbf{u}$ and strains $\boldsymbol{\varepsilon}$:

$$\varepsilon_x = \frac{\partial u_x}{\partial x}, \qquad \varepsilon_y = \frac{\partial u_y}{\partial y}, \qquad \varepsilon_z = \frac{\partial u_z}{\partial z}$$
$$\gamma_{xy} = \frac{\partial u_x}{\partial y} + \frac{\partial u_y}{\partial x}, \qquad \gamma_{xz} = \frac{\partial u_x}{\partial z} + \frac{\partial u_z}{\partial x}, \qquad \gamma_{yz} = \frac{\partial u_y}{\partial z} + \frac{\partial u_z}{\partial y} \tag{2.1}$$

which, in matrix notation, are reduced to one equation

$$\boldsymbol{\varepsilon}_{(6\times1)} = \mathbf{L}_{(6\times3)}\ \mathbf{u}_{(3\times1)} \tag{2.2}$$

where

$$\mathbf{L} = \begin{bmatrix} \partial/\partial x & 0 & 0 \\ 0 & \partial/\partial y & 0 \\ 0 & 0 & \partial/\partial z \\ \partial/\partial y & \partial/\partial x & 0 \\ \partial/\partial z & 0 & \partial/\partial x \\ 0 & \partial/\partial z & \partial/\partial y \end{bmatrix} \tag{2.3}$$

(II) equilibrium equations, that is differential equations relating stresses $\boldsymbol{\sigma}$ and body forces $\hat{\mathbf{b}}$:

$$\frac{\partial \sigma_x}{\partial x} + \frac{\partial \tau_{yx}}{\partial y} + \frac{\partial \tau_{zx}}{\partial z} + \hat{b}_x = 0$$
$$\frac{\partial \tau_{xy}}{\partial x} + \frac{\partial \sigma_y}{\partial y} + \frac{\partial \tau_{zy}}{\partial z} + \hat{b}_y = 0 \tag{2.4}$$
$$\frac{\partial \tau_{xz}}{\partial x} + \frac{\partial \tau_{yz}}{\partial y} + \frac{\partial \sigma_z}{\partial z} + \hat{b}_z = 0$$

which can be written as the following matrix equation

$$\mathbf{L}^{\mathrm{T}}_{(3\times6)}\ \boldsymbol{\sigma}_{(6\times1)} + \hat{\mathbf{b}}_{(3\times1)} = \mathbf{0}_{(3\times1)} \tag{2.5}$$

(III) constitutive equations, that is algebraic equations transforming stresses $\boldsymbol{\sigma}$ into strains $\boldsymbol{\varepsilon}$:

$$\varepsilon_x = \frac{1}{E}\left[\sigma_x - \nu(\sigma_y + \sigma_z)\right], \qquad \varepsilon_y = \frac{1}{E}\left[\sigma_y - \nu(\sigma_x + \sigma_z)\right]$$
$$\varepsilon_z = \frac{1}{E}\left[\sigma_z - \nu(\sigma_x + \sigma_y)\right] \tag{2.6}$$
$$\gamma_{xy} = \frac{1}{G}\tau_{xy}, \quad \gamma_{xz} = \frac{1}{G}\tau_{xz}, \quad \gamma_{yz} = \frac{1}{G}\tau_{yz}, \quad \text{where} \quad G = \frac{E}{2(1+\nu)}$$

which correspond to one matrix equation

$$\boldsymbol{\epsilon}_{(6\times1)} = \mathbf{C}_{(6\times6)}\ \boldsymbol{\sigma}_{(6\times1)} \tag{2.7}$$

where

$$\mathbf{C} = \frac{1}{E}\begin{bmatrix} 1.0 & -\nu & -\nu & 0 & 0 & 0 \\ -\nu & 1.0 & -\nu & 0 & 0 & 0 \\ -\nu & -\nu & 1.0 & 0 & 0 & 0 \\ 0 & 0 & 0 & 2(1+\nu) & 0 & 0 \\ 0 & 0 & 0 & 0 & 2(1+\nu) & 0 \\ 0 & 0 & 0 & 0 & 0 & 2(1+\nu) \end{bmatrix} \tag{2.8}$$

The equations in which stresses are calculated on the basis of strains are also often used and then the following matrix relation can be written

$$\boldsymbol{\sigma}_{(6\times1)} = \mathbf{E}_{(6\times6)}\, \boldsymbol{\varepsilon}_{(6\times1)} \tag{2.9}$$

As we can see, both the flexibility matrix $\mathbf{C}$ and the rigidity matrix $\mathbf{E} = \mathbf{C}^{-1}$ are used.

Figure 2.1 illustrates the notation and sign convention for the stresses revealed on the walls of elementary volumetric segment $d\Omega = dx\, dy\, dz$.

The information about boundary conditions at every point on the surface must be provided in addition to equations (I)–(III). The static boundary conditions resulting from zero or nonzero boundary loads $\hat{\boldsymbol{\sigma}}_b$ are given on the part of boundary $\partial\Omega_\sigma$ and formulated as the relation

$$\boldsymbol{\sigma}_b = \hat{\boldsymbol{\sigma}}_b \tag{2.10}$$

The kinematic boundary conditions, imposing given boundary displacements $\hat{\mathbf{u}}$ on the part of boundary $\partial\Omega_u$ (including zero ones resulting from rigid supports) are written as the relation

$$\mathbf{u}_b = \hat{\mathbf{u}}_b \tag{2.11}$$

The stress vector $\boldsymbol{\sigma}_b$ and the displacement $\mathbf{u}_b$ vector, written with respect to a boundary vector base, can be identified according to transformation rules, known from the *theory of elasticity*, which in matrix notation reads:

$$\boldsymbol{\sigma}_{b(3\times1)} = \mathbf{T}^{\sigma}_{(3\times6)}\, \boldsymbol{\sigma}_{(6\times1)}, \qquad \mathbf{u}_{b(3\times1)} = \mathbf{T}^{u}_{(3\times3)}\, \mathbf{u}_{(3\times1)} \tag{2.12}$$

where $\mathbf{T}^{\sigma}$ and $\mathbf{T}^{u}$ are the transformation matrices for stress and displacement, respectively.

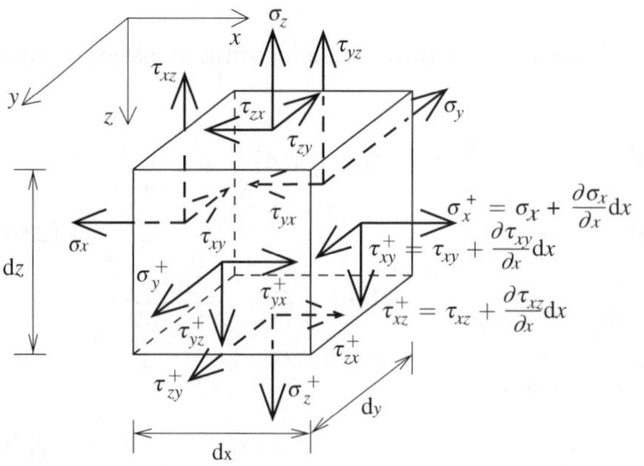

Figure 2.1 Notation and sign convention for stresses in a 3D body

Box 2.1 Equations describing boundary value problems for a 3D body

Local formulation

a) Matrix notation for three groups of equations:

$$\boldsymbol{\varepsilon}_{(6\times1)} = \mathbf{L}_{(6\times3)}\,\mathbf{u}_{(3\times1)}, \quad \mathbf{L}^{\mathrm{T}}_{(3\times6)}\,\boldsymbol{\sigma}_{(6\times1)} + \hat{\mathbf{b}}_{(3\times1)} = \mathbf{0}_{(3\times1)}, \quad \boldsymbol{\varepsilon}_{(6\times1)} = \mathbf{C}_{(6\times6)}\,\boldsymbol{\sigma}_{(6\times1)}$$

Boundary conditions:

$$\text{kinematic} \quad \mathbf{u}_b = \hat{\mathbf{u}}_b \quad \text{on } \partial\Omega_u \quad \text{and/or} \quad \text{static} \quad \boldsymbol{\sigma}_b = \hat{\boldsymbol{\sigma}}_b \quad \text{on } \partial\Omega_\sigma$$

b) Engineering notation of 15 equations:

(I) kinematic equations (6):

$$\varepsilon_x = \frac{\partial u_x}{\partial x}, \qquad \varepsilon_y = \frac{\partial u_y}{\partial y}, \qquad \varepsilon_z = \frac{\partial u_z}{\partial z}$$

$$\gamma_{xy} = \frac{\partial u_x}{\partial y} + \frac{\partial u_y}{\partial x}, \qquad \gamma_{xz} = \frac{\partial u_x}{\partial z} + \frac{\partial u_z}{\partial x}, \qquad \gamma_{yz} = \frac{\partial u_y}{\partial z} + \frac{\partial u_z}{\partial y}$$

(II) equilibrium equations (3):

$$\frac{\partial \sigma_x}{\partial x} + \frac{\partial \tau_{yx}}{\partial y} + \frac{\partial \tau_{zx}}{\partial z} + \hat{b}_x = 0, \qquad \frac{\partial \tau_{xy}}{\partial x} + \frac{\partial \sigma_y}{\partial y} + \frac{\partial \tau_{zy}}{\partial z} + \hat{b}_y = 0$$

$$\frac{\partial \tau_{xz}}{\partial x} + \frac{\partial \tau_{yz}}{\partial y} + \frac{\partial \sigma_z}{\partial z} + \hat{b}_z = 0$$

(III) constitutive equations (6):

$$\varepsilon_x = \frac{1}{E}[\sigma_x - \nu(\sigma_y + \sigma_z)], \quad \varepsilon_y = \frac{1}{E}[\sigma_y - \nu(\sigma_x + \sigma_z)]$$

$$\varepsilon_z = \frac{1}{E}[\sigma_z - \nu(\sigma_x + \sigma_y)], \quad \gamma_{xy} = \frac{1}{G}\tau_{xy}, \quad \gamma_{xz} = \frac{1}{G}\tau_{xz}, \quad \gamma_{yz} = \frac{1}{G}\tau_{yz}$$

Global formulation

Total potential energy

$$\Pi = U - W$$

Internal strain energy

$$U = \frac{1}{2}\int_\Omega \boldsymbol{\sigma}^{\mathrm{T}}\boldsymbol{\varepsilon}\,\mathrm{d}\Omega$$

External work of body and surface loads

$$W = \int_\Omega \hat{\mathbf{b}}^{\mathrm{T}}\mathbf{u}\,\mathrm{d}\Omega + \int_{\partial\Omega_\sigma} \hat{\boldsymbol{\sigma}}_b^{\mathrm{T}}\mathbf{u}_b\,\mathrm{d}\partial\Omega$$

In the present chapter and the subsequent ones, the sets of equations will be gathered in boxes in order to confront the equations describing various typical boundary value problems. Next to the general equations, the relations for particular cases will be given. The sets of equations for local and global formulations will always have a similar structure. As the first one, the set of equations describing the BVP for a 3D body in the Cartesian coordinate system is presented in Box 2.1.

The system of 15 equations with boundary conditions is equivalent to the condition of minimum total potential energy Π of the deformable body with assumed kinematic constraints. The total energy is a functional composed of the strain energy U and the work of external body and surface loads W (the latter with a minus sign). This is the so-called global formulation of the problem.

Reference

Timoshenko S and Goodier JN 1951 *Theory of Elasticity* 2nd edn. McGraw-Hill, New York-Toronto-London.

3

Equations of Thin Shells According to the Three-Parameter Kirchhoff–Love Theory

3.1 General Equations for Thin Shells

The geometry of a shell that is represented by its middle surface is described in a system of curvilinear coordinates ξ_α ($\alpha = 1, 2$) by means of two Lame coefficients A_α and two principal curvature radii R_α. In order to analyse a particular surface the geometrical parameters mentioned should be identified, including the equation of the surface in a corresponding coordinate system.

To begin with, the equations of the three-parameter Kirchhoff–Love thin shell theory (local formulation) according to Sanders (1959) are presented in this section. They include:

(I) kinematic equations (Box 3.1)
(II) equilibrium equations (Box 3.2)
(III) constitutive equations (Box 3.3)
(IV) boundary conditions (Box 3.4)

The energy-based global formulation of the theory is given in Box 3.5.

Box 3.1 Equations of the thin shell theory by Sanders (1959)

(I) Kinematic equations (6):

$$\varepsilon_{11} = \frac{1}{A_1}\frac{\partial u_1}{\partial \xi_1} + \frac{1}{A_1 A_2}\frac{\partial A_1}{\partial \xi_2} u_2 + \frac{w}{R_1}$$

$$\varepsilon_{22} = \frac{1}{A_2}\frac{\partial u_2}{\partial \xi_2} + \frac{1}{A_1 A_2}\frac{\partial A_2}{\partial \xi_1} u_1 + \frac{w}{R_2}$$

$$\gamma_{12} = \frac{A_2}{A_1}\frac{\partial}{\partial \xi_1}\left(\frac{u_2}{A_2}\right) + \frac{A_1}{A_2}\frac{\partial}{\partial \xi_2}\left(\frac{u_1}{A_1}\right)$$

$$\kappa_{11} = -\frac{1}{A_1}\frac{\partial}{\partial \xi_1}\left(\frac{1}{A_1}\frac{\partial w}{\partial \xi_1} - \frac{u_1}{R_1}\right) - \frac{1}{A_1 A_2}\left(\frac{1}{A_2}\frac{\partial w}{\partial \xi_2} - \frac{u_2}{R_2}\right)\frac{\partial A_1}{\partial \xi_2}$$

Plate and Shell Structures: Selected Analytical and Finite Element Solutions, First Edition.
Maria Radwańska, Anna Stankiewicz, Adam Wosatko and Jerzy Pamin.
© 2017 John Wiley & Sons Ltd. Published 2017 by John Wiley & Sons Ltd.

$$\kappa_{22} = -\frac{1}{A_2}\frac{\partial}{\partial \xi_2}\left(\frac{1}{A_2}\frac{\partial w}{\partial \xi_2} - \frac{u_2}{R_2}\right) - \frac{1}{A_1 A_2}\left(\frac{1}{A_1}\frac{\partial w}{\partial \xi_1} - \frac{u_1}{R_1}\right)\frac{\partial A_2}{\partial \xi_1}$$

$$\chi_{12} = -\frac{A_2}{A_1}\frac{\partial}{\partial \xi_1}\left[\frac{1}{A_2}\left(\frac{1}{A_2}\frac{\partial w}{\partial \xi_2} - \frac{u_2}{R_2}\right)\right] - \frac{A_1}{A_2}\frac{\partial}{\partial \xi_2}\left[\frac{1}{A_1}\left(\frac{1}{A_1}\frac{\partial w}{\partial \xi_1} - \frac{u_1}{R_1}\right)\right]$$
$$+ \frac{1}{2A_1 A_2}\left(\frac{1}{R_2} - \frac{1}{R_1}\right)\left[\frac{\partial (A_2 u_2)}{\partial \xi_1} - \frac{\partial (A_1 u_1)}{\partial \xi_2}\right]$$

Box 3.2 Equations of the thin shell theory by Sanders

(II) Equilibrium equations (5):

$$\frac{\partial (A_2 n_{11})}{\partial \xi_1} + \frac{\partial (A_1 n_{12})}{\partial \xi_2} + \frac{\partial A_1}{\partial \xi_2} n_{12} - \frac{\partial A_2}{\partial \xi_1} n_{22} + \frac{A_1 A_2}{R_1} t_1$$
$$+ \frac{A_1}{2}\frac{\partial}{\partial \xi_2}\left[\left(\frac{1}{R_1} - \frac{1}{R_2}\right) m_{12}\right] + A_1 A_2 \hat{p}_1 = 0$$

$$\frac{\partial (A_2 n_{12})}{\partial \xi_1} + \frac{\partial (A_1 n_{22})}{\partial \xi_2} + \frac{\partial A_2}{\partial \xi_1} n_{12} - \frac{\partial A_1}{\partial \xi_2} n_{11} + \frac{A_1 A_2}{R_2} t_2$$
$$+ \frac{A_2}{2}\frac{\partial}{\partial \xi_1}\left[\left(\frac{1}{R_2} - \frac{1}{R_1}\right) m_{12}\right] + A_1 A_2 \hat{p}_2 = 0$$

$$\frac{\partial (A_2 t_1)}{\partial \xi_1} + \frac{\partial (A_1 t_2)}{\partial \xi_2} - A_1 A_2\left(\frac{n_{11}}{R_1} + \frac{n_{22}}{R_2}\right) + A_1 A_2 \hat{p}_n = 0$$

$$\frac{\partial (A_2 m_{11})}{\partial \xi_1} + \frac{\partial (A_1 m_{12})}{\partial \xi_2} + \frac{\partial A_1}{\partial \xi_2} m_{12} - \frac{\partial A_2}{\partial \xi_1} m_{22} - A_1 A_2 t_1 = 0$$

$$\frac{\partial (A_2 m_{12})}{\partial \xi_1} + \frac{\partial (A_1 m_{22})}{\partial \xi_2} + \frac{\partial A_2}{\partial \xi_1} m_{12} - \frac{\partial A_1}{\partial \xi_2} m_{11} - A_1 A_2 t_2 = 0$$

Box 3.3 Equations of the thin shell theory by Sanders

(III) Constitutive equations (6):

$$n_{11} = D^n(\varepsilon_{11} + \nu\varepsilon_{22}), \quad n_{22} = D^n(\varepsilon_{22} + \nu\varepsilon_{11}), \quad n_{12} = D^n\frac{(1-\nu)}{2}\gamma_{12}$$

$$m_{11} = D^m(\kappa_{11} + \nu\kappa_{22}), \quad m_{22} = D^m(\kappa_{22} + \nu\kappa_{11}), \quad m_{12} = D^m\frac{(1-\nu)}{2}\chi_{12}$$

where: $D^n = \dfrac{E\,h}{1-\nu^2}, \quad D^m = \dfrac{E\,h^3}{12(1-\nu^2)}$

Box 3.4 Equations of the thin shell theory by Sanders

(IV) Boundary conditions:

a) kinematic: $u_\nu = \hat{u}_\nu, \quad u_s = \hat{u}_s, \quad w = \hat{w}, \quad \vartheta_\nu = \hat{\vartheta}_\nu$

b) static:

$$n_\nu = \hat{n}_\nu, \quad \tilde{n}_{\nu s} = n_{\nu s} + \left(\frac{3}{2R_s} - \frac{1}{2R_\nu}\right) m_{\nu s} = \hat{n}_{\nu s}$$

$$\tilde{t}_\nu = t_\nu + \frac{1}{A_s}\frac{\partial m_{\nu s}}{\partial \xi_s} = \hat{t}_\nu, \quad m_\nu = \hat{m}_\nu$$

c) mixed

Box 3.5 Global formulation of the thin shell theory by Sanders

Total potential energy $\Pi = U^n + U^m - W$

Internal energy

$$U^n = \frac{D^n}{2}\int_\Omega \left(n_{11}\varepsilon_{11} + n_{12}\gamma_{12} + n_{22}\varepsilon_{22}\right) \mathrm{d}\Omega$$

$$U^m = \frac{D^m}{2}\int_\Omega \left(m_{11}\kappa_{11} + m_{12}\chi_{12} + m_{22}\kappa_{22}\right) \mathrm{d}\Omega$$

External load work

$$W = \int_\Omega \left(\hat{p}_1 u_1 + \hat{p}_2 u_2 + \hat{p}_n w\right) \mathrm{d}\Omega + \int_{\partial\Omega_\sigma} \left(\hat{n}_\nu u_\nu + \hat{n}_{\nu s} u_s + \hat{t}_\nu w + \hat{m}_\nu \vartheta_\nu\right) \mathrm{d}\partial\Omega$$

The general shell equations will be followed by three sets of equations for three types of surface structures: flat membrane, see Subsection 3.3.1 and Box 3.6, a plate in bending, see Subsection 3.3.2 and Box 3.7, cylindrical shell, see Subsection 3.3.3 and Box 3.8.

The static analysis begins with the definition of known external actions, represented by three vectors:

- surface load vector $\hat{\mathbf{p}}_{(3\times1)} = [\hat{p}_1, \hat{p}_2, \hat{p}_n]^{\mathrm{T}}$ [N/m²]
- boundary load vector $\hat{\mathbf{p}}_{b(4\times1)} = [\hat{n}_\nu, \hat{n}_{\nu s}, \hat{t}_\nu \mid \hat{m}_\nu]^{\mathrm{T}}$ [N/m], [Nm/m]
- vector of generalized boundary displacements $\hat{\mathbf{u}}_{b(4\times1)} = [\hat{u}_\nu, \hat{u}_s, \hat{w} \mid \hat{\vartheta}_\nu]^{\mathrm{T}}$ [m], [−]

Box 3.6 Equations for flat rectangular membranes (in a Cartesian coordinate system)

Local formulation

a) Matrix notation:

$$\mathbf{e}^n_{(3\times1)} = \mathbf{L}^n_{(3\times2)}\mathbf{u}^n_{(2\times1)}, \quad \mathbf{L}^{n\,\mathrm{T}}_{(2\times3)}\,\mathbf{s}^n_{(3\times1)} + \hat{\mathbf{p}}^n_{(2\times1)} = \mathbf{0}_{(2\times1)}, \quad \mathbf{s}^n_{(3\times1)} = \mathbf{D}^n_{(3\times3)}\,\mathbf{e}^n_{(3\times1)}$$

$$\text{where} \quad \mathbf{L}^n_{(3\times2)} = \begin{bmatrix} \partial/\partial x & 0 \\ 0 & \partial/\partial y \\ \partial/\partial y & \partial/\partial x \end{bmatrix}, \quad \mathbf{D}^n_{(3\times3)} = \frac{Eh}{1-\nu^2}\begin{bmatrix} 1 & \nu & 0 \\ \nu & 1 & 0 \\ 0 & 0 & \frac{1-\nu}{2} \end{bmatrix}$$

boundary conditions: kinematic $\mathbf{u}^n_b = \hat{\mathbf{u}}^n_b$ and/or static $\mathbf{s}^n_b = \hat{\mathbf{p}}^n_b$

b) Traditional notation – eight equations and boundary conditions

(I) kinematic equations (3):

$$\varepsilon_x = \frac{\partial u}{\partial x}, \quad \varepsilon_y = \frac{\partial v}{\partial y}, \quad \gamma_{xy} = \frac{\partial u}{\partial y} + \frac{\partial v}{\partial x}$$

(II) equilibrium equations (2):

$$\frac{\partial n_x}{\partial x} + \frac{\partial n_{yx}}{\partial y} + \hat{p}_x = 0, \quad \frac{\partial n_{xy}}{\partial x} + \frac{\partial n_y}{\partial y} + \hat{p}_y = 0$$

(III) constitutive equations (3):

$$n_x = D^n(\varepsilon_x + \nu\varepsilon_y), \quad n_y = D^n(\varepsilon_y + \nu\varepsilon_x), \quad n_{xy} = D^n(1-\nu)\,\gamma_{xy}/2$$

where $D^n = E\,h/(1-\nu^2)$

(IV) boundary conditions:

a) kinematic: $u_\nu = \hat{u}_\nu$, $u_s = \hat{u}_s$

b) static: $n_\nu = \hat{n}_\nu$, $n_s = \hat{n}_s$

c) mixed (one kinematic and one static)

Global formulation

Total potential energy $\Pi^n = U^n - W^n$

Internal strain energy

$$U^n = \frac{D^n}{2}\int_A \left[\left(\frac{\partial u}{\partial x}\right)^2 + \left(\frac{\partial v}{\partial y}\right)^2 + 2\nu\frac{\partial u}{\partial x}\frac{\partial v}{\partial y} + \frac{1-\nu}{2}\left(\frac{\partial u}{\partial y} + \frac{\partial v}{\partial x}\right)^2\right] \mathrm{d}x\,\mathrm{d}y$$

External load work

$$W^n = \int_A [\hat{p}_x u + \hat{p}_y v]\,\mathrm{d}x\,\mathrm{d}y + \int_{\partial A_\sigma} [\hat{n}_\nu u_\nu + \hat{n}_{\nu s} u_s]\,\mathrm{d}s$$

Box 3.7 Equations for thin rectangular plates according to Kirchhoff–Love theory

Local formulation

a) Matrix notation of kinematic and constitutive equations:

$$\mathbf{e}^m_{(3\times1)} = \mathbf{L}^m_{(3\times1)}\mathbf{u}^m_{(1\times1)}, \quad \mathbf{m}_{(3\times1)} = \mathbf{D}^m_{(3\times3)}\,\mathbf{e}^m_{(3\times1)}$$

$$\text{where:}\quad \mathbf{L}^m_{(3\times1)} = \begin{bmatrix} -\partial^2/\partial x^2 \\ -\partial^2/\partial y^2 \\ -2\partial^2/\partial x\partial y \end{bmatrix}, \quad \mathbf{D}^m_{(3\times3)} = \frac{E\,h^3}{12(1-\nu^2)}\begin{bmatrix} 1 & \nu & 0 \\ \nu & 1 & 0 \\ 0 & 0 & \frac{1-\nu}{2} \end{bmatrix}$$

boundary conditions: kinematic $\mathbf{u}^m_b = \hat{\mathbf{u}}^m_b$ and/or static $\mathbf{s}^m_b = \hat{\mathbf{p}}^m_b$

b) Traditional notation – nine equations and boundary conditions

(I) kinematic equations (3):

$$\kappa_x = -\frac{\partial^2 w}{\partial x^2}, \quad \kappa_y = -\frac{\partial^2 w}{\partial y^2}, \quad \chi_{xy} = -2\frac{\partial^2 w}{\partial x\partial y}$$

(II) equilibrium equations (3):

$$\frac{\partial t_x}{\partial x} + \frac{\partial t_y}{\partial y} + \hat{p}_z = 0, \quad \frac{\partial m_x}{\partial x} + \frac{\partial m_{yx}}{\partial y} - t_x = 0, \quad \frac{\partial m_{xy}}{\partial x} + \frac{\partial m_y}{\partial y} - t_y = 0$$

(III) constitutive equations (3):

$$m_x = D^m(\kappa_x + \nu\kappa_y), \quad m_y = D^m(\kappa_y + \nu\kappa_x), \quad m_{xy} = D^m(1-\nu)\,\chi_{xy}/2$$

where $D^m = E\,h^3/[12(1-\nu^2)]$

(IV) conditions

- on boundary lines:
  a) kinematic: $w = \hat{w}$, $\vartheta_\nu = \hat{\vartheta}_\nu$
  b) static: $\tilde{t}_\nu = t_\nu + \dfrac{\partial m_{\nu s}}{\partial s} = \hat{t}_\nu$, $m_\nu = \hat{m}_\nu$
  c) mixed (one kinematic and one static)
- in corners: kinematic $w_i = \hat{w}_i$ or static $T_i = 2\,m_{xy} = \hat{P}_i$

Global formulation

Total potential energy $\Pi^m = U^m - W^m$

Internal strain energy

$$U^m = \frac{D^m}{2}\int_A \left[\left(\frac{\partial^2 w}{\partial x^2}\right)^2 + \left(\frac{\partial^2 w}{\partial y^2}\right)^2 + 2\nu\frac{\partial^2 w}{\partial x^2}\frac{\partial^2 w}{\partial y^2} + 2(1-\nu)\left(\frac{\partial^2 w}{\partial x\partial y}\right)^2\right] \mathrm{d}x\,\mathrm{d}y$$

External load work

$$W^m = \int_A \hat{p}_z w\,\mathrm{d}x\,\mathrm{d}y + \int_{\partial A_\sigma} (\hat{t}_\nu w + \hat{m}_\nu \vartheta_\nu)\,\mathrm{d}s$$

Box 3.8 Equations for cylindrical shells of revolution in an axisymmetric membrane-bending state

Local formulation

Traditional notation – 11 equations and boundary conditions

(I) kinematic equations (4):

$$\varepsilon_x = u', \quad \varepsilon_\theta = \frac{w}{R}, \quad \kappa_x = -w'', \quad \kappa_\theta = 0, \quad (\,)' = \frac{\mathrm{d}(\,)}{\mathrm{d}x}$$

(II) equilibrium equations (3):

$$n_x' + \hat{p}_x = 0, \quad t_x' - \frac{n_\theta}{R} + \hat{p}_n = 0, \quad m_x' - t_x = 0$$

(III) constitutive equations (4):

$$n_x = D^n(\varepsilon_x + \nu\varepsilon_\theta), \qquad n_\theta = D^n(\varepsilon_\theta + \nu\varepsilon_x)$$

$$m_x = D^m\kappa_x, \qquad m_\theta = D^m\nu\kappa_x$$

where: $D^n = \dfrac{Eh}{1-\nu^2}$, $\quad D^m = \dfrac{Eh^3}{12(1-\nu^2)}$

(IV) conditions on boundary line $x_b = \text{const.}$:

a) kinematic:

$$u_x = \hat{u}_x, \quad \vartheta_x = \hat{\vartheta}_x, \quad w = \hat{w}$$

b) static:

$$n_x = \hat{n}_x, \quad m_x = \hat{m}_x, \quad t_x = \hat{t}_x$$

c) mixed

Global formulation

Total potential energy $\quad \Pi = U^n + U^m - W$

Internal strain energy

$$U^n = \frac{D^n}{2}\int_\Omega \left(n_x\,\varepsilon_x + n_\theta\,\varepsilon_\theta\right)\mathrm{d}\Omega$$

$$U^m = \frac{D^m}{2}\int_\Omega \left(m_x\,\kappa_x + m_\theta\,\kappa_\theta\right)\mathrm{d}\Omega$$

External load work

$$W = \int_\Omega \left(\hat{p}_x\,u + \hat{p}_n\,w\right)\mathrm{d}\Omega + \int_{\partial\Omega_\sigma} \left(\hat{n}_x\,u_x + \hat{t}_x\,w + \hat{m}_x\,\vartheta_x\right)\mathrm{d}\partial\Omega$$

Next, the behaviour of shell structures is described by the following groups of vectors at a point on the middle surface with curvilinear coordinates ξ_α

- translation vector $\mathbf{u}_{(3\times1)}(\xi_1, \xi_2) = [u_1, u_2, w]^{\mathrm{T}}$ [m]
- generalized strain vector $\mathbf{e}_{(6\times1)}(\xi_1, \xi_2) = [\varepsilon_{11}, \varepsilon_{22}, \gamma_{12} \mid \kappa_{11}, \kappa_{22}, \chi_{12}]^{\mathrm{T}}$ [−], [m^{-1}]
- generalized stress vector (stress resultants: forces and moments)
$\mathbf{s}_{(8\times1)}(\xi_1, \xi_2) = [n_{11}, n_{22}, n_{12} \mid m_{11}, m_{22}, m_{12} \mid t_1, t_2]^{\mathrm{T}}$, [N/m], [Nm/m], [N/m]

as well as rotations ϑ_1, ϑ_2 of the normal $\mathbf{n}$ and rotation ϑ_n about the normal, which depend on the translations and their derivatives, see Equations (1.43) and (1.44).

The adopted notation and sign convention are shown for:

- surface loading and generalized displacements (translations and rotations) – Figure 3.1
- forces and moments in shell cross-sections – Figure 3.2
- kinematic and static boundary quantities – Figure 3.3.

A short review of theories describing elastic plates and shells is given in Section 1.2. More details of the solutions of thin shell problems can be found in Başar and Krätzig (2001), Borkowski et al. (2001), Girkmann (1956), Nowacki (1980), Timoshenko and Woinowsky-Krieger (1959) and Waszczyszyn and Radwańska (1995).

For thin shells, subjected to Kirchhoff–Love constraints, the best shell theory equations with respect to the consistence of their local formulation with the virtual work principle (global form) were given by Sanders (1959). The Sanders equations

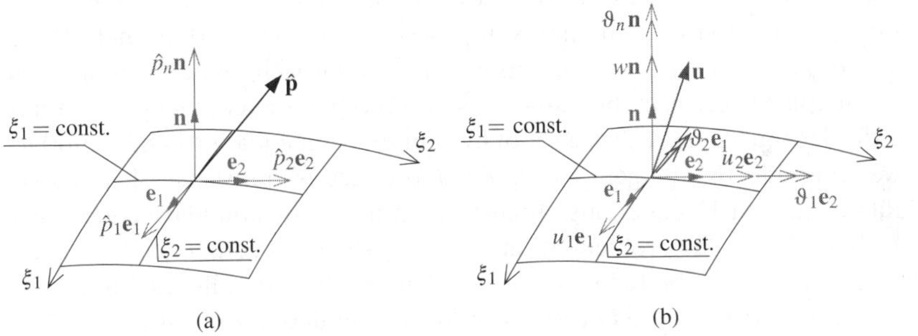

Figure 3.1 (a) Surface loading and (b) generalized displacements

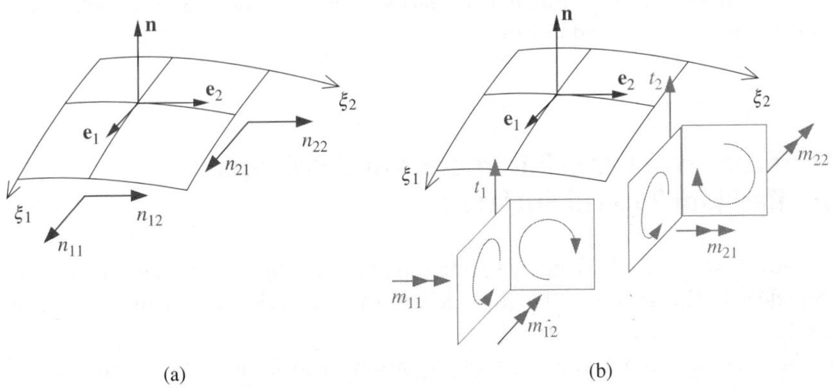

Figure 3.2 (a) Forces in membrane state and (b) forces and moments in bending state

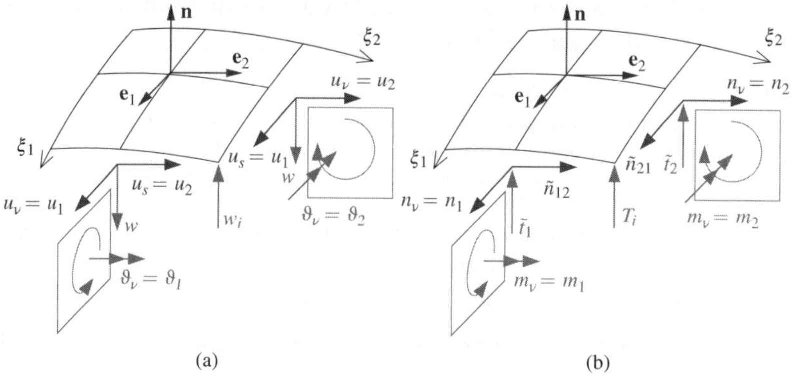

Figure 3.3 Boundary quantities: (a) kinematic and (b) static

are an improved version of the equations of the first-order approximation for thin shells given by Love, in which transverse shear $\gamma_{\alpha z}$ and normal strains ε_{zz} along the thickness are assumed to be zero. The Sanders equations deal with two-dimensional middle surfaces using curvilinear coordinates. Forces and moments are defined on the basis of stresses integrated along the shell thickness (stress resultants). Kinematic relations are improved compared with Love's first approximation theory, as all strains vanish during the motion of the shell as a rigid body. In the Sanders formulation the generalized forces are reduced from 10 to 8 by averaging the tangential forces $n_{12} = (\overline{n}_{12} + \overline{n}_{21})/2$ and twisting moments $m_{12} = (\overline{m}_{12} + \overline{m}_{21})/2$. Additional strains $\gamma_{12} = \varepsilon_{12} + \varepsilon_{21}$ and $\chi_{12} = \kappa_{12} + \kappa_{21}$ are also used. In his approach, Sanders adopts the virtual work principle to derive, in the framework of three-parameter theory (with three independent displacements u_1, u_2, w), a set of five equilibrium equations and boundary conditions defining effective tangent forces $\tilde{n}_{\nu s}$ and effective transverse shear forces $\tilde{t}_\nu$. The 17 unknowns and 17 equations obtained (completed by boundary conditions) constitute the boundary value problem for an arbitrary shell. All equations (kinematic, equilibrium, constitutive and boundary conditions) are presented in Boxes 3.1–3.4.

At this point of our analysis the equations of local formulation for a shell are given only in the traditional notation, although all the relations can also be written in the matrix form.

Energy-based expressions for a global formulation that takes into account membrane and bending effects are presented in Box 3.5.

3.2 Specification of Lame Parameters and Principal Curvature Radii for Typical Surfaces

In three-dimensional space, elementary surface segments lying in the shell middle surface are considered. The shape of the shell determines the relevant set of curvilinear coordinates ξ_α.

Defining the surface geometry, its general equation must be written. In the case of axisymmetric shells described in a spherical or cylindrical system (see Figure 3.4), the meridian equation is given as $r = r(\xi_1)$.

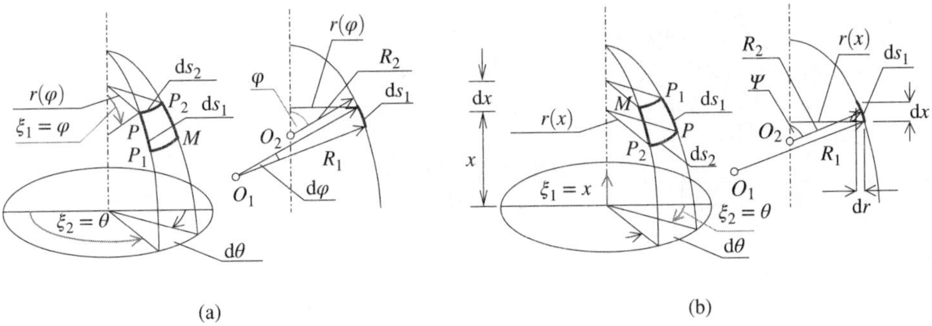

Figure 3.4 Elementary segments of axisymmetric shells described in a coordinate system: (a) spherical and (b) cylindrical

In Subsections 1.3.2 and 1.3.4, expressions for Lame coefficients A_α and principal curvature radii R_α are specified. The relevant information is quoted next in the description of typical surfaces.

Lame coefficients A_α can be calculated as the lengths of vectors tangent to the surface $A_\alpha = |\mathbf{g}_\alpha|$ or from geometric considerations of the elementary surface segment on the basis of the formulae

$$ds_\alpha = A_\alpha \, d\xi_\alpha \tag{3.1}$$

The principal curvature radii R_α result directly from the surface equation.

3.2.1 Shells of Revolution in a Spherical Coordinate System

Once curvilinear coordinates $\xi_1 = \varphi, \xi_2 = \theta$ have been selected, an elementary segment P–P_1–M–P_2 in the shell middle surface defined by the meridian equation $r = r(\varphi)$ is considered. From the geometric analysis the following relations are derived:

$$\begin{aligned} ds_1 &= A_1 \, d\xi_1 = R_1 \, d\varphi, \quad A_1 = R_1(\varphi) \\ ds_2 &= A_2 \, d\xi_2 = r \, d\theta, \quad A_2 = r(\varphi) \end{aligned} \tag{3.2}$$

The curvature radii depend on the meridian shape:

$$R_1 = R_1(\varphi), \quad R_2 = r(\varphi)/\sin\varphi \tag{3.3}$$

Example 3.1 For a spherical shell (Figure 3.5) the meridian equation

$$r(\varphi) = R \sin\varphi \tag{3.4}$$

and four geometric parameters:

$$A_1 = R = \text{const.}, \quad A_2 = r(\varphi) = R \sin\varphi, \quad R_1 = R_2 = R = \text{const.} \tag{3.5}$$

must be known.

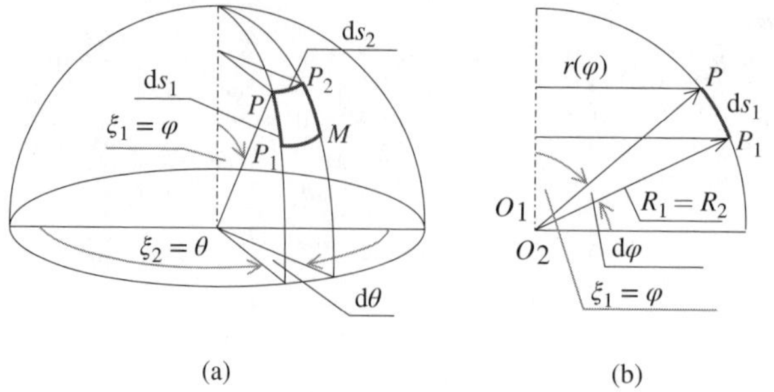

Figure 3.5 Geometry of a spherical shell

3.2.2 Shells of Revolution in a Cylindrical Coordinate System

In the analysis of a shell described by cylindrical coordinates $\xi_1 = x$, $\xi_2 = \theta$, the meridian equation is $r = r(x)$ and angle ψ is additionally introduced to describe the direction tangential to the meridian.

From the geometric analysis the following formulae are derived:

$$\begin{aligned} &ds_1 = A_1\, d\xi_1 = (1/\sin\psi)\, dx = \sqrt{1+(r_{,x})^2}\, dx, \quad A_1 = \sqrt{1+(r_{,x})^2} \\ &ds_2 = A_2\, d\xi_2 = r\, d\theta, \quad A_2 = r(x) \\ &R_1 = R_1(x), \quad R_2 = r(x)/\sin\psi = r(x)\sqrt{1+(r_{,x})^2} \end{aligned} \tag{3.6}$$

Example 3.2 The cylindrical shell in Figure 3.6a with the meridian defined by $r(x) = R = \text{const.}$ is described by:

$$\begin{aligned} &r_{,x} = 0, \quad A_1 = 1 = \text{const.}, \quad A_2 = r(x) = R = \text{const.} \\ &R_1 = \infty, \quad R_2 = R = \text{const.} \end{aligned} \tag{3.7}$$

Example 3.3 The conical shell in Figure 3.6b is defined by the meridian equation $r(x) = x\ \tan\alpha$, where angle $\alpha = \pi/2 - \psi$ is given. The following geometric parameters

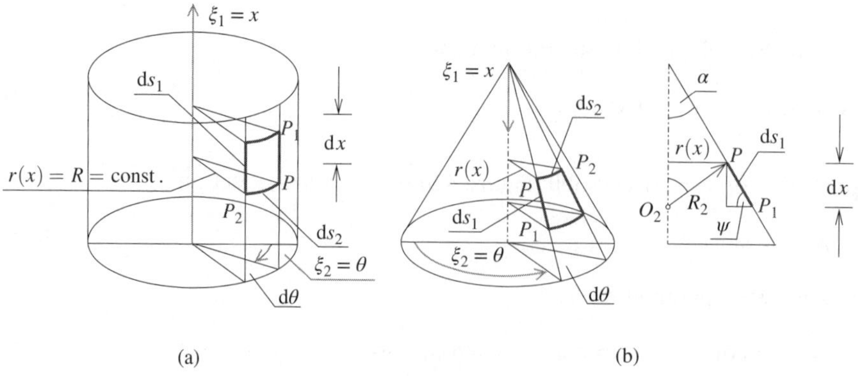

Figure 3.6 (a) Cylindrical shell and (b) conical shell

can be specified:

$$r_{,x} = \tan\,\alpha = \text{const.},\quad A_1 = 1/\cos\,\alpha = \text{const.},\quad A_2 = r(x) = x\,\tan\,\alpha$$
$$R_1 = \infty,\quad R_2 = r(x)/\cos\,\alpha = (x\,\tan\,\alpha)/\cos\,\alpha \tag{3.8}$$

3.2.3 Shallow Shells Represented by Rectangular Projection

Using the Cartesian coordinates $\xi_1 = x$, $\xi_2 = y$, the surface equation $z = z(x, y)$ (Figure 3.7) and taking into account the simplification resulting from inequalities $z_{,x} \ll 1$, $z_{,y} \ll 1$, the Lame coefficients are:

$$A_1 = \sqrt{1 + (z_{,x})^2} \approx 1,\quad A_2 = \sqrt{1 + (z_{,y})^2} \approx 1 \tag{3.9}$$

while the curvature radii R_α are large and result from the surface equation $z = z(x, y)$.

Example 3.4 A paraboloidal shell (Figure 3.7a) with rectangular projection is defined by the equation

$$z(x, y) = \frac{f}{2a^2}(x^2 + y^2) \tag{3.10}$$

while a hyperbolic paraboloid (Figure 3.7b) is defined by

$$z(x, y) = \frac{f}{ab}xy \tag{3.11}$$

3.2.4 Flat Membranes and Plates in Cartesian or Polar Coordinate System

Flat shell structures frequently take the form of rectangular, circular or annular membranes and plates (Figure 3.8).

Example 3.5 Rectangular membranes and plates (Figure 3.8a) described by Cartesian coordinates $\xi_1 = x$, $\xi_2 = y$ are characterized by the following parameters:

$$ds_1 = dx,\quad ds_2 = dy,\quad A_1 = A_2 = 1 = \text{const.},\quad R_1 = R_2 = \infty \tag{3.12}$$

Example 3.6 Circular or annular membranes and plates (Figure 3.8b) parametrized by polar coordinates $\xi_1 = r$, $\xi_2 = \theta$ can be identified by parameters:

$$ds_1 = dr,\quad ds_2 = rd\theta,\quad A_1 = 1 = \text{const.},\quad A_2 = r,\quad R_1 = R_2 = \infty \tag{3.13}$$

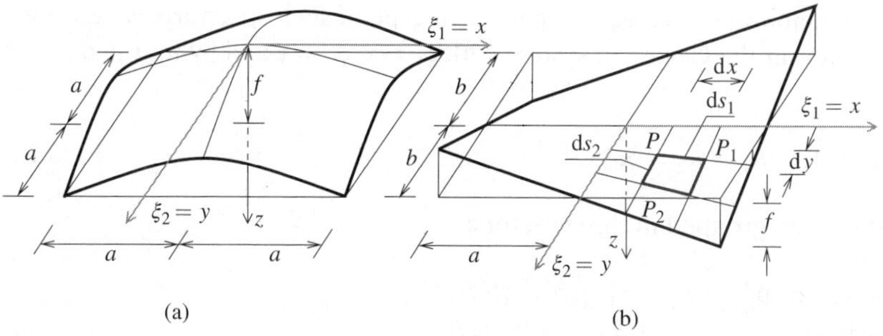

Figure 3.7 Shallow shells: (a) paraboloid and (b) hyperbolic paraboloid

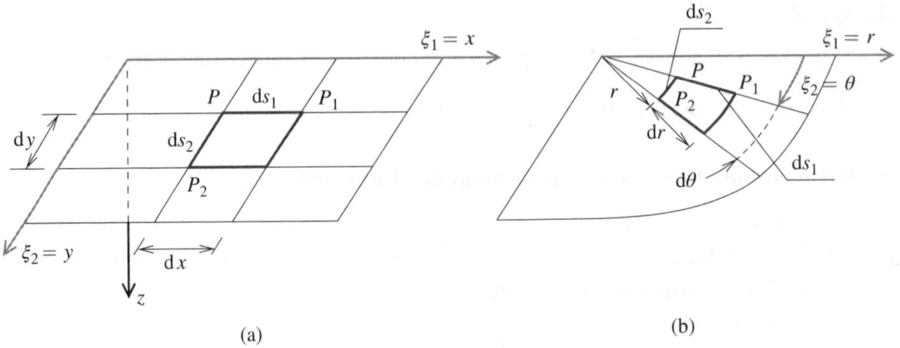

Figure 3.8 Segments from middle surface with two types of coordinates: (a) Cartesian and (b) polar

3.3 Transition from General Shell Equations to Particular Cases of Plates and Shells

In Section 3.1 the general equations for thin shells have been presented. These equations can be treated as a starting point for the derivation of equations describing a particular type of shell in a certain stress state. Then, the coordinate system, Lame coefficients and principal curvature radii are specified and relevant assumptions are made. This approach is more universal and coherent with the formulation of some general finite elements, which are used to discretize shells, as well as, in particular cases, flat membrane and plates. The equations of rectangular membranes and plates under bending are presented in traditional and matrix notation in Subsections 3.3.1, 3.3.2 and, additionally, taking into account FEM, in Subsections 5.2.1 and 5.2.2.

The next subsections contain a brief description of geometry and external actions as well all unknowns determined by solving the boundary value problems for three types of shell structures: rectangular membranes, thin rectangular plates under bending and cylindrical shells in axisymmetric membrane-bending state. Finally, three Boxes 3.6–3.8 are given to illustrate the local and global formulations of the models for the three shell structure types mentioned.

3.3.1 Equations of Rectangular Flat Membranes

The rectangular membrane, treated as a particular type of shell structure, is described in a natural manner in the Cartesian system with $\xi_1 = x$, $\xi_2 = y$ and by four geometric parameters:

$$A_1 = A_2 = 1, \quad R_1 = R_2 = \infty \tag{3.14}$$

External actions are grouped in three vectors:

- surface load vector $\hat{\mathbf{p}}^n_{(2\times1)}(x, y) = [\hat{p}_x, \hat{p}_y]^{\mathrm{T}}$ [N/m^2]
- boundary load vector $\hat{\mathbf{p}}^n_{b(2\times1)}(s) = [\hat{p}_v, \hat{p}_s]^{\mathrm{T}}$ [N/m]
- boundary displacement vector $\hat{\mathbf{u}}^n_{b(2\times1)}(s) = [\hat{u}_v, \hat{u}_s]^{\mathrm{T}}$ [m]

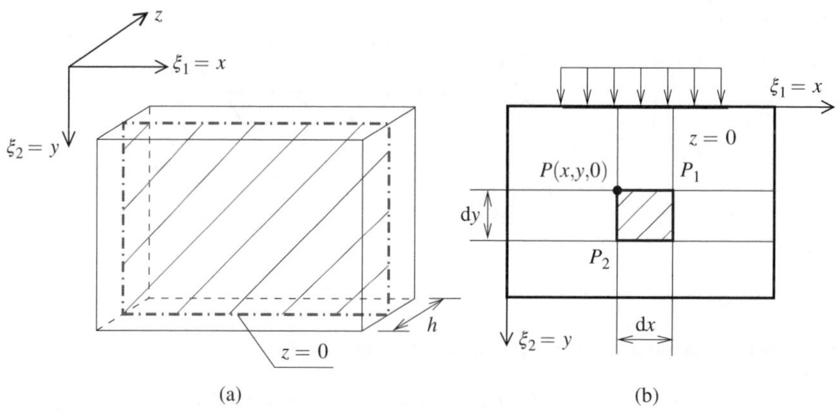

Figure 3.9 (a) Rectangular flat membrane and (b) its geometric 2D model

At any point $P(x, y, 0)$ on the middle surface (Figure 3.9b) eight fields (functions of two coordinates) are introduced. They are included in unknown vectors:

- displacement vector $\mathbf{u}^n_{(2\times1)}(x, y) = [u_x, u_y]^{\mathrm{T}} = [u, v]^{\mathrm{T}}$ [m]
- strain vector $\mathbf{e}^n_{(3\times1)}(x, y) = [\varepsilon_x, \varepsilon_y, \gamma_{xy}]^{\mathrm{T}}$ [–]
- membrane force vector $\mathbf{s}^n_{(3\times1)}(x, y) = \mathbf{n}(x, y) = [n_x, n_y, n_{xy}]^{\mathrm{T}}$ [N/m]

These fields are found from the solution of the boundary value problem, formulated in Box 3.6 in the form of eight equations and suitable boundary conditions.

3.3.2 Equations of Rectangular Plates in Bending

In transition from shells to rectangular plates (modelled by the middle plane) the Cartesian coordinate system is used and, consequently, geometric parameters:

$$A_1 = A_2 = 1, \quad R_1 = R_2 = \infty \tag{3.15}$$

For plates the external actions are grouped in three vectors:

- surface load vector $\hat{\mathbf{p}}^m_{(1\times1)}(x, y) = [\hat{p}_z]$ [N/m^2]
- boundary load vector $\hat{\mathbf{p}}^m_{b(2\times1)}(s) = [\hat{t}_\nu \mid \hat{m}_\nu]^{\mathrm{T}}$ [N/m], [Nm/m]
- boundary displacement vector $\hat{\mathbf{u}}^m_{b(2\times1)}(s) = [\hat{w} \mid \hat{\vartheta}_\nu]^{\mathrm{T}}$ [m], [–]

At any point $P(x, y, 0)$ on the middle plane of the plate in Figure 3.10b the following vectors are defined:

- one-component displacement vector $\mathbf{u}^m_{(1\times1)}(x, y) = [w(x, y)]$ [m]
- strain vector $\mathbf{e}^m_{(3\times1)}(x, y) = [\kappa_x, \kappa_y, \chi_{xy}]^{\mathrm{T}}$ [1/m], $\mathbf{e}^t_{(2\times1)}(x, y) = [\gamma_{xz}, \gamma_{yz}]^{\mathrm{T}} = [0, 0]^{\mathrm{T}}$
- generalized resultant force vector $\mathbf{s}^m = [\mathbf{m}^{\mathrm{T}}, \mathbf{t}^{\mathrm{T}}]^{\mathrm{T}}$:
  - moments $\mathbf{m}_{(3\times1)}(x, y) = [m_x, m_y, m_{xy}]^{\mathrm{T}}$ [Nm/m]
  - transverse shear forces $\mathbf{t}_{(2\times1)}(x, y) = [t_x, t_y]^{\mathrm{T}}$ [N/m]

which are unknown in the presented mathematical model and are calculated from the set of nine equations shown in Box 3.7.

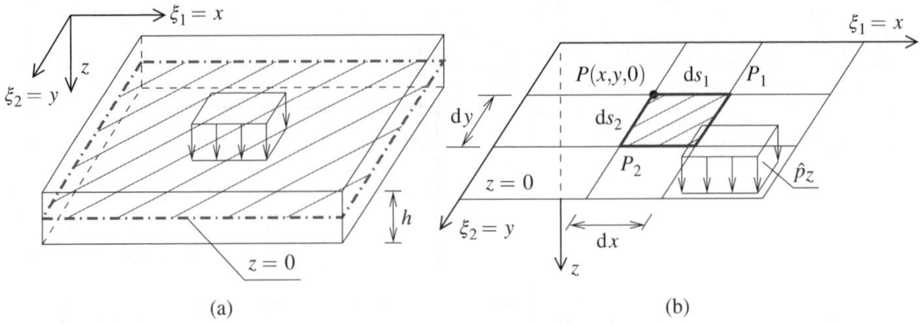

Figure 3.10 (a) Rectangular plate and (b) its geometric 2D model

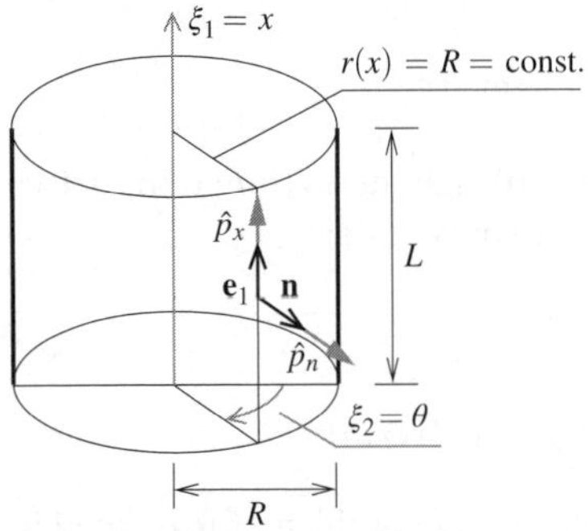

Figure 3.11 Geometry of cylindrical shell

3.3.3 Equations of Cylindrical Shells in an Axisymmetric Membrane-Bending State

The cylindrical shell (Figure 3.11) is naturally described in the cylindrical coordinates system with $\xi_1 = x$, $\xi_2 = \theta$ by the meridian equation $r(x) = R = \text{const.}$, and four geometrical parameters:

$$A_1 = 1, \quad A_2 = R, \quad R_1 = \infty, \quad R_2 = R \tag{3.16}$$

In the analysis of the axisymmetric cylindrical surface, the relations $\partial A_\alpha / \partial \xi_2 = 0$ and $R_1 = \infty$, are taken into account and the following additional dependencies hold:

$$\begin{aligned} &\hat{p}_2 = \hat{p}_\theta = 0, \quad u_2 = u_\theta = 0, \qquad \gamma_{12} = \gamma_{x\theta} = 0, \quad \chi_{12} = \chi_{x\theta} = 0 \\ &n_{12} = n_{x\theta} = 0, \quad m_{12} = m_{x\theta} = 0, \quad t_2 = t_\theta = 0 \end{aligned} \tag{3.17}$$

Notice that partial derivatives with respect to $\xi_2 = \theta$ vanish and hence only ordinary derivatives with respect to $\xi_1 = x$ denoted by $(\ )' = \mathrm{d}(\ )/\mathrm{d}x$ occur in the set of equations,

External actions are represented by vectors: surface load vector, boundary loads and given displacements, respectively: $\hat{\mathbf{p}}(x) = [\hat{p}_x, \hat{p}_n]^{\mathrm{T}}$, $\hat{\mathbf{p}}_b(s) = [\hat{n}_x, \hat{m}_x, \hat{t}_x]^{\mathrm{T}}$, $\hat{\mathbf{u}}_b(s) = [\hat{u}_x, \hat{\vartheta}_x, \hat{w}]^{\mathrm{T}}$.

At any point on the middle surface, 11 fields (functions dependent only on variable x in the axisymmetric state) grouped into three vectors are introduced:

- displacement vector $\mathbf{u}_{(2\times1)}(x) = [u_x, w]^{\mathrm{T}} = [u, w]^{\mathrm{T}}$ [m]
- vectors of strains
  - membrane strains $\mathbf{e}^n_{(2\times1)}(x) = [\varepsilon_x, \varepsilon_\theta]^{\mathrm{T}}$ [–]
  - bending strains $\mathbf{e}^m_{(2\times1)}(x) = [\kappa_x, \kappa_\theta]^{\mathrm{T}}$ [1/m]

- vector of generalized forces
  - membrane forces $\mathbf{n}_{(2\times1)}(x) = [n_x, n_\theta]^{\mathrm{T}}$ [N/m]
  - moments $\mathbf{m}_{(2\times1)}(x) = [m_x, m_\theta]^{\mathrm{T}}$ [Nm/m]
  - transverse forces $\mathbf{t}_{(1\times1)}(x) = [t_x]$ [N/m]

Taking into account this information significantly simplifies the description of the boundary value problem – the number of the unknowns and equations is reduced from 17 to 11. The Sanders equations, specified for the cylindrical shell in an axisymmetric state, take the particular form given in Box 3.8 in traditional notation.

3.4 Displacement Equations for Multi-Parameter Plate and Shell Theories

In previous sections we have presented the mathematical models for plates and shells in the form of a set of kinematic (I), equilibrium (II) and constitutive (III) equations, complete with information about boundary constrains and loading. In the context of local formulation a second type of mathematical model is considered in the form of displacement differential equation (or system of equations) of the so-called multi-parameter shell theory. The specification of the theory results from the number of independent parameters (fields) that are generalized displacements determined on the middle surface of a plate or shell. In addition, this number is also linked with the order of differential equation (or system of equations) of the theory. Moreover, the order of the equation shows the number of required boundary conditions.

Next, the details describing the most interesting five cases (information about the number of parameters, order of theory and number of boundary conditions) are given.

- **Two-parameter theory of rectangular membranes**:
  (i) Two translations in middle plane $u(x, y)$ and $v(x, y)$ are treated as two parameters, (ii) fourth-order theory is considered, because the set of displacement equations contains two second-order differential equations with partial derivatives of $u(x, y)$ and $v(x, y)$, see Chapter 6, Equation (6.2), (iii) on each membrane edge two kinematic and/or static boundary conditions (the latter expressed by displacements) are adopted.
- **One-parameter theory of rectangular thin K–L plates in bending**:
  (i) only one function of deflection (one parameter) is sufficient to determine the plate behaviour, (ii) fourth-order theory is connected with one fourth-order differential equation with partial derivatives of $w(x, y)$, see Chapter 8, Equation (8.19), (iii) for a plate under bending we must formulate two boundary conditions for each edge.
- **Three-parameter theory of rectangular moderately thick M–R plates in bending and transverse shear states:**
  (i) the deflection $w(x, y)$ and two rotation angles $\vartheta_x(x, y)$ and $\vartheta_y(x, y)$ are independent functions, (ii) sixth-order theory with three displacement equations contain up to second-order partial derivatives of three generalized displacements, see Chapter 8, Equations (8.41), (iii) three boundary conditions with in terms of generalized displacement have to be specified on every edge.
- **Three-parameter theory of thin K–L shells in membrane-bending states** (e.g. cylindrical shell segment):
  (i) the deformed state of the shell is described using three translations $u(x, \theta)$, $v(x, \theta)$ and $w(x, \theta)$ at every point on the middle surface, (ii) eighth-order theory based on a set

Box 3.9 Displacement differential equations for selected types of plates and shells

- Displacement equations for rectangular membranes

$$\left(\frac{\partial^2}{\partial x^2}+\frac{1-\nu}{2}\frac{\partial^2}{\partial y^2}\right)u+\frac{1+\nu}{2}\frac{\partial^2}{\partial x\partial y}v=-\frac{1}{D^n}\hat{p}_x$$

$$\frac{1+\nu}{2}\frac{\partial^2}{\partial x\partial y}u+\left(\frac{\partial^2}{\partial y^2}+\frac{1-\nu}{2}\frac{\partial^2}{\partial x^2}\right)v=-\frac{1}{D^n}\hat{p}_y$$

- Displacement equation for rectangular plates (K–L theory)

$$\nabla^2\nabla^2 w(x,y)=\frac{\partial^4 w}{\partial x^4}+2\frac{\partial^4 w}{\partial x^2\,\partial y^2}+\frac{\partial^4 w}{\partial y^4}=\frac{\hat{p}_z}{D^m}$$

- Displacement equations for rectangular plates (M–R theory)

$$D^t\left[\frac{\partial}{\partial x}\left(\frac{\partial w}{\partial x}+\vartheta_x\right)+\frac{\partial}{\partial y}\left(\frac{\partial w}{\partial y}+\vartheta_y\right)\right]+\hat{p}_z=0$$

$$D^m\left[\frac{\partial^2\vartheta_x}{\partial x^2}+\nu\frac{\partial^2\vartheta_y}{\partial x\,\partial y}+\frac{(1-\nu)}{2}\frac{\partial^2\vartheta_x}{\partial y^2}+\frac{(1-\nu)}{2}\frac{\partial^2\vartheta_y}{\partial x\,\partial y}\right]-D^t\left(\frac{\partial w}{\partial x}+\vartheta_x\right)=0$$

$$D^m\left[\frac{\partial^2\vartheta_y}{\partial y^2}+\nu\frac{\partial^2\vartheta_x}{\partial x\,\partial y}+\frac{(1-\nu)}{2}\frac{\partial^2\vartheta_x}{\partial x\,\partial y}+\frac{(1-\nu)}{2}\frac{\partial^2\vartheta_y}{\partial x^2}\right]-D^t\left(\frac{\partial w}{\partial y}+\vartheta_y\right)=0$$

- Displacement equations for cylindrical shells

$$\frac{\partial^2 u}{\partial x^2}+\frac{1-\nu}{2\,R^2}\frac{\partial^2 u}{\partial\theta^2}+\frac{1+\nu}{2\,R}\frac{\partial^2 v}{\partial x\,\partial\theta}+\frac{\nu}{R}\frac{\partial w}{\partial x}=-\frac{\hat{p}_x}{D^n}$$

$$\frac{1+\nu}{2\,R}\frac{\partial^2 u}{\partial x\,\partial\theta}+\frac{1-\nu}{2}\frac{\partial^2 v}{\partial x^2}+\frac{1}{R^2}\frac{\partial^2 v}{\partial\theta^2}+\frac{1}{R^2}\frac{\partial w}{\partial\theta}=-\frac{\hat{p}_\theta}{D^n}$$

$$\frac{\nu}{R}\frac{\partial u}{\partial x}+\frac{1}{R^2}\frac{\partial v}{\partial\theta}+\frac{w}{R^2}+\frac{h^2}{12}\left(\frac{\partial^4 w}{\partial x^4}+\frac{2}{R^2}\frac{\partial^4 w}{\partial x^2\,\partial\theta^2}+\frac{1}{R^4}\frac{\partial^4 w}{\partial\theta^4}\right)=-\frac{\hat{p}_n}{D^n}$$

- Displacement equations for cylindrical shells in axisymmetric membrane-bending states

$$u''+\frac{\nu}{R}w'=-\frac{\hat{p}_x}{D^n},\qquad \frac{\nu\,u'}{R}+\frac{h^2}{12}w^{\mathrm{IV}}+\frac{w}{R^2}=\frac{\hat{p}_n}{D^n}$$

$$D^n=(E\,h)/(1-\nu^2),\quad h^2/12=D^m/D^n,\quad (\,)'=\mathrm{d}(\,)/\mathrm{d}x$$

of three differential equations with up to second-order partial derivatives of two functions $u(x, \theta)$, $v(x, \theta)$ and up to fourth-order partial derivatives of function $w(x, \theta)$, see Chapter 11, Equation (11.73), (iii) on all boundary lines we must write four conditions: one for each of two functions $u(x, \theta)$, $v(x, \theta)$ and two for function $w(x, \theta)$.

- **Two-parameter theory of thin K–L shells in axisymmetric membrane-bending states** (e.g. cylindrical shell):
  (i) displacements are described by two functions $u(x)$, $w(x)$ with one independent cylindrical coordinate x, (ii) sixth-order theory contains a set of two differential equations with first and second-order derivatives of function $u(x)$ and first- and four-order derivatives of $w(x)$, see Chapter 11, Equations (11.5), (iii) on each boundary line we write one boundary condition for function $u(x)$ and two conditions for function $w(x)$.

This summary is given to explain the names of applied theories, their orders and the associated number of boundary conditions. The displacement differential equations for each of these theories are also presented in Box 3.9.

3.5 Remarks

The contents of this chapter can be treated as a brief encyclopaedic lecture on the *shell theory*. Starting from thin shell general equations, three sets of equations for typical shells, namely rectangular membranes, plates under bending and cylindrical shells in the axisymmetric membrane-bending state have been presented. It seems advisable to include, at the beginning of the lecture, the contents of Sections 1.1 and 1.5 on the general classification of shell structures.

In this chapter we have made an attempt to treat complicated theoretical issues in a simple manner, referring to many non-trivial terms from the *theory of elasticity, strength of materials* and *structural mechanics*.

References

Başar Y and Krätzig WB 2001 *Theory of Shell Structures* Fortschritt Berichte 18, Nr. 258 2nd edn. Düsseldorf: VDI Verlag Gmbh, Ruhr–Universität Bochum.

Borkowski A, Jemielita G, Michalak B, Nagórski R, Pietraszkiewicz W, Rudnicki M, Woźniak M and Woźniak C 2001 *Mechanics of Elastic Plates and Shells*, vol. 8 of *Technical Mechanics*. PWN, Warsaw (in Polish).

Girkmann K 1956 *Flächentragwerke*. Springer-Verlag, Wien.

Nowacki W 1980 *Plates and Shells*. PWN, Warsaw (in Polish).

Sanders JL 1959 An improved first-approximation theory of thin shells. Technical Report TR-R24, NASA.

Timoshenko S and Woinowsky-Krieger S 1959 *Theory of Plates and Shells*. McGraw-Hill, New York-Auckland.

Waszczyszyn Z and Radwańska M 1995 Basic equations and calculations methods for elastic shell structures. In Borkowski A, Cichoń C, Radwańska M, Sawczuk A and Waszczyszyn Z, *Structural Mechanics: Computer Approach*, vol. **3**. Arkady, Warsaw, chapter 9, pp. 11–190 (in Polish).

4

General Information about Models and Computational Aspects

This chapter is an introduction to the modelling of mechanical problems of plates and shells and to computational methods applied in linear analysis. The details can be found in the following chapters and in the book by Radwańska (2009).

The problems of structural mechanics can be divided into so-called initial-boundary value problems (IBVP) and boundary value problems (BVP). Moreover, two kinds of models of a structure can be distinguished: continuous models (CM) and discrete models (DM).

In the case of an initial-boundary value problem for a continuous model, one deals with a process that is dependent on both spatial coordinates (two in a 2D problem) and time. The mathematical model is formulated as a set of partial differential equations with space and time derivatives, completed with boundary and initial conditions.

The boundary value problem for a continuous model is described by a set of partial differential equations with spatial derivatives, which needs to be supplemented by boundary conditions resulting from prescribed generalized displacements and/or loads on the boundary of the domain. Solving the problem analytically for a CM, a set of continuous functions is obtained as a solution.

In structural mechanics we generally deal with continuous structures. In many cases it is necessary to replace a continuous systems with a discrete model of the structure. When a computational method is applied, the discretization process converts CM into DM. The mathematical model is then (in most cases) a set of linear algebraic equations completed with boundary and initial conditions for a nonstationary problem (IBVP) or only by boundary conditions in the case of a stationary problem (BVP). As a result of the solution of a problem for a DM, a set of discrete values the so-called generalized coordinates, is obtained. Generalized coordinates are variables that describe the behaviour of the discrete system (structure).

In this chapter three mechanical problems are briefly reviewed:

- statics
- buckling
- free vibrations

These problems correspond to two mathematical models:

- linear boundary value problem
- eigenvalue problem

The selected problems can be solved using different approaches, for example:

- analytical solution

Plate and Shell Structures: Selected Analytical and Finite Element Solutions, First Edition.
Maria Radwańska, Anna Stankiewicz, Adam Wosatko and Jerzy Pamin.
© 2017 John Wiley & Sons Ltd. Published 2017 by John Wiley & Sons Ltd.

- finite difference method (FDM)
- finite element method (FEM)

Omitting a detailed derivation of mathematical relations, a confrontation of computational models for three different mechanical problems is performed for a thin plate (details are given in Chapters 8, 14 and 15). In particular, a simply supported square plate with the input data given in Table 4.1 is going to be considered. The loading configurations for the static and buckling analyses are depicted in Figure 4.1.

4.1 Analytical Approach to Statics, Buckling and Free Vibrations

4.1.1 Statics of a Thin Plate in Bending

In order to find the deflection function for a thin plate in bending, one has to solve the following inhomogeneous fourth-order partial differential equation for deflection (transverse displacement) with the right-hand side dependent on the intensity of transverse load

$$D^m \nabla^2 \nabla^2 w(x, y) = \hat{p}_z(x, y) \tag{4.1}$$

Table 4.1 Input data for the simply supported square plate in Figure 4.1

Geometry:	length $L = a = b = 1.0$ m
	thickness $h = 0.01$ m
Material:	Young's modulus $E = 2 \times 10^8$ kN/m^2
	Poisson's ratio $\nu = 0.3$
	density $\rho = 0.787$ kN s^2/m^4
Calculated parameters:	cross-sectional stiffness $D^m = E\,h^3 \;/\; [12(1 - \nu^2)] = 18.315$ kNm
	surface density $\mu = \rho\, h = 0.00787$ kN s^2/m^3
Loading:	transverse surface load $\hat{p}_z = 1.0$ kN/m^2
	reference load $p_x^* = 1.0$ kN/m

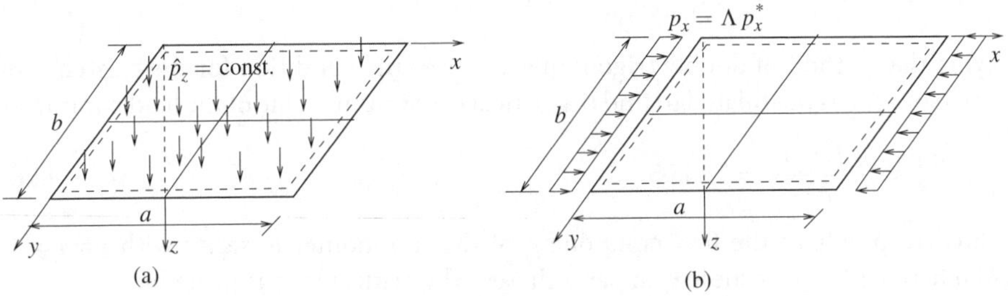

Figure 4.1 Loading configuration for: (a) statics and (b) buckling

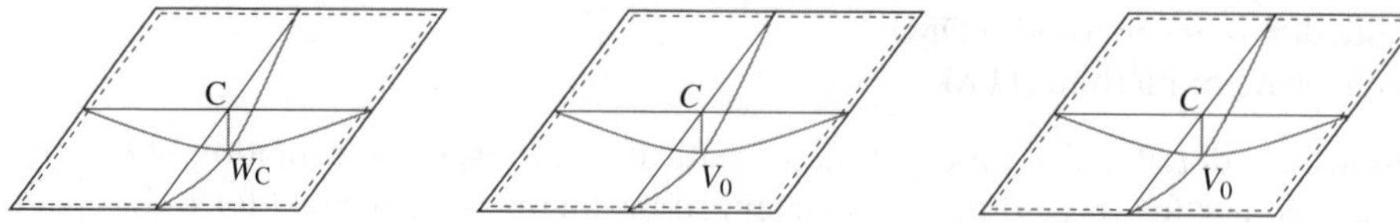

(a) (b) (c)

Figure 4.2 Results for simply supported square plate: (a) static deflection mode, (b) first buckling mode and (c) first free vibration mode

Equation (4.1) has to be completed with boundary conditions. For each simply supported edge the following mixed boundary conditions have to be satisfied:

$$w = 0, \qquad m_\nu = 0 \tag{4.2}$$

Using the method of double trigonometric series (see Section 8.8) the maximum deflection at the centre of the plate is found as

$$w_C^{\text{anal}} = 0.00406 \frac{\hat{p}_z L^4}{D^m} = 0.00022168 \text{ m} \tag{4.3}$$

The special feature of the obtained deflection mode of the square plate is that it has four planes of symmetry (see Figure 4.2a).

4.1.2 Buckling of a Plate

In linear buckling analysis a single parameter load is considered with scalar factor Λ and reference load p^*. The loading process is realized by a monotonic increase of parameter Λ. The plate shown in Figure 4.1b, subjected to in-plane load $p_x = \Lambda\, p_x^*$ inducing a unidirectional compression, is considered.

For a sufficiently small value of Λ the plate behaves as a membrane with the nonzero force:

$$n_x = \Lambda\, n_x^*, \quad n_x^* = -1.0\text{kN/m} \tag{4.4}$$

When stability is lost due to buckling then bending occurs (see Section 14.2).

The phenomenon is described by a fourth-order partial differential equation. In a considered case it is homogeneous and includes the following unknowns: buckling load parameter Λ and function $w(x, y)$, which describes the postbuckling deflection mode with a specific shape admitted by the simply supported boundary

$$D^m\, \nabla^2\nabla^2 w(x, y) - \Lambda\, n_x^* \frac{\partial^2 w(x, y)}{\partial x^2} = 0 \tag{4.5}$$

Applying the method of double trigonometric series the condition for the existence of nonzero solution is postulated to find the critical (minimum) value of the load parameter

$$\Lambda_{\text{cr}}^{(1,1)} = \frac{4\, D^m\, \pi^2}{a^2} = 723.05 \tag{4.6}$$

which corresponds to the first component of the trigonometric series with $i = j = 1$, marked here with upper indices in parentheses. The critical load is given by

$$p_{\text{cr}}^{(1,1)\text{ anal}} = \Lambda_{\text{cr}}^{(1,1)}\, p_x^* = 723.05 \text{ kN/m} \tag{4.7}$$

and the function of buckling deflection $w(x, y) = V^{(1,1)}(x, y)$ for a square plate with the number of half-waves $i = j = 1$, see Figure 4.2b, reads

$$V^{(1,1)}(x, y) = V_0 \sin \frac{\pi x}{a} \sin \frac{\pi y}{a} \tag{4.8}$$

with arbitrary amplitude V_0.

4.1.3 Transverse Free Vibrations of a Plate

For the case of transverse vibrations, taking the inertia forces into account the equilibrium equation for a plate in bending, written in terms of the deflection, has the form (see Section 15.2)

$$D^m \nabla^2\nabla^2 w(x, y, t) + \mu \frac{\partial^2 w(x, y, t)}{\partial t^2} = 0 \tag{4.9}$$

After separation of variables

$$w(x, y, t) = V(x, y)\, T(t) \tag{4.10}$$

a homogeneous fourth-order differential equation with partial derivatives with respect to two space coordinates and unknown angular frequency of vibration ω and the amplitude of the deflection $V(x, y)$ is obtained

$$\nabla^2\nabla^2 V(x, y) - \tilde{\omega}\, V(x, y) = 0 \tag{4.11}$$

where

$$\tilde{\omega} = \frac{\mu\, \omega^2}{D^m} \tag{4.12}$$

From the condition for the existence of nonzero solution the fundamental (minimum) angular frequency of vibration ω is computed according to

$$\omega^{(1,1)\text{anal}} = 2\, \frac{\pi^2}{a^2} \sqrt{\frac{D^m}{\mu}} = 952.40\, \frac{\text{rad}}{\text{s}} \tag{4.13}$$

Next the frequency f is found

$$f = \frac{\omega}{2\,\pi} = 151.58 \text{ Hz} \tag{4.14}$$

The form of vibration for $i = j = 1$ (see Figure 4.2c) is given by

$$V^{(1,1)}(x, y) = V_0 \sin \frac{\pi x}{a} \sin \frac{\pi y}{a} \tag{4.15}$$

with arbitrary coefficient V_0.

4.2 Approximate Approach According to the Finite Difference Method

One of the known discretization methods is the Finite Difference Method (FDM) in which:

- the domain of the problem is replaced by a finite set of grid points
- derivatives are substituted by approximating difference quotients
- a set of linear algebraic equations is formulated and solved

Details can be found in Section 8.11.

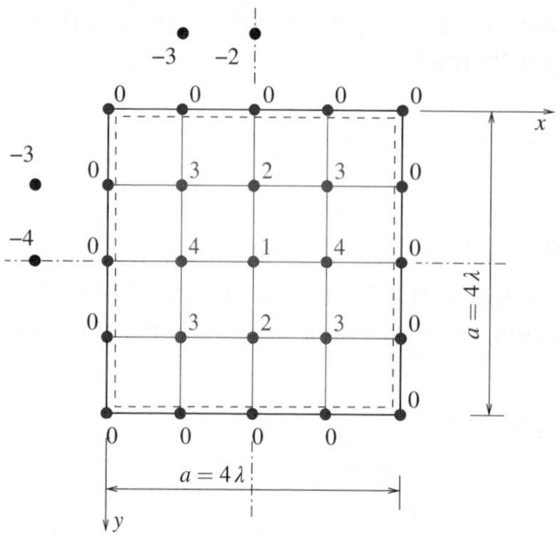

Figure 4.3 FDM discretization for simply supported square plate

4.2.1 Set of Algebraic Equations for Statics of a Plate in Bending

In Figure 4.3 the configuration of the considered plate after FDM discretization (with a four internal points with nonzero deflection) is shown. If we take into account Equation (4.1) and omit the detailed derivations, the set of four inhomogeneous linear equations for computing the values of nonzero deflections for four internal grid points of the plate is given:

$$\mathbf{AW} = \mathbf{B} \tag{4.16}$$

$$\begin{bmatrix} 20 & -16 & 8 & -16 \\ -8 & 20 & -16 & 4 \\ 2 & -8 & 20 & -8 \\ 8 & 4 & -16 & 20 \end{bmatrix} \begin{bmatrix} w_1 \\ w_2 \\ w_3 \\ w_4 \end{bmatrix} = B \begin{bmatrix} 1 \\ 1 \\ 1 \\ 1 \end{bmatrix} \tag{4.17}$$

$$\begin{bmatrix} 10 & -8 & 4 & -8 \\ -8 & 20 & -16 & 4 \\ 4 & -16 & 40 & -16 \\ -8 & 4 & -16 & 20 \end{bmatrix} \begin{bmatrix} w_1 \\ w_2 \\ w_3 \\ w_4 \end{bmatrix} = B \begin{bmatrix} 0.5 \\ 1 \\ 2 \\ 1 \end{bmatrix} \tag{4.18}$$

where

$$B = \frac{\hat{p}_z \lambda^4}{D^m} = 0.00021328 \text{ m} \tag{4.19}$$

and the last form of this equation contains a symmmetric matrix. The solution of this system of equations for $B = 1$ is

$$\mathbf{W} = \begin{bmatrix} w_1 \\ w_2 \\ w_3 \\ w_4 \end{bmatrix} = \begin{bmatrix} 1.0313 \\ 0.7500 \\ 0.5469 \\ 0.7500 \end{bmatrix} \tag{4.20}$$

The maximum value of deflection of the plate at point C (element w_1 of solution vector $\mathbf{W}$ multiplied by B) is compared with the exact analytical result

$$w_C^{\mathrm{FDM}} = B\, w_1 = 0.00021966 \text{ m} = 0.99089\, w_C^{\mathrm{anal}} \tag{4.21}$$

The approximate drawing of the plate deformation, made on the basis of discrete deflection values, is analogous to the static displacement mode presented in Figure 4.2a.

4.2.2 Set of Homogeneous Algebraic Equations for Plate Buckling

Taking Equation (4.5) and adopting the FDM algorithm (compare Section 8.11 and Subsection 14.2.2) one can obtain a set of equations representing the nonstandard algebraic eigenvalue problem that can be written in the following matrix form:

$$(\mathbf{A} - \tilde{\Lambda}\ \mathbf{N})\ \mathbf{V} = \mathbf{0} \tag{4.22}$$

$$\left(\begin{bmatrix} 10 & -8 & 4 & -8 \\ -8 & 20 & -16 & 4 \\ 4 & -16 & 40 & -16 \\ -8 & 4 & -16 & 20 \end{bmatrix} - \tilde{\Lambda} \begin{bmatrix} 1 & 0 & 0 & -1 \\ 0 & 2 & -2 & 0 \\ 0 & -2 & 4 & 0 \\ -1 & 0 & 0 & 2 \end{bmatrix}\right) \begin{bmatrix} v_1 \\ v_2 \\ v_3 \\ v_4 \end{bmatrix} = \begin{bmatrix} 0 \\ 0 \\ 0 \\ 0 \end{bmatrix} \tag{4.23}$$

From the solution that is composed of four coupled pairs of eigenvalues and eigenvectors $(\tilde{\Lambda}_i, \mathbf{V}_i)$, $i = 1, 2, 3, 4$ the minimum eigenvalue is chosen and the critical load is calculated and compared with the exact analytical value:

$$(\tilde{\Lambda}_1, \mathbf{V}_1) = (2.3431, [0.6667, 0.4714, 0.3333, 0.4714]^{\mathrm{T}}) \tag{4.24}$$

$$\tilde{\Lambda}_1 = \left(\frac{\lambda^2}{D^m}\right) \Lambda_{\mathrm{cr}} \rightarrow \Lambda_{\mathrm{cr}} = \left(\frac{D^m}{\lambda^2}\right) \tilde{\Lambda}_1 = 293.04 \times 2.3431 = 687.46 \tag{4.25}$$

$$p_{\mathrm{cr}}^{\mathrm{FDM}} = \Lambda_{\mathrm{cr}}\, p_x^* = 687.46\ \mathrm{kN}/m = 0.9508 p_{\mathrm{cr}}^{\mathrm{anal}} \tag{4.26}$$

The first eigenvector is a discrete representation of the buckling form with four planes of symmetry (similar to the mode shown in Figure 4.2b).

4.2.3 Set of Homogeneous Algebraic Equations for Transverse Free Vibrations of a Plate

The formulation of the nonstandard algebraic eigenvalue problem is based on Equation (4.11) with partial derivatives of deflection amplitude function. Applying the algorithm of FDM, the homogeneous matrix equation is written for four specified grid points:

$$(\mathbf{A} - \tilde{\omega}\ \mathbf{M})\ \mathbf{V} = \mathbf{0} \tag{4.27}$$

$$\left(\begin{bmatrix} 10 & -8 & 4 & -8 \\ -8 & 20 & -16 & 4 \\ 4 & -16 & 40 & -16 \\ -8 & 4 & -16 & 20 \end{bmatrix} - \tilde{\omega} \begin{bmatrix} 0.5 & 0 & 0 & 0 \\ 0 & 1 & 0 & 0 \\ 0 & 0 & 2 & 0 \\ 0 & 0 & 0 & 1 \end{bmatrix}\right) \begin{bmatrix} v_1 \\ v_2 \\ v_3 \\ v_4 \end{bmatrix} = \begin{bmatrix} 0 \\ 0 \\ 0 \\ 0 \end{bmatrix} \tag{4.28}$$

The goal of the calculations is to determine the lowest eigenvalue and associated eigenvector

$$(\tilde{\omega}_1, \mathbf{V}_1) = (1.3726, [0.6667, 0.4714, 0.3333, 0.4714]^{\mathrm{T}}) \tag{4.29}$$

that are used to find the fundamental angular frequency compared to the exact value:

$$\tilde{\omega}_1 = \left(\frac{\mu\, \lambda^4}{D^m}\right) \omega^2, \quad \omega = \sqrt{\left(\frac{D^m}{\mu\, \lambda^4}\right) \tilde{\omega}_1} \tag{4.30}$$

$$\omega^{\text{FDM}} = \sqrt{595\,761 \times 1.3726} = 904.29 \text{ rad}/s = 0.94949\ \omega^{\text{anal}} \tag{4.31}$$

The basic form of vibration has four planes of symmetry as shown in Figure 4.2c.

4.3 Computational Analysis by Finite Element Method

The reader of this book is expected to know the basis of the Finite Element Method (FEM). In this section, FEM algorithms for the problems of statics, buckling and free vibrations are presented in Boxes 4.1–4.3 to complete the discussion of different computational models.

Box 4.1 FEM – algorithm for linear statics

1) Domain discretization
2) Calculation of linear stiffness matrix $\mathbf{k}^e$ and substitute nodal load vector $\mathbf{f}^e$ for each FE
3) Transformation of $\mathbf{k}^e$, $\mathbf{f}^e$ from local to global coordinate system to obtain $\mathbf{K}^e$, $\mathbf{F}^e$
4) Assembly of global stiffness matrix $\mathbf{K}$ and load vector $\mathbf{F}$ (adding concentrated loads, if any, to the load vector)
5) Formation of system of equations $\mathbf{K\,Q} = \mathbf{F} + \mathbf{R}$ with unknown vector of nodal degrees of freedom $\mathbf{Q}$ and reaction forces $\mathbf{R}$
6) Modification of the system of equations by imposing kinematic boundary conditions
7) Calculation of generalized nodal displacement vector $\mathbf{Q}$
8) Computation of reaction force vector $\mathbf{R}$
9) Calculation of resultant forces and generalized strains for each FE

Box 4.2 FEM – algorithm for buckling analysis

Stage I: Static analysis of prebuckling state for reference load p^* (and Λ=1)

1) Domain discretization
2) Calculation of linear stiffness matrix $\mathbf{k}^e$ and load configuration vector $\mathbf{p}^{*e}$ for each FE and transformation to global coordinate system
3) Assembly of global matrix $\mathbf{K}$ and vector $\mathbf{P}^*$ and imposing boundary constraints
4) Calculation of generalized nodal displacement vector $\mathbf{Q}^*$
5) Computation of generalized stress vector σ^* for each FE

Stage II: Buckling analysis – eigenvalue problem

1) Calculation of initial stress stiffness matrix $\mathbf{k}_\sigma^*$ for each FE, transformation
2) Assembly of global matrix $\mathbf{K}_\sigma^*$
3) Formation of system of equations for eigenvalue problem $(\mathbf{K} + \Lambda\mathbf{K}_\sigma^*)\ \mathbf{V} = \mathbf{0}$ and incorporation of kinematic boundary conditions
4) Solution of system of equations: computation of eigenvalues (buckling load parameters) Λ_i and eigenvectors (buckling modes) $\mathbf{V}_i$

Box 4.3 FEM – algorithm for free vibrations analysis

1) Domain discretization
2) Calculation of linear stiffness matrix $\mathbf{k}^e$ and mass matrix $\mathbf{m}^e$ for each FE
3) Assembly of global matrices $\mathbf{K}$ and $\mathbf{M}$
4) Formation of system of equations for eigenvalue problem $(\mathbf{K} - \omega^2\mathbf{M})\mathbf{V} = \mathbf{0}$ and its modification due to kinematic boundary conditions
5) Solution of system of equations: computation of eigenvalues (squares of angular frequency values) ω_i^2 and respective eigenvectors (vibration modes) $\mathbf{V}_i$

Table 4.2 Equations of mathematical and numerical models

Problem \ Method	Analytical	FDM	FEM
Statics	$D^m\,\nabla^2\nabla^2 w(x,y) = \hat{p}_z(x,y)$	$\mathbf{A}\,\mathbf{W} = \mathbf{B}$	$\mathbf{K}\,\mathbf{Q} = \mathbf{F} + \mathbf{P}$
Linear buckling	$D^m\,\nabla^2\nabla^2 w(x,y) - \Lambda\, n_x^* \dfrac{\partial^2 w(x,y)}{\partial x^2} = 0$	$(\mathbf{A} - \tilde{\Lambda}\,\mathbf{N})\,\mathbf{V} = \mathbf{0}$	$(\mathbf{K} + \Lambda\mathbf{K}_\sigma^*)\,\mathbf{V} = \mathbf{0}$
Free vibrations	$D^m\,\nabla^2\nabla^2 w(x,y,t) + \mu\, \dfrac{\partial^2 w(x,y,t)}{\partial t^2} = 0$	$(\mathbf{A} - \tilde{\omega}\,\mathbf{M})\,\mathbf{V} = \mathbf{0}$	$(\mathbf{K} - \omega^2\mathbf{M})\mathbf{V} = \mathbf{0}$

4.4 Computational Models – Summary

To summarize, the analytical models in the form of differential equations and the numerical models (FDM, FEM) in the form of matrix equations are shown for the three previously-mentioned mechanical problems in Table 4.2.

Reference

Radwańska M 2009 *Shell Structures*. Cracow University of Technology, Cracow (in Polish).

5

Description of Finite Elements for Analysis of Plates and Shells

5.1 General Information on Finite Elements

As presented in Chapter 3, three typical states of shell structures are distinguished:

- membrane state, analogous to plane state in a flat membrane
- bending state, which in thin plates and shells is accompanied by zero transverse shear strains and passive transverse forces
- transverse shear state, must be taken into account in the analysis of moderately thick shell structures under bending

A wide variety of finite elements have been worked out for the analysis of shell structures using FEM. They have different application ranges, from very general to specific, see MacNeal (1998), Yang et al. (1990, 2000). There are finite elements exclusively for the discretization of a particular type of shell structures. There are also FEs that are formulated in a very universal manner, with approximation of geometry and displacement fields that cover all the three mentioned states and whose aim is the discretization of thin and moderately thick shells. However, beside problems related to satisfactory approximation purely numerical problems can occur, especially when we want to consider extremely thin shell structures. In fact, there is no FE that enables numerical analysis with sufficient accuracy and efficiency in a wide range of mechanical problems for shell structures. However, Section 18.3 contains the description of advanced FEs that have quite universal applications.

Significant progress in FE formulation is associated with the research on (materially and geometrically) nonlinear problems. In this respect, the reader is referred to the works mentioned in Chapter 1 and the literature dealing with this particular issue.

The authors strongly advise the users of computer programs based on FEM to study the description of FEs included in manuals to be aware of the capabilities and limitations of FEs.

Taking into account the classification of shell structures proposed in Section 1.5 and the general knowledge of FEM, the following types of FEs can be distinguished:

- FEs for flat membranes (plates with in-plane load)
- FEs for thin plates under bending, based on one-parameter Kirchhoff–Love (K–L) theory

Plate and Shell Structures: Selected Analytical and Finite Element Solutions, First Edition.
Maria Radwańska, Anna Stankiewicz, Adam Wosatko and Jerzy Pamin.
© 2017 John Wiley & Sons Ltd. Published 2017 by John Wiley & Sons Ltd.

- FEs for moderately thick plates under bending and transverse shear, based on three-parameter Mindlin–Reissner (M–R) theory
- FEs for curved thin shells, based on three-parameter K–L shell theory
- FEs for curved moderately thick shells, based on five(six)-parameter M–R shell theory
- degenerated FEs based on 3D continuum equations, modified by shell hypotheses, compatible with five-parameter M–R moderately thick shell theory
- solid FEs used for discretization of thick shells and based on 3D continuum equations

FEs with zero curvature (flat) are treated as a special type of curved shell FEs. This is possible when the isoparametric approach is applied, that is the same finite element approximation for the description of both the geometry and deformation is used. The information about zero curvatures of element middle surface implies the separation of membrane and bending effects.

With respect to various applications, the displacement-based FEs are classified according to the following strict criteria:

- number of approximated fields within FE (displacements and in some cases independent rotations of the normal to the middle surface and about the normal)
- number and type of nodal degrees of freedom (dofs)
- formulae based on the relevant shell theory, namely with kinematic and constitutive relations, on the basis of which **B** and **D** matrices necessary for the calculation of FE stiffness matrix $\mathbf{k}^e$ are derived
- formulae used for calculating the potential energy of the shell structures or for writing the virtual work principle
- quadrature of integration over the element area and edge, used for calculating the matrices and vectors of a shell FE

The next classification of FEs applied to discretization of shell structures with geometry-oriented approximation includes:

- geometrically one-dimensional FEs used for the description of shell structures of axisymmetric geometry, with FE approximation in the meridional direction and with expansion of each function into a trigonometric series with respect to circumferential coordinate
- two-dimensional flat three-, four-node FEs (discretized membranes, plates and shells)
- two-dimensional curved six-, eight-, nine-node FEs, related to the classical three-parameter K–L or five(six)-parameter M–R theories

A wide application of the one-dimensional shell FEs (conical two-node or curved three-node) mentioned previously should be emphasized. They are used in the numerical analysis of axisymmetric shell structures: circular or annular membranes or plates, cylindrical, conical or spherical shells. The numerical results obtained with these FEs are very often confronted with analytical solutions and are satisfactory, see Chapters 7 and 9–11.

From the abundant bibliography dealing with the basis of FEM the authors would like to point out the following works ordered chronologically: Zienkiewicz and

Cheung (1967), Ashwell and Gallagher (1976), Hinton and Owen (1979, 1984), Irons and Ahmad (1980), Bathe (1982), Crisfield (1986), Yang (1986), Cook et al. (1989), Batoz and Dhatt (1990), Cook (1995), Bischoff et al. (2004), Zienkiewicz et al. (2005), Hartmann and Katz (2007), Oñate (2013).

5.2 Description of Selected FEs

In further description of FEs, the following indices will be used to highlight the typical states:

- *n* denotes membrane (in-plane) state
- *m* denotes bending
- *t* denotes transverse shear

For the description of FEs we define:

- *NNDOF* – number of nodal degrees of freedom (dofs)
- *NEN* – number of element nodes
- *NEDOF* – number of element dofs

5.2.1 Flat Rectangular Four-Node Membrane FE

The membrane FE in question (see Figure 5.1) is characterized by the following numbers: $NNDOF = 2$, $NEN = 4$, $NEDOF = NNDOF \times NEN = 8$ and by the vectors of node and element displacements:

$$\mathbf{q}_{node} = [u_{node}, v_{node}]^{\mathrm{T}}, \quad \mathbf{q}^{en} = [u_1, v_1 | u_2, v_2 | u_3, v_3 | u_4, v_4]^{\mathrm{T}} \tag{5.1}$$

for $node = 1, \ldots, NEN$ and $e = 1, \ldots, NSE$. Index n is used to mark the membrane state.

To approximate the displacements $u(\xi, \eta)$ and $v(\xi, \eta)$, shape functions $N_i(\xi, \eta)$, where $i = 1, 2, 3, 4$, bilinear with respect to dimensionless normalized coordinates $\xi, \eta \in [-1, +1]$ are used (see Figure 5.2):

$$\begin{aligned} \mathbf{u}^n(\xi,\eta)_{(2\times1)} &= [u(\xi,\eta), v(\xi,\eta)]^{\mathrm{T}} = \mathbf{N}^n_{(2\times8)}\ \mathbf{q}^{en}_{(8\times1)} \\ u(\xi,\eta) &= N_1\ u_1 + N_2\ u_2 + N_3\ u_3 + N_4\ u_4 \\ v(\xi,\eta) &= N_1\ v_1 + N_2\ v_2 + N_3\ v_3 + N_4\ v_4 \end{aligned} \tag{5.2}$$

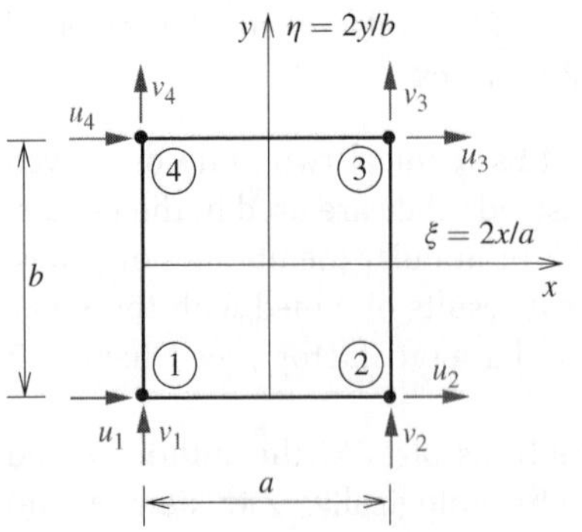

Figure 5.1 Flat rectangular four-node membrane FE

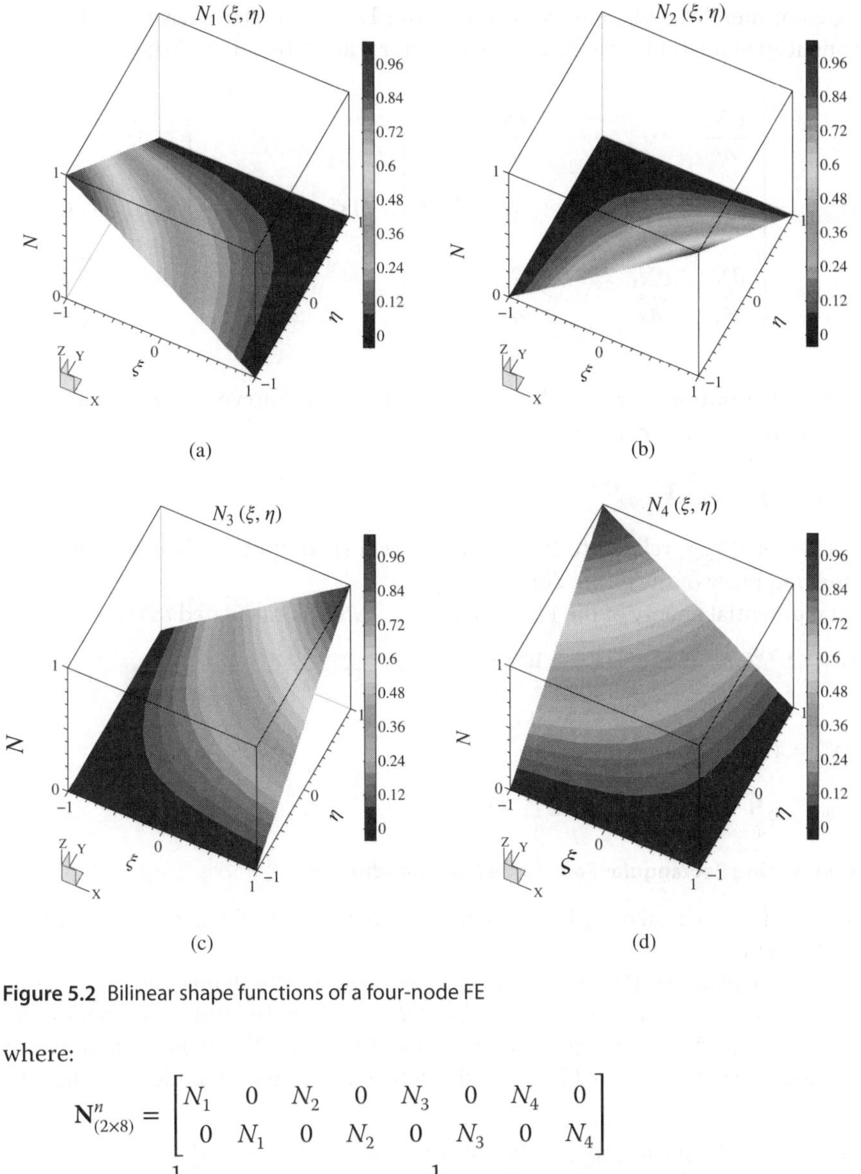

Figure 5.2 Bilinear shape functions of a four-node FE

where:

$$\mathbf{N}^n_{(2\times8)} = \begin{bmatrix} N_1 & 0 & N_2 & 0 & N_3 & 0 & N_4 & 0 \\ 0 & N_1 & 0 & N_2 & 0 & N_3 & 0 & N_4 \end{bmatrix}$$

$$N_1 = \frac{1}{4}(1-\xi)(1-\eta), \quad N_2 = \frac{1}{4}(1+\xi)(1-\eta) \qquad (5.3)$$

$$N_3 = \frac{1}{4}(1+\xi)(1+\eta), \quad N_4 = \frac{1}{4}(1-\xi)(1+\eta)$$

According to the fundamental reference works on FEM the known formula for the linear stiffness matrix of a membrane four-node FE with $NEDOF = 8$ reads

$$\mathbf{k}^{en}_{(8\times8)} = \int_{A^e} \mathbf{B}^{n\,\mathrm{T}}_{(8\times3)} \mathbf{D}^n_{(3\times3)} \mathbf{B}^n_{(3\times8)} \mathrm{d}A \qquad (5.4)$$

Calculation of stiffness matrix requires integration over area, using the so-called full numerical integration (FI) with 2×2 Gauss points or reduced integration (RI) with one Gauss point. In the integrand the following matrices occur: $\mathbf{B}^n = \mathbf{L}^n\, \mathbf{N}^n$, the matrix resulting from membrane kinematic relations and $\mathbf{D}^n$, the matrix derived from membrane constitutive relations (see Box 3.6 in Chapter 3 and Box 6.1 in Chapter 6):

$$\mathbf{B}^n_{(3\times8)} = \begin{bmatrix} \dfrac{\partial N_1}{\partial x} & 0 & \cdots & \dfrac{\partial N_4}{\partial x} & 0 \\ 0 & \dfrac{\partial N_1}{\partial y} & \cdots & 0 & \dfrac{\partial N_4}{\partial y} \\ \dfrac{\partial N_1}{\partial y} & \dfrac{\partial N_1}{\partial x} & \cdots & \dfrac{\partial N_4}{\partial y} & \dfrac{\partial N_4}{\partial x} \end{bmatrix}, \quad \mathbf{D}^n_{(3\times3)} = \frac{E\,h}{1-\nu^2}\begin{bmatrix} 1 & \nu & 0 \\ \nu & 1 & 0 \\ 0 & 0 & \dfrac{1-\nu}{2} \end{bmatrix} \tag{5.5}$$

When the distribution of surface load $\hat{\mathbf{p}}^{en}_{(2\times1)} = [\hat{p}_x, \hat{p}_y]^{\mathrm{T}}$ is known, the vector of nodal substitutes of the surface load is calculated as

$$\mathbf{f}^{en}_{(8\times1)} = \int_{A^e} \mathbf{N}^{en\,\mathrm{T}}_{(8\times2)}\,\hat{\mathbf{p}}^{en}_{(2\times1)}\,\mathrm{d}A \tag{5.6}$$

Similarly, vector $\mathbf{f}^{en}_{b\,(8\times1)}$ related to the boundary load is computed when edge ∂A^e is a part of the boundary of the entire domain ∂A_σ.

The total potential energy of the FE set ($e = 1, \ldots, NSE$) is calculated as the sum

$$\Pi^n = \Sigma_{e=1}^{NSE}\,\Pi^{en} = \Sigma_{e=1}^{NSE}(U^{en}_{\mathrm{int}} - W^{en}) \tag{5.7}$$

where:

$$\begin{aligned} \Pi^{en} &= U^{en}_{\mathrm{int}} - W^{en} = U^{en}_{\mathrm{int}} + U^{en}_{\mathrm{ext}} \\ U^{en}_{\mathrm{int}} &= \frac{1}{2}\mathbf{q}^{en\,\mathrm{T}}\,\mathbf{k}^{en}\,\mathbf{q}^{en}, \quad U^{en}_{\mathrm{ext}} = -W^{en} = -(\mathbf{f}^{en\mathrm{T}} + \mathbf{f}_b^{en\mathrm{T}})\,\mathbf{q}^{en} \end{aligned} \tag{5.8}$$

5.2.2 Conforming Rectangular Four-Node Plate Bending FE

The description of a typical plate FE is based on the mechanics of plates under bending, see also Chapter 8.

The presented plate FE (Figure 5.3) is characterized by the following parameters: $NNDOF = 4$, $NEN = 4$ and $NEDOF = NNDOF \times NEN = 16$. Index m marks the bending state related to moments (bending and twisting). The deflection function $w(x, y)$ is approximated within FE using the following vector of node and element dofs:

$$\begin{aligned} \mathbf{q}^m_{node(4\times1)} &= [w, \varphi_x, \varphi_y, \chi]^{\mathrm{T}}_{node} = [w, \partial w/\partial y, -\partial w/\partial x, \partial^2 w/\partial x\partial y]^{\mathrm{T}}_{node} \\ \mathbf{q}^{em}_{(16\times1)} &= [\mathbf{q}_1, \mathbf{q}_2, \mathbf{q}_3, \mathbf{q}_4]^{\mathrm{T}} = [w_1, \varphi_{x1}, \varphi_{y1}, \chi_1 | \ldots . | \ldots . | \ldots \chi_4]^{\mathrm{T}} \end{aligned} \tag{5.9}$$

The application of $NEDOF = 16$ and Hermitian (bicubic) shape functions (Figure 5.4) leads to a conforming plate bending FE, in which C^1-continuity is guaranteed on interelement lines (continuity of deflection function and its two first derivatives $w_{,\nu}$ and $w_{,s}$).

Adopting the dimensionless coordinates:

$$\xi = 2\left(\frac{x}{a}\right) - 1, \quad \eta = 2\left(\frac{y}{b}\right) - 1 \tag{5.10}$$

we write the polynomial base for the deflection approximation (according to the Pascal triangle):

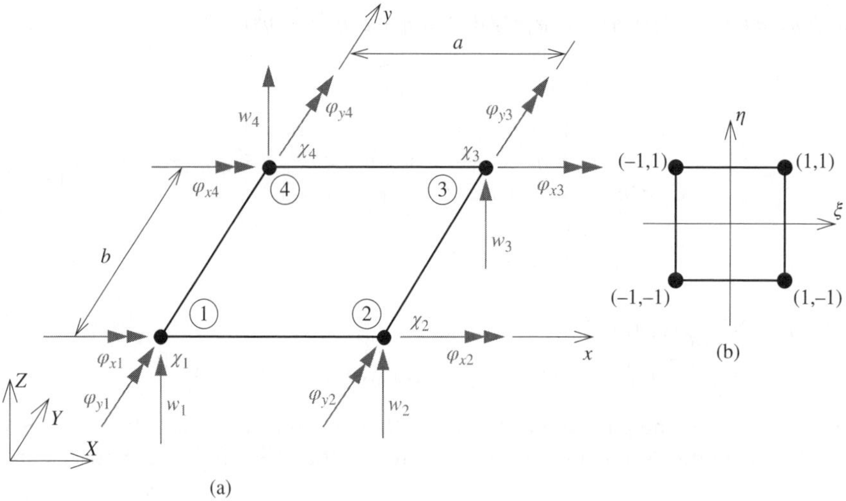

Figure 5.3 (a) Conforming rectangular four-node plate bending FE and (b) reference FE

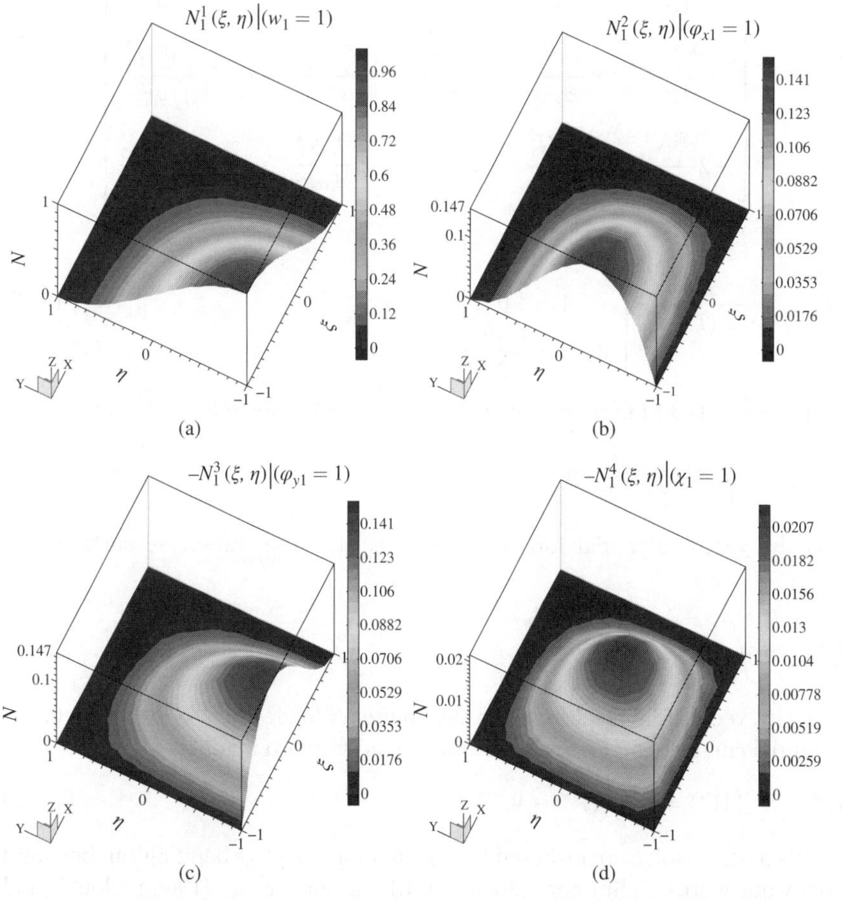

Figure 5.4 Hermitian (bicubic) shape functions $N_1^1, N_1^2, N_1^3, N_1^4$ related to node number 1 in conforming plate bending element

$$w(\xi,\eta) = (\alpha_1 + \alpha_2\xi + \alpha_3\xi^2 + \alpha_4\xi^3)(\beta_1 + \beta_2\eta + \beta_3\eta^2 + \beta_4\eta^3)$$

$$= \begin{array}{ccccccccccccc} & & & & & & C_1 & & + & & & & \\ & & & & C_2\xi & & + & & C_3\eta & & + & & \\ & & C_4\xi^2 & & + & & C_5\xi\eta & & + & & C_6\eta^2 & & + \\ C_7\xi^3 & & + & & C_8\xi^2\eta & & + & & C_9\xi\eta^2 & & + & & C_{10}\eta^3 \; + \\ & & C_{11}\xi^3\eta & & + & & C_{12}\xi^2\eta^2 & & + & & C_{13}\xi\eta^3 & & + \\ & & & & C_{14}\xi^3\eta^2 & & + & & C_{15}\xi^2\eta^3 & & + & & \\ & & & & & & C_{16}\xi^3\eta^3 & & & & & & \end{array} \tag{5.11}$$

$$\begin{aligned} w(\xi,\eta) &= \mathbf{N}^m_{(1\times16)}(\xi,\eta)\,\mathbf{q}^{em}_{(16\times1)} \\ &= [N_1^1, N_1^2, N_1^3, N_1^4, \mid N_2^1 \ldots N_4^4]\;[w_1, \varphi_{x1}, \varphi_{y1}, \chi_1 \mid w_2 \ldots \chi_4]^{\mathrm{T}} \end{aligned}$$

On the basis of the kinematic equations (K–L theory) and the constitutive relations (cf. Box 3.7 in Chapter 3 and Box 8.1 in Chapter 8) the following two matrices are constructed:

$$\mathbf{B}^m_{(3\times16)} = \begin{bmatrix} -\dfrac{\partial^2 N_1^1}{\partial x^2} & -\dfrac{\partial^2 N_1^2}{\partial x^2} & -\dfrac{\partial^2 N_1^3}{\partial x^2} & -\dfrac{\partial^2 N_1^4}{\partial x^2} & \cdots & -\dfrac{\partial^2 N_4^4}{\partial x^2} \\ -\dfrac{\partial^2 N_1^1}{\partial y^2} & -\dfrac{\partial^2 N_1^2}{\partial y^2} & -\dfrac{\partial^2 N_1^3}{\partial y^2} & -\dfrac{\partial^2 N_1^4}{\partial y^2} & \cdots & -\dfrac{\partial^2 N_4^4}{\partial y^2} \\ -2\dfrac{\partial^2 N_1^1}{\partial x \partial y} & -2\dfrac{\partial^2 N_1^2}{\partial x \partial y} & -2\dfrac{\partial^2 N_1^3}{\partial x \partial y} & -2\dfrac{\partial^2 N_1^4}{\partial x \partial y} & \cdots & -2\dfrac{\partial^2 N_4^4}{\partial x \partial y} \end{bmatrix} \tag{5.12}$$

$$\mathbf{D}^m_{(3\times3)} = \frac{E\,h^3}{12\,(1-\nu^2)} \begin{bmatrix} 1 & \nu & 0 \\ \nu & 1 & 0 \\ 0 & 0 & \dfrac{1-\nu}{2} \end{bmatrix} \tag{5.13}$$

and the stiffness matrix of four-node plate bending FE with $NEDOF = 16$

$$\mathbf{k}^{em}_{(16\times16)} = \int_{A^e} \mathbf{B}^{m\,\mathrm{T}}_{(16\times3)}\mathbf{D}^m_{(3\times3)}\mathbf{B}^m_{(3\times16)}\mathrm{d}A \tag{5.14}$$

as well as the vector of nodal forces which substitute the transverse surface load $\hat{\mathbf{p}}^{em}_{(1\times1)} = [\hat{p}_z]$

$$\mathbf{f}^{em}_{(16\times1)} = \int_{A^e} \mathbf{N}^{em\,\mathrm{T}}_{(16\times1)}\,\hat{p}_z\mathrm{d}A \tag{5.15}$$

and, optionally, vector $\mathbf{f}_b^{em}$ resulting from the boundary load, when $\partial A^e \cap \partial A_\sigma \neq 0$.

The potential energy of an FE set is assembled in a standard manner

$$\Pi^m = \Sigma_{e=1}^{NSE}\Pi^{em} = \Sigma_{e=1}^{NSE}(U^{em} - W^{em}) \tag{5.16}$$

The details of the displacement-based FE formulation for plate bending can be found in the following works: Adini and Clough (1961); Bogner et al. (1966); Clough and Felippa (1968); Fraeijs de Veubeke (1968); Melosh (1963).

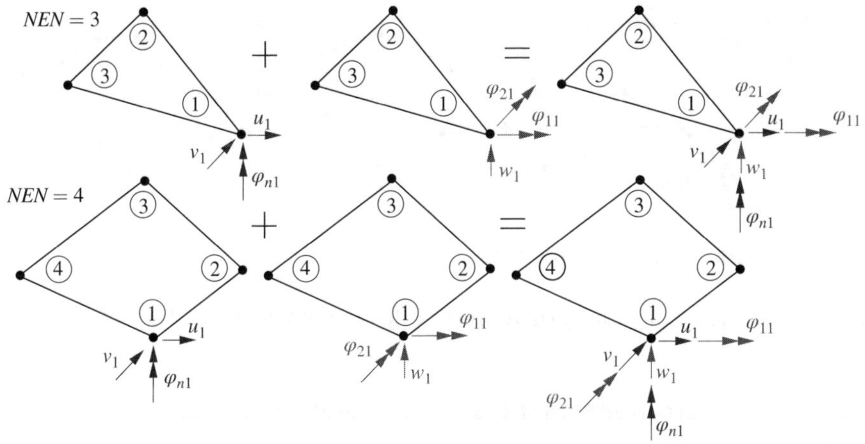

Figure 5.5 Flat shell three- and four-node elements composed of membrane and plate bending FEs

5.2.3 Nonconforming Flat Three- and Four-Node FEs for Thin Shells

Flat shell FEs with $NEN = 3$ or $NEN = 4$ (Figure 5.5) are often applied. They are obtained by combining flat membrane and plate bending FEs.

In these FEs the following nodal membrane (n) and bending (m) dofs are gathered in the vectors

$$\mathbf{q}^{\mathrm{T}}_{node} = [\mathbf{q}^{n}, \mathbf{q}^{m}]_{node} \tag{5.17}$$

resulting in $NNDOF = 2 + 3 = 5$:

$$\begin{aligned} \mathbf{q}^{n}_{node} &= [u, v]^{\mathrm{T}}_{node}, \quad \mathbf{q}^{m}_{node} = [w, \varphi_1, \varphi_2]^{\mathrm{T}}_{node} \\ \mathbf{q}_{node} &= [u, v, w, \varphi_1, \varphi_2]^{\mathrm{T}}_{node} \end{aligned} \tag{5.18}$$

In some cases it is necessary to introduce a third rotational dof – the so-called drilling rotation φ_n. Then, it is incorporated into the description of the membrane state and for $NNDOF = 3 + 3 = 6$ we obtain:

$$\begin{aligned} \mathbf{q}^{n}_{node} &= [u, v, \varphi_n]^{\mathrm{T}}_{node}, \quad \mathbf{q}^{m}_{node} = [w, \varphi_1, \varphi_2]^{\mathrm{T}}_{node} \\ \mathbf{q}_{node} &= [u, v, w, \varphi_1, \varphi_2, \varphi_n]^{\mathrm{T}}_{node} \end{aligned} \tag{5.19}$$

This complement of approximation by the third rotational dof is indispensable when branching shells, folded plates or bars with thin-walled cross sections are discretized with flat FEs, see Subsection 18.3.2.

The approximation of three displacement fields is expressed by

$$\mathbf{u}(\xi_1, \xi_2)_{(3\times1)} = [u(\xi_1, \xi_2), v(\xi_1, \xi_2), w(\xi_1, \xi_2)]^{\mathrm{T}} = \mathbf{N}_{(3\times NEDOF)}\, \mathbf{q}^{e}_{(NEDOF\times1)} \tag{5.20}$$

where $NEDOF = NEN \times NNDOF$, and the membrane and bending dofs are included in vector $\mathbf{q}^e$.

The total internal energy of the presented FE is given as the sum

$$U^e = U^{en} + U^{em}, \quad \text{where: } U^{en} = \frac{1}{2}\mathbf{q}^{e\,\mathrm{T}}\mathbf{k}^{en}\mathbf{q}^{e}, \quad U^{em} = \frac{1}{2}\mathbf{q}^{e\mathrm{T}}\mathbf{k}^{em}\mathbf{q}^{e} \tag{5.21}$$

Element stiffness matrices $\mathbf{k}^{en}$ and $\mathbf{k}^{em}$ have zero submatrices, resulting from the lack of coupling of membrane and bending contributions. Before the assembly process it is

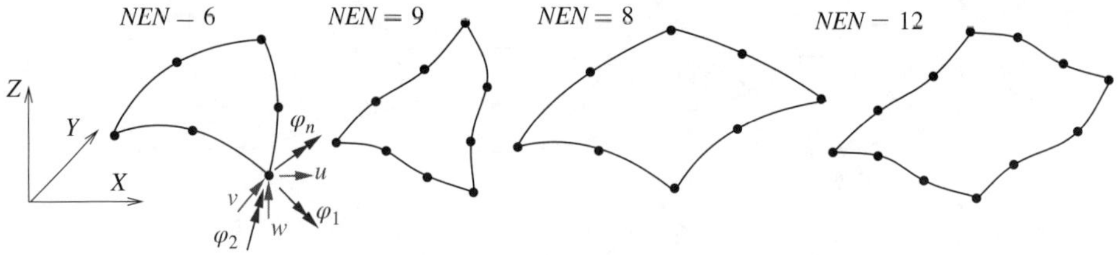

Figure 5.6 Curved shell FEs with $NEN = 6, 9, 8, 12$

necessary to make relevant transformations to calculate the global stiffness matrix for the FE set.

5.2.4 Two-Dimensional Curved Shell FE based on the Kirchhoff–Love Thin Shell Theory

Figure 5.6 illustrates curved triangular six- and nine-node FEs and quadrilateral eight- and 12-node FEs. Usually isoparametric formulation is applied, using identical interpolation functions to describe both FE geometry and displacement fields.

In the displacement-based FE model based on three-parameter K–L shell theory three translations are approximated

$$\mathbf{u}(\xi_1, \xi_2)_{(3\times1)} = [u(\xi_1, \xi_2), v(\xi_1, \xi_2), w(\xi_1, \xi_2)]^{\mathrm{T}} = \mathbf{N}_{(3\times NEDOF)}\, \mathbf{q}^{e}_{(NEDOF\times1)} \tag{5.22}$$

The following dofs for $NNDOF = 5$ or $NNDOF = 6$ are used

$$\mathbf{q}_{node} = [u, v, w, \varphi_1, \varphi_2]^{\mathrm{T}}_{node} \quad \text{or} \quad \mathbf{q}_{node} = [u, v, w, \varphi_1, \varphi_2, \varphi_n]^{\mathrm{T}}_{node} \tag{5.23}$$

For instance, for the FE with $NEN = 6$ and $NNDOF = 6$ we obtain $NEDOF = NNDOF \times NEN = 6 \times 6 = 36$.

The translational nodal dofs are often referred to the directions of the global coordinate system (X, Y, Z), while the directions of the rotational dofs result from the axes of the local nodal base vectors tangent and normal to the surface. When the K–L theory is applied, fields $u_1(\xi_1, \xi_2)$ and $u_2(\xi_1, \xi_2)$ require, in general, C^0-continuity, while the displacement normal to the shell $w(\xi_1, \xi_2)$ – C^1-continuity.

In the formulation of the FE model based on the global energetic approach, the formula for internal energy of thin shells is used, summing the effects of membrane and bending states. In the calculations of the FE stiffness matrix the following matrices are used: $\mathbf{B}^n$, $\mathbf{B}^m$ (resulting from the kinematic relations for membrane and bending states), and $\mathbf{D}^n$, $\mathbf{D}^m$ (computed on the basis of constitutive equations of these two states).

5.2.5 Curved FE based on the Mindlin–Reissner Moderately Thick Shell Theory

In this case, frequently encountered in FE codes, the approximation specifies five (or six) independent fields: three translations u_1, u_2, w and two rotations of the normal to the middle surface φ_1, φ_2 (and additionally the third rotation φ_n around the normal)

$$\begin{aligned}\mathbf{u}(\xi_1, \xi_2)_{(5\times1)} &= [u(\xi_1, \xi_2), v(\xi_1, \xi_2), w(\xi_1, \xi_2), \varphi_1(\xi_1, \xi_2), \varphi_2(\xi_1, \xi_2)]^{\mathrm{T}} \\ &= \mathbf{N}_{(5\times NEDOF)}\, \mathbf{q}^{e}_{(NEDOF\times1)}\end{aligned} \tag{5.24}$$

The following vectors of node dofs are used for $NNDOF = 5$ or $NNDOF = 6$

$$\mathbf{q}_{node} = [u, v, w, \varphi_1, \varphi_2]^{\mathrm{T}}_{node} \quad \text{or} \quad \mathbf{q}_{node} = [u, v, w, \varphi_1, \varphi_2, \varphi_n]^{\mathrm{T}}_{node} \tag{5.25}$$

For independent approximation all fields have C^0-continuity guaranteed.

In the M–R theory describing moderately thick shells, not only are the membrane and bending effects considered, but also the transverse shear ones. On the basis of the kinematic and constitutive relations the corresponding matrices: $\mathbf{B}^n$, $\mathbf{B}^m$, $\mathbf{B}^t$ and $\mathbf{D}^n$, $\mathbf{D}^m$, $\mathbf{D}^t$, respectively, are formed (see Section 8.5). In the FE description the membrane U^n, bending U^m and transverse shear U^t energies are summed. The stiffness matrix has three components resulting from the three states mentioned previously.

5.2.6 Degenerated Shell FE

The FEs described in this section do not refer immediately to the shell theory equations. They are based on 3D continuum equations, modified by the kinematic M–R hypothesis about the behaviour of a straight fibre (director), which does not preserve the normality to the deformed middle surface.

In degenerated FEs, used for the discretization of both thin and moderately thick shells, the stresses and strains related to transverse shear are considered next to membrane and bending ones. The applied so-called degeneration consists in the transition from a three-dimensional FE to a corresponding two-dimensional FE, using the finite element approximation of:

- geometry – on the basis of vectors of node position $\mathbf{x}_{node}$ and vectors representing node thickness $\mathbf{t}_{node}$
- translation fields $\mathbf{u}(\mathbf{x}) = [u, v, w]^{\mathrm{T}}$ using nodal translations $[u, v, w]_{node}$ and nodal rotations $[\varphi_1, \varphi_2]_{node}$

For example, for element FE8 shown in Figure 5.7c, of $NNDOF = 5$, $NEN = 8$, $NEDOF = NNDOF \times NEN = 40$ are obtained.

Three fields $u(\xi_1, \xi_2)$, $v(\xi_1, \xi_2)$, $w(\xi_1, \xi_2)$ have C^0-continuity on interelement boundaries in the middle surface and linear distribution in the thickness direction guaranteed.

The details of the degenerated FEs can be found in Ahmad et al. (1970). As already mentioned, these elements can be applied in the discretization of shell structures with small and moderately thickness. Moreover, eight- and nine-node elements can be applied in the approximation of the geometry and deformation of shells of any shape, including curved shells and flat membranes or plates.

5.2.7 Three-Dimensional Solid FE for Thick Shells

Standard solid FEs are applied in the analysis of thick membranes, plates and shells (Figure 5.8).

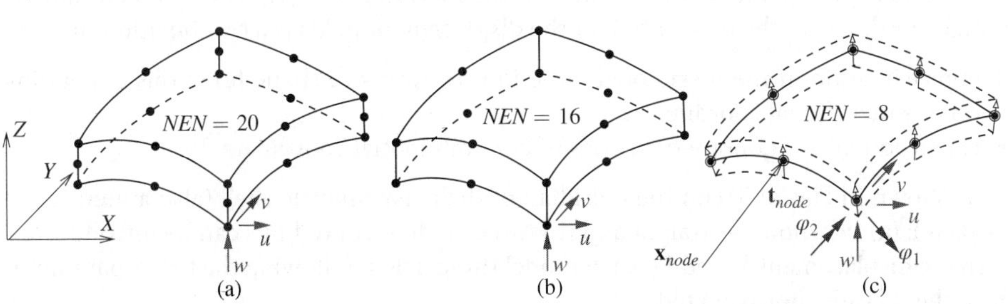

Figure 5.7 Idea of transition from 3D FE: (a) FE20, (b) FE16 to (c) eight-noded 2D degenerated FE8

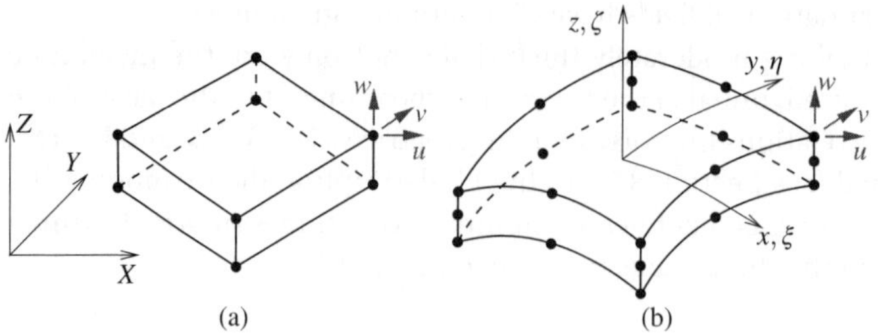

Figure 5.8 Solid 3D FEs: (a) eight-node and (b) 20-node

Three displacement fields are approximated

$$\mathbf{u}(\xi,\eta,\zeta)_{(3\times1)} = [u(\xi,\eta,\zeta), v(\xi,\eta,\zeta), w(\xi,\eta,\zeta)]^{\mathrm{T}} = \mathbf{N}_{(3\times NEDOF)}\,\mathbf{q}^{e}_{(NEDOF\times1)} \tag{5.26}$$

using three nodal dofs

$$\mathbf{q}_{node} = [u, v, w]^{\mathrm{T}}_{node} \tag{5.27}$$

The applied shape functions depend on three coordinates ξ, η, ζ. Three translation fields $u(\xi,\eta,\zeta), v(\xi,\eta,\zeta), w(\xi,\eta,\zeta)$ have C^0-continuity guaranteed on interelement boundaries.

The starting point for the 3D solid FE formulation is the energy expression

$$U^e = \frac{1}{2}\int_{V^e} \varepsilon^{\mathrm{T}}_{(1\times6)}\,\sigma_{(6\times1)}\mathrm{d}V = \frac{1}{2}\mathbf{q}^{e\,\mathrm{T}}_{(1\times NEDOF)}\,\mathbf{k}^{e}_{(NEDOF\times NEDOF)}\,\mathbf{q}^{e}_{(NEDOF\times1)} \tag{5.28}$$

which contains volume integral, and strains $\varepsilon_{(6\times1)}$ and stresses $\sigma_{(6\times1)}$ characteristic for a 3D-body are taken into account, see Chapter 2.

The Gauss numerical integration in three directions must be performed to calculate the FE stiffness matrix according to the formula

$$\mathbf{k}^{e}_{(NEDOF\times NEDOF)} = \int_{V^e} \mathbf{B}^{\mathrm{T}}_{(NEDOF\times6)}\,\mathbf{D}_{(6\times6)}\,\mathbf{B}_{(6\times NEDOF)})\,\mathrm{d}V \tag{5.29}$$

5.2.8 Geometrically One-Dimensional FE for Thin Shell Structures

Axisymmetric shell surfaces are formed by rotation of a relevant line around the symmetry axis. Composing an FE model a set of nodes is introduced on the meridian. Different numbers of nodes are used, which result in straight (conical) or curved FEs. The nodes occurring on the meridian represent parallel circles, giving finally an axisymmetric surface domain.

Introducing a separation with respect to the meridional coordinate ξ and circumferential coordinate θ, the description of the displacement fields is a combination of:

- approximation in the meridional direction by means of shape functions depending on one meridional coordinate ξ
- expansion into trigonometric series of circumferential coordinate θ

As shown in Figure 5.9 on the meridian of the axisymmetric shell (also a membrane or plate): (a) two-node – straight and (b) three-node – curved FEs can be introduced.

In the displacement-based FE shell model three fields (following the three-parameter K–L theory) are approximated

$$\mathbf{u}(\xi,\theta) = [u(\xi,\theta), v(\xi,\theta), w(\xi,\theta)]^{\mathrm{T}} \tag{5.30}$$

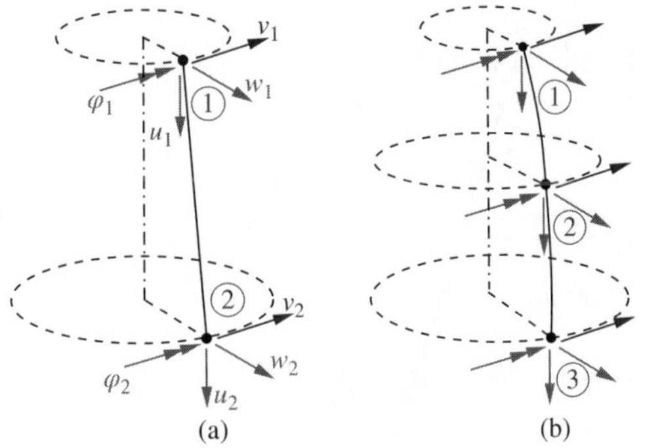

Figure 5.9 One-dimensional shell FEs: (a) straight and (b) curved

After the expansion of the three displacement functions into trigonometric series of the circumferential coordinate θ:

$$u(\xi,\theta) = \sum_{j=0}^{J} u^{(j)}(\xi)\ \cos(j\theta)$$
$$v(\xi,\theta) = \sum_{j=0}^{J} v^{(j)}(\xi)\ \sin(j\theta) \tag{5.31}$$
$$w(\xi,\theta) = \sum_{j=0}^{J} w^{(j)}(\xi)\ \cos(j\theta)$$

amplitude functions $u^{(j)}(\xi)$, $v^{(j)}(\xi)$, $w^{(j)}(\xi)$ appear, attributed to each number j (upper index). These functions of displacement amplitudes are approximated using geometrically one-dimensional shape functions and nodal dofs for each number $j = 0, 1, 2, \ldots, J$

$$\mathbf{u}^{(j)}_{(3\times1)} = [u^{(j)}(\xi), v^{(j)}(\xi), w^{(j)}(\xi)]^{\mathrm{T}} = \mathbf{N}_{(3\times NEDOF)}\,\mathbf{q}^{(j)\,e}_{(NEDOF\times1)} \tag{5.32}$$

In the shell FE, four dofs for a node are adopted ($NNDOF = 4$). For example, in a conical FE with $NEN = 2$ we obtain the following nodal and element displacement vectors:

$$\mathbf{q}^{(j)}_{node} = [u^{(j)}, v^{(j)}, w^{(j)}, \varphi^{(j)}]^{\mathrm{T}}_{node}, \quad \mathbf{q}^{(j)\,e\,\mathrm{T}}_{(8\times1)} = [\mathbf{q}^{(j)}_1, \mathbf{q}^{(j)}_2] \tag{5.33}$$

and consequently $NEDOF = NNDOF \times NEN = 8$. For the description of displacement amplitudes $u^{(j)}(\xi)$ and $v^{(j)}(\xi)$ (tangent to the middle surface of a shell) two linear shape functions are applied that are associated with two dof pairs: $u^{(j)}_1$, $u^{(j)}_2$ and $v^{(j)}_1$, $v^{(j)}_2$. On the other hand, the amplitude of displacement $w^{(j)}(\xi)$, normal to the middle surface, is described by four Hermitian cubic functions and four dofs: $w^{(j)}_1$, $\varphi^{(j)}_1$, $w^{(j)}_2$, $\varphi^{(j)}_2$. The application of nodal rotations $\varphi^{(j)}_{node}$ next to nodal displacements $w^{(j)}_{node}$ raises the order of approximation of displacement amplitude $w^{(j)}(\xi)$.

The description of generalized strains and resultant forces includes membrane and bending effects in thin shells, and uses corresponding kinematic and constitutive relations. Calculating the stiffness matrices different integration methods are applied: numerical in the meridional direction and analytical in the circumferential direction.

In the two-node conical FE the direction of the element meridian is described using angle $\psi =$ const., which is automatically calculated on the basis of the coordinates of

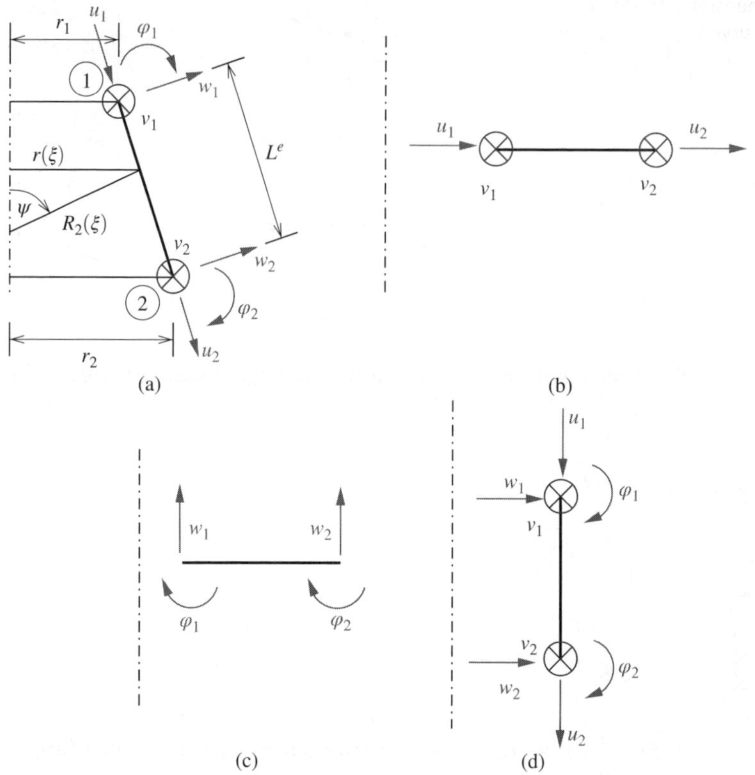

Figure 5.10 Transition from: (a) conical shell FE to (b) membrane, (c) plate and (d) cylindrical shell FEs

the two nodes. To define the FE geometry the following quantities are necessary:

$$\begin{aligned} A_1 &= A = L^e, \qquad & A_2 &= r = r_1 + \xi L^e \cos\psi \\ 1/R_1 &= 0, \qquad & R_2 &= \frac{r_1}{\sin\psi} + \xi L^e \cot\psi \end{aligned} \tag{5.34}$$

This FE called SRK (Shell element Rotational Kirchhoff–Love theory) was implemented in FEM package ANKA (1993) and proved to be very effective in the solution of problems for axisymmetric membranes, plates and shells. This package was developed at Cracow University of Technology and used among others for computations presented in the book of Waszczyszyn et al. (1994).

The determination of the coordinates of both FE nodes enables the adaptation of the described shell element to particular cases (Figure 5.10), for example to the approximation of:

- circular or annular membrane
- circular or annular plate under bending
- cylindrical or conical shells
- any axisymmetric shell, whose meridian is approximated by a segment line

The following parameters occur in the kinematic relations and in the FE description: radii of two parallel circles r_1, r_2, length L^e, angle ψ, curvature radius R_2. In the ANKA computer code the values of these parameters control the incorporation of appropriate components of the general model. In other words, the respective components of the model are activated when appropriate:

- membrane state for membranes
- bending state for plates
- both membrane and bending effects for shells

The axisymmetric load for membranes, plates and shells corresponds to a single component of the expansion with $j = 0$. In the case of standard wind load of a cylindrical shell, the analysis of load and response for each number $j = 0, 1, \ldots, 6$ is performed and finally the solutions are summed.

5.3 Remarks on Displacement-based FE Formulation

Flat FEs demonstrate weaknesses that, in some cases, can become essential:

- the geometry of a curved shell is represented only in a rough way and the discretization introduces sharp bends (fold lines), inducing bending moments not present in the real structure (the use of a sufficiently large number of FEs for a better representation of the surface can result in a reasonable solution)
- flat FEs are not able to couple membrane and bending states
- the lack of drilling rotation φ_n prevents the proper modelling of the connection of FEs lying in different planes; this dof enriches the approximation of the membrane state and facilitates the transformation of a nodal rotation vector from the local coordinate system to the global one

Curved FEs that are based on the Kirchhoff–Love theory for shells can reproduce an arbitrary geometry of a shell; however, they require a formulation of (quite complicated in this case), kinematic equations relating the strains to the derivatives of displacements. The description of membrane, bending and transverse shear strain fields requires an approximation of translations and their first and second gradients. An important issue is the necessity to guarantee an appropriate continuity of the approximated fields and satisfaction of the condition of strain-free rigid motions.

Solid FEs have a wide application range, especially in the case of isoparametric (linear, parabolic, cubic) FEs, though 3D discretization leads to a large number of dofs and thus a large system of equations. Due to the use of numerical integration in three dimensions the cost of computations grows significantly. Large differences between the dimensions in the middle surface and along the thickness induce a strong disproportion in energy and stiffness balance and thus can result in an ill-conditioned system of equations. To avoid this problem the order of approximation in the thickness direction should be lower in comparison with the two other directions. This has recently become possible by using highly specialized *hp*-adaptive techniques of approximation, see Demkowicz et al. (2007).

Summarizing, the complexity of the description of various shell structures in linear as well as nonlinear analysis sets very high demands on the FE approximation. The computation method must be efficient in terms of accuracy and robustness. The approach should be general enough to take different cases into account. The formulation should be based on the best available theory.

Further information on the FE formulation strategy for shell structures and the discussion of more advanced approximation techniques, including numerous references to the literature, can be found in Chapter 18.

References

Adini A and Clough RW 1961 Analysis of plate bending by the finite element method. Technical Report G-7337, University of California, Berkeley.

Ahmad S, Irons BM and Zienkiewicz OC 1970 Analysis of thick and thin shell structures by curved finite element. *International Journal for Numerical Methods in Engineering* **2**(3), 419–451.

ANKA 1993 ANKA – computer code for nonlinear analysis of structures: User's manual. Technical report, Cracow University of Technology, Cracow (in Polish).

Ashwell DG and Gallagher RH (eds) 1976 *Finite Elements for Thin Shells and Curved Members*. John Wiley & Sons, Ltd, Chichester, UK.

Bathe KJ 1982 *Finite Element Procedures in Engineering Analysis*. Prentice-Hall.

Batoz JL and Dhatt G 1990 *Modélisation des Structures par Élément Finis*. Hermes, Paris.

Bischoff M, Wall WA, Bletzinger KU and Ramm E 2004 Models and finite elements for thin-walled structures. In Stein E, de Borst R and Hughes TJR (eds), *Encyclopedia of Computational Mechanics: Solids and Structures*, vol. 2. John Wiley & Sons, Ltd, Chichester, UK. chapter 3, pp. 59–137.

Bogner FK, Fox RL and Schmit LA 1966 The generation of interelement-compatible stiffness and mass matrices by the use of interpolation formulas *Proceedings of the Conference on Matrix Methods in Structural Mechanics*, pp. 397–443, report AFFDL TR-66-80. Wright-Patterson Air Force Base, Ohio.

Clough RW and Felippa CA 1968 A refined quadrilateral element for analysis of plate bending *Proceedings of 2nd Conference on Matrix Methods in Structural Mechanics*, pp. 399–440, report AFFDL TR-68-150. Wright-Patterson Air Force Base.

Cook RD 1995 *Finite Element Modeling for Stress Analysis*. John Wiley & Sons, New York-Chichester.

Cook RD, Malkus DS and Plesha ME 1989 *Concepts and Applications of Finite Element Analysis* 3rd edn. John Wiley & Sons, New York.

Crisfield MA 1986 *Finite Elements and Solution Procedures for Structural Analysis – Linear Analysis* vol. 1. Pineridge Press, Swansea.

Demkowicz L, Kurtz J, Pardo D, Paszyński M, Rachowicz W and Zdunek A 2007 *Computing with hp-Adaptive Finite Elements: Frontiers Three Dimensional Elliptic and Maxwell Problems with Applications*, vol. 2, 1st edn. Chapman & Hall/CRC.

Fraeijs de Veubeke B 1968 A conforming finite element for plate bending. *International Journal of Solids and Structures* **4**(1), 95–108.

Hartmann F and Katz C 2007 *Structural Analysis with Finite Elements* 2nd edn. Springer.

Hinton E and Owen DR 1979 *An Introduction to Finite Element Computations*. Pineridge Press, Swansea.

Hinton E and Owen DR 1984 *Finite Element Software for Plates and Shells*. Pineridge Press.

Irons B and Ahmad S 1980 *Techniques of Finite Elements*. Ellis Horwood, Chichester.

MacNeal RH 1998 Perspective on finite element for shell analysis. *Finite Elements in Analysis and Design* **30**(3), 175–186.

Melosh RJ 1963 Basis for derivation of matrices for the direct stiffness method. *AIAA Journal* **1**(7), 1631–1637.

Oñate E 2013 *Structural Analysis with the Finite Element Method. Linear Statics. Volume 2: Beams, Plates and Shells,* Lecture Notes on Numerical Methods in Engineering and Sciences. CIMNE Springer.

Waszczyszyn Z, Cichoń C and Radwańska M 1994 *Stability of Structures by Finite Element Methods.* Elsevier, Amsterdam.

Yang H, Saigal S and Liaw DG 1990 Advances of thin shell finite elements and some applications – version i. *Computers & Structures* **35**(4), 481–504.

Yang HTY, Saigal S, Masud A and Kapania RK 2000 A survey of recent shell finite elements. *International Journal for Numerical Methods in Engineering* **47**(1–3), 101–127.

Yang TY 1986 *Finite Element Structural Analysis.* Prentice-Hall, Englewood Cliffs, New York.

Zienkiewicz OC and Cheung YK 1967 *The Finite Element Method in Structural and Continuum Mechanics.* McGraw-Hill, London.

Zienkiewicz OC, Taylor RL and Zhu JZ 2005 *The Finite Element Method: Its Basis and Fundamentals,* 6th edn. Elsevier/Butterworth-Heinemann.

Part 2

Plates

6

Flat Rectangular Membranes

6.1 Introduction

A flat 3D body with thickness $h \ll (L_x, L_y)$, shown in Figure 6.1a, treated as a set of layers in plane stress state is often called a flat membrane. The membrane is represented by its middle plane and all applied loads are reduced to this plane. The distribution of stresses along the thickness is constant (Figure 6.1b).

Thus, the middle plane of the membrane ($z = 0$) is its two-dimensional geometric model (Figure 6.1c). In our analysis a constant thickness $h(x, y) = h =$ const. is assumed. Extended information about rectangular membranes can be found in Andermann (1966); Girkmann (1956); Waszczyszyn and Radwańska (1995).

The following vectors are known at the beginning of the analysis:

- vector of load reduced to the middle plane $\hat{\mathbf{p}}^n_{(2\times1)}(x, y) = [\hat{p}_x, \hat{p}_y]^\mathrm{T}$ [N/m^2]
- boundary load vector $\hat{\mathbf{p}}^n_{b(2\times1)}(s) = [\hat{p}_\nu, \hat{p}_s]^\mathrm{T}$ [N/m]
- vector of known displacements on boundary lines $\hat{\mathbf{u}}^n_{b(2\times1)}(s) = [\hat{u}_\nu, \hat{u}_s]^\mathrm{T}$ [m]

The boundary line is paramterized by coordinate s, and at each boundary point one can define base directions in the middle plane: normal ν and tangent s. Note that all known quantities are indicated by the hat mark ($\hat{}$).

The stress state is described by membrane forces (stress resultants) defined per unit length of a cross-sectional line in the middle plane (Figure 6.3). The forces are calculated as integrals of corresponding stresses over the thickness of the membrane:

$$
\begin{aligned}
n_x &= \int_{-h/2}^{+h/2} \sigma_x \, \mathrm{d}z = \sigma_x \, h, \qquad n_y = \int_{-h/2}^{+h/2} \sigma_y \, \mathrm{d}z = \sigma_y \, h \\
n_{xy} &= \int_{-h/2}^{+h/2} \tau_{xy} \, \mathrm{d}z = \tau_{xy} \, h = \tau_{yx} \, h = n_{yx}
\end{aligned}
\tag{6.1}
$$

In a solution of the boundary value problem (BVP) eight fields (functions of two coordinates x, y) are introduced at any point of the middle plane. They can be grouped in three vectors:

- vector of displacements (translations) $\mathbf{u}^n_{(2\times1)}(x, y) = [u_x, u_y]^\mathrm{T} = [u, v]^\mathrm{T}$ [m]
- vector of membrane strains $\mathbf{e}^n_{(3\times1)}(x, y) = [\varepsilon_x, \varepsilon_y, \gamma_{xy}]^\mathrm{T}$ [−]
- vector of membrane forces $\mathbf{s}^n_{(3\times1)}(x, y) = \mathbf{n}(x, y) = [n_x, n_y, n_{xy}]^\mathrm{T}$ [N/m]

Graphical interpretation of strains in the membrane state is presented in Figure 6.2.

Plate and Shell Structures: Selected Analytical and Finite Element Solutions, First Edition.
Maria Radwańska, Anna Stankiewicz, Adam Wosatko and Jerzy Pamin.
© 2017 John Wiley & Sons Ltd. Published 2017 by John Wiley & Sons Ltd.

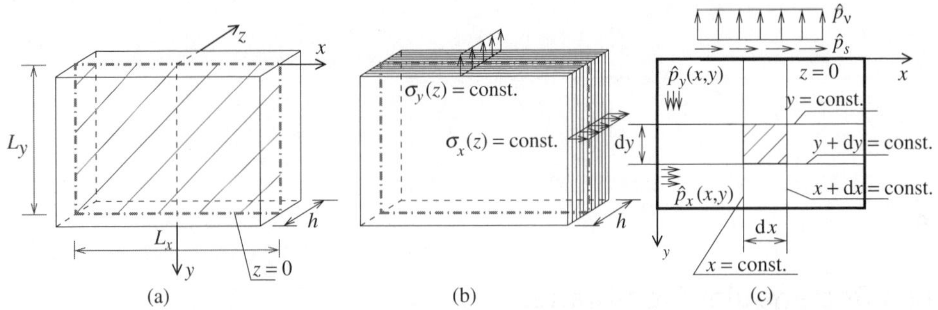

Figure 6.1 (a) Membrane as a 3D body, (b) membrane as a set of layers and (c) middle plane as a 2D model of a membrane

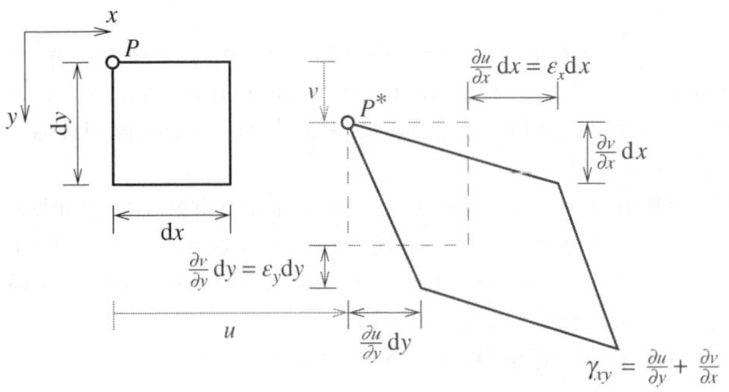

Figure 6.2 Graphical interpretation of displacements and strains for membranes

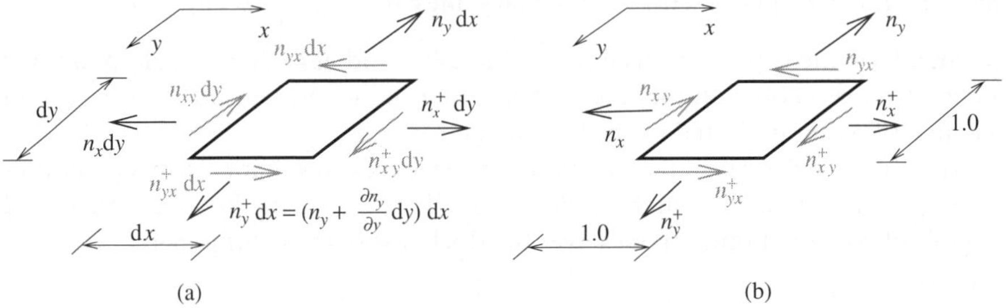

Figure 6.3 Sign convention of membrane forces for: (a) elementary segment $dx \times dy$ and (b) segment with sides of unit length

6.2 Governing Equations

6.2.1 Local Formulation

The mathematical model of a flat membrane includes: three kinematic (I), two equilibrium (II) and three constitutive equations (III) (in total, eight equations), valid for any point on the middle plane of the membrane. It is worth noting that the equilibrium and constitutive equations for flat membranes are obtained by multiplying the equations describing the case of a plane stress state, known from the *theory of elasticity*

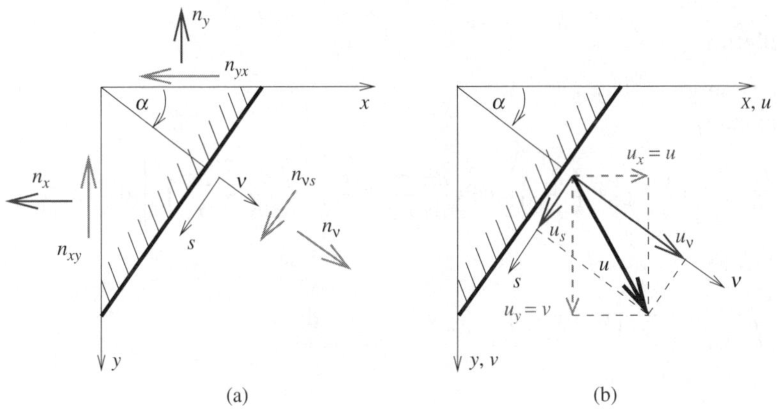

Figure 6.4 Notation of boundary quantities related to line α = const.: (a) static and (b) kinematic; for rectangular domain angle α has the following values: 0°, 90°, 180° and 270°

(Timoshenko and Goodier 1951), by the thickness h. Thus, the stresses are substituted by membrane forces and a factor called membrane stiffness $D^n = E\,h/(1-\nu^2)$ is introduced into the constitutive equations. Equations (I)–(III) must be supplemented by relations describing the boundary conditions (IV) formulated using quantities shown in Figure 6.4. In the membrane state two conditions are required for each boundary line.

The mathematical model (local and global formulation) is defined by equations presented in Box 6.1.

Box 6.1 Equations for flat rectangular membranes (in a Cartesian coordinate system)

Local formulation – eight equations and boundary conditions

(I) kinematic equations (3):

$$\varepsilon_x = \frac{\partial u}{\partial x}, \qquad \varepsilon_y = \frac{\partial v}{\partial y}, \qquad \gamma_{xy} = \frac{\partial u}{\partial y} + \frac{\partial v}{\partial x}$$

(II) equilibrium equations (2):

$$\frac{\partial n_x}{\partial x} + \frac{\partial n_{yx}}{\partial y} + \hat{p}_x = 0, \qquad \frac{\partial n_{xy}}{\partial x} + \frac{\partial n_y}{\partial y} + \hat{p}_y = 0$$

(III) constitutive equations (3):

$$n_x = D^n(\varepsilon_x + \nu\varepsilon_y), \qquad n_y = D^n(\varepsilon_y + \nu\varepsilon_x), \qquad n_{xy} = D^n(1-\nu)\,\gamma_{xy}/2$$

where $D^n = E\,h/(1-\nu^2)$

(IV) boundary conditions:

a) kinematic: $u_\nu = \hat{u}_\nu$, $u_s = \hat{u}_s$

b) static: $n_\nu = \hat{n}_\nu$, $n_{\nu s} = \hat{n}_{\nu s}$

c) mixed (one kinematic and one static)

Global formulation

Total potential energy $\Pi^n = U^n - W^n$

Internal strain energy

$$U^n = \frac{D^n}{2}\int_A \left[\left(\frac{\partial u}{\partial x}\right)^2 + \left(\frac{\partial v}{\partial y}\right)^2 + 2\nu\frac{\partial u}{\partial x}\frac{\partial v}{\partial y} + \frac{1-\nu}{2}\left(\frac{\partial u}{\partial y} + \frac{\partial v}{\partial x}\right)^2\right] \mathrm{d}x\,\mathrm{d}y$$

External load work

$$W^n = \int_A \left[\hat{p}_x u + \hat{p}_y v\right] \mathrm{d}x\,\mathrm{d}y + \int_{\partial A_\sigma} \left[\hat{n}_\nu u_\nu + \hat{n}_{\nu s} u_s\right] \mathrm{d}s$$

6.2.2 Equilibrium Equations in Terms of In-Plane Displacements

After coupling the kinematic and constitutive relations, two equilibrium equations can be formulated as two second-order displacement partial differential equations with two unknown displacement functions $u(x, y)$, $v(x, y)$

$$\begin{aligned} &\left(\frac{\partial^2}{\partial x^2} + \frac{1-\nu}{2}\frac{\partial^2}{\partial y^2}\right)u + \frac{1+\nu}{2}\frac{\partial^2}{\partial x \partial y}v = -\frac{1}{D^n}\hat{p}_x \\ &\frac{1+\nu}{2}\frac{\partial^2}{\partial x \partial y}u + \left(\frac{\partial^2}{\partial y^2} + \frac{1-\nu}{2}\frac{\partial^2}{\partial x^2}\right)v = -\frac{1}{D^n}\hat{p}_y \end{aligned} \tag{6.2}$$

In the solution of the two equations integration constants appear, which are calculated from the boundary conditions (two conditions for each side of the rectangular domain).

In Figure 6.5, different variants of boundary conditions are shown, depending on the applied type of kinematic constraints and boundary loads:

- line A–B ($y = 0, x \in [0, a], \nu = -y, s = +x$): $n_\nu = n_y = 0,\ -n_{\nu s} = n_{yx} = 0$
- line B–C ($x = a, y \in [0, b], \nu = +x, s = +y$): $n_\nu = n_x = \hat{p},\ n_{\nu s} = n_{xy} = 0$
- line C–D ($y = b, x \in [0, a], \nu = +y, s = -x$): $u_\nu = u_y = 0,\ -n_{\nu s} = n_{yx} = \hat{p}$
- line D–A ($x = 0, y \in [0, b], \nu = -x, s = -y$): $-u_\nu = u_x = 0,\ -u_s = u_y = 0$

6.2.3 Principal Membrane Forces and their Directions

In the analysis of membranes (as in the *theory of elasticity* and *strength of materials*) the principal membrane forces and their directions can be calculated at any point on

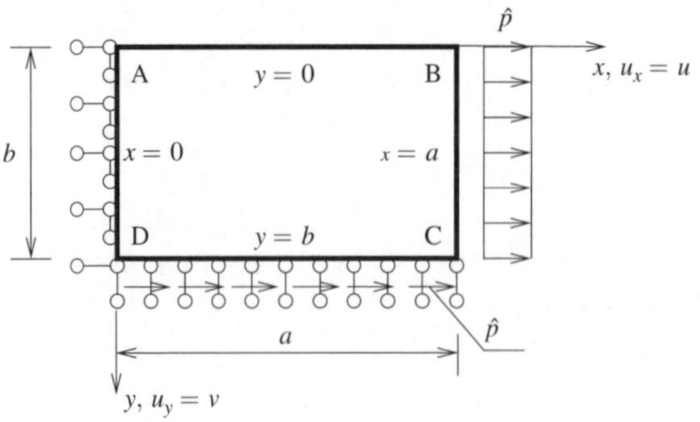

Figure 6.5 Examples of boundary constraints and loads applied to a rectangular flat membrane

the middle plane using formulae

$$n_{I,II} = \frac{1}{2}(n_x + n_y) \pm \frac{1}{2}\sqrt{(n_x - n_y)^2 + 4n_{xy}^2} \tag{6.3}$$

Angle α between axis x and the direction of the maximum force n_I is calculated from relation

$$\tan(2\alpha) = \frac{2\, n_{xy}}{n_x - n_y} \tag{6.4}$$

After the directions of the principal membrane forces have been determined, the so-called trajectories of the principal forces can be traced. Along these lines (tangentially to them) the extreme forces occur. The trajectories connected with tensile forces are particularly important since they help one to design the arrangement of steel bars in reinforced concrete structures. The presentation of directions and magnitudes of principal membrane forces is one of the most illustrative ways of result visualization in FEM packages, see Figures 6.12g or 6.15f.

6.2.4 Equations for a Flat Membrane Formulated using Airy's Stress Function

The strain compatibility equation is included in the problem formulation

$$\frac{\partial^2 \varepsilon_x}{\partial y^2} + \frac{\partial^2 \varepsilon_y}{\partial x^2} - \frac{\partial^2 \gamma_{xy}}{\partial x \partial y} = 0 \tag{6.5}$$

Using three constitutive equations, it can be expressed in terms of membrane forces

$$\frac{\partial^2 n_x}{\partial y^2} + \frac{\partial^2 n_y}{\partial x^2} - \nu\left(\frac{\partial^2 n_x}{\partial x^2} + \frac{\partial^2 n_y}{\partial y^2}\right) - 2(1+\nu)\frac{\partial^2 n_{xy}}{\partial x \partial y} = 0 \tag{6.6}$$

The elimination of the tangential force leads to a new form of the strain compatibility condition

$$\nabla^2\, (n_x + n_y) = -(1+\nu)\left(\frac{\partial \hat{p}_x}{\partial x} + \frac{\partial \hat{p}_y}{\partial y}\right) \tag{6.7}$$

which together with two equilibrium equations:

$$\frac{\partial n_x}{\partial x} + \frac{\partial n_{yx}}{\partial y} = -\hat{p}_x, \qquad \frac{\partial n_{xy}}{\partial x} + \frac{\partial n_y}{\partial y} = -\hat{p}_y \tag{6.8}$$

describes a statically determinate flat membrane (three equations with three unknown forces). However, they must be supplemented with static boundary conditions.

Now, Equations (6.7) and (6.8) will be compressed to one equation by introducing Airy's stress function $F(x, y)$ [Nm], which appears in the following relations with the membrane forces [N/m]:

$$n_x = \frac{\partial^2 F}{\partial y^2} - \int_0^x \hat{p}_x\, \mathrm{d}x, \qquad n_y = \frac{\partial^2 F}{\partial x^2} - \int_0^y \hat{p}_y\, \mathrm{d}y, \qquad n_{xy} = -\frac{\partial^2 F}{\partial x \partial y} \tag{6.9}$$

and satisfies two equilibrium equations.

For the special case of constant surface loads $\hat{p}_x(x, y) = c_x$, $\hat{p}_y(x, y) = c_y$ the relations between the forces and the stress function can be modified:

$$n_x = \frac{\partial^2 F}{\partial y^2} - c_x\, x, \qquad n_y = \frac{\partial^2 F}{\partial x^2} - c_y\, y, \qquad n_{xy} = -\frac{\partial^2 F}{\partial x \partial y} \tag{6.10}$$

Finally, one fundamental flat membrane biharmonic equation is obtained with one unknown Airy's stress function

$$\nabla^2\nabla^2 F = 0 \tag{6.11}$$

In the analytical solution of simple examples the stress function $F(x, y)$ is employed in a polynomial form

$$\begin{aligned} F(x,y) = {} & a_1 x + b_1 y + \\ & a_2 x^2 + b_2 xy + c_2 y^2 + \\ & a_3 x^3 + b_3 x^2 y + c_3 xy^2 + d_3 y^3 + \\ & a_4 x^4 + b_4 x^3 y + c_4 x^2 y^2 + d_4 xy^3 + e_4 y^4 \end{aligned} \tag{6.12}$$

It can be noticed that the stress function adopted as the first-degree polynomial corresponds to the zero stress field. Moreover, the 2nd and 3rd degree polynomials satisfy the biharmonic equation for any values of the coefficients. In the case of 4th and higher degree polynomials, relations between coefficients and their values are obtained on the basis of the problem definition.

6.2.5 Global Formulation

A functional describing the potential energy and written with respect to the middle plane can be expressed by membrane forces and strains in the following form

$$U^n = \int_A (n_x\, \varepsilon_x + n_y\, \varepsilon_y + n_{xy}\, \gamma_{xy})\mathrm{d}x\, \mathrm{d}y \tag{6.13}$$

In order to eliminate the membrane forces, the constitutive equations are used and the elastic energy of a membrane is obtained, which depends on the membrane stiffness and squares of strains

$$U^n = \frac{D^n}{2}\int_A \left(\varepsilon_x^2 + \varepsilon_y^2 + 2\nu\, \varepsilon_x \varepsilon_y + \frac{1-\nu}{2}\gamma_{xy}^2\right) \mathrm{d}x\, \mathrm{d}y \tag{6.14}$$

Taking the kinematic equations into account, the strain energy U^n can also be expressed as a surface integral, with the integrand that depends on the first derivatives of membrane displacements

$$U^n = \frac{D^n}{2}\int_A \left\{ \left(\frac{\partial u}{\partial x} + \frac{\partial v}{\partial y}\right)^2 - 2(1-\nu)\left[\frac{\partial u}{\partial x}\frac{\partial v}{\partial y} - \frac{1}{4}\left(\frac{\partial u}{\partial y} + \frac{\partial v}{\partial x}\right)^2\right]\right\} \mathrm{d}x\, \mathrm{d}y \tag{6.15}$$

This form of U^n is the basis for development of the displacement-based finite element model.

The work of external surface and boundary loads W^n has the form

$$W^n = \int_A \left[\hat{p}_x u + \hat{p}_y v\right] \mathrm{d}x\, \mathrm{d}y + \int_{\partial A_\sigma} \left[\hat{n}_\nu u_\nu + \hat{n}_{\nu s} u_s\right] \mathrm{d}s \tag{6.16}$$

The total potential energy Π^n is formulated as the difference of these integrals

$$\Pi^n = U^n - W^n \tag{6.17}$$

The equations of the global formulation are also included in Box 6.1.

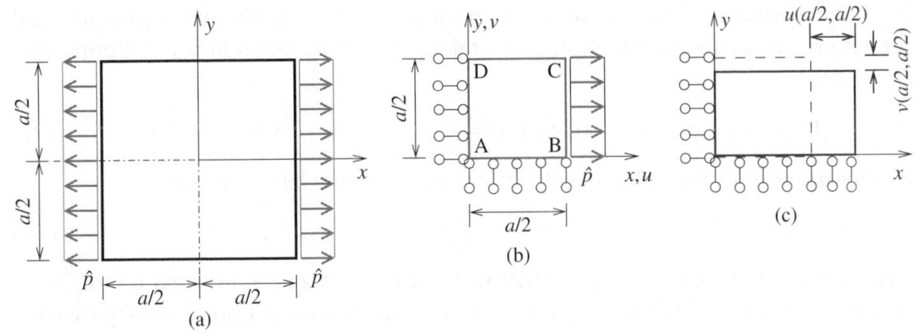

Figure 6.6 Square membrane under unidirectional tension: (a) whole configuration, (b) quarter of the domain with relevant boundary conditions and (c) subdomain before and after deformation

6.3 Square Membrane under Unidirectional Tension

The unidirectional tension of a square membrane with contour load $p_\nu = \hat{p} = \text{const.}$ on two opposite edges is the simplest case of a two-dimensional plane problem (Figure 6.6).

The calculation model (set of equations) and its solution is presented. The set of eight equations (three kinematic, two equilibrium, three constitutive) is formulated and supplemented with static and/or kinematic boundary conditions. The solution of the problem in local formulation is obtained by finding the distributions of three membrane forces, three strains and two displacements on the middle plane of the membrane. In the numerical example, a quarter of the membrane domain is analysed.

6.3.1 Analytical Solution

For a full description of the problem the set of eight equations is rewritten (see Box 6.1):

(I) kinematic equations (3):

$$\varepsilon_x = u_{,x}, \qquad \varepsilon_y = v_{,y}, \qquad \gamma_{xy} = u_{,y} + v_{,x} \tag{6.18}$$

(II) equilibrium equations (2):

$$n_{x,x} + n_{yx,y} + \hat{p}_x = 0, \qquad n_{xy,x} + n_{y,y} + \hat{p}_y = 0 \tag{6.19}$$

(III) constitutive equations (3):

$$\varepsilon_x = (n_x - \nu n_y)/(Eh), \quad \varepsilon_y = (n_y - \nu n_x)/(Eh), \quad \gamma_{xy} = 2(1+\nu)n_{xy}/(Eh) \tag{6.20}$$

For four boundary lines of the membrane quarter, the following boundary conditions are formulated:

- line A–B ($y = 0, x \in [0, a/2]$): $\quad -u_\nu = v = 0, \ -n_{\nu s} = n_{yx} = 0$
- line B–C ($x = a/2, y \in [0, a/2]$): $\quad n_\nu = n_x = \hat{p}, \ n_{\nu s} = n_{xy} = 0$
- line C–D ($y = a/2, x \in [0, a/2]$): $\quad n_\nu = n_y = 0, \ -n_{\nu s} = n_{yx} = 0$
- line D–A ($x = 0, y \in [0, a/2]$): $\quad -u_\nu = u = 0, \ -n_{\nu s} = n_{xy} = 0$

In the analytical solution, three groups of functions are determined in three steps of the consideration:

- membrane forces, obtained after integration of two equilibrium equations and the strain compatibility equation (6.7), taking the static boundary conditions into account:

$$n_x(x,y) = \hat{p} = \text{const.}, \qquad n_y(x,y) = 0, \qquad n_{xy}(x,y) = 0 \tag{6.21}$$

- membrane strains, derived from algebraic constitutive relations (6.20):

$$\varepsilon_x(x,y) = \hat{p}/(Eh), \qquad \varepsilon_y(x,y) = -\nu\hat{p}/(Eh), \qquad \gamma_{xy}(x,y) = 0 \tag{6.22}$$

- displacement functions (translations) on the middle plane, found from integrated kinematic equations (6.18) using of kinematic boundary conditions – see Figure 6.6c:

$$u(x,y) = \frac{\hat{p}}{E\,h}\,x, \qquad v(x,y) = -\frac{\nu\hat{p}}{E\,h}\,y \tag{6.23}$$

The values of the following input parameters must be specified to obtain the numerical results: Young's modulus E, Poisson's ratio ν, length a, thickness h, surface loads $\hat{p}_x$, $\hat{p}_y$ and normal boundary load $\hat{p}_\nu = \hat{p}$. Their values are given in Box 6.2.

The verification values of two translations $u(a/2, y)$ and $v(x, a/2)$ for two boundary lines as well as the computed membrane forces and strains are presented in Box 6.2.

It should be emphasized that in the case of nonzero Poisson's ratio, the following relations are valid (see Figure 6.6c):

$$\varepsilon_y(x,y) = -\nu\;\varepsilon_x(x,y), \qquad v(a/2, a/2) = -\nu\;u(a/2, a/2) \tag{6.24}$$

Box 6.2 Square membrane under unidirectional tension

Data

$E = 2.0 \times 10^8$ kN/m², $\nu = 0.3$, $a = 1.0$ m, $h = 0.01$ m
$\hat{p}_x(x,y) = \hat{p}_y(x,y) = 0.0$, $\hat{p}_\nu = \hat{p} = 100.0$ kN/m

---

Check values from analytical solution

$u(a/2, y) = u(0.5, y) = 2.50 \times 10^{-5}$ m, $v(x, a/2) = v(x, 0.5) = -0.75 \times 10^{-5}$ m
$\varepsilon_x(x,y) = 5.0 \times 10^{-5} = \text{const.}$, $\varepsilon_y(x,y) = -1.5 \times 10^{-5} = \text{const.}$, $\gamma_{xy}(x,y) = 0.0$
$n_x(x,y) = \hat{p} = 100$ kN/m, $n_y(x,y) = n_{xy}(x,y) = 0.0$

Check values from FEM solution using ANKA

$NSE = 8 \times 8$ (quarter of domain), $NEN = 4$, $NNDOF = 2$, $NEDOF = 8$
$u^{\text{FEM}}(a/2, a/2) = u^{\text{FEM}}(0.5, 0.5) = 2.50 \times 10^{-5}$ m
$v^{\text{FEM}}(a/2, a/2) = v^{\text{FEM}}(0.5, 0.5) = -0.75 \times 10^{-5}$ m
$n_x^{\text{FEM}}(x,y) = 100$ kN/m, $n_y^{\text{FEM}}(x,y) = n_{xy}^{\text{FEM}}(x,y) = 0.0$

Membrane configuration shown in Figure 6.6

6.3.2 Analytical Solution with Airy's Stress Function

In unidirectional tension or compression of a square membrane, the constant stress σ_x as well as constant membrane force n_x occur in the considered domain. The following boundary conditions are formulated for the whole membrane, taking two planes of symmetry into account (Figure 6.6a):

- lines $x = \pm a/2$: $n_x = \hat{p}, \quad n_{xy} = 0$
- lines $y = \pm a/2$: $n_y = 0, \quad n_{yx} = 0$

On the basis of the expected stress distribution resulting from the applied load the following (polynomial) stress function $F(x, y)$ [Nm] can be assumed

$$F(x, y) = c_2\, y^2 \tag{6.25}$$

Concluding from the relations between membrane forces and Airy's function, one obtains:

$$n_x = F_{,yy} = 2\, c_2, \quad n_y = F_{,xx} = 0, \quad n_{xy} = -F_{,xy} = 0 \tag{6.26}$$

From the nonzero static boundary condition

$$n_x(\pm a/2, y) = 2\, c_2 = \hat{p} \tag{6.27}$$

the value of constant c_2 is calculated and as a result a uniform distribution of the only nonzero force n_x is obtained:

$$c_2 = \hat{p}/2, \qquad n_x(x, y) = \hat{p} = \text{const.} \tag{6.28}$$

6.3.3 Numerical Solution

The detailed numerical data for the considered case are as in Subsection 6.3.1. The domain of square membrane is discretized with $NSE = 8 \times 8$ four-node FEs with $NNDOF = 2$ and $NEDOF = 8$. The boundary normal load for $x = \pm a/2$ is replaced by substitute nodal forces.

Using FEM package ANKA (1993) the nonzero membrane force $n_x^{\text{FEM}}(x, y) = \hat{p}$ and $n_y^{\text{FEM}}(x, y) = n_{xy}^{\text{FEM}}(x, y) = 0$ are reproduced exactly, which should be the case for any FE code. The values of two translations of corner $x = y = +a/2$ are included in Box 6.2.

The reproduction of even simple unidirectional tension/compression conditions can lead to unexpected numerical results of FEM analysis. It happens when spurious deformation modes occur. For example, this phenomenon can take place if the problem is discretized with finite elements in which reduced numerical integration is applied. For more advanced information the reader is referred to Subsection 14.4.2 and Section 17.2, dealing with problem of the influence of numerical integration on the properties of FEs.

6.4 Square Membrane under Uniform Shear

6.4.1 Analytical Solution

Pure uniform shear state with the only nonzero force $n_{xy}(x, y)$ and shear strain $\gamma_{xy}(x, y)$ (see Figure 6.7) is treated as one of the basic membrane strain states apart from the unidirectional tension/compression.

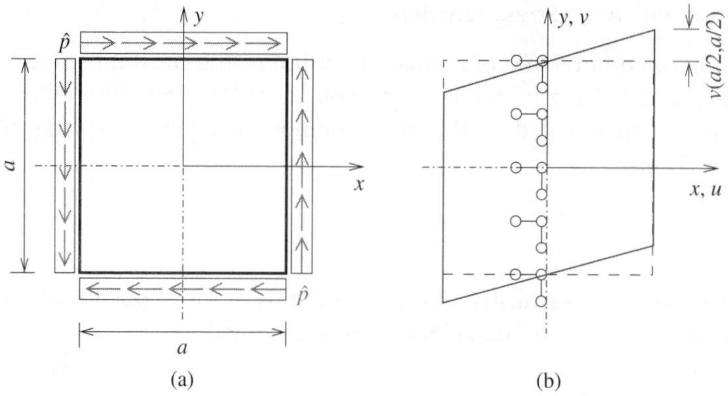

Figure 6.7 Square membrane under uniform shear: (a) configuration and (b) deformation

For analytical solution of the problem, the eight equations given in Box 6.1 are used, completed by adequate boundary conditions (Figure 6.7b):

- line $x = 0, y \in [-a/2, +a/2]$: $\quad u = v = 0$
- lines $x = \pm a/2, y \in [-a/2, +a/2]$: $\quad n_{xy} = \hat{p},\ n_x = 0$
- lines $y = \pm a/2, x \in [-a/2, +a/2]$: $\quad n_{yx} = \hat{p},\ n_y = 0$

The following distributions are obtained as a solution of the problem:

- membrane forces:

$$n_x(x, y) = n_y(x, y) = 0, \qquad n_{xy}(x, y) = \hat{p} = \text{const.} \tag{6.29}$$

- membrane strains:

$$\varepsilon_x(x, y) = \varepsilon_y(x, y) = 0, \qquad \gamma_{xy}(x, y) = \frac{2(1+v)\,\hat{p}}{E\,h} = \text{const.} \tag{6.30}$$

- displacements functions:

$$u(x, y) = 0, \qquad v(x, y) = \frac{2(1+v)\,\hat{p}}{E\,h}\,x \tag{6.31}$$

For detailed calculations the values of the following parameters must be specified: Young's modulus E, Poisson's ratio v, length a, thickness h, surface loads $\hat{p}_x, \hat{p}_y$ and tangential boundary load $\hat{p}_s = \hat{p}$. The input and output data are included in Box 6.3. The obtained results show that the static and kinematic constraints are satisfied.

6.4.2 FEM Results

The membrane considered in the previous subsection is now discretized with $NSE = 8 \times 8$ four-node FEs with $NNDOF = 2$ and $NEDOF = 8$. The numerical solution is obtained using FEM program ANKA (1993). The uniform, tangential boundary load is replaced with substitute tangential nodal forces.

The results for the membrane under uniform shear with boundary conditions, shown in Figure 6.7b, are consistent with the analytical solution (see Box 6.3):

Box 6.3 Square membrane under uniform shear

Data

$E = 2.0 \times 10^8$ kN/m^2, $\nu = 0.3$, $a = 1.0$ m, $h = 0.01$ m
$\hat{p}_x(x, y) = \hat{p}_y(x, y) = 0.0$, $\hat{p}_s = \hat{p} = 100.0$ kN/m

---

Check values from analytical solution

$u(x, y) = 0.0$, $v(0, y) = 0.0$, $v(\pm a/2, y) = v(\pm 0.5, y) = \pm 0.65 \times 10^{-4}$ m
$\varepsilon_x(x, y) = \varepsilon_y(x, y) = 0.0$, $\gamma_{xy}(x, y) = 1.3 \times 10^{-4} = \text{const.}$
$n_x(x, y) = n_y(x, y) = 0.0$, $n_{xy}(x, y) = 100$ kN/m

Check values from FEM solution using ANKA

$NSE = 8 \times 8$ (whole domain), $NEN = 4$, $NNDOF = 2$, $NEDOF = 8$
$v^{FEM}(a/2, a/2) = v^{FEM}(0.5, 0.5) = 0.65 \times 10^{-4}$ m
$n_x^{FEM}(x, y) = n_y^{FEM}(x, y) = 0.0$, $n_{xy}^{FEM}(x, y) = 100$ kN/m

---

Membrane configuration shown in Figure 6.7

- zero horizontal translation in the whole domain $u(x, y) = 0$
- vertical translation v depends linearly on the x-coordinate and for all points lying on line $x = \pm a/2$ its values are consistent with the analytical solution
$v^{FEM} = v^{anal} = \pm 0.65 \times 10^{-4}$ m
- membrane forces correspond exactly to uniform shear state
$n_{xy}(x, y) = \hat{p} = 100.0$ kN/m $=$ const., $n_x(x, y) = n_y(x, y) = 0$

6.5 Pure In-Plane Bending of a Square Membrane

In-plane bending of a membrane is imposed by a boundary load defined by the linear function $p_x(y) = -2\hat{p}\, y/a$, where $\hat{p} = p_x(-a/2)$. The load is applied on two parallel lines $x = \pm a/2$ as shown in Figure 6.8.

The kinematic constraints are imposed at three points: F(0, 0), E(0, $+a/2$), D(0, $-a/2$) on symmetry axis $x = 0$ and shown in Figure 6.8b.

In order to get the unique solution of the set of equations describing the membrane state the following static boundary conditions are formulated:

- lines $x = \pm a/2, y \in [-a/2, +a/2]$: $n_x = -2\hat{p}y/a$, $n_{xy} = 0$
- lines $y = \pm a/2, x \in [-a/2, +a/2]$: $n_y = 0$, $n_{yx} = 0$

In the case of pure in-plane bending the only nonzero force is $n_x(x, y) = -2\hat{p}\, y/a$ while $n_y(x, y) = n_{xy}(x, y) = 0$.

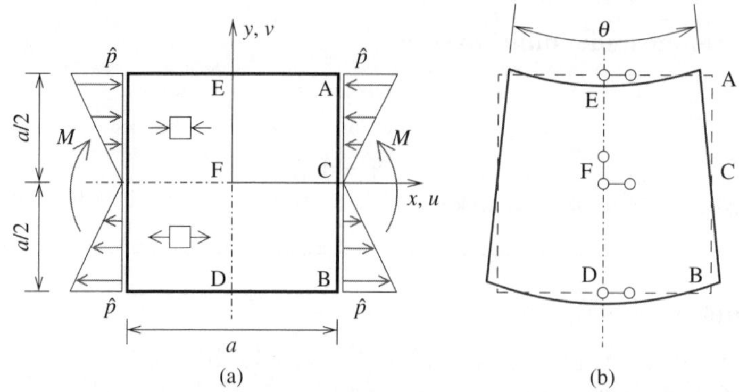

Figure 6.8 Square membrane under pure in-plane bending: (a) geometry and tractions with equivalent moments and (b) configuration before and after deformation

From constitutive relations the strains are determined:

$$\varepsilon_x = \frac{1}{E\,h}\,n_x = -2\,\frac{\hat{p}}{E\,h}\,\frac{y}{a} = \frac{\partial u}{\partial x}, \qquad \varepsilon_y = -\frac{\nu}{E\,h}\,n_x = 2\nu\frac{\hat{p}}{E\,h}\,\frac{y}{a} = \frac{\partial v}{\partial y}$$
$$\gamma_{xy} = 0 = \frac{\partial u}{\partial y} + \frac{\partial v}{\partial x} \tag{6.32}$$

Subsequently, the kinematic relations are integrated to find the functions of translations:

$$u(x,y) = -2\,\frac{\hat{p}}{E\,h\,a}\,xy + f_1(y), \qquad v(x,y) = -2\nu\frac{\hat{p}}{E\,h\,a}\,\frac{y^2}{2} + f_2(x) \tag{6.33}$$

where $f_1(y), f_2(x)$ are integration functions.

Taking relation $\gamma_{xy} = 0$ and kinematic constraints $u(0,0) = 0$ and $v(0,0) = 0$ into account, the final forms of functions $u(x,y)$ and $v(x,y)$ are obtained:

$$u(x,y) = -2\frac{\hat{p}}{Eha}xy = -\frac{4}{a^2}\,\overline{u}\,xy, \quad v(x,y) = \frac{\hat{p}}{Eha}(x^2 + \nu y^2) = \frac{2}{a^2}\,\overline{u}(x^2 + \nu y^2) \tag{6.34}$$

where

$$\overline{u} = \frac{\hat{p}\,a}{2\,E\,h} = u_B \tag{6.35}$$

The displacement values are calculated at six characteristic points – see Figure 6.8 and Box 6.4:

- $A(a/2, a/2)$: $u_A = -\overline{u},\ v_A = 0.5(1+\nu)\overline{u}$
- $B(a/2, -a/2)$: $u_B = \overline{u},\ v_B = 0.5(1+\nu)\overline{u}$
- $C(a/2, 0)$: $u_C = 0,\ v_C = 0.5\overline{u}$
- $D(0, -a/2)$: $u_D = 0,\ v_D = 0.5\nu\overline{u}$
- $E(0, a/2)$: $u_E = 0,\ v_E = 0.5\nu\overline{u}$
- $F(0, 0)$: $u_F = 0,\ v_F = 0$

Angle θ defined between the lines $x = \pm a/2$ in the deformed configuration (Figure 6.8b) is given by

$$\theta^{\text{anal}} = \frac{2\hat{p}}{E\,h} = 4\frac{\overline{u}}{a} \tag{6.36}$$

Now, strain components can be written as functions of θ as:

$$\begin{aligned} &\varepsilon_x(y) = -4\,\frac{\overline{u}}{a}\,\frac{y}{a} = -\theta\,\frac{y}{a}, \qquad \varepsilon_y(y) = 4\nu\,\frac{\overline{u}}{a}\,\frac{y}{a} = \nu\theta\,\frac{y}{a} = -\nu\varepsilon_x(y) \\ &\varepsilon_{x(\text{A,E})} = -2\,\frac{\overline{u}}{a}, \qquad \varepsilon_{x(\text{C,F})} = 0, \qquad \varepsilon_{x(\text{B,D})} = 2\,\frac{\overline{u}}{a} \end{aligned} \tag{6.37}$$

Taking these formulae into account and adopting the following input values: Young's modulus E, Poisson's ratio ν, length a, thickness h and the intensity of boundary load $\hat{p}$ at the corners, the output values of displacements, strains and forces are calculated. The input data and verification results of analytical and numerical calculations are included in Box 6.4.

It is worth noticing that in the case of pure in-plane bending shear strain $\gamma_{xy}^{\text{anal}}$ is equal to zero in the whole domain. The knowledge of exact solution is useful in the assessment of correctness of some FEs. The problem of spurious reproduction of strain $\gamma_{xy}^{\text{FEM}} \neq 0$ in pure in-plane bending by four-node bilinear FEs is indicated in Section 17.3.1.

Box 6.4 Pure in-plane bending of a square membrane

Data

$E = 2.0 \times 10^8$ kN/m^2, $\nu = 0.3$, $a = 1.0$ m, $h = 0.01$ m
$\hat{p}_x(x, y) = \hat{p}_y(x, y) = 0.0$, $\hat{p} = 100.0$ kN/m

---

Check values from analytical solution

$u_{\text{A}} = -2.5 \times 10^{-5}$ m, $v_{\text{A}} = 1.625 \times 10^{-5}$ m,
$u_{\text{B}} = 2.5 \times 10^{-5}$ m, $v_{\text{B}} = 1.625 \times 10^{-5}$ m
$u_{\text{C}} = 0$, $v_{\text{C}} = 1.25 \times 10^{-5}$ m, $u_{\text{D}} = 0$, $v_{\text{D}} = 0.375 \times 10^{-5}$ m
$u_{\text{E}} = 0$, $v_{\text{E}} = 0.375 \times 10^{-5}$ m, $u_{\text{F}} = 0$, $v_{\text{F}} = 0$
$\varepsilon_{x(\text{A,E})} = -5.0 \times 10^{-5}$, $\varepsilon_{x(\text{C,F})} = 0$, $\varepsilon_{x(\text{B,D})} = 5.0 \times 10^{-5}$, $\gamma_{xy}(x, y) = 0.0$
$n_{x(\text{A,E})} = -100.0$ kN/m, $n_{x(\text{C,F})} = 0$, $n_{x(\text{B,D})} = 100.0$ kN/m
$n_y(x, y) = n_{xy}(x, y) = 0.0$

Check values from FEM solution using ANKA

$NSE = 16 \times 16$ (whole domain), $NEN = 4$, $NNDOF = 2$, $NEDOF = 8$
$u_{\text{A}}^{\text{FEM}} = -2.474 \times 10^{-5}$ m $= 0.990u_{\text{A}}^{\text{anal}}$, $v_{\text{A}}^{\text{FEM}} = 1.611 \times 10^{-5}$ m $= 0.991v_{\text{A}}^{\text{anal}}$

Membrane configuration shown in Figure 6.8

Moreover, in the considered case, the buckling phenomena can occur due to compressive force n_x in the upper half of the membrane. The issue of determination of the critical load is discussed in Section 14.4.2.

6.6 Cantilever Beam with a Load on the Free Side

In this section the problem of bending of a cantilever beam (shown in Figure 6.9) is considered. Bending of the beam is induced by a parabolic tangential (shear) traction applied to the free end of the cantilever ($x = 0$). The resultant (integral) of the applied load is denoted by $\tilde{P}$. The top and bottom boundary of the beam ($y = \pm c$) are unloaded. The kinematic constraints are imposed on line $x = L$ and are going to be discussed in detail later on. The adopted unit width of the beam ($b = 1$) corresponds to the unit thickness of membrane ($h = 1$) and thus the values of membrane forces and corresponding stress components are equal. To solve the problem, the equations of two-dimensional plane stress state are used.

6.6.1 Analytical Solution

In the analytical solution the strain compatibility equation expressed by Airy's function is used with the assumption of zero surface loads

$$\nabla^2\nabla^2 F = 0 \tag{6.38}$$

The solution of this problem can be found in Timoshenko and Goodier (1951) or Girkmann (1956). The relations between stress components and Airy's function F [N] are given by:

$$\sigma_x = \frac{\partial^2 F}{\partial y^2}, \qquad \sigma_y = \frac{\partial^2 F}{\partial x^2}, \qquad \tau_{xy} = -\frac{\partial^2 F}{\partial x \partial y} \tag{6.39}$$

Stress function $F(x, y)$ is adopted in a form of polynomials (of second, third or fourth degree) depending on the expected stress distribution that results from loading and boundary conditions. In the considered case the following function F is proposed

$$F = b_2\, xy + \frac{d_4}{6}\, xy^3 \tag{6.40}$$

and resulting stress distributions are given by:

$$\sigma_x = F_{,yy} = d_4\, xy, \qquad \sigma_y = F_{,xx} = 0, \qquad \tau_{xy} = -F_{,xy} = -b_2 - \frac{d_4}{2}\, y^2 \tag{6.41}$$

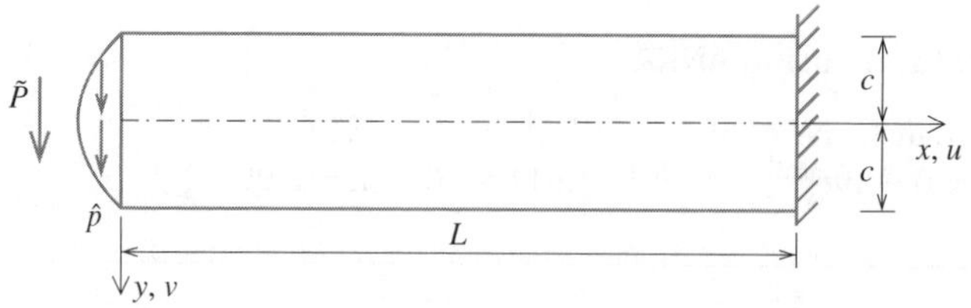

Figure 6.9 Cantilever beam modelled as a flat rectangular membrane

Constants b_2 and d_4 are derived from the boundary conditions. Taking the lack of tangential load on two lines $y = \pm c$ into account

$$\tau_{xy}(x, \pm c) = -b_2 - \frac{d_4}{2}\, c^2 = 0 \tag{6.42}$$

since the integral relation between resultant force P and shear stress τ_{xy} on line $x = 0$ is

$$\tilde{P} = \frac{P}{b} = -\int_{-c}^{+c} \tau_{xy}\, \mathrm{d}y = \int_{-c}^{+c} \left(b_2 + \frac{d_4}{2}\, y^2\right) \mathrm{d}y \tag{6.43}$$

one obtains the relations:

$$b_2 = \frac{3\,P}{4\,c\,b}, \qquad d_4 = -\frac{2}{c^2}\, b_2 = -\frac{3\,P}{2\,c^3\,b} \tag{6.44}$$

and finally the stress functions read:

$$\sigma_x(x, y) = -\frac{3\,P}{2\,c^3\,b}\, xy, \qquad \sigma_y(x, y) = 0, \qquad \tau_{xy}(x, y) = -\frac{3\,P}{4\,c\,b}\left(1 - \frac{y^2}{c^2}\right) \tag{6.45}$$

Adopting $a = 2\,c$ and $b = 1$ the cross-sectional inertia moment I given by formula $I = ba^3/12$ is calculated here as $I = 2bc^3/3$ and introduced into Equations (6.45) leading to:

$$\sigma_x(x, y) = -\frac{P}{I}\, xy, \qquad \sigma_y(x, y) = 0, \qquad \tau_{xy}(x, y) = -\frac{P}{2\,I}(c^2 - y^2) \tag{6.46}$$

Combining the constitutive and kinematic equations the three strain components can be written as follows:

$$\begin{aligned}
\varepsilon_x(x, y) &= \frac{\sigma_x}{E} = -\frac{P}{E\,I}\, xy = \frac{\partial u}{\partial x}, \qquad \varepsilon_y(x, y) = -\nu\,\frac{\sigma_x}{E} = \frac{\nu\,P}{E\,I}\, xy = \frac{\partial v}{\partial y} \\
\gamma_{xy}(x, y) &= \frac{\tau_{xy}}{G} = -\frac{P}{2\,G\,I}(c^2 - y^2) = \frac{\partial u}{\partial y} + \frac{\partial v}{\partial x}
\end{aligned} \tag{6.47}$$

After integration (see Timoshenko and Goodier 1951) the formulae for translation functions, expressed using four unknown constants, can be obtained:

$$\begin{aligned}
u &= -\frac{P}{2\,E\,I}\, x^2 y - \frac{\nu\,P}{6\,E\,I}\, y^3 + \frac{P}{6\,G\,I}\, y^3 + C_1\, y + C_2 \\
v &= \frac{\nu\,P}{2\,E\,I}\, xy^2 + \frac{P}{6\,E\,I}\, x^3 + C_3\, x + C_4
\end{aligned} \tag{6.48}$$

The constants C_i, where $i = 1, \ldots, 4$, are determined on the basis of the following consideration. Firstly, we substitute Equations (6.48) into the kinematic equation for the shear strain $(6.47)_3$ to obtain the relation between constants: $C_1 + C_3 = -Pc^2/(2GI)$.

For the point with coordinates $x = L$, $y = 0$ on the axis of the beam (the right-hand clamped end) we introduce constraints of the motion of the beam in the (x, y) plane, restraining there two translations and one rotation. Please note that the shear strain in Equation $(6.47)_3$, which has the negative sign, is related to two rotation angles expressed by the following derivatives: $\partial v/\partial x$ (rotation of horizontal element of beam axis) and $\partial u/\partial y$ (rotation of vertical element along beam cross section). This means that the clamped boundary conditions on the right-hand side (i.e. at point $x = L$, $y = 0$) can be imposed in the following two manners:

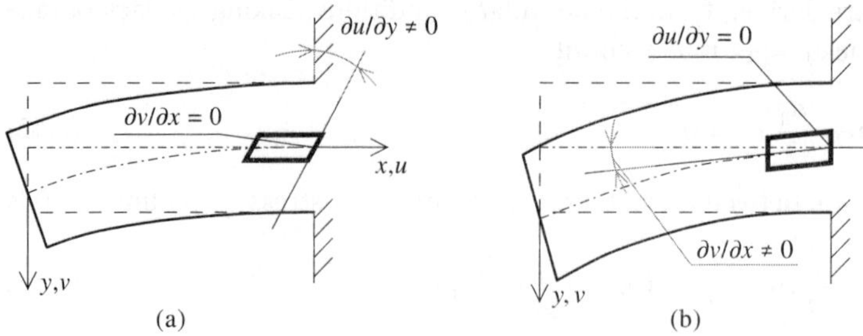

Figure 6.10 Cantilever beam – two cases of boundary conditions on a clamped beam end

Case 1 – horizontal element of beam axis is fixed – Figure 6.10a:

$$u = v = 0, \qquad \frac{\partial v}{\partial x} = 0 \tag{6.49}$$

Case 2 – vertical element of beam cross section is fixed – Figure 6.10b:

$$u = v = 0, \qquad \frac{\partial u}{\partial y} = 0 \tag{6.50}$$

Omitting the detailed calculations of constants C_i, where $i = 1, \ldots, 4$, the two sets of relations describing the distributions of horizontal and vertical translations are given:

Case 1 – The influence of shear stresses on beam translations is neglected and the Bernoulli–Euler theory for beams is used. Vertical translation is denoted by top index (B–E). The displacements are specified as follows:

$$\begin{aligned} u(x,y) &= -\frac{P}{2EI}x^2y + \left(-\frac{\nu P}{6EI} + \frac{P}{6GI}\right)y^3 + \left(\frac{PL^2}{6EI} - \frac{Pc^2}{2GI}\right)y \\ v^{(\mathrm{B-E})}(x,y) &= \frac{\nu P}{2EI}xy^2 + \frac{P}{6EI}x^3 - \frac{PL^2}{2EI}x + \frac{PL^3}{3EI} \end{aligned} \tag{6.51}$$

with the equation of curved beam axis (for $y = 0$) and the formula for maximum deflection (for $x = 0, y = 0$):

$$v^{(\mathrm{B-E})}(x,0) = \frac{P}{6EI}x^3 - \frac{PL^2}{2EI}x + \frac{PL^3}{3EI}, \qquad v^{(\mathrm{B-E})}(0,0) = \frac{PL^3}{3EI} \tag{6.52}$$

Case 2 – The influence of shear stresses on the beam deflection is included as in the Timoshenko beam theory, which is marked by top index (T). In Figure 6.10b the rotation angle $\partial v/\partial x = -Pc^2/(2GI)$ represents the influence of shear stresses on the deflection and results in an additional component of deflection at the left-hand free end, equal to $\Delta v^{(\mathrm{T})}(0,0) = P\,L\,c^2/(2\,G\,I)$:

$$\begin{aligned} v^{(\mathrm{T})}(x,0) &= \frac{P}{6\,EI}x^3 - \frac{P\,L^2}{2\,EI}x + \frac{P\,L^3}{3\,EI} + \underline{\frac{P\,c^2}{2\,GI}(L-x)}, \\ v^{(\mathrm{T})}(0,0) &= \frac{P\,L^3}{3\,EI} + \underline{\frac{P\,L\,c^2}{2\,GI}} \end{aligned} \tag{6.53}$$

The second option gives more a precise description of transverse bending of the beam. The underlined factors take the influence of shear stresses into account.

Box 6.5 Cantilever beam with a load on the free side

Data

$E = 2.0 \times 10^4$ kN/cm^2, $\nu = 0.25$
$L = 48.0$ cm, $a = 2\,c = 12.0$ cm, $b = 1.0$ cm, $I = 144.0$ cm^4
$\tilde{P} = 40.0$ kN/cm

Check values from analytical solution

$\sigma_x(L, \pm c) = \mp 800.0$ MPa, $\tau_{xy}(x, 0) = -50.0$ MPa
$u(0, \pm c) = \pm 0.0934$ cm, $u(0, 0) = 0.0$
$v^{(B-E)}(0, 0) = 0.512$ cm, $v^{(T)}(0, 0) = 0.542$ cm

Check values from FEM solution using ROBOT

$NSE^{I} = 4 \times 16$, $NSE^{II} = 16 \times 64$
$v^{II,FEM}(0, 0) = UY(0, 0) = 0.535$ cm, $v^{(B-E)}(0, 0) < v^{II,FEM}(0, 0) < v^{(T)}(0, 0)$

Membrane configuration shown in Figure 6.9

We emphasize that in general continuum modelling (2D or 3D) the clamped edge can be represented by different constraints and usually cannot deform. Therefore, these solutions and, in particular, the stress distributions in Equations (6.46) are valid sufficiently far from the clamped edge according to the de Saint-Venant principle.

In order to obtain the numerical results, the values of Young's modulus E, Poisson's ratio ν, dimensions L, c, b and load resultant $\tilde{P}$ must be specified, see Box 6.5.

The following formulae are used to find:

- cross-sectional moment of inertia $I = 2\,c^3\,b\,/3$
- normal stresses at extreme points ($y = \pm c$) on clamped cross section $x = L$
  $\sigma_x(L, \pm c) = \mp P\,L\,c/I$
- tangential stress on beam axis $y = 0$ for $x \in [0, L]$ $\tau_{xy}(x, 0) = -P\,c^2/(2\,I)$
- horizontal translations at extreme points ($y = \pm c$) at the free end $x = 0$
  $u(0, +c) = -\nu\,P\,c^3/(6\,EI) + P\,L^2\,c/(6\,EI) + P\,c^3/(6\,GI) = -u(0, -c)$
- horizontal translation of the point $x = y = 0$ on the beam axis at the free end $u(0, 0) = 0$
- vertical deflection of the same point, calculated considering the following two cases:
  Case 1 – neglecting shear stresses (transverse force) in the beam
  $v^{(B-E)}(0, 0) = P\,L^3/(3\,EI)$
  Case 2 – taking into account shear stresses (transverse force)
  $v^{(T)}(0, 0) = P\,L^3/(3\,EI) + P\,L\,c^2/(2\,GI)$

The input data and check results are included in Box 6.5. The deformation and distributions of normal and shear stress components are shown in Figure 6.11.

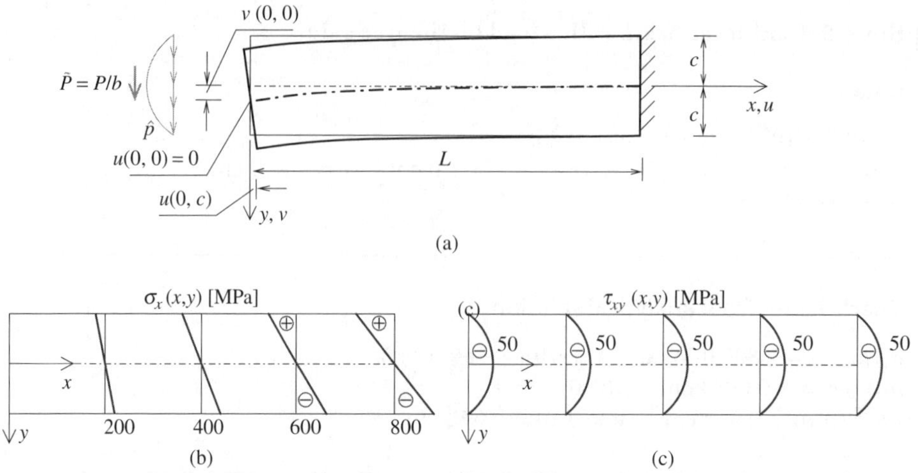

Figure 6.11 Cantilever beam – analytical results: (a) deformation, diagrams of (b) normal stresses and (c) in-plane shear stresses and along sections x = const.

The problem of bending of a cantilever beam is a typical example discussed within the *strength of materials* or *structural mechanics* course. On the basis of the solution for the beam described by the equations for a two-dimensional plane problem the following remarks can be formulated:

- pure tension or compression occurs on extreme lines ($y = \pm c$) with $n_x = \sigma_x\, h \neq 0$ and $n_{xy} = \tau_{xy}\, h = 0$
- pure shear occurs on the middle line ($y = 0$) with $n_x = \sigma_x\, h = 0$ and $n_{xy} = \tau_{xy}\, h \neq 0$
- bending with shear effects takes place on intermediate lines with $n_x = \sigma_x\, h \neq 0$ and $n_{xy} = \tau_{xy}\, h \neq 0$
- linear distribution of normal stresses σ_x on cross-sectional line $x =$ const. and parabolic distribution of shear stresses τ_{xy} (analogous to tangential traction on free edge $x = 0$) are noted

6.6.2 FEM Results

The problem of bending of a cantilever beam is one of well-known benchmarks used for FEM code verification. In Subsection 17.5.1 the remarks on numerical simulations performed using different FEs are presented.

The computations are carried out with FEM package ROBOT (2006). The beam is discretized with four-node FEs and two meshes with $NSN^{\mathrm{I}} = 4 \times 16 = 64$ and $NSN^{\mathrm{II}} = 16 \times 64 = 1024$ FEs are considered. The maximum value of deflection obtained for the fine mesh (see Box 6.5) is between the (analytical) results obtained using Bernoulli–Euler and Timoshenko theories.

The most important results of computations are shown in Figure 6.12:

- plot of the deformed configuration (coarse mesh)
- contour plots of translations and selected stresses for the fine mesh
- visualization of principal membrane stresses for the coarse mesh

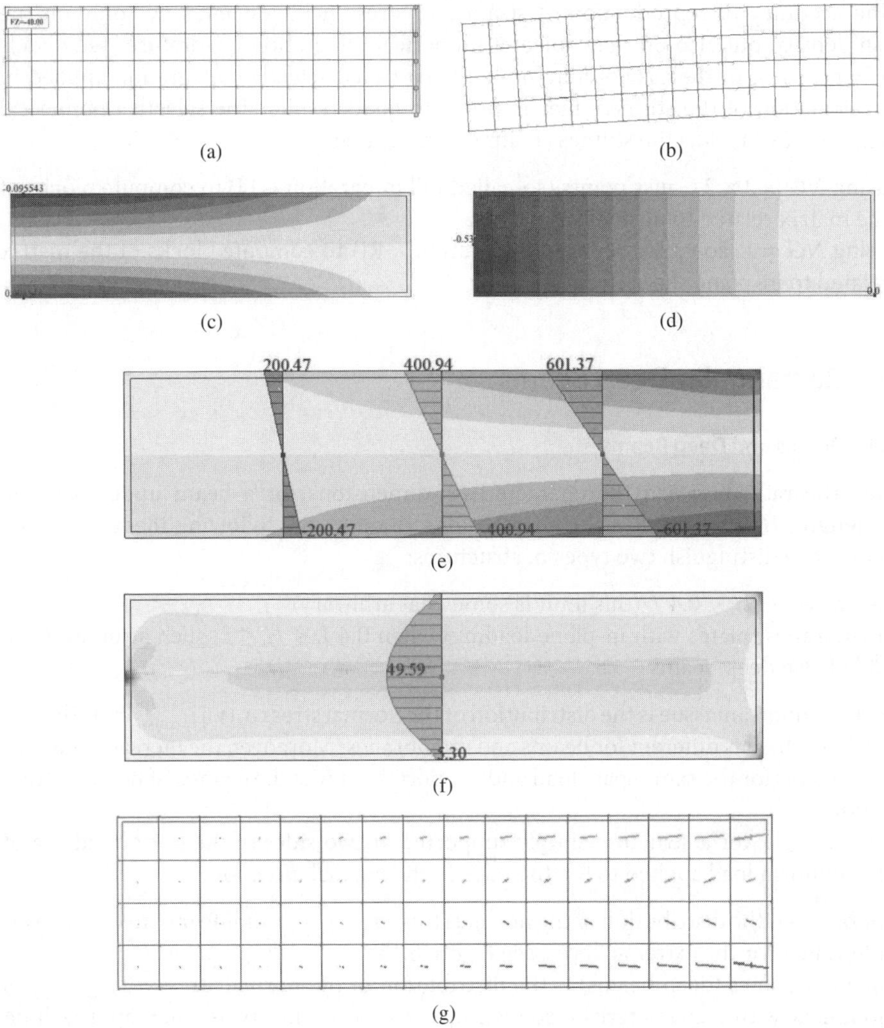

Figure 6.12 Cantilever beam – numerical results: (a) and (b) beam before and after deformation for a coarse FE mesh ($NSN^{I} = 4 \times 16$), contour plots of translations [cm]: (c) u_x, (d) u_y, stresses [MPa]: (e) σ_x, (f) τ_{xy} for fine FE mesh ($NSN^{II} = 16 \times 64$); (g) directions and magnitudes of principal membrane stresses for coarse mesh; FEM results from ROBOT. *Subfigures (c) and (e) shown in Plate section for color representation of this figure.*

Special attention is focused on the visualization of principal membrane stresses for the coarse mesh. On the basis of the results presented in Figure 6.12g two remarks can be formulated:

- the stresses of opposite sign are observed in the neighbourhood of top and bottom edges
- an increase of the absolute value of stress when approaching the supported right vertical edge is noticed – which is in compliance with Figure 6.12e

The accuracy of reproduction of strain and stress fields depends not only on the mesh density, but also on the choice of numerical integration quadrature (see details in Chapter 17). In the case of four-node FE, so-called selective integration applied in the calculation of the stiffness matrix provides good results. The selective numerical integration (SI) means the stiffness matrix is integrated:

- using $NG = 2 \times 2$ Gauss points (so-called full integration – FI) to compute blocks of the matrix related to normal strains ε_x, ε_y
- using $NG = 1$ (so-called reduced integration – RI) to compute blocks of the matrix related to in-plane shear strain γ_{xy}

6.7 Rectangular Deep Beams

6.7.1 Beams and Deep Beams

When the ratio between the characteristic dimensions for a beam under bending (L – length, H – height, h – thickness) is described by the following inequalities, we can formally distinguish two types of structures:

- beams when $H < 0.4\,L$ (this limit is somewhat arbitrary)
- membranes (plates with in-plane loading) when $0.4\,L \leq H \leq L$; such structures are called also deep beams

A very important issue is the distribution of the normal stress $\sigma_x(y)|_{x=\text{const.}}$ in the height direction, which is different for beams and membranes. Moreover, the increase of membrane height (for the same span, load and supports) induces the change of normal stress function.

For example, for a structure simply supported at two sides of the bottom edge and with a uniform load applied to the top edge in the normal direction:

- for beams – the distribution of the normal stress $\sigma_x(y)|_{x=\text{const.}}$ is linear with equal absolute values for the extreme fibres, see Figure 6.13a
- for membranes (deep beams) – the distribution of the normal stress $\sigma_x(y)|_{x=\text{const.}}$ is nonlinear with a characteristic sudden increase of tensile stress when approaching the bottom edge of the membrane, see Figure 6.13b

The membrane forces (stresses) depend strongly on the load location (top or bottom edge) and supports (with the horizontal displacement prevented or allowed and at a point or on a line), see Andermann (1966).

6.7.2 Square Membrane with a Uniform Load on the Top Edge, Supported on Two Parts of the Bottom Edge – FDM and FEM Results

In this section a square membrane with uniform load on the top boundary line is analysed. The membrane is supported on two parts of the bottom edge as shown in Figure 6.13b.

The solution is obtained:

- solving a differential equation with Airy's function using the finite difference method (FDM)
- using the finite element method (FEM)

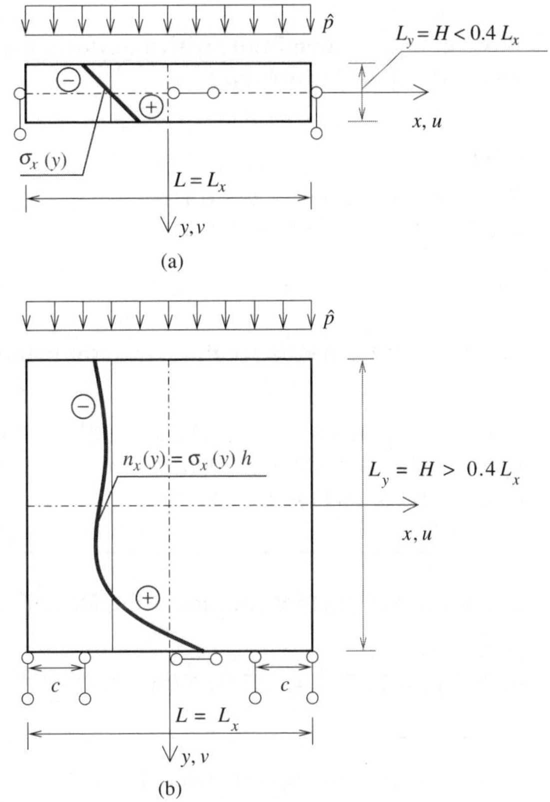

Figure 6.13 Model of: (a) beam and (b) deep beam – with characteristic distribution of stress σ_x along the vertical line $x = 0$

In previous years, in static calculations, engineers and students tended to use the FDM (see Andermann 1966). Then, the computational effort was focused on the formulation and solution of large sets of algebraic equations resulting from the methodology of FDM. When FDM is used, the results of calculations are obtained as a set of discrete values for a finite number of points in the structure. The obtained values are used to plot the distribution of essential functions along selected section lines.

For detailed calculations, the values of Young's modulus E and Poisson's ratio ν, dimensions L, H and the unit thickness of membrane ($h = 1$), the lengths of supported edge segments c and load $\hat{p}$ are specified in Box 6.6 (see Andermann 1966).

Selected results of calculations, presented in Figure 6.14 are reprinted from Andermann (1966), pages 53–54. The distributions of membrane forces have been plotted (along the corresponding lines) on the basis of values of three membrane forces n_x, n_y, n_{xy}. The normal force fields $n_x(x, y)$ and $n_y(x, y)$ are symmetric about the line $x = 0$, while the tangential forces $n_{xy}(x, y)$ are antisymmetric and thus $n_{xy}(0, y) = 0$ and the sign of the force is opposite for the right and left part of the domain.

In the vicinity of the top edge of the deep beam with a constant distributed load $p_v = -\hat{p}$ applied to this edge the membrane force n_y and stress σ_y are constant along horizontal section lines and for the unit thickness h of the beam $n_y = -\hat{p} = \sigma_y h$. In Figure 6.14a the characteristic nonlinear distribution of $n_x(y) = \sigma_x(y)h$ is shown along the height for the line $x = 0$. One should pay attention to the increase of the tensile force at the bottom edge.

Box 6.6 Square membrane with a uniform load on the top edge, supported on two parts of the bottom edge

Data

$E = 1.0 \times 10^7\ \text{kN/m}^2, \quad \nu = 0.1667$
$L = L_x = L_y = H = 1.0\ \text{m}, \quad h = 1, \quad c = 0.1\ \text{m}, \quad \hat{p} = 100.0\ \text{kN/m}$

Check values from FDM solution by Andermann (1966)

$\Delta_x = \Delta_y = L/10$
$n_x(0, H) = -0.515\ \hat{p} = -51.5\ \text{kN/m}, \quad n_x^{\text{FDM}}(0,0) = 1.390\ \hat{p} = 139.0\ \text{kN/m}$
$n_y(0, 0.9\ H) = -0.987\ \hat{p} = -98.7\ \text{kN/m}$
$n_x(0, 0.1\ H) = 0.023\ \hat{p} = 2.3\ \text{kN/m}, \quad n_x(\pm L, 0.1\ H) = -3.489\ \hat{p} = -348.9\ \text{kN/m}$

Check values from FEM solution using ROBOT

$NSE = 20 \times 20$
$n_x^{\text{FEM}}(0,0) = 1.550\ \hat{p} = 155.0\ \text{kN/m} = 1.12\ n_x^{\text{FDM}}(0,0)$

Membrane configuration shown in Figure 6.13b

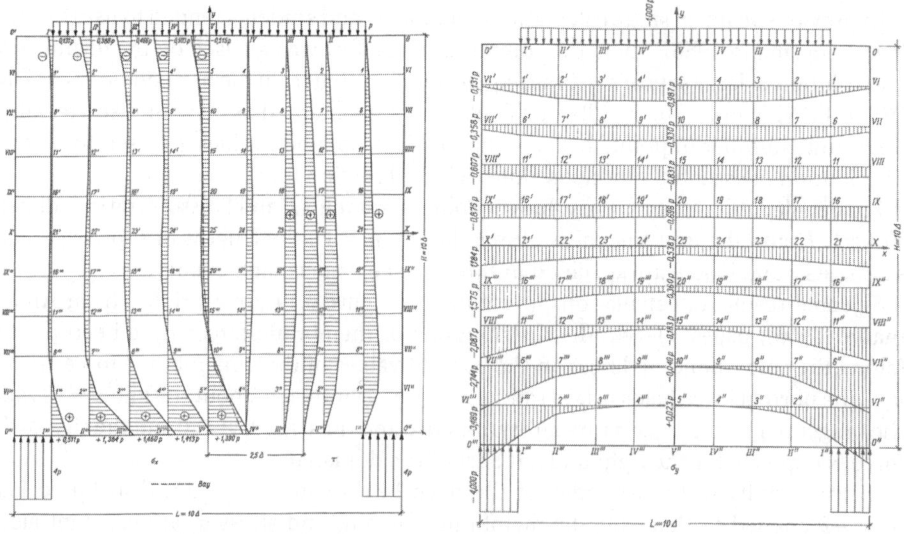

(a) (b)

Figure 6.14 Square deep beam – diagrams of membrane forces: (a) normal $n_x(x, y)$ in the left-hand part of the domain, tangential $n_{xy}(x, y)$ in the right-hand part of the domain and (b) normal $n_y(x, y)$ in whole domain; results obtained using FDM. Source: Andermann (1966).

The selected values of membrane forces read from Figure 6.14 are included in Box 6.6.

In Figure 6.15 the results of FEM analysis using package ROBOT (2006) are depicted. The attention should be focused on Figure 6.15e where the distribution of normal force $n_x(0, y)$ along the vertical symmetry axis is shown with the maximum value of tensile force at point $x = y = 0$. The comparison of $n_x(0, 0)$ obtained from FEM and FDM is included in Box 6.6.

Figure 6.15f with directions and magnitudes of principal membrane forces shows the distribution of compressive force n_{II} in the form of stress fluxes going from the loaded top edge to narrow support zones at the bottom edge.

6.8 Membrane with Variable Thicknesses or Material Parameters

6.8.1 Introduction

The issue of continuity of the displacement and strain fields is an essential aspect of approximated finite element method. For instance, when displacement-based (one-field) FEs are used, the displacement field is the primary one and strains and stresses are secondary fields. Depending on the order of derivatives appearing in the kinematic equations, the degree of polynomial representation of the displacement field is reduced in the description of strain and stress fields. Thereby, the continuous approximation of generalized displacements is usually associated with discontinuities in the description of strains and stresses.

However, there are cases where the jumps of strains and/or stresses are admissible or physically justified. One then has to take this issue into account in the procedure of stress smoothing at the postprocessing stage.

The algorithm of FEM computations usually contains the following sequence of operations:

- approximation of displacements within FE
- calculation of stress components at Gauss points
- extrapolation of values from Gauss points to nodes (on interelement lines)
- averaging of stresses in common nodes of neighbouring FEs

The set of averaged nodal values of stresses is used to create the contour plots of selected components of the stress vector. The contour maps obtained in this way do not show the discontinuities and provide properly looking results. Thus, some users of computer programs can be unaware of the quality of representation of strain or stress fields for different types of FEs.

Two examples in which discontinuity of a particular field of strain or stress is physically justifiable are discussed in the next subsections and in Hartmann and Katz (2007).

6.8.2 Membrane with Different Thicknesses in Three Subdomains – FEM Solution

A membrane with three horizontal subdomains and jumps of the thickness is considered, see Figure 6.16. Attention is focused on the distribution of stress σ_y and membrane force n_y normal to the boundaries of subdomains and their values on both sides of the interface, denoted by indices (i) and (j).

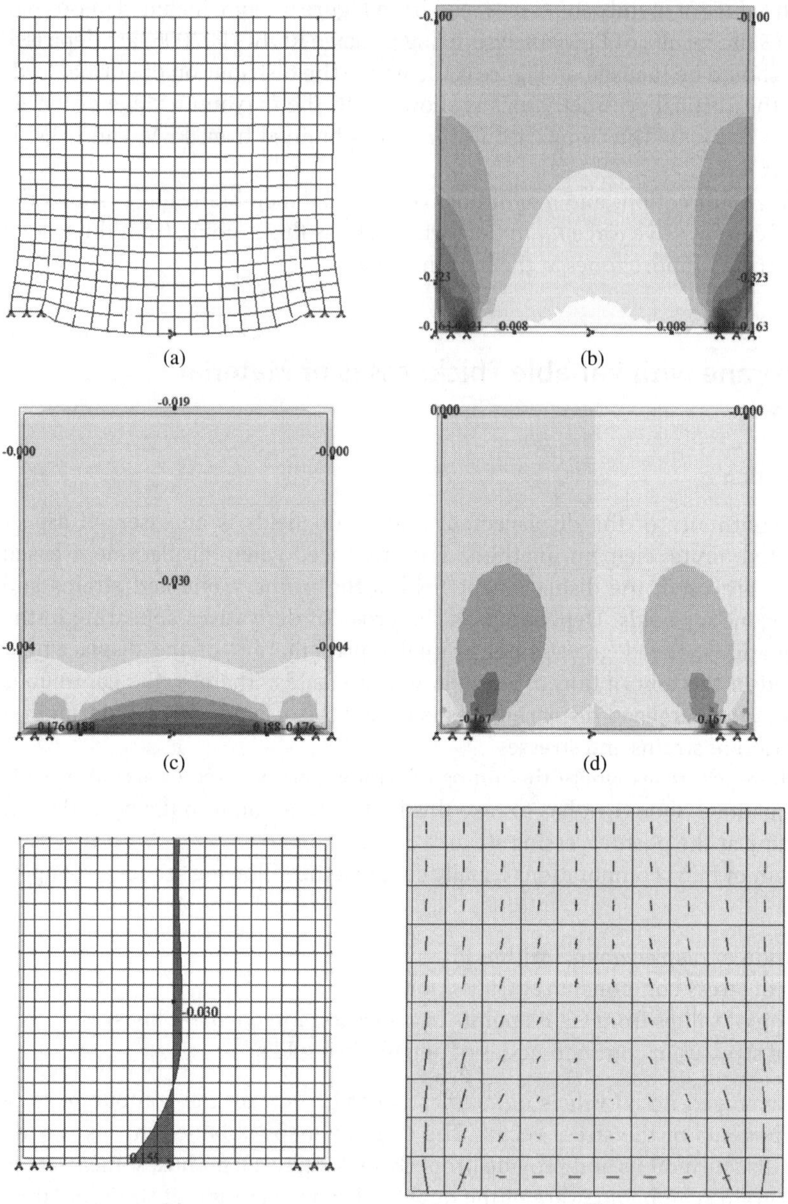

Figure 6.15 Square deep beam: (a) deformation, contour plots of membrane forces: (b) $n_y(x, y)$, (c) $n_x(x, y)$, (d) $n_{xy}(x, y)$, (e) diagram of distribution of $n_x(0, y)$ [MN/m] along central vertical line $x = 0$ with maximum value of $n_x(0, 0) = 0.155$ [MN/m] $\simeq 1.5\,\hat{p}$ and (f) vizualization of directions and magnitudes of principal membrane forces; FEM results from ROBOT

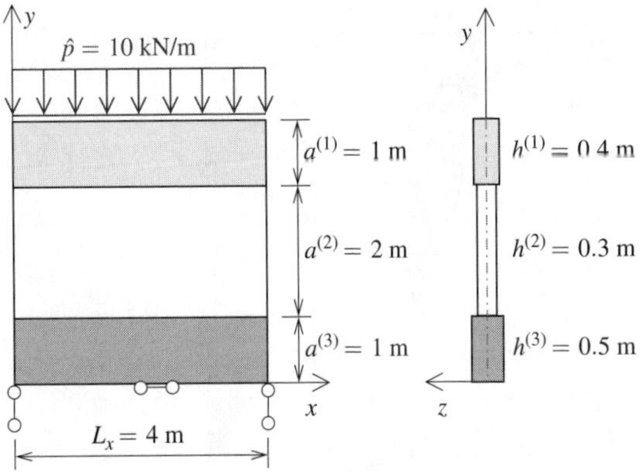

Figure 6.16 Example of membrane with thickness discontinuity

In the two-dimensional problem the equilibrium requirement on interfaces of membrane subdomains leads to the equality of normal membrane forces for $\nu = y$

$$n_\nu^{(i)} = n_\nu^{(j)} \quad \rightarrow \quad n_y^{(i)} = n_y^{(j)} \tag{6.54}$$

Writing the relations between forces and stress components:

$$n_y^{(i)} = \sigma_y^{(i)}\, h^{(i)}, \qquad n_y^{(j)} = \sigma_y^{(j)}\, h^{(j)} \tag{6.55}$$

and taking the thickness change into account $h^{(i)} \neq h^{(j)}$, the discontinuity of normal stress and normal strain is proven:

$$\sigma_y^{(i)} \neq \sigma_y^{(j)}, \qquad \epsilon_y^{(i)} \neq \epsilon_y^{(j)} \tag{6.56}$$

The following data are adopted for numerical computations (see also Figure 6.16):

- Young's modulus $E = 2.0 \times 10^7$ kPa, Poisson's ratio $\nu = 0.1$
- length $L_x = 4.0$ m, subdomain heights: $a^{(1)} = 1.0$ m, $a^{(2)} = 2.0$ m, $a^{(3)} = 1.0$ m
- respective thicknesses: $h^{(1)} = 0.4$ m, $h^{(2)} = 0.3$ m, $h^{(3)} = 0.5$ m

When the considered problem is solved the analysis of contour plots of characteristic stresses and membrane forces should be performed (see Figure 6.17). Additionally, the distribution of relevant functions along a central vertical cross-sectional line is plotted. It should be emphasized that at the boundary of subdomains with different thicknesses the continuity of normal force $n_\nu = n_y$ is preserved, but the discontinuity of normal stress $\sigma_\nu = \sigma_y$ occurs.

6.8.3 Membrane with Different Material Parameters in Three Subdomains – FEM Solution

A membrane with three subdomains with different values of material parameters is considered. The membrane deformation involves the continuous distribution of strain field tangential to the interface of the subdomains for $s = x$:

$$\varepsilon_s^{(i)} = \varepsilon_s^{(j)} \quad \rightarrow \quad \varepsilon_x^{(i)} = \varepsilon_x^{(j)} \tag{6.57}$$

The stress σ_x is proportional to strains using the constitutive relations, in which different values of Young's modulus and Poisson's ratio appear:

$$\sigma_x^{(i)} = \frac{E^{(i)}}{1-(\nu^{(i)})^2}\left(\varepsilon_x^{(i)} + \nu^{(i)}\varepsilon_y^{(i)}\right), \qquad \sigma_x^{(j)} = \frac{E^{(j)}}{1-(\nu^{(j)})^2}\left(\varepsilon_x^{(j)} + \nu^{(j)}\varepsilon_y^{(j)}\right) \tag{6.58}$$

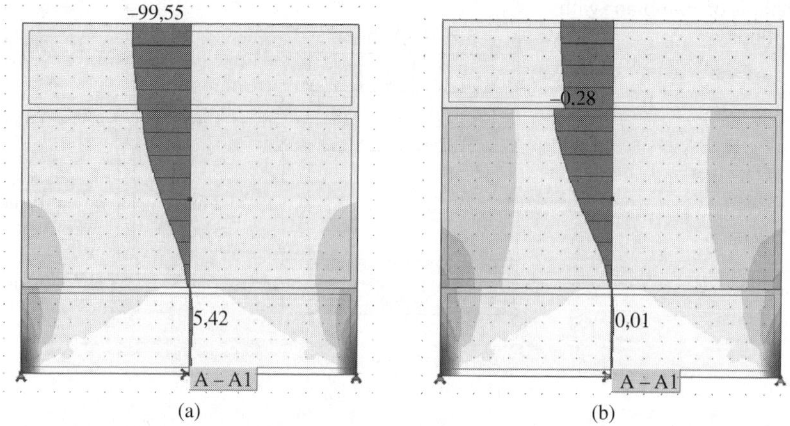

Figure 6.17 Membrane with thickness discontinuities – contour plots of: (a) force n_y [kN/m] and (b) stress σ_y [MPa], with diagrams of distributions along vertical line $x = L_x/2$ (denoted as A-A1); FEM results from ROBOT

The different values of $E^{(i)} \neq E^{(j)}$ and $\nu^{(i)} \neq \nu^{(j)}$ in spite of the equality of strain ϵ_x result in the discontinuity of stress σ_x:

$$\epsilon_x^{(i)} = \epsilon_x^{(j)} \quad \rightarrow \quad \sigma_x^{(i)} \neq \sigma_x^{(j)} \tag{6.59}$$

The following data are adopted for numerical computations:

- Young's moduli: $E^{(1)} = 30$ MPa, $E^{(2)} = 20$ MPa, Poisson's ratios: $\nu^{(1)} = 0.16$, $\nu^{(2)} = 0.1$
- length $L_x = 4.0$ m, subdomain heights: $a^{(1)} = 1.0$ m, $a^{(2)} = 2.0$ m, $a^{(3)} = 1.0$ m
- thickness: $h = 0.3$ m

The results of the (sudden) change of material parameters are visible as discontinuities in the contour plot of stress σ_x in Figure 6.18b. Additionally, the plot of this stress along a central vertical cross-sectional line is included.

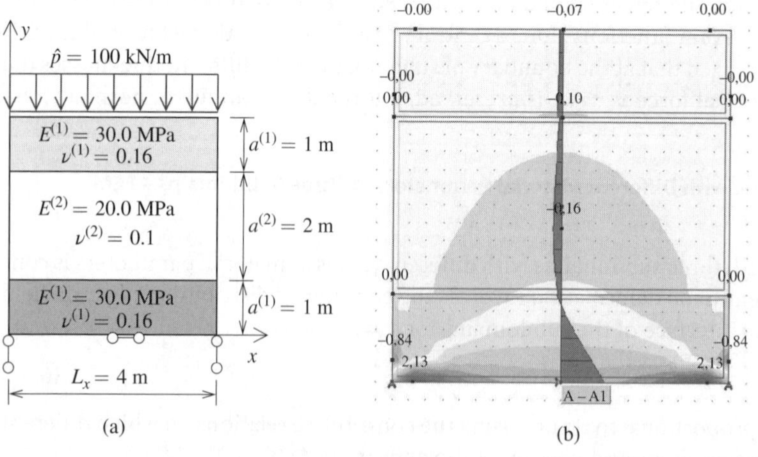

Figure 6.18 Membrane with different elastic constants in subdomains: (a) configuration and (b) contour plot of stress σ_x [MPa] with diagram of distribution along central vertical line $x = L_x/2$ (denoted as A-A1); FEM results from ROBOT

The main conclusion is that the field of stress tangential to the boundary of two subdomains with different material parameters is discontinuous. This fact requires consideration while the reinforcement of concrete structure is designed (reinforcement bars parallel to the interface of subdomains should have different section areas). On the basis of the examples described in this section, one can conclude that the averaging of stress along the interface of subdomains with different thicknesses or material parameters is not permitted.

References

Andermann F 1966 *Rectangular Membranes: Static Calculations.* Arkady, Warsaw (in Polish).

ANKA 1993 ANKA – computer code for nonlinear analysis of structures: User's manual. Technical report, Cracow University of Technology, Cracow (in Polish).

Girkmann K 1956 *Flächentragwerke.* Springer-Verlag, Wien.

Hartmann F and Katz C 2007 *Structural Analysis with Finite Elements* 2nd edn. Springer.

ROBOT 2006 ROBOT Millennium: User's Guide. Technical report, RoboBAT, Cracow (in Polish).

Timoshenko S and Goodier JN 1951 *Theory of Elasticity* 2nd edn. McGraw-Hill, New York-Toronto-London.

Waszczyszyn Z and Radwańska M 1995 Basic equations and calculations methods for elastic shell structures. In Borkowski A, Cichoń C, Radwańska M, Sawczuk A and Waszczyszyn Z, *Structural Mechanics: Computer Approach*, vol. **3**. Arkady, Warsaw, chapter 9, pp. 11–190 (in Polish).

7

Circular and Annular Membranes

7.1 Equations of Membranes – Local and Global Formulation

In a general case of circular and annular membranes described in a polar coordinate system and shown in Figure 7.1, the following prescribed external actions (loads and constraints) written in a vector form are taken into account:

- vector of surface load $\hat{\mathbf{p}}^n_{(2\times1)}(r,0) = \left[\hat{p}_r, \hat{p}_\theta\right]^{\mathrm{T}}$ $\left[\mathrm{N/m^2}\right]$
- vector of boundary tractions $\hat{\mathbf{p}}^n_{b(2\times1)}(s) = \left[\hat{p}_v, \hat{p}_s\right]^{\mathrm{T}}$ $\left[\mathrm{N/m}\right]$
- vector of constrained boundary displacements (translations) $\hat{\mathbf{u}}^n_{b(2\times1)}(s) = \left[\hat{u}_v, \hat{u}_s\right]^{\mathrm{T}}$[m]

These external actions generate the following fields of unknown quantities:

- vector of displacements $\mathbf{u}^n_{(2\times1)}(r,\theta) = [u_r, u_\theta]^{\mathrm{T}} = [u, v]^{\mathrm{T}}$ [m]
- vector of membrane strains $\mathbf{e}^n_{(3\times1)}(r,\theta) = [\varepsilon_r, \varepsilon_\theta, \gamma_{r\theta}]^{\mathrm{T}}$ [–]
- vector of membrane forces $\mathbf{s}^n_{(3\times1)}(r,\theta) = \mathbf{n}(r,\theta) = [n_r, n_\theta, n_{r\theta}]^{\mathrm{T}}$ [N/m]

In Figure 7.2 an elementary surface segment with normal membrane forces: radial n_r, circumferential n_θ and shear $n_{r\theta}$ is depicted. In Box 7.1 the local and global formulations of the boundary value problem are given. The membrane stiffness is defined as $D^n = E\,h/(1-\nu^2)$ and occurs in the constitutive equations and the internal energy functional.

The local formulation of the problem is expressed in the form of a set of eight (differential and algebraic) equations complemented by relevant boundary conditions. Combining kinematic and constitutive equations and introducing them into equilibrium equations, one can obtain two second-order partial differential equations with partial derivatives of two unknown displacements: $u_r(r,\theta)$, $u_\theta(r,\theta)$. These two equations written in polar coordinate system are analogous to Equations (6.2), describing rectangular membranes and indicated in Subsection 6.2.2.

The global formulation is written as a sum of membrane strain energy and work of external loads.

Plate and Shell Structures: Selected Analytical and Finite Element Solutions, First Edition.
Maria Radwańska, Anna Stankiewicz, Adam Wosatko and Jerzy Pamin.
© 2017 John Wiley & Sons Ltd. Published 2017 by John Wiley & Sons Ltd.

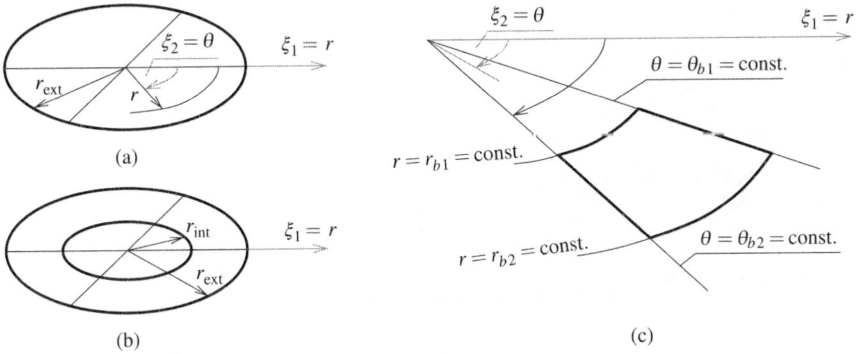

Figure 7.1 Membranes described in a polar coordinate system: (a) circular, (b) annular and (c) segment

Box 7.1 Equations for circular and annular membranes

Local formulation – set of eight equations and boundary conditions

(I) kinematic equations (3):

$$\varepsilon_r = \frac{\partial u_r}{\partial r}, \quad \varepsilon_\theta = \frac{u_r}{r} + \frac{1}{r}\frac{\partial u_\theta}{\partial \theta}, \quad \gamma_{r\theta} = \frac{1}{r}\frac{\partial u_r}{\partial \theta} + \frac{\partial u_\theta}{\partial r} - \frac{u_\theta}{r}$$

(II) equilibrium equations (2):

$$\frac{\partial n_r}{\partial r} + \frac{1}{r}\frac{\partial n_{\theta r}}{\partial \theta} + \frac{n_r - n_\theta}{r} + \hat{p}_r = 0, \qquad \frac{\partial n_{r\theta}}{\partial r} + \frac{1}{r}\frac{\partial n_\theta}{\partial \theta} + \frac{2n_{r\theta}}{r} + \hat{p}_\theta = 0$$

(III) constitutive equations (3):

$$n_r = D^n(\varepsilon_r + \nu\varepsilon_\theta), \quad n_\theta = D^n(\varepsilon_\theta + \nu\varepsilon_r), \quad n_{r\theta} = D^n(1-\nu)\,\gamma_{r\theta}/2$$

where $D^n = E\,h/(1-\nu^2)$

(IV) boundary conditions:

a) kinematic: $u_\nu = \hat{u}_\nu$, $u_s = \hat{u}_s$

b) static: $n_\nu = \hat{n}_\nu$, $n_{\nu s} = \hat{n}_{\nu s}$

c) mixed (one kinematic and one static)

Global formulation

Total potential energy $\Pi^n = U^n - W^n$

Internal strain energy

$$U^n = \frac{D^n}{2}\int_A \left\{ \left(\frac{\partial u}{\partial r} + \frac{u}{r} + \frac{1}{r}\frac{\partial v}{\partial \theta} \right)^2 - 2(1-\nu)\left[\frac{1}{r}\frac{\partial u}{\partial r}\left(\frac{\partial v}{\partial \theta} + u \right) - \frac{1}{4}\left(\frac{1}{r}\frac{\partial u}{\partial \theta} + \frac{\partial v}{\partial r} - \frac{v}{r} \right)^2 \right] \right\} r\,\mathrm{d}r\,\mathrm{d}\theta$$

External load work

$$W^n = \int_A [\hat{p}_r u + \hat{p}_\theta v] r\,\mathrm{d}r\,\mathrm{d}\theta + \int_{\partial A_\sigma} [\hat{n}_\nu u_\nu + \hat{n}_{\nu s} u_s]\,\mathrm{d}s$$

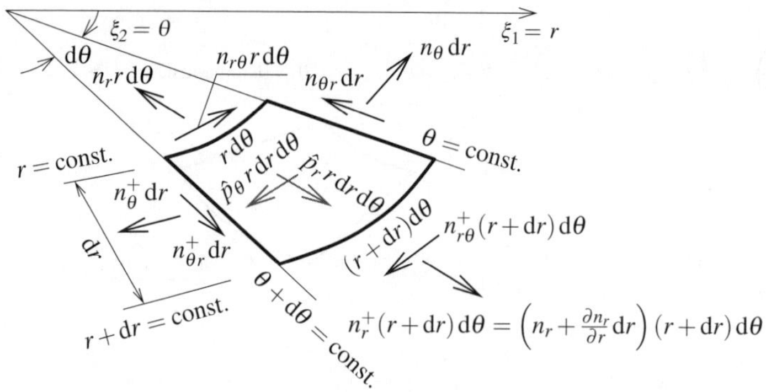

Figure 7.2 Membrane forces in a general case of deformation described in a polar coordinate system, shown on the boundary of an elementary surface segment

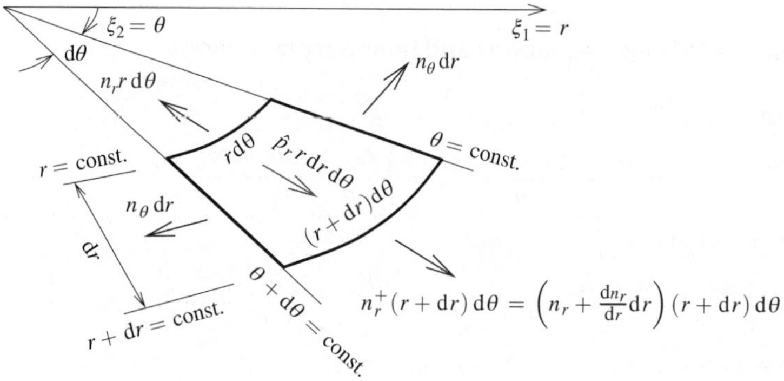

Figure 7.3 Membrane forces in the axisymmetric state

7.2 Equations for the Axisymmetric Membrane State

Geometrically, the one-dimensional boundary value problem of a membrane in the axisymmetric state is described by the following functions written as components of three vectors:

- radial translation $\mathbf{u}^n_{(1\times1)}(r) = [u_r(r)]$ [m]
- membrane (radial and circumferential) strains $\mathbf{e}^n_{(2\times1)}(r) = [\varepsilon_r, \varepsilon_\theta]^{\mathrm{T}}$ [−]
- membrane (radial and circumferential) forces shown in Figure 7.3
  $\mathbf{s}^n_{(2\times1)}(r) = \mathbf{n}(r) = [n_r, n_\theta]^{\mathrm{T}}$ [N/m]

The mathematical model of the considered problem in local and global formulation is written in Box 7.2.

In the case of local formulation of the problem, the set of five equations can be reduced by the coupling of kinematic and constitutive relations (equations I and III in Box 7.2). The equilibrium equation expressed by radial translation $u_r(r)$ and its (first and second) ordinary derivatives now becomes a second-order differential equation

$$\frac{\mathrm{d}^2 u_r}{\mathrm{d}r^2} + \frac{1}{r}\,\frac{\mathrm{d}u_r}{\mathrm{d}r} - \frac{1}{r^2}\,u_r = \frac{\hat{p}_r(r)}{D^n} \tag{7.1}$$

Box 7.2 Equations for membranes in an axisymmetric state

Local formulation – set of five equations and boundary conditions

(I) kinematic equations (2):

$$\varepsilon_r = \frac{\mathrm{d}u_r}{\mathrm{d}r}, \quad \varepsilon_\theta = \frac{u_r}{r}$$

(II) equilibrium equations (1):

$$\frac{\mathrm{d}n_r}{\mathrm{d}r} + \frac{n_r - n_\theta}{r} + \hat{p}_r = 0$$

(III) constitutive equations (2):

$$n_r = D^n(\varepsilon_r + \nu\varepsilon_\theta), \quad n_\theta = D^n(\varepsilon_\theta + \nu\varepsilon_r)$$

where $D^n = E\,h/(1-\nu^2)$

(IV) boundary conditions formulated for each edge $r = r_\mathrm{b} = \mathrm{const}$ – one kinematic or one static:

$$u_\nu = \hat{u}_\nu \quad \text{or} \quad n_\nu = \hat{n}_\nu$$

Global formulation

Total potential energy $\Pi^n = U^n - W^n$
Internal strain energy

$$U^n = \frac{D^n}{2}\, 2\pi \int_A \left[\left(\frac{\mathrm{d}u_r}{\mathrm{d}r} + \frac{u}{r} \right)^2 - 2(1-\nu)\left(\frac{\mathrm{d}u_r}{\mathrm{d}r}\,\frac{u_r}{r} \right) \right] r\,\mathrm{d}r$$

External load work

$$W^n = 2\pi \int_A [\hat{p}_r u_r]\, r\,\mathrm{d}r + \int_{\partial A_\sigma} [\hat{n}_\nu u_\nu]\,\mathrm{d}s$$

The unique solution of this equation requires appropriate boundary conditions. For example, in the case of an annular membrane, two boundary conditions are written (one for the internal and one for the external contour).

7.3 Annular Membrane

In this section the analytical and numerical (FEM) solutions for an annular membrane shown in Figure 7.4 are discussed. The considered structure is loaded only on its external edge by uniform normal load $\hat{p}_\nu$. Surface load $\hat{p}_r(r)$ is zero in the whole domain. The internal edge is supported in such a way that radial translation is prevented.

Five functions describing the behaviour of the membrane: $u_r(r)$, $\varepsilon_r(r)$, $\varepsilon_\theta(r)$, $n_r(r)$ and $n_\theta(r)$ are searched for. The complete definition of the problem requires the specification of material constants: Young's modulus E, Poisson's ratio ν and dimensions: radii of internal r_int and external r_ext boundaries, thickness h and boundary traction $\hat{p}_\nu$. The input values are included in Box 7.3.

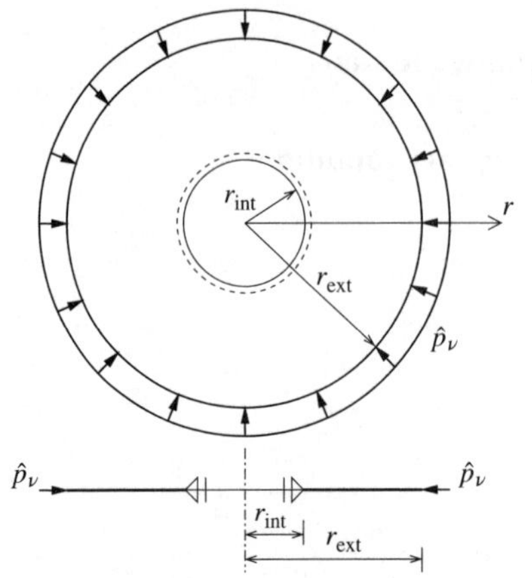

Figure 7.4 Configuration of axisymmetric annular membrane: dimensions, boundary load and constraints

Box 7.3 Annular membrane in an axisymmetric state

Data

$E = 2.0 \times 10^7$ kN/m^2, $\quad \nu = 0.1, \quad r_{int} = 1.0$ m, $\quad r_{ext} = 3.0$ m, $\quad h = 0.1$ m
$\hat{p}_r(r) = 0.0, \quad \hat{p}_\nu = \hat{p} = 1.0$ kN/m for $r_{ext} = 3.0$ m
$D^n = 2.02 \times 10^6$ kN/m

Check values from analytical solution

$u_r(1.0) = 0.0,$	$u_r(3.0) = -1.10 \times 10^{-6}$ m
$\varepsilon_r(1.0) = -0.825 \times 10^{-6},$	$\varepsilon_r(3.0) = -0.458 \times 10^{-6}$
$\varepsilon_\theta(1.0) = 0.0,$	$\varepsilon_\theta(3.0) = -0.366 \times 10^{-6}$
$n_r(1.0) = -1.667$ kN/m,	$n_r(3.0) = -1.00$ kN/m
$n_\theta(1.0) = -0.1667$ kN/m,	$n_\theta(3.0) = -0.8327$ kN/m

---

Check values from FEM solution using ANKA (SRK FEs)

$NSE = 16, \quad NEN = 2, \quad NNDOF = 4, \quad NEDOF = 8$
$u_r^{FEM}(3.0) = -1.099 \times 10^{-6}$ m $\approx u_r^{anal}(1.0)$
$n_r^{FEM}(1.0) = -1.667$ kN/m $= n_r^{anal}(1.0)$
$n_r^{FEM}(3.0) = -1.00$ kN/m $= n_r^{anal}(3.0)$
$n_\theta^{FEM}(1.0) = -0.1573$ kN/m $= 0.942\ n_\theta^{anal}(1.0)$
$n_\theta^{FEM}(3.0) = -0.8337$ kN/m $\approx n_\theta^{anal}(3.0)$

Membrane configuration shown in Figure 7.4

7.3.1 Analytical Solution

The boundary value problem of the membrane in the axisymmetric state is geometrically a one-dimensional problem. It can be solved analytically starting from the second-order displacement differential equation

$$\frac{d^2u_r}{dr^2} + \frac{1}{r}\frac{du_r}{dr} - \frac{1}{r^2}u_r = \frac{\hat{p}_r(r)}{D^n} \tag{7.2}$$

with appropriate kinematic (for r_{int}) and static (for r_{ext}) boundary conditions:

$$u_r(r_{int}) = 0, \quad n_r(r_{ext}) = -\hat{p}_v \tag{7.3}$$

Solution of differential equation (7.2) is a sum of general and particular integrals

$$u_r(r) = u_r^o(r) + \overline{u}_r(r) \tag{7.4}$$

Due to the lack of surface load, the particular integral $\overline{u}_r(r) = 0$. The general integral includes two constants of integration and has the following form

$$u_r(r) = u_r^o(r) = C_1\, r + C_2\, \frac{1}{r} \tag{7.5}$$

Now, using the kinematic and constitutive equations, the solution functions can be written as:

- radial and circumferential strains:

$$\varepsilon_r(r) = \frac{du_r}{dr} = C_1 - C_2\,\frac{1}{r^2}, \quad \varepsilon_\theta(r) = \frac{u_r}{r} = C_1 + C_2\,\frac{1}{r^2} \tag{7.6}$$

- radial and circumferential membrane forces:

$$\begin{aligned} n_r(r) &= D^n(\varepsilon_r + \nu\varepsilon_\theta) = D^n\left[(1+\nu)\,C_1 - \frac{(1-\nu)}{r^2}\,C_2\right] \\ n_\theta(r) &= D^n(\varepsilon_\theta + \nu\varepsilon_r) = D^n\left[(1+\nu)\,C_1 + \frac{(1-\nu)}{r^2}\,C_2\right] \end{aligned} \tag{7.7}$$

To find the unknown integration constants C_1 and C_2 the set of two equations resulting from kinematic and static boundary conditions expressed in Equation (7.3):

$$C_1\, r_{int} + C_2\,\frac{1}{r_{int}} = 0, \quad D^n\left[(1+\nu)\,C_1 - \frac{(1-\nu)}{r_{ext}^2}\,C_2\right] = -\hat{p}_v \tag{7.8}$$

is solved. For example, for input data listed in Box 7.3, one obtains $C_1 = -0.4125 \times 10^{-6}$ [–], $C_2 = 0.4125 \times 10^{-6}$ m^2.

Finally, the five functions describing the behaviour of the considered membrane, for $r \in [\,1.0, 3.0\,]$ are given next and shown in Figure 7.5:

- radial translation

$$u_r(r) = C_1\, r + C_2\,\frac{1}{r} = 0.4125\left(-r + \frac{1}{r}\right) \times 10^{-6}\ [\text{m}] \tag{7.9}$$

- radial and circumferential strains:

$$\begin{aligned} \varepsilon_r(r) &= \frac{du_r}{dr} = C_1 - C_2\,\frac{1}{r^2} = 0.4125\left(-1.0 - \frac{1}{r^2}\right) \times 10^{-6}\ [-] \\ \varepsilon_\theta(r) &= \frac{u_r}{r} = C_1 + C_2\,\frac{1}{r^2} = 0.4125\left(-1.0 + \frac{1}{r^2}\right) \times 10^{-6}\ [-] \end{aligned} \tag{7.10}$$

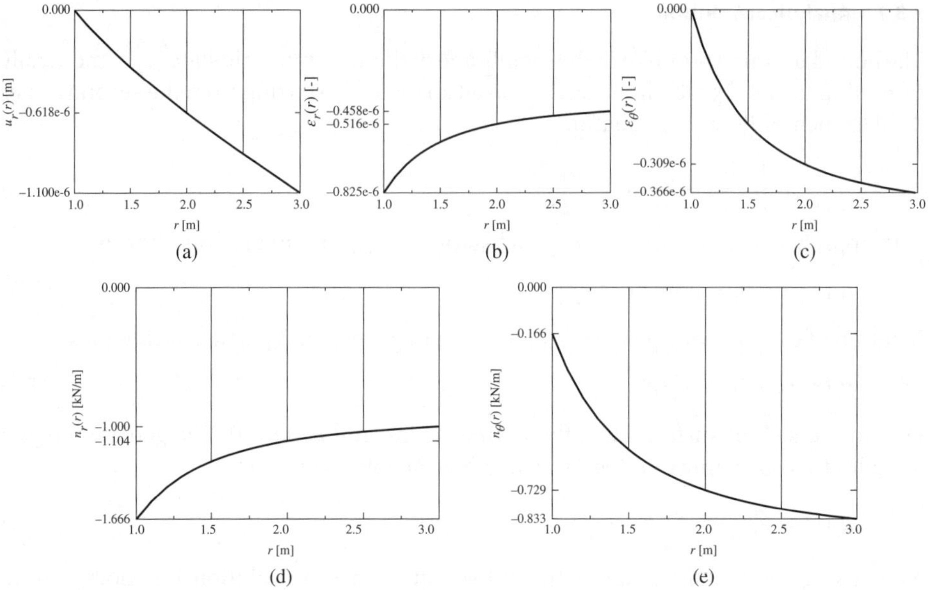

Figure 7.5 Annular membrane: functions obtained from analytical solution: (a) $u_r(r)$, (b) $\varepsilon_r(r)$, (c) $\varepsilon_\theta(r)$ and (d) $n_r(r)$, (e) $n_\theta(r)$

- radial and circumferential membrane forces:

$$
\begin{aligned}
n_r(r) &= D^n(\varepsilon_r + \nu\varepsilon_\theta) = D^n\left[(1+\nu)\,C_1 - \frac{(1-\nu)}{r^2}\,C_2\right] \\
&= 2.02\times10^6\left[-1.1\times0.4125 - \frac{0.9}{r^2}\,0.4125\right]\times10^{-6} = -0.916 - \frac{0.750}{r^2}\ \text{[kN/m]} \\
n_\theta(r) &= D^n(\varepsilon_\theta + \nu\varepsilon_r) = D^n\left[(1+\nu)\,C_1 + \frac{(1-\nu)}{r^2}\,C_2\right] \\
&= 2.02\times10^6\left[-1.1\times0.4125 + \frac{0.9}{r^2}\,0.4125\right]\times10^{-6} = -0.916 + \frac{0.750}{r^2}\ \text{[kN/m]}
\end{aligned}
\tag{7.11}
$$

Check results are represented by values of radial displacement, two membrane strains and two membrane forces calculated for the internal and external boundaries of the annular membrane. Output values are included in Box 7.3. It is worth noticing that two values $u_r(1.0) = 0$ and $n_r(3.0) = -1.0$ kN/m result from the strict fulfilment of boundary conditions.

7.3.2 FEM Solution

In the FEM analysis of an annular membrane in axisymmetric state, two ways of discretization are possible. The first option is to take the axisymmetric state of the symmetric annular membrane into account and discretize the line $\theta =$ const. with geometrically one-dimensional, two-node elements SRK (see Figure 7.6b). The description of this FE can be found in Subsection 5.2.8 or in ANKA (1993) and Waszczyszyn et al. (1994). The second way is to discretize the whole domain or an appropriate segment of the membrane using two-dimensional FEs.

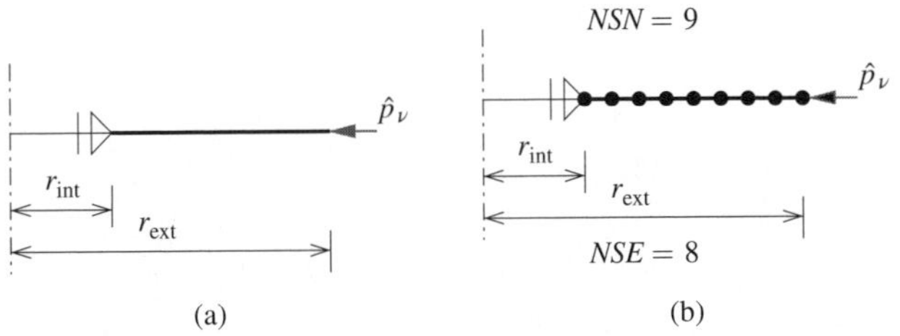

Figure 7.6 Annular membrane: (a) radial cross section (for θ = const.) and (b) configuration discretized with SRK FEs

In this example, the first option has been applied. The determination of coordinates of nodes from a radial section line θ = const. automatically enforces the modification of shell equations and their transformation into relations describing circular membranes. Then the option of axisymmetric state is set by taking $j = 0$ in the input data. For $j = 0$, $\cos j\,\theta = 1$ and functions describing the problem do not depend on circumferential coordinate θ.

It should be mentioned that, for this kind of one-dimensional FE, each node is in fact treated as a circle of radius defined by its radial coordinate. Thus, the boundary load applied to the external contour is defined as $2\pi r_{ext}\hat{p}$ and assigned to external, boundary node. Additionally, the value of horizontal reaction calculated at the internal node must be divided by $2\pi r_{int}$ to obtain uniformly distributed traction along the whole constrained contour.

Calculations have been repeated for $NSE = 4, 8, 16$. A comparison of selected values obtained from analytical and numerical solution for internal and external boundaries, $r = r_{int}$ and $r = r_{ext}$, respectively, leads to the conclusion that the results computed for $NSE = 16$ are in a very good compliance with exact solution. The check values are included in Box 7.3.

References

ANKA 1993 ANKA – computer code for nonlinear analysis of structures: User's manual. Technical report, Cracow University of Technology, Cracow (in Polish).

Waszczyszyn Z, Cichoń C and Radwańska M 1994 *Stability of Structures by Finite Element Methods*. Elsevier, Amsterdam.

8

Rectangular Plates under Bending

8.1 Introduction

Analysis is carried out for a rectangular plate of constant thickness h, in the Cartesian coordinate system. The plate is represented by the middle plane ($z = 0$) and two limiting planes $z = \pm h/2$. The axis z is usually directed downwards (Figure 8.1). The loads that can be applied to the plate are: surface transverse load $\hat{p}_z$ [N/m^2], boundary loads $\hat{t}_\nu$ [N/m] and/or $\hat{m}_\nu$ [Nm/m] on arbitrary contour lines.

The most commonly used theories for plates under bending are:

- the Kirchhoff–Love (K–L) theory for thin plates (see Section 8.2)
- the Mindlin–Reissner (M–R) theory for moderately thick plates (see Section 8.5)

The choice of appropriate theory depends on the value of h/L ratio, where L denotes the smaller of in-plane plate dimensions L_x and L_y, see Box 1.1. In a mathematical model of appropriate theory different kinematic and static constraints (which are in fact hypotheses) are imposed on the plate behaviour. For these two theories different kinematic equations hold, combining the fields of generalized displacements and generalized strains, see the books by Girkmann (1956), Timoshenko and Woinowsky-Krieger (1959), Kączkowski (1980) and Reddy (1999, 2007).

8.2 Equations for the Classical Kirchhoff–Love Thin Plate Theory

8.2.1 Assumptions and Basic Relations

The Kirchhoff–Love theory, describing thin plates ($h/L < 1/10$), is called a one-parameter theory, because only one field – deflection $w(x, y)$ – is the essential function for the middle plane, while the rotations of the normal to the middle plane are expressed by means of the first derivatives of the deflection function with respect to coordinates x and y. The K–L thin plate theory is coherent with the three-parameter

Plate and Shell Structures: Selected Analytical and Finite Element Solutions, First Edition.
Maria Radwańska, Anna Stankiewicz, Adam Wosatko and Jerzy Pamin.
© 2017 John Wiley & Sons Ltd. Published 2017 by John Wiley & Sons Ltd.

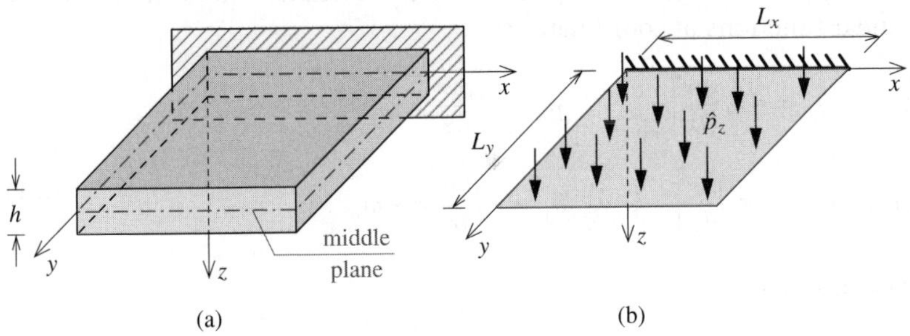

Figure 8.1 Plate under bending: (a) 3D geometrical model and (b) middle plane ($z = 0$) as a 2D geometrical model

theory of thin shells, in which membrane and bending effects are combined. The hypotheses of K–L theory for both plates and shells read:

Kinematic hypothesis: *a straight fibre, located at the intersection of two cross-planes, normal to the undeformed (initial) middle surface of shell structure, after the application of external actions remains straight, with unchanged length and normal to the deformed (current) middle surface.*

Static hypothesis: *stress σ_z, compared with other stress tensor components, is so small that for all points of a thin shell structure it can be omitted in constitutive relations.*

The segment of the plate with dimensions $dx \times h$, shown in the (x, z) plane (Figure 8.2a), changes its position during loading, with the rectangular shape preserved, as a result of zero transverse shear strain $\gamma_{xz} = 0$. The same holds true for the (y, z) plane (Figure 8.2b) with $\gamma_{yz} = 0$. For thin plates, disregarding the deformation due to transverse shear is justified.

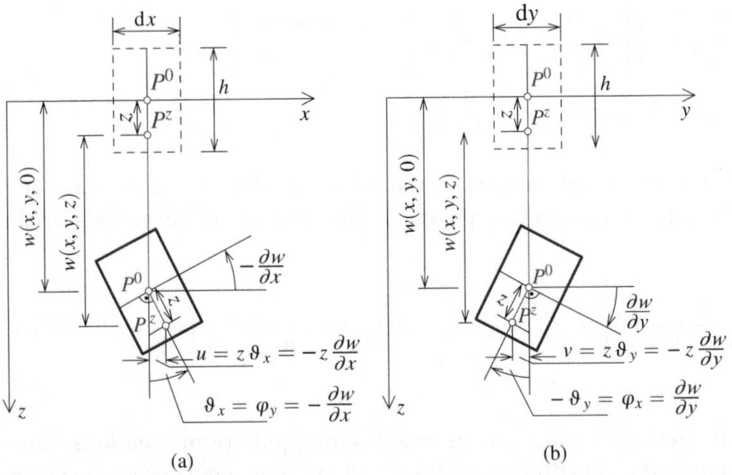

Figure 8.2 Description of plate behaviour following K–L constraints – respective segments ($dx \times h$) and ($dy \times h$) before and after deformation: (a) in the (x, z) plane and (b) in the (y, z) plane

Translations of any point $P(x, y, z)$ from plane $z \neq 0$ equidistant from the middle plane are written as linear functions of coordinate z:

$$\begin{aligned} &u(x,y,z) = z\,\vartheta_x = -z\left.\frac{\partial w}{\partial x}\right|_{z=0}, \qquad \vartheta_x = -\frac{\partial w}{\partial x} = \varphi_y \\ &v(x,y,z) = z\,\vartheta_y = -z\left.\frac{\partial w}{\partial y}\right|_{z=0}, \qquad \vartheta_y = -\frac{\partial w}{\partial y} = -\varphi_x \\ &w(x,y,z) = w(x,y,0) \end{aligned} \tag{8.1}$$

using two rotations of normal ϑ_x and ϑ_y, related to respective deflection derivatives $w_{,x}$ and $w_{,y}$, or two angles denoted by φ_x and φ_y (note that we use two symbols for partial derivatives interchangeably, e.g. $\partial w/\partial x$ and $w_{,x}$). The deflection $w(x, y)$ does not depend on z.

The components of strain tensor are linear functions of the z coordinate, thus at any point $P(x, y, z)$:

$$\begin{aligned} &\varepsilon_x(x,y,z) = -z\left.\frac{\partial^2 w}{\partial x^2}\right|_{z=0} = z\,\kappa_x(x,y,0), \qquad \kappa_x = -\frac{\partial^2 w}{\partial x^2} \\ &\varepsilon_y(x,y,z) = -z\left.\frac{\partial^2 w}{\partial y^2}\right|_{z=0} = z\,\kappa_y(x,y,0), \qquad \kappa_y = -\frac{\partial^2 w}{\partial y^2} \\ &\gamma_{xy}(x,y,z) = -2\,z\left.\frac{\partial^2 w}{\partial x \partial y}\right|_{z=0} = z\,\chi_{xy}(x,y,0), \qquad \chi_{xy} = -2\frac{\partial^2 w}{\partial x \partial y} \end{aligned} \tag{8.2}$$

In these relations two curvatures occur, $\kappa_x(x, y, 0)$, $\kappa_y(x, y, 0)$, and warping $\chi_{xy}(x, y, 0)$, defined by second partial derivatives of the deflection function $w_{,xx}$, $w_{,yy}$, $w_{,xy}$.

Assuming an isotropic material and taking the static K–L hypothesis into account, the constitutive equations at a point on an equidistant plane specified by coordinate z can be written as for plane stress state, in the following form:

$$\begin{aligned} &\begin{bmatrix}\sigma_x\\ \sigma_y\end{bmatrix} = \frac{E}{1-\nu^2}\begin{bmatrix}1 & \nu\\ \nu & 1\end{bmatrix}\begin{bmatrix}\varepsilon_x\\ \varepsilon_y\end{bmatrix} = -z\,\frac{E}{1-\nu^2}\begin{bmatrix}1 & \nu\\ \nu & 1\end{bmatrix}\begin{bmatrix}\partial^2 w/\partial x^2\\ \partial^2 w/\partial y^2\end{bmatrix} \\ &\tau_{xy} = G\,\gamma_{xy} = z\,G\,\chi_{xy} = -2\,z\,G\,\frac{\partial^2 w}{\partial x\,\partial y} \end{aligned} \tag{8.3}$$

The change of $\sigma_x(z)$, $\sigma_y(z)$, $\tau_{xy}(z)$ stresses in the thickness direction is linear, while the distribution of τ_{xz}, τ_{yz} components is parabolic (Figure 8.3a) and described by the formulae:

$$\tau_{xz}(z) = -\frac{E}{2(1-\nu^2)}\left(\frac{h^2}{4} - z^2\right)(\nabla^2 w)_{,x}, \quad \tau_{yz}(z) = -\frac{E}{2(1-\nu^2)}\left(\frac{h^2}{4} - z^2\right)(\nabla^2 w)_{,y} \tag{8.4}$$

The transverse shear stresses τ_{xz} and τ_{yz} are included in the equilibrium equations. However, they are omitted in the constitutive relations and internal energy since the corresponding strains γ_{xz}, γ_{yz} are zero, which is consistent with the adopted kinematic K–L hypothesis.

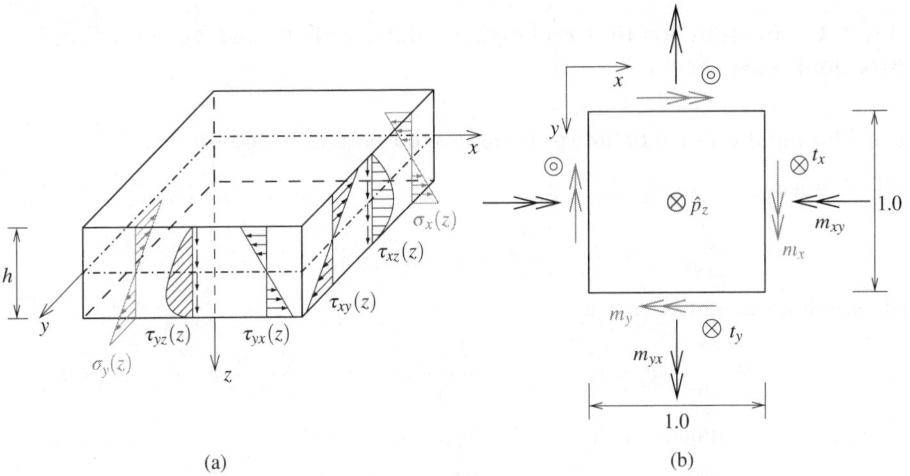

Figure 8.3 (a) Stress distribution along the thickness of plate under bending and (b) the sign convention of surface load and generalized resultant forces

The stresses are the basis for the definition of generalized resultant forces, related to the middle plane $z = 0$ and to the unit length from cross-sectional lines: two bending moments, a twisting moment and two transverse shear forces (Figure 8.3):

$$m_x = \int_{-h/2}^{+h/2} \sigma_x z \, \mathrm{d}z, \qquad m_y = \int_{-h/2}^{+h/2} \sigma_y z \, \mathrm{d}z, \qquad m_{xy} = \int_{-h/2}^{+h/2} \tau_{xy} z \, \mathrm{d}z = m_{yx} \tag{8.5}$$

$$t_x = \int_{-h/2}^{+h/2} \tau_{xz} \, \mathrm{d}z, \qquad t_y = \int_{-h/2}^{+h/2} \tau_{yz} \, \mathrm{d}z \tag{8.6}$$

Figure 8.3 illustrates: (i) linear distributions of stresses $\sigma_x(z)$, $\sigma_y(z)$, $\tau_{xy}(z)$ along the plate thickness, (ii) quadratic functions $\tau_{xz}(z)$, $\tau_{yz}(z)$ and (iii) positive generalized resultant forces for bending state: bending and twisting moments m_x, m_y, m_{xy}, and transverse shear forces t_x, t_y.

Summing up, at any point on the middle plane the following vectors are introduced:

- one-component displacement vector (deflection) $\mathbf{u}^m_{(1\times1)}(x,y) = [w(x,y)]$ [m]
- strain vectors $\mathbf{e}^m_{(3\times1)}(x,y) = [\kappa_x, \kappa_y, \chi_{xy}]^{\mathrm{T}}$ [1/m], $\mathbf{e}^t_{(2\times1)}(x,y) = [\gamma_{xz}, \gamma_{yz}]^{\mathrm{T}} = [0,0]^{\mathrm{T}}$
- vector of generalized resultant forces $\mathbf{s}^m_{(5\times1)}(x,y) = [\mathbf{m}^{\mathrm{T}}, \mathbf{t}^{\mathrm{T}}]^{\mathrm{T}}$:
  - moments $\mathbf{m}_{(3\times1)}(x,y) = [m_x, m_y, m_{xy}]^{\mathrm{T}}$ [Nm/m]
  - transverse shear forces $\mathbf{t}_{(2\times1)}(x,y) = [t_x, t_y]^{\mathrm{T}}$ [N/m]

In Box 8.1 the set of nine differential and algebraic equations, which is a mathematical model in the local formulation for the rectangular K–L plate bending problem, is given, including the information about boundary conditions. The model has previously been given in Box 3.7 in the general Chapter 3 dealing with shells and plates considered as a particular type of shell. Moreover, the global formulation represented by the internal strain energy dependent on partial derivatives of the deflection function and the external load work is included.

Box 8.1 Equations for thin rectangular plates under bending according to Kirchhoff–Love theory

Local formulation – set of nine equations and boundary conditions

(I) kinematic equations (3):

$$\kappa_x = -\frac{\partial^2 w}{\partial x^2}, \quad \kappa_y = -\frac{\partial^2 w}{\partial y^2}, \quad \chi_{xy} = -2\frac{\partial^2 w}{\partial x \partial y}$$

(II) equilibrium equations (3):

$$\frac{\partial t_x}{\partial x} + \frac{\partial t_y}{\partial y} + \hat{p}_z = 0, \quad \frac{\partial m_x}{\partial x} + \frac{\partial m_{yx}}{\partial y} - t_x = 0, \quad \frac{\partial m_{xy}}{\partial x} + \frac{\partial m_y}{\partial y} - t_y = 0$$

(III) constitutive equations (3):

$$m_x = D^m(\kappa_x + \nu\kappa_y), \quad m_y = D^m(\kappa_y + \nu\kappa_x), \quad m_{xy} = D^m(1-\nu)\,\chi_{xy}/2$$

where $D^m = E\,h^3/[12(1-\nu^2)]$

(IV) conditions

- on boundary lines:
  a) kinematic: $w = \hat{w}$, $\vartheta_\nu = -w_{,\nu} = \hat{\vartheta}_\nu$
  b) static: $\tilde{t}_\nu = t_\nu + \frac{\partial m_{\nu s}}{\partial s} = \hat{t}_\nu$, $m_\nu = \hat{m}_\nu$
  c) mixed (one kinematic and one static)
- at corners: kinematic $w_i = \hat{w}_i$ or static $T_i = 2\,m_{xy} = \hat{P}_i$

Global formulation

Total potential energy $\Pi^m = U^m - W^m$
Internal strain energy

$$U^m = \frac{D^m}{2}\int_A \left[\left(\frac{\partial^2 w}{\partial x^2}\right)^2 + \left(\frac{\partial^2 w}{\partial y^2}\right)^2 + 2\nu\,\frac{\partial^2 w}{\partial x^2}\,\frac{\partial^2 w}{\partial y^2} + 2(1-\nu)\left(\frac{\partial^2 w}{\partial x \partial y}\right)^2\right] \mathrm{d}x\,\mathrm{d}y$$

External load work

$$W^m = \int_A \hat{p}_z w\;\mathrm{d}x\,\mathrm{d}y + \int_{\partial A_\sigma} (\hat{t}_\nu w + \hat{m}_\nu \vartheta_\nu)\;\mathrm{d}s + \sum_i \hat{P}_i\,w_i$$

Taking the constitutive and kinematic equations into account the following relations between the moments and the deflection function are obtained:

$$m_x = -D^m(w_{,xx} + \nu\,w_{,yy}), \quad m_y = -D^m(w_{,yy} + \nu\,w_{,xx}), \quad m_{xy} = -D^m(1-\nu)w_{,xy} \tag{8.7}$$

where $D^m = E\,h^3/[12(1-\nu^2)]$ is the bending stiffness of the plate.

In the K–L theory transverse shear forces t_x and t_y are called inactive, since transverse shear strains $\gamma_{xz} = \gamma_{yz} = 0$. These forces are coupled with the first derivatives of moments using the two equilibrium equations, see Box 8.1, Equations $(\mathrm{II})_{2,3}$:

$$t_x = m_{x,x} + m_{yx,y}, \qquad t_y = m_{xy,x} + m_{y,y} \tag{8.8}$$

and furthermore with third derivatives of the deflection function:

$$t_x = -D^m(\nabla^2 w)_{,x}, \qquad t_y = -D^m(\nabla^2 w)_{,y} \tag{8.9}$$

What is an essential novelty in the K–L thin plate theory is the reduction of three boundary quantities t_ν, m_ν, $m_{\nu s}$ to two $\tilde{t}_\nu$, m_ν. It is possible due to the coupling of transverse force t_ν and twisting moment $m_{\nu s}$ into the so-called effective transverse shear force, given by definition and then the relation with third derivatives of the deflection function

$$\tilde{t}_\nu = t_\nu + \frac{\partial m_{\nu s}}{\partial s} = -D^m\left[w_{,\nu\nu\nu} + (2-\nu)w_{,ss\nu}\right] \tag{8.10}$$

Besides, the twisting moments at the plate corner produce a concentrated force and an effective corner force is calculated by the following formula

$$T_i = 2\, m_{xy} = -2\, D^m\,(1-\nu)\, w_{,xy} \tag{8.11}$$

under the assumption that there is a right angle between the free edges crossing at corner i. In fact, the definitions of the effective transverse shear force in plates and shells, as well as the effective tangent force in shells and the effective corner force occur in the principle of virtual work (to be precise in the derivation of boundary conditions) used in the Sanders shell theory, see Chapter 3.

The essential information about the behaviour of a plate under bending is given by the trajectories of the principal moments; the magnitudes and directions of extreme moments $m_{\mathrm{I}} = m_{\max}$ and $m_{\mathrm{II}} = m_{\min}$ are calculated with the following formulae:

$$m_{\mathrm{I,\,II}} = \frac{1}{2}(m_x + m_y) \pm \frac{1}{2}\sqrt{(m_x - m_y)^2 + 4\,(m_{xy})^2}, \qquad \tan(2\alpha) = \frac{2\, m_{xy}}{m_x - m_y} \tag{8.12}$$

They are very helpful in the qualitative and quantitative analysis of the behaviour of a plate under bending.

Engineering design requires checking of extreme stresses. Taking the relations between the moments and stresses into account:

$$\sigma_x(z) = \frac{12\, m_x}{h^3}\, z, \qquad \sigma_y(z) = \frac{12\, m_y}{h^3}\, z, \qquad \tau_{xy}(z) = \frac{12\, m_{xy}}{h^3}\, z \tag{8.13}$$

we can calculate extreme normal stresses σ_x and σ_y as well as shear stresses $\tau_{xy} = \tau_{yx}$ that appear on the bottom and top limiting planes for $z = \pm h/2$:

$$\sigma_x(\pm h/2) = \pm\frac{6\, m_x}{h^2}, \quad \sigma_y(\pm h/2) = \pm\frac{6\, m_y}{h^2}, \quad \tau_{xy}(\pm h/2) = \tau_{yx}(\pm h/2) = \pm\frac{6\, m_{xy}}{h^2} \tag{8.14}$$

The distribution of normal stress $\sigma_z(z)$ along the thickness is described by the third-order function of z and is connected with the deflection function by the following dependence, resulting from the equilibrium equation

$$\sigma_z(z) = -\frac{E\,(h+z)}{6(1-\nu^2)}\left(\frac{h}{2} - z\right)^2 \nabla^2\nabla^2 w \tag{8.15}$$

after fulfilling the static condition

$$\sigma_z(x, y, -h/2) = -\hat{p}_z(x, y) \tag{8.16}$$

related to transverse load $\hat{p}_z(x, y)$ applied at the top surface.

To make this section complete, the set of nine equations for K–L thin plates is listed in Box 8.1. Moreover, the global formulation described further in Subsection 8.2.4 is rewritten.

8.2.2 Equilibrium Equation for a Plate Expressed by Moments

After modification of two equilibrium equations $(\mathrm{II})_{2,3}$ in Box 8.1 in order to relate the two transverse shear forces with moments:

$$t_x = \frac{\partial m_x}{\partial x} + \frac{\partial m_{yx}}{\partial y}, \qquad t_y = \frac{\partial m_{xy}}{\partial x} + \frac{\partial m_y}{\partial y} \tag{8.17}$$

we obtain an equilibrium equation $(\mathrm{II})_1$ of a plate under bending, with physical interpretation $\Sigma P_z = 0$, expressed by second partial derivatives of the three moments

$$\frac{\partial^2 m_x}{\partial x^2} + 2\,\frac{\partial^2 m_{xy}}{\partial x\,\partial y} + \frac{\partial^2 m_y}{\partial y^2} = -\hat{p}_z(x, y) \tag{8.18}$$

8.2.3 Displacement Differential Equation for a Thin Rectangular Plate According to the Kirchhoff–Love Theory

Introducing the combined constitutive equations (III) and kinematic relations (I) from Box 8.1 into the plate Equation (8.18), we obtain the displacement equation for a bending plate – a fourth-order partial differential equation with unknown deflection function $w(x, y)$

$$\nabla^2\nabla^2 w(x, y) = \frac{\partial^4 w}{\partial x^4} + 2\,\frac{\partial^4 w}{\partial x^2\,\partial y^2} + \frac{\partial^4 w}{\partial y^4} = \frac{\hat{p}_z}{D^m} \tag{8.19}$$

This equation must be complemented by the information about boundary loads and kinematic constraints. Since the displacement differential equation of bending plate is expressed only by the derivatives of deflection function $w(x, y)$ all boundary conditions (both kinematic and static) have to be written in terms of function w and its appropriate derivatives.

Examples of boundary conditions are written for the plate shown in Figure 8.4:

- edge $x = 0, y \in [0, b]$: $w = 0, \qquad \vartheta_x = -w_{,x} = 0$
- edge $x = a, y \in [0, b]$: $m_x = -D^m(w_{,xx} + \nu\, w_{,yy}) = -\hat{m}$; $\tilde{t}_x = -D^m\,[w_{,xxx} + (2 - \nu)w_{,yyx}] = 0$
- edge $y = 0, x \in [0, a]$: $w = 0, \qquad m_y = -D^m(w_{,yy} + \nu\, w_{,xx}) = 0$
- edge $y = b, x \in [0, a]$: $m_y = -D^m(w_{,yy} + \nu\, w_{,xx}) = 0$; $\tilde{t}_y = -D^m\,[w_{,yyy} + (2 - \nu)w_{,xxy}] = \hat{t}$
- corner $x = a,\ y = b$: $T = -2\,D^m(1 - \nu)w_{,xy} = 0$

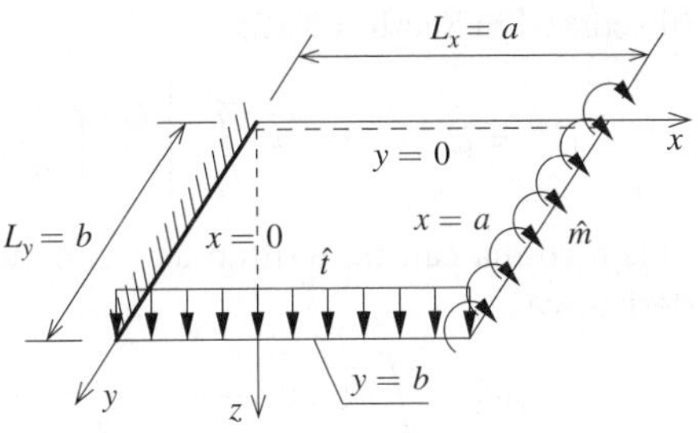

Figure 8.4 Example of rectangular plate under bending with various boundary constraints and loads

8.2.4 Global Formulation for a Kirchhoff–Love Thin Plate

Taking bending and twisting moments and caused by them curvatures and warping into account, the potential energy U^m of a plate in bending can be written as follows

$$U^m = \int_A [m_x\ \kappa_x + m_{xy}\ \chi_{xy} + m_y\ \kappa_y]\,\mathrm{d}x\,\mathrm{d}y \tag{8.20}$$

Using the relations between the moments and bending strains the elastic energy U^m can be expressed using bending stiffness D^m, and the squares of curvatures and warping

$$U^m = \frac{D^m}{2}\int_A [\kappa_x^2 + \kappa_y^2 + 2\nu\ \kappa_x\ \kappa_y + (1-\nu)\ \chi_{xy}^2/2]\,\mathrm{d}x\,\mathrm{d}y \tag{8.21}$$

The third possible form of elastic energy U^m includes the second partial derivatives of the deflection function in the integrand expression and can be written in two ways:

$$U^m = \frac{D^m}{2}\int_A \left[\left(\frac{\partial^2 w}{\partial x^2}\right)^2 + \left(\frac{\partial^2 w}{\partial y^2}\right)^2 + 2\nu\,\frac{\partial^2 w}{\partial x^2}\,\frac{\partial^2 w}{\partial y^2} + 2\,(1-\nu)\left(\frac{\partial^2 w}{\partial x\,\partial y}\right)^2\right]\mathrm{d}x\,\mathrm{d}y \tag{8.22}$$

$$U^m = \frac{D^m}{2}\int_A \left[(\nabla^2 w)^2 - 2\,(1-\nu)\left(\frac{\partial^2 w}{\partial x^2}\,\frac{\partial^2 w}{\partial y^2} - \left(\frac{\partial^2 w}{\partial x\,\partial y}\right)^2\right)\right]\mathrm{d}x\,\mathrm{d}y \tag{8.23}$$

The work of surface and boundary loads in the bending state is given by the following formula

$$W^m = \int_A \hat{p}_z\ w\,\mathrm{d}x\,\mathrm{d}y + \int_{\partial A_\sigma} (\hat{t}_\nu\ w + \hat{m}_\nu\ \vartheta_\nu)\,\mathrm{d}s + \sum_i \hat{P}_i\ w_i \tag{8.24}$$

These formulae enabling the determination of the functional of total potential energy $\Pi^m = U^m - W^m$ are going to be used in Section 8.12 dealing with an approximation method known in the literature as the Ritz method.

8.3 Derivation of Displacement Equation for a Thin Plate from the Principle of Minimum Potential Energy

The general form of total potential energy Π^m of a rectangular bending plate is given in the previous section. The energy takes a simplified form for certain boundary conditions

(discussed in Section 8.12)

$$\Pi^m = U^m - W^m = \int_A \left[\frac{D^m}{2} \left(\frac{\partial^2 w}{\partial x^2} + \frac{\partial^2 w}{\partial y^2} \right)^2 - \hat{p}_z \, w \right] \mathrm{d}x \, \mathrm{d}y \tag{8.25}$$

This formula can be written as a functional of the deflection function and its second derivatives

$$\Pi^m[w] = \int_A F(x, y, w, w_{,xx}, w_{,yy}) \, \mathrm{d}x \, \mathrm{d}y = \int_A F(x, y, w, f, g) \, \mathrm{d}x \, \mathrm{d}y \tag{8.26}$$

where the following new functions are introduced:

$$F = \frac{D^m}{2} \left(\frac{\partial^2 w}{\partial x^2} + \frac{\partial^2 w}{\partial y^2} \right)^2 - \hat{p}_z \, w, \qquad f = \frac{\partial^2 w}{\partial x^2}, \qquad g = \frac{\partial^2 w}{\partial y^2} \tag{8.27}$$

Now the total potential energy of a plate under bending can be written as

$$\Pi^m = \int_A F \, \mathrm{d}x \, \mathrm{d}y = \int_A \left[\frac{D^m}{2} (f + g)^2 - \hat{p}_z \, w \right] \mathrm{d}x \, \mathrm{d}y \tag{8.28}$$

The Euler equation derived from the minimum condition $\delta \, \Pi^m = 0$ for the integrand F can be written as

$$\frac{\partial F}{\partial w} + \frac{\partial^2}{\partial x^2} \left(\frac{\partial F}{\partial w_{,xx}} \right) + \frac{\partial^2}{\partial y^2} \left(\frac{\partial F}{\partial w_{,yy}} \right) = 0 \quad \text{or} \quad F_{,w} + \frac{\partial^2}{\partial x^2} F_{,f} + \frac{\partial^2}{\partial y^2} F_{,g} = 0 \tag{8.29}$$

Let us first calculate the derivatives:

$$\begin{aligned} F_{,w} &= -\hat{p}_z, \qquad F_{,f} = D^m \, (f + g) = D^m \, (w_{,xx} + w_{,yy}) \\ F_{,g} &= D^m \, (f + g) = D^m \, (w_{,xx} + w_{,yy}) \end{aligned} \tag{8.30}$$

in order to finally present the Euler equation (8.29) in a particular form, which is known as the displacement differential equation for thin rectangular plates under bending

$$-\hat{p}_z + D^m \left[\frac{\partial^2}{\partial x^2} (w_{,xx} + w_{,yy}) + \frac{\partial^2}{\partial y^2} (w_{,xx} + w_{,yy}) \right] = 0 \tag{8.31}$$

or in a concise form

$$\nabla^2 \nabla^2 w(x, y) = \frac{\hat{p}_z(x, y)}{D^m} \tag{8.32}$$

8.4 Equation for a Plate under Bending Resting on a Winkler Elastic Foundation

The simplest model for the one-parameter action of the foundation on a structure was created by Winkler. In this model an assumption is made that soil resistance is proportional to deflection, works in the opposite direction to deflection and depends on foundation stiffness, denoted in this section by k [kN/m^3]. The resistance can be treated as an additional load applied to the plate

$$\tilde{p}_z(x, y; k) = -k \, w(x, y) \tag{8.33}$$

which is introduced into the equilibrium equation and into the known displacement differential equation for a plate

$$\nabla^2\nabla^2 w(x,y) + \frac{k}{D^m}\, w(x,y) = \frac{\hat{p}_z(x,y)}{D^m} \tag{8.34}$$

and, moreover, into work W^m in the global formulation. In Subsection 8.11.4 the approximate solution of this problem using the Finite Difference Method is discussed.

8.5 Equations of Mindlin–Reissner Moderately Thick Plate Theory

8.5.1 Kinematics and Fundamental Relations for Mindlin–Reissner Plates

Moderately thick plates under bending are described by the equations of the so-called Mindlin–Reissner (M–R) three-parameter plate theory, with three independent generalized displacements: deflection $w(x,y)$ and two rotations of normal to middle plane $\vartheta_x(x,y)$, $\vartheta_y(x,y)$. Moderately thick shells are the subject of Mindlin–Reissner five-parameter shell theory.

Now, in the consideration, transverse shear deformation effects must be taken into account beside bending effects (contrary to K–L theory). The straight fibre, initially normal to the middle plane, is subjected to two rotations ϑ_x, ϑ_y that are different from the rotations of the tangent to the deformed middle surface (expressed by derivatives $\partial w/\partial x$, $\partial w/\partial y$). Note that the length of the fibre does not change. In the equations written next rotations φ_x and φ_y are used alternatively; in this case indices come from the names of coordinate axes to which rotation vectors are parallel (Figure 8.5):

$$\varphi_y = \vartheta_x, \qquad \varphi_x = -\vartheta_y \tag{8.35}$$

The introduction of the additional symbols for rotation angles φ_x, φ_y is compatible with the notation used for rotational dofs in FEM.

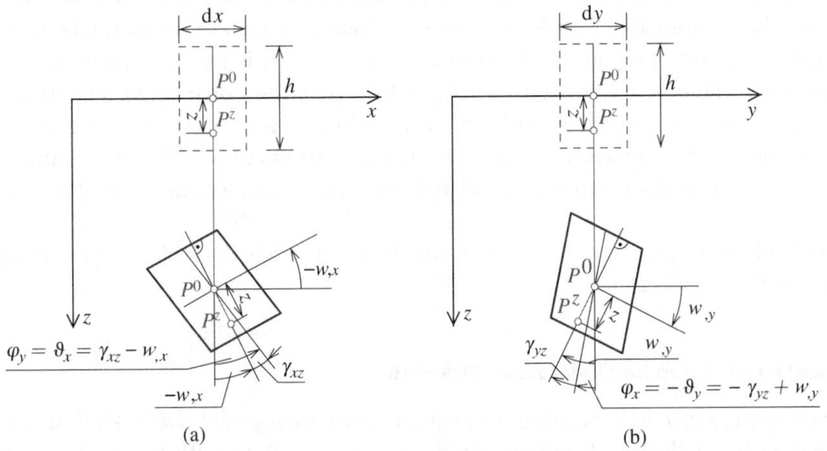

Figure 8.5 Description of plate behaviour following M–R constraints – respective segments (dx × *h*) and (dy × *h*) before and after deformation: (a) in the (*x*, *z*) plane and (b) in the (*y*, *z*) plane

According to the Mindlin–Reissner theory, plate deformation is described by the following formulae:

$$
\begin{aligned}
&u(x,y,z) = z\,\vartheta_x = z\,\varphi_y, \qquad v(x,y,z) = z\,\vartheta_y = -z\,\varphi_x \\
&\varepsilon_x(x,y,z) = z\,\frac{\partial \vartheta_x}{\partial x} = z\,\frac{\partial \varphi_y}{\partial x} = z\,\kappa_x(x,y,0) \\
&\varepsilon_y(x,y,z) = z\,\frac{\partial \vartheta_y}{\partial y} = -z\,\frac{\partial \varphi_x}{\partial y} = z\,\kappa_y(x,y,0) \\
&\gamma_{xy}(x,y,z) = z\left(\frac{\partial \vartheta_x}{\partial y} + \frac{\partial \vartheta_y}{\partial x}\right) = z\left(\frac{\partial \varphi_y}{\partial y} - \frac{\partial \varphi_x}{\partial x}\right) = z\,\chi_{xy}(x,y,0) \\
&\gamma_{xz}(x,y,z) = \frac{\partial w}{\partial x} + \vartheta_x = \frac{\partial w}{\partial x} + \varphi_y \neq 0 \\
&\gamma_{yz}(x,y,z) = \frac{\partial w}{\partial y} + \vartheta_y = \frac{\partial w}{\partial y} - \varphi_x \neq 0
\end{aligned}
\tag{8.36}
$$

The behaviour of the moderately thick plate, at any point on its middle plane, is described using:

- one-component displacement vector (deflection) $\mathbf{u}^m_{(1\times1)}(x,y) = [w(x,y)]$ [m]
- vector of rotations of normal to middle surface (independent of deflection) $\vartheta_{(2\times1)}(x,y) = [\vartheta_x, \vartheta_y]^{\mathrm{T}} = [\varphi_y, -\varphi_x]^{\mathrm{T}}$ [–]
- strain vector $\mathbf{e}_{(5\times1)}(x,y) = [\mathbf{e}^{m\,\mathrm{T}}, \mathbf{e}^{t\,\mathrm{T}}]^{\mathrm{T}}$:
  - for bending $\mathbf{e}^m_{(3\times1)}(x,y) = [\kappa_x, \kappa_y, \chi_{xy}]^{\mathrm{T}}$ [1/m]
  - for transverse shear $\mathbf{e}^t_{(2\times1)}(x,y) = [\gamma_{xz}, \gamma_{yz}]^{\mathrm{T}}$ [–]
- generalized resultant force vector $\mathbf{s}^m_{(5\times1)}(x,y) = [\mathbf{m}^{\mathrm{T}}, \mathbf{t}^{\mathrm{T}}]^{\mathrm{T}}$:
  - moments $\mathbf{m}_{(3\times1)}(x,y) = [m_x, m_y, m_{xy}]^{\mathrm{T}}$ [Nm/m]
  - transverse shear forces $\mathbf{t}_{(2\times1)}(x,y) = [t_x, t_y]^{\mathrm{T}}$ [N/m]

These 13 quantities are unknowns that are calculated from the set of 13 equations that compose the mathematical model of Mindlin–Reissner theory for moderately thick bending plates. The set of equations is given in Box 8.2 and it is the local formulation of the boundary value problem. It is worth noting that the constitutive equations are formulated for both bending and transverse shear. The correction coefficient denoted commonly by k appears in the formula for stiffness $D^t = kGh$ and is introduced in order to consider the parabolic distribution of transverse shear stresses in z direction.

An example of analytical and numerical analysis for moderately thick plates in bending is discussed in Section 8.14.

8.5.2 Global Formulation for Moderately Thick Plates

The strain energy is a sum of bending and transverse shear energies $U = U^m + U^t$. It can be described with a deflection function $w(x,y)$ and two rotations with two variants of notation: $\vartheta_x(x,y), \vartheta_y(x,y)$ or $\varphi_x(x,y), \varphi_y(x,y)$. The two alternative groups of equations are:

$$U^m[\vartheta_x, \vartheta_y] = \frac{D^m}{2}\int_A [\vartheta_{x,x}^2 + \vartheta_{y,y}^2 + 2\nu(\vartheta_{x,y} + \vartheta_{y,x})^2]\,\mathrm{d}x\,\mathrm{d}y$$
$$U^t[w, \vartheta_x, \vartheta_y] = \frac{D^t}{2}\int_A [(w_{,x} + \vartheta_x)^2 + (w_{,y} + \vartheta_y)^2]\,\mathrm{d}x\,\mathrm{d}y \qquad (8.37)$$

$$U^m[\varphi_x, \varphi_y] = \frac{D^m}{2}\int_A [\varphi_{y,x}^2 + \varphi_{x,y}^2 + 2\nu(\varphi_{y,y} - \varphi_{x,x})^2]\,\mathrm{d}x\,\mathrm{d}y$$
$$U^t[w, \varphi_x, \varphi_y] = \frac{D^t}{2}\int_A [(w_{,x} + \varphi_y)^2 + (w_{,y} - \varphi_x)^2]\,\mathrm{d}x\,\mathrm{d}y \qquad (8.38)$$

Box 8.2 Equations for moderately thick rectangular plates according to Mindlin–Reissner theory

Local formulation – set of 13 equations and boundary conditions

(I) kinematic equations (5):

$$\kappa_x = \frac{\partial \vartheta_x}{\partial x}, \quad \kappa_y = \frac{\partial \vartheta_y}{\partial y}, \quad \chi_{xy} = \frac{\partial \vartheta_x}{\partial y} + \frac{\partial \vartheta_y}{\partial x}, \quad \gamma_{xz} = \frac{\partial w}{\partial x} + \vartheta_x, \quad \gamma_{yz} = \frac{\partial w}{\partial y} + \vartheta_y$$

(II) equilibrium equations (3):

$$\frac{\partial t_x}{\partial x} + \frac{\partial t_y}{\partial y} + \hat{p}_z = 0, \quad \frac{\partial m_x}{\partial x} + \frac{\partial m_{yx}}{\partial y} - t_x = 0, \quad \frac{\partial m_{xy}}{\partial x} + \frac{\partial m_y}{\partial y} - t_y = 0$$

(III) constitutive equations (5):

$$m_x = D^m(\kappa_x + \nu\kappa_y), \quad m_y = D^m(\kappa_y + \nu\kappa_x), \quad m_{xy} = D^m(1-\nu)\,\chi_{xy}/2$$
$$t_x = D^t\,\gamma_{xz}, \quad t_y = D^t\,\gamma_{yz}$$

where: $D^m = E\,h^3/[12(1-\nu^2)]$, $D^t = k\,G\,h = k\,E\,h/[2(1+\nu)]$, $k = 5/6$

(IV) boundary conditions:

a) kinematic: $w = \hat{w}$, $\vartheta_\nu = \hat{\vartheta}_\nu$, $\vartheta_s = \hat{\vartheta}_s$

b) static: $t_\nu = \hat{t}_\nu$, $m_\nu = \hat{m}_\nu$, $m_{\nu s} = \hat{m}_{\nu s}$

c) mixed

Global formulation

Total potential energy $\Pi^m = U^m + U^t - W^m$

Internal strain energy

$$U^m[\vartheta_x, \vartheta_y] = \frac{D^m}{2}\int_A \left[\left(\frac{\partial \vartheta_x}{\partial x}\right)^2 + \left(\frac{\partial \vartheta_y}{\partial y}\right)^2 + 2\nu\left(\frac{\partial \vartheta_x}{\partial y} + \frac{\partial \vartheta_y}{\partial x}\right)^2\right]\mathrm{d}x\,\mathrm{d}y$$

$$U^t[w, \vartheta_x, \vartheta_y] = \frac{D^t}{2}\int_A \left[\left(\frac{\partial w}{\partial x} + \vartheta_x\right)^2 + \left(\frac{\partial w}{\partial y} + \vartheta_y\right)^2\right]\mathrm{d}x\,\mathrm{d}y$$

External load work

$$W^m = \int_A \hat{p}_z w\,\mathrm{d}x\,\mathrm{d}y + \int_{\partial A_\sigma} (\hat{t}_\nu w + \hat{m}_\nu \vartheta_\nu + \hat{m}_{\nu s}\vartheta_s)\,\mathrm{d}s + \sum_i \hat{P}_i\,w_i$$

The work of external load is as follows

$$W^m = \int_A \hat{p}_z \; w \; dx\,dy + \int_{\partial A_\sigma} (\hat{t}_\nu \; w + \hat{m}_\nu \; \vartheta_\nu + \hat{m}_{\nu s} \; \vartheta_s) \; ds + \sum_i \hat{P}_i \; w_i \tag{8.39}$$

Consequently, the total potential energy of a moderately thick plate has three components

$$\Pi = U^m + U^t - W^m \tag{8.40}$$

8.5.3 Equations for Mindlin–Reissner Moderately Thick Plates Expressed by Generalized Displacements

In order to find the three fields (deflection and two rotations) $w(x,y)$, $\vartheta_x(x,y)$, $\vartheta_y(x,y)$ the following three displacement equations are derived:

$$D^t \left[\frac{\partial}{\partial x}\left(\frac{\partial w}{\partial x} + \vartheta_x\right) + \frac{\partial}{\partial y}\left(\frac{\partial w}{\partial y} + \vartheta_y\right) \right] + \hat{p}_z = 0$$

$$D^m \left[\frac{\partial^2 \vartheta_x}{\partial x^2} + \nu \; \frac{\partial^2 \vartheta_y}{\partial x \, \partial y} + \frac{(1-\nu)}{2} \; \frac{\partial^2 \vartheta_x}{\partial y^2} + \frac{(1-\nu)}{2} \; \frac{\partial^2 \vartheta_y}{\partial x \, \partial y} \right] - D^t \left(\frac{\partial w}{\partial x} + \vartheta_x\right) = 0$$

$$D^m \left[\frac{\partial^2 \vartheta_y}{\partial y^2} + \nu \; \frac{\partial^2 \vartheta_x}{\partial x \, \partial y} + \frac{(1-\nu)}{2} \; \frac{\partial^2 \vartheta_x}{\partial x \, \partial y} + \frac{(1-\nu)}{2} \; \frac{\partial^2 \vartheta_y}{\partial x^2} \right] - D^t \left(\frac{\partial w}{\partial y} + \vartheta_y\right) = 0 \tag{8.41}$$

The solution of these differential equations requires information about boundary conditions written as functions of deflection, two rotations and their derivatives.

8.6 Analytical Solution of a Sinusoidally Loaded Rectangular Plate

To show the main features of an analytical solution for a 2D domain (the middle plane of a bending plate) a rectangular, the simply supported plate shown in Figure 8.6 is considered.

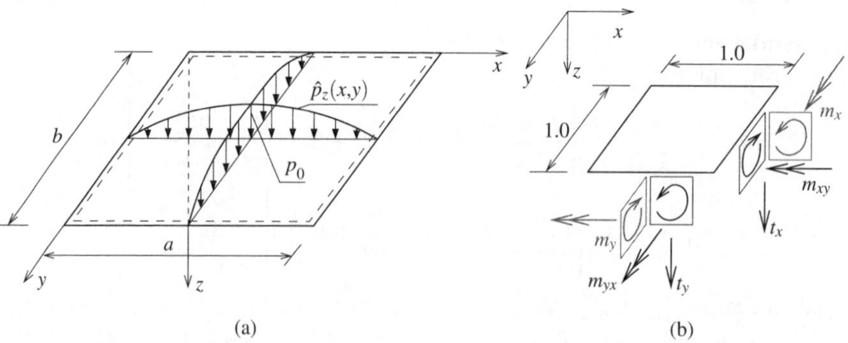

Figure 8.6 Rectangular, simply supported plate under sinusoidal load: (a) configuration and (b) criterion of notation and sign convention for moments and transverse shear forces – repeated from Figure 8.3

This example is a fundamental one in every *plates* and *shells* course analogously to a simply supported beam (1D) discussed in a *structural mechanics* course.

To solve the problem, all functions describing the behaviour of the structure under the given load have to be determined. Finally, the diagrams of deflection function $w(x, y)$, two bending moments $m_x(x, y)$, $m_y(x, y)$, twisting moment $m_{xy}(x, y)$ and two transverse shear forces $t_x(x, y)$, $t_y(x, y)$ are presented in the 2D domain. The distribution of edge reactions is shown along the contour lines.

The load of considered plate is described by the following two sinusoidal functions

$$\hat{p}_z(x, y) = p_0 \sin\frac{\pi x}{a} \sin\frac{\pi y}{b} \tag{8.42}$$

For such a surface, the load and boundary conditions, related to four simply supported edges, the deflection function can be predicted in the following form

$$w(x, y) = w_0 \sin\frac{\pi x}{a} \sin\frac{\pi y}{b} \tag{8.43}$$

This function $w(x, y)$ fulfils mixed boundary conditions on each simply supported line:

- kinematic condition – zero value of the deflection $w = 0$
- static condition – zero moment $m_\nu = -D^m(w_{,\nu\nu} + \nu w_{,ss}) = 0$

For a supported edge with $w = 0$ and thus $w_{,ss} = 0$ the second static boundary condition is reduced to the following kinematic constraint $w_{,\nu\nu} = 0$. As a consequence we obtain:

$$\begin{aligned} &w(0, y) = w(a, y) = w(x, 0) = w(x, b) = 0 \\ &w_{,xx}|_{x=0} = w_{,xx}|_{x=a} = 0, \quad w_{,yy}|_{y=0} = w_{,yy}|_{y=b} = 0 \end{aligned} \tag{8.44}$$

Considering the left-hand and right-hand sides of the displacement equation of a thin plate under bending, deflection amplitude w_0 can be related to load amplitude p_0:

$$\begin{aligned} &\nabla^2\nabla^2 w(x, y) = \frac{\hat{p}_z}{D^m} \\ &\text{LHS} = w_{,xxxx} + 2\, w_{,xxyy} + w_{,yyyy} = w_0\, \pi^4 \left(\frac{1}{a^2} + \frac{1}{b^2}\right)^2 \sin\frac{\pi x}{a} \sin\frac{\pi y}{b} \\ &\text{RHS} = \frac{p_0}{D^m} \sin\frac{\pi x}{a} \sin\frac{\pi y}{b} \end{aligned} \tag{8.45}$$

$$w_0 = \frac{p_0}{D^m\, \pi^4 \left(\frac{1}{a^2} + \frac{1}{b^2}\right)^2} \tag{8.46}$$

For a square plate ($a = b$) the maximum deflection at the centre is given by

$$w_{\max} = w_0 = \frac{p_0\, a^4}{4\, D^m\, \pi^4} \tag{8.47}$$

The resultant of surface load is denoted by P and calculated as

$$P = \int_0^a \int_0^b p_0 \sin\frac{\pi\, x}{a} \sin\frac{\pi y}{b}\, \mathrm{d}x\, \mathrm{d}y = \frac{4\, a\, b\, p_0}{\pi^2} \tag{8.48}$$

Now, using the combined constitutive and kinematic equations (8.7) and (8.9), the distributions of three moments $m_x(x, y)$, $m_y(x, y)$, $m_{xy}(x, y)$ and two transverse shear forces

$t_x(x,y)$, $t_y(x,y)$ are determined. In the formulae next, amplitudes M_x, M_y, M_{xy}, T_x, T_y (extreme values appearing at characteristic points) are written with capital letters:

$$\begin{aligned}
m_x(x,y) &= -D^m(w_{,xx} + \nu\, w_{,yy}) = D^m\, w_0\, \pi^2 \left(\frac{1}{a^2} + \frac{\nu}{b^2}\right) \sin\frac{\pi x}{a} \sin\frac{\pi y}{b} \\
&= M_x \sin\frac{\pi x}{a} \sin\frac{\pi y}{b} \\
m_y(x,y) &= -D^m(w_{,yy} + \nu\, w_{,xx}) = D^m\, w_0\, \pi^2 \left(\frac{\nu}{a^2} + \frac{1}{b^2}\right) \sin\frac{\pi x}{a} \sin\frac{\pi y}{b} \\
&= M_y \sin\frac{\pi x}{a} \sin\frac{\pi y}{b} \\
m_{xy}(x,y) &= -(1-\nu)D^m\, w_{,xy} = -D^m\, w_0\, \pi^2 \frac{(1-\nu)}{ab} \cos\frac{\pi x}{a} \cos\frac{\pi y}{b} \\
&= -M_{xy} \cos\frac{\pi x}{a} \cos\frac{\pi y}{b}
\end{aligned} \tag{8.49}$$

$$\begin{aligned}
t_x(x,y) &= -D^m(w_{,xxx} + w_{,yyx}) = D^m\, w_0 \pi^3 \frac{1}{a} \left(\frac{1}{a^2} + \frac{1}{b^2}\right) \cos\frac{\pi x}{a} \sin\frac{\pi y}{b} \\
&= T_x \cos\frac{\pi x}{a} \sin\frac{\pi y}{b} \\
t_y(x,y) &= -D^m(w_{,xxy} + w_{,yyy}) = D^m\, w_0 \pi^3 \frac{1}{b} \left(\frac{1}{a^2} + \frac{1}{b^2}\right) \sin\frac{\pi x}{a} \cos\frac{\pi y}{b} \\
&= T_y \sin\frac{\pi x}{a} \cos\frac{\pi y}{b}
\end{aligned} \tag{8.50}$$

Corresponding diagrams and contour plots are depicted in Figures 8.7–8.11. Figure 8.7 illustrates deflection w in the plate domain. Bending moments m_x and m_y are presented in Figure 8.8. In Figure 8.9a the real directions of the corner twisting moments are also highlighted. Next Figures 8.10 and 8.11 show transverse shear forces t_x and t_y.

Then, the boundary reactions (Figure 8.12) are derived from the relation $\tilde{t}_\nu = t_\nu + \frac{\partial m_{\nu s}}{\partial s}$ for effective transverse shear force (valid for the K–L theory). The adopted sign convention is like that for the transverse shear forces (see Figure 8.12a). The boundary and corner reactions are positive when their directions are compatible with the direction of the z-axis (Figure 8.12b).

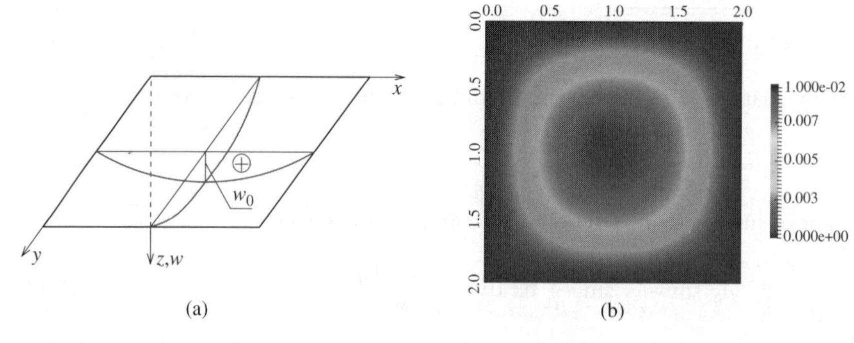

Figure 8.7 Deflection w for square simply supported plate under sinusoidal load: (a) diagram and (b) contour plot

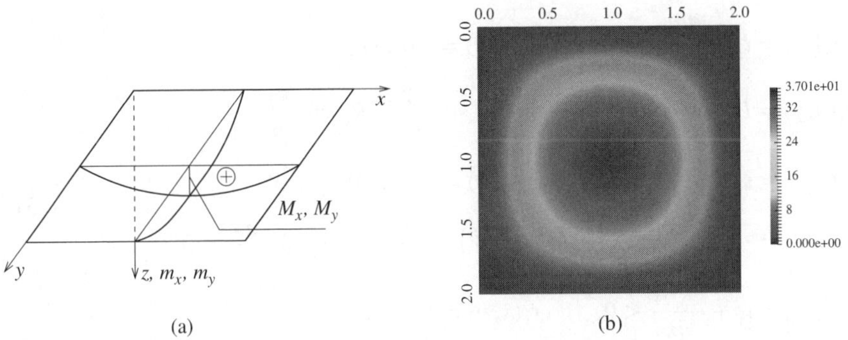

Figure 8.8 Bending moments m_x, m_y for a square simply supported plate under sinusoidal load: (a) diagram and (b) contour plot (the same for m_x and m_y)

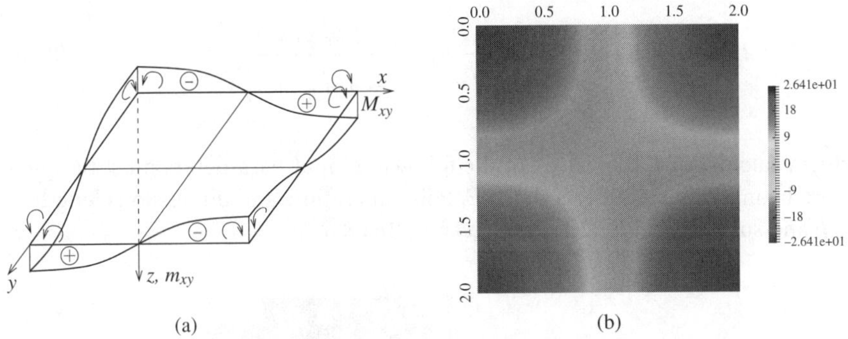

Figure 8.9 Twisting moment m_{xy} for a square simply supported plate under sinusoidal load: (a) diagram and (b) contour plot

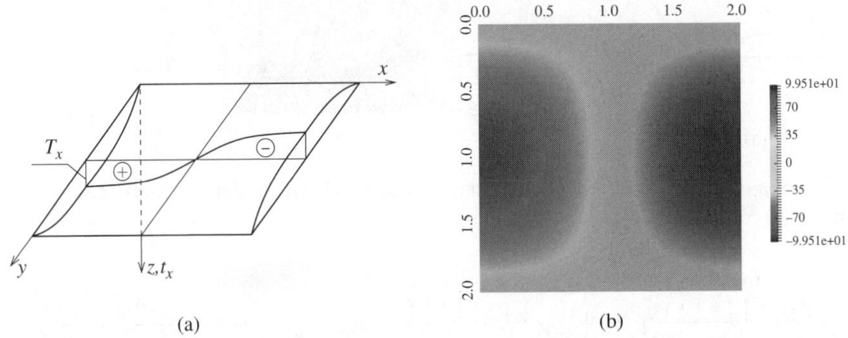

Figure 8.10 Transverse shear force t_x for a square simply supported plate under sinusoidal load: (a) diagram and (b) contour plot

For four boundary lines the boundary reactions are calculated from effective transverse shear forces according to the following relations:

$$\tilde{t}_x|_{x=0} = t_x + \frac{\partial m_{xy}}{\partial y} > 0, \qquad r_x(0,y) = -\tilde{t}_x(0,y)$$

$$r_x(0,y) = -D^m\, w_0\, \pi^3 \frac{1}{a}\left(\frac{2-\nu}{b^2} + \frac{1}{a^2}\right)\sin\frac{\pi y}{b} = -R_x\, \sin\frac{\pi y}{b} < 0$$

$$
\begin{aligned}
&\tilde{t}_x|_{x=a} = t_x + \frac{\partial m_{xy}}{\partial y} < 0, \qquad r_x(a,y) = \tilde{t}_x(a,y)\\
&r_x(a,y) = D^m\ w_0\ \pi^3 \frac{1}{a}\left(\frac{2-\nu}{b^2} + \frac{1}{a^2}\right)\sin\frac{\pi y}{b} = R_x\ \sin\frac{\pi y}{b} < 0\\
&\tilde{t}_y|_{y=0} = t_y + \frac{\partial m_{yx}}{\partial x} > 0, \qquad r_y(x,0) = -\tilde{t}_y(x,0)\\
&r_y(x,0) = -D^m\ w_0\ \pi^3 \frac{1}{b}\left(\frac{2-\nu}{a^2} + \frac{1}{b^2}\right)\sin\frac{\pi x}{a} = -R_y\ \sin\frac{\pi x}{a} < 0\\
&\tilde{t}_y|_{y=b} = t_y + \frac{\partial m_{yx}}{\partial x} < 0, \qquad r_y(x,b) = \tilde{t}_y(x,b)\\
&r_y(x,b) = D^m\ w_0\ \pi^3 \frac{1}{b}\left(\frac{2-\nu}{a^2} + \frac{1}{b^2}\right)\sin\frac{\pi x}{a} = R_y\ \sin\frac{\pi x}{a} < 0
\end{aligned}
\tag{8.51}
$$

The concentrated reactions at four corners are calculated as

$$
T_i = R_i = 2\ |m_{xy}| = 2(1-\nu)D^m\ \pi^2 \frac{1}{ab}\ w_0 = \frac{2\ p_0\ (1-\nu)}{\pi^2\, a\, b\left(\frac{1}{a^2} + \frac{1}{b^2}\right)^2}
\tag{8.52}
$$

To perform calculations, the values of the following input parameters must be specified: material constants Young's modulus E, Poisson's ratio ν and dimensions length a, thickness h and surface load p_0. They are given in Box 8.3.

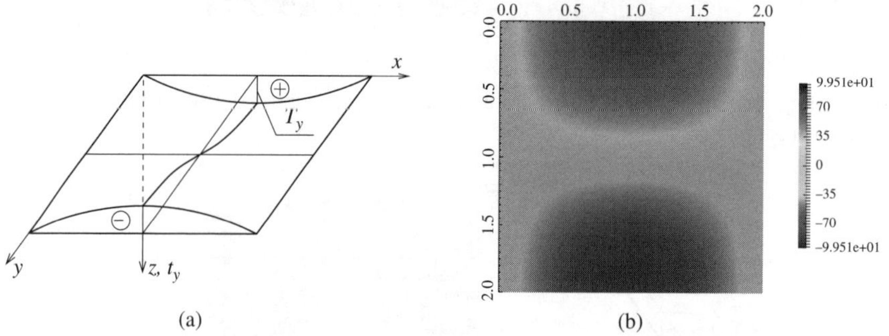

Figure 8.11 Transverse shear force t_y for a square simply supported plate under sinusoidal load: (a) diagram and (b) contour plot

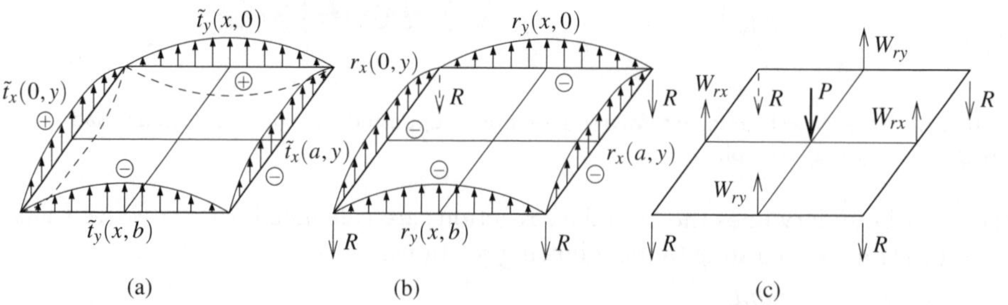

Figure 8.12 Rectangular, simply supported plate under sinusoidal load: diagrams of distributions (along plate edges) of: (a) effective transverse shear forces, (b) linear boundary and corner reactions and (c) presentation of all actions in the analysed plate as concentrated forces, see also Equation (8.53)

Box 8.3 Rectangular simply supported plate under a sinusoidal load

Data

$E = 1.5 \times 10^7\ \mathrm{kN/m^2}$, $\nu = 1/6$, $a = b = 2.0\ \mathrm{m}$, $h = 0.1\ \mathrm{m}$
$D^m = 1285.71\ \mathrm{kNm}$
$p_0 = 313.1\ \mathrm{kN/m^2}$, $P = 507.6\ \mathrm{kN}$

Check values from analytical solution

$w_0 = 0.01\ \mathrm{m}$, $M_x = M_y = 37.01\ \mathrm{kNm/m}$, $M_{xy} = \pm 26.41\ \mathrm{kNm/m}$
$T_x = T_y = \pm 99.51\ \mathrm{kN/m}$, $W_r = -179.9\ \mathrm{kN}$, $R = 52.9\ \mathrm{kN}$

Plate configuration shown in Figure 8.6

The following values describing the solution of the considered square plate: load resultant P, maximum deflection w_0, maximum values of two bending moments $M_x = M_y$, extreme value of twisting moment M_{xy}, extreme transverse shear forces $T_x = T_y$, resultants of boundary reactions at each edge W_r, concentrated reactions in each corner R can also be found in Box 8.3.

To verify the calculations, condition $\Sigma P_z = 0$ (Figure 8.12c) which includes: load resultant $P > 0$, four resultants of edge reactions $2W_{rx} < 0$, $2W_{ry} < 0$ and four corner reactions $4R > 0$ is checked and it is satisfied

$$\begin{aligned} &P + \int_0^a r_y(x,0)\,\mathrm{d}x + \int_0^a r_y(x,b)\,\mathrm{d}x + \int_0^b r_x(0,y)\,\mathrm{d}y + \int_0^b r_x(a,0)\,\mathrm{d}y + 4\,R \\ &= P + 2\,W_{ry} + 2\,W_{rx} + 4\,R = 507.6 - 4 \times 179.9 + 4 \times 52.9 = 0 \end{aligned} \tag{8.53}$$

8.7 Analysis of Plates under Bending Using Expansions in Double or Single Trigonometric Series

8.7.1 Application of Navier's Method – Double Trigonometric Series

The method using double trigonometric series (DTSM) is applied in the analysis of a rectangular plate simply supported on all edges (Figure 8.13a). In this particular case the deflection function is adopted as the double sum for $i, j = 1, \ldots, \infty$ of products of unknown coefficients W_{ij} and sine functions with respect to both coordinates x, y

$$w(x,y) = \sum_{i=1}^{\infty}\sum_{j=1}^{\infty} W_{ij}\,\sin(\alpha_i x)\,\sin(\beta_j y)\,, \quad \text{where:} \quad \alpha_i = \frac{i\pi}{a}, \quad \beta_j = \frac{j\pi}{b} \tag{8.54}$$

The adopted function satisfies the boundary conditions on the simply supported contour:

- kinematic condition $w = 0$
- static condition $m_\nu = -D^m(w_{,\nu\nu} + \nu w_{,ss}) = 0$,
  which, taking into account $w_{,ss} = 0$, is reduced to the requirement that $w_{,\nu\nu} = 0$

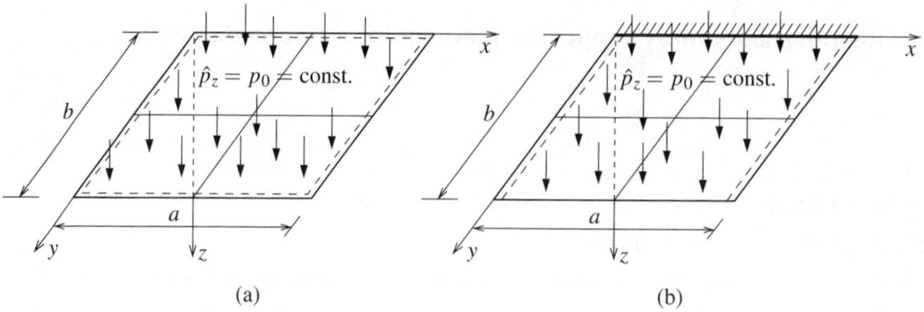

Figure 8.13 Rectangular plates with two types of boundary conditions analysed using two different methods: (a) double and (b) single trigonometric series

Thus the boundary conditions are employed in the following form:

$$\begin{aligned} &w(0,y) = w(a,y) = w(x,0) = w(x,b) = 0 \\ &w_{,xx}|_{x=0} = w_{,xx}|_{x=a} = 0, \quad w_{,yy}|_{y=0} = w_{,yy}|_{y=b} = 0 \end{aligned} \tag{8.55}$$

The analysis starts with the known displacement differential equation for a rectangular plate under bending

$$\nabla^2\nabla^2 w(x,y) = \frac{\hat{p}_z(x,y)}{D^m} \tag{8.56}$$

in which function $w(x,y)$ given by Equation (8.54) is substituted, including unknown coefficients W_{ij}, $i,j = 1,2,\ldots$. This gives

$$\sum_{i=1}^{\infty}\sum_{j=1}^{\infty} (\alpha_i^2 + \beta_j^2)^2\, W_{ij}\, \sin(\alpha_i\, x)\, \sin(\beta_j\, y) = \frac{\hat{p}_z(x,y)}{D^m} \tag{8.57}$$

In order to calculate a single coefficient W_{ij} from the double sum for any combination of $i,j = 1,2,\ldots$, let us use the orthogonality of the sine functions, expressed by:

$$\begin{aligned} &\int_0^a \sin(\alpha_i\, x)\, \sin(\alpha_k\, x)\, \mathrm{d}x = \frac{a}{2} \quad \text{for } i = k \quad \text{and} \\ &\qquad \int_0^a \sin(\alpha_i\, x)\, \sin(\alpha_k\, x)\, \mathrm{d}x = 0 \quad \text{for } i \neq k \\ &\int_0^b \sin(\beta_j\, y)\, \sin(\beta_l\, y)\, \mathrm{d}y = \frac{b}{2} \quad \text{for } j = l \quad \text{and} \\ &\qquad \int_0^b \sin(\beta_j\, y)\, \sin(\beta_l\, y)\, \mathrm{d}y = 0 \quad \text{for } j \neq l \end{aligned} \tag{8.58}$$

The single coefficient W_{ij} can thus be calculated using the following sequence of operations:

i) multiply Equation (8.57) by factor $\sin(\alpha_k\, x)$ and integrate in the interval $[0, a]$
ii) multiply the obtained equation again by $\sin(\beta_l\, y)$ and integrate in the interval $[0, b]$

$$\left[\sum_{i=1}^{\infty}\sum_{j=1}^{\infty}(\alpha_i^2+\beta_j^2)^2\, W_{ij}\, \sin(\alpha_i x)\, \sin(\beta_j y)\right] \Bigg| \cdot \sin(\alpha_k x) \Bigg| \int_0^a \cdots \mathrm{d}x \Bigg| \cdot \sin(\beta_l y) \Bigg| \int_0^b \cdots \mathrm{d}y$$

$$= \frac{\hat{p}_z(x,y)}{D^m} \Bigg| \cdot \sin(\alpha_k x) \Bigg| \int_0^a \cdots \mathrm{d}x \Bigg| \cdot \sin(\beta_l y) \Bigg| \int_0^b \cdots \mathrm{d}y \tag{8.59}$$

$$(\alpha_i^2+\beta_j^2)^2\, W_{ij}\, \frac{a}{2}\, \frac{b}{2} = \frac{1}{D^m} \int_0^a \int_0^b \hat{p}_z(x,y)\, \sin(\alpha_i\, x)\, \sin(\beta_j\, y)\, \mathrm{d}x\, \mathrm{d}y \tag{8.60}$$

$$W_{ij} = \frac{4}{D^m(\alpha_i^2+\beta_j^2)^2\, a\, b} \int_0^a \int_0^b \hat{p}_z(x,y)\, \sin(\alpha_i\, x)\, \sin(\beta_j\, y)\, \mathrm{d}x\, \mathrm{d}y \tag{8.61}$$

The standard example of the use of double trigonometric series is the solution for a simply supported plate subjected to uniformly distributed transverse load $\hat{p}_z = p_0 = \text{const.}$

For such a load the following formula for the coefficients W_{ij} is obtained

$$\begin{aligned} W_{ij} &= \frac{4\, p_0}{D^m(\alpha_i^2+\beta_j^2)^2 a\, b} \int_0^a \int_0^b \sin(\alpha_i\, x)\, \sin(\beta_j\, y)\, \mathrm{d}x\, \mathrm{d}y \\ &= \frac{4\, p_0}{D^m(\alpha_i^2+\beta_j^2)^2 a\, b} \cdot \frac{4\, a\, b}{i\, j\, \pi^2} \end{aligned} \tag{8.62}$$

for $i, j = 1, 3, 5, \ldots$ and the final deflection function has the form

$$w(x,y) = \frac{16\, p_0}{D^m\, \pi^6} \sum_{i=1,3,\ldots}^{\infty} \sum_{j=1,3,\ldots}^{\infty} \frac{1}{i\, j \left(\frac{i^2}{a^2}+\frac{j^2}{b^2}\right)^2} \sin(\alpha_i\, x)\, \sin(\beta_j\, y) \tag{8.63}$$

The square bending plate is described by the following parameters: material constants E, ν, dimensions a, h and load p_0. Their values are listed in Box 8.4.

Box 8.4 Square simply supported plate with a uniform load

Data

$E = 1.5 \times 10^7\ \text{kN/m}^2$, $\nu = 0.15$, $a = b = 2.0\ \text{m}$, $h = 0.1\ \text{m}$
$D^m = 1278.77\ \text{kNm}$, $p_0 = 100.0\ \text{kN/m}^2$

Check values from analytical solution

$w_{\max} = 0.00508\ \text{m}$, $m_{x,\max} = m_{y,\max} = 17.28\ \text{kNm/m}$

Plate configuration shown in Figure 8.14a

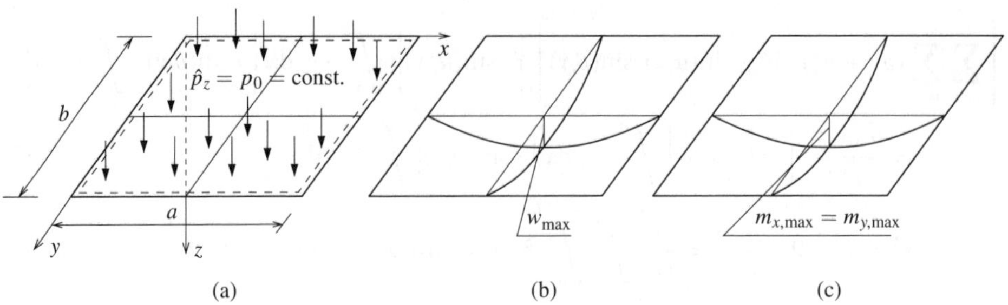

Figure 8.14 Simply supported square plate with uniformly distributed load: (a) configuration, (b) middle surface of plate after deformation and (c) distribution of bending moments

For the square plate the maximum deflection marked in Figure 8.14b is determined from the formula

$$w_{\text{max}} = W_{11} = 0.00406 \, \frac{p_0 \, a^4}{D^m} \tag{8.64}$$

and then two equal maximum bending moments, as shown in Figure 8.14c, at the central point of the plate are

$$m_{x,\text{max}} = m_{y,\text{max}} = 0.0432 \, p_0 \, a^2 \tag{8.65}$$

The values of maximum deflection w_{max} and two bending moments $m_{x,\text{max}} = m_{y,\text{max}}$ calculated for the adopted input data are also included in Box 8.4.

8.7.2 Idea of Levy's Method – Single Trigonometric Series

Single trigonometric series method (STSM) can be applied to find a solution for many cases of rectangular plates. When two parallel edges of a rectangular plate (Figure 8.13b) are simply supported (e.g. the ones parallel to axis y) and the other two are arbitrarily supported, the deflection function is adopted in the form of a single sine series of variable x, with functional factors $Y_j(y)$, for $j = 1, \ldots, \infty$

$$w(x, y) = \sum_{j=1}^{\infty} Y_j(y) \, \sin(\alpha_j \, x) \, , \quad \text{where} \quad \alpha_j = j \, \pi / a \tag{8.66}$$

Each function $Y_j(y)$ is determined by solving the fourth-order differential equation with ordinary derivatives with respect to coordinate y

$$Y_j^{\text{IV}} - 2 \, \alpha_j^2 \, Y_j^{\text{II}} + \alpha_j^4 \, Y_j = \frac{2}{D^m \, a} \int_0^a \hat{p}_z(x, y) \, \sin(\alpha_j \, x) \, \mathrm{d}x \quad \text{for} \quad j = 1, 2, \ldots \tag{8.67}$$

which is obtained by a modification of Equation (8.56) leading to the elimination of coordinate x. After integration of Equation (8.67), four constants must be calculated from two boundary conditions on each line $y = 0$ and $y = b$.

Based on this method, which takes into account different variants of boundary conditions for edges $y = 0$ and $y = b$, and using various combinations of solutions for numerous cases of loads and supports, tables for engineers have been worked out and are available in relevant books. The information given in these tables enables the calculation of extreme values of deflection and bending moments, as well as reactions at characteristic points of a plate. It should be noted that the dedicated tables are prepared for reinforced concrete, see for example Starosolski (2009), and steel plates, for example Timoshenko and Woinowsky-Krieger (1959), in which different values of Poisson's ratio are adopted. One should also pay attention to different directions of the z-axis and thus different signs of loads, deflection and bending or twisting moments.

8.8 Simply Supported or Clamped Square Plate with Uniform Load

In this section two square plates with uniform load and two cases of boundary conditions on the whole contour are analysed:

case A – simply supported plate, see Figure 8.15a
case B – clamped plate, see Figure 8.15b

The adopted material constants require the use of the tables for steel plates (see Timoshenko and Woinowsky-Krieger 1959), which takes into account the proper value of Poisson's ratio ($\nu = 0.3$). Note that the z-axis is directed downwards according to the sign convention in Figure 8.3, which affects the signs of the deflection and bending moments. Note that in Figures 8.15a,b the origins of Cartesian coordinate systems are located at different points for cases A and B.

The considered plate is described by the following parameters: material constants E, ν, dimensions $L_x = L_y = L$, two alternative values the thickness $h^{(1)}$, $h^{(2)}$ and surface load p_0. Their values are listed in Box 8.5.

For the two cases of boundary conditions A and B, appropriate formulae are presented in the following two subsections. However, the results of calculations are included in one Box 8.5. Two check values: maximum deflection w_C and bending moments $m_{xC} = m_{yC}$ (at the centre C of the plate) are presented in Box 8.5 and can be used to compare analytical and FEM solutions for the two cases of boundary conditions.

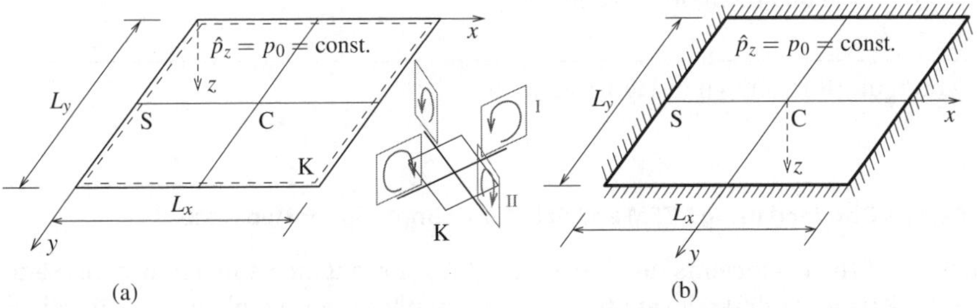

Figure 8.15 Two benchmarks for bending analysis – square plate with uniform load and different boundary conditions: (a) simply supported plate (case A) with principal bending moments at corner K and (b) clamped plate (case B)

Box 8.5 Simply supported (case A) and clamped (case B) square plate with a uniform load

Data

$E = 2.0 \times 10^8 \text{kN/m}^2, \quad \nu = 0.3, \quad L = L_x = L_y = 1.0\,\text{m}$

$h^{(1)} = 0.004\ \text{m}, \quad D^{m(1)} = 1.1722\,\text{kNm}$

$h^{(2)} = 0.008\ \text{m}, \quad D^{m(2)} = 9.3773\,\text{kNm}$

$p_0 = 10.0\,\text{kN/m}^2$

Check values from analytical solution

A $\quad w_{\text{C}}^{(1)} = 0.0345\ \text{m} \approx 8.6h^{(1)}, \quad w_{\text{C}}^{(2)} = 0.00433\ \text{m} \approx 0.54h^{(2)}$

$m_{x\text{C}} = m_{y\text{C}} = 0.479\,\text{kNm/m}$

$\sigma_{x\text{C}}^{(1)}|_{\pm h^{(1)}/2} = \sigma_{y\text{C}}^{(1)}|_{\pm h^{(1)}/2} = \pm 179 \times 10^3\ \text{MPa}$

$\sigma_{x\text{C}}^{(2)}|_{\pm h^{(2)}/2} = \sigma_{y\text{C}}^{(2)}|_{\pm h^{(2)}/2} = \pm 44 \times 10^3\ \text{MPa}$

$t_{x\text{S}} = 3.38\,\text{kN/m}, \quad r_{x\text{S}} = \tilde{t}_{x\text{S}} = 4.20\,\text{kN/m}, \quad R_{\text{K}} = 0.65\,\text{kN}$

$m_{x\text{K}} = m_{y\text{K}} = 0.0\,\text{kNm/m} \quad m_{\text{I, II K}} = \pm 0.394\,\text{kNm/m}$

B $\quad w_{\text{C}}^{(1)} = 0.0107\ \text{m}$

$m_{x\text{C}} = m_{y\text{C}} = 0.231\,\text{kNm/m} \quad m_{\nu\text{S}} = -0.513\,\text{kNm/m}$

Check values from FEM solution using ROBOT

$NSE = 100, 286, \quad NEN = 4$

A $\quad w_{\text{C}}^{(1)\,\text{FEM}} = -0.0346\ \text{m} = -1.003 w_{\text{C}}^{(1)\,\text{anal}}$

$m_{x\text{C}}^{\text{FEM}} = m_{y\text{C}}^{\text{FEM}} = -0.482\,\text{kNm/m} = -1.006 m_{x\,\text{C}}^{\text{anal}}$

B $\quad w_{\text{C}}^{(1)\,\text{FEM}} = -0.0109\ \text{m} = -1.006 w_{\text{C}}^{(1)\,\text{anal}}$

$m_{x\,\text{C}}^{\text{FEM}} = m_{y\,\text{C}}^{\text{FEM}} = -0.238\,\text{kNm/m} = -1.030 m_{x\,\text{C}}^{\text{anal}}$

$m_{\nu\,\text{S}}^{\text{FEM}} = 0.509\,\text{kNm/m} = -0.992 m_{\nu\text{S}}^{\text{anal}}$

Plate configurations shown in Figure 8.15a,b

8.8.1 Results Obtained using DTSM and FEM for a Simply Supported Plate

On the basis of the coefficients taken from the table for engineers in Timoshenko and Woinowsky-Krieger (1959) relevant for a square simply supported plate, one can calculate the values of characteristic quantities that describe the behaviour of the plate. The factors given in the table come from the DTSM method. In Figure 8.15a the points C, S and K are shown, for which these computations have been performed.

The essential formulae rewritten from the tables are given next:

$$\begin{aligned}
&w_{\mathrm{C}}^{\mathrm{DTSM\,(A)}} = 0.00406\,\hat{p}\,L^4/D^m \\
&m_{x\,\mathrm{C}}^{\mathrm{DTSM\,(A)}} = m_{y\,\mathrm{C}}^{\mathrm{DTSM\,(A)}} = 0.0479\,\hat{p}\,L^2 \\
&t_{x\,\mathrm{S}}^{\mathrm{DTSM\,(A)}} = 0.338\,\hat{p}\,L, \qquad r_{x\,\mathrm{S}}^{\mathrm{DTSM\,(A)}} = \tilde{t}_{x\,\mathrm{S}}^{\mathrm{DTSM\,(A)}} = 0.420\,\hat{p}\,L \\
&R_{\mathrm{K}}^{\mathrm{DTSM\,(A)}} = 0.065\,\hat{p}\,L
\end{aligned} \tag{8.68}$$

Two values of plate thickness $h^{(1)}$ and $h^{(2)}$ are adopted in calculations. It should be noted that the maximum deflection $w_{\mathrm{C}}^{(1)}$ obtained for thickness $h^{(1)}$ exceeds the thickness over eight times, see Box 8.5. In geometrically linear plate theory the condition $w_{\max} < h$ must be satisfied. Moreover, it is known that bending stiffness $D^{m\ (i)}$, deflection w_{C} and extreme normal stress on planes $z = \pm h/2$ depend on thickness $h^{(i)}$ while the bending moments, transverse shear forces and reactions do not, see Equation (8.68). The results of calculations for the two values of thickness lead to the following conclusions. When the thickness for a bending plate is doubled, then:

- bending stiffness D^m increases eight times and thus maximum deflection w_{C} decreases eight times
- bending moments, transverse shear forces and reactions do not change
- normal stresses σ_x and σ_y at top and bottom surfaces (for $z = \pm h/2$) decrease four times

The results of analytical and numerical computations (see Box 8.5) can be summarized as follows. For the z-axis and load directed downwards, the plate deflects down and this involves the extension of bottom fibres in both directions. In Figure 8.16 we notice that the surface of the deformed plate and the distributions of bending moments have two planes of symmetry. The distribution of the twisting moment is antisymmetric.

At the corners, where $m_{x\mathrm{K}} = m_{y\mathrm{K}} = 0$, the extreme values of twisting moments appear, which are principal moments at these points and are calculated using formulae:

$$m_{\mathrm{I\,K}} = 0.0394\,\hat{p}\,L^2, \qquad m_{\mathrm{II\,K}} = -0.0394\,\hat{p}\,L^2 \tag{8.69}$$

Their values are included in Box 8.5. In Figure 8.15a the real senses of principal moments at corner K are shown. Note that in the vicinity of corners: (i) on the top surface tensile stress occurs in the direction of bisector of the right angle and (ii) on the bottom surface tensile stress appears in the direction perpendicular to the bisector of the right angle. These facts have to be taken into account when the reinforcement of a concrete plate is designed.

The twisting moments at the corners produce the concentrated reactions R_{K} that result in large stress and strain related to transverse shear. According to the K–L theory, these effects are neglected. Thus, the more precise description of stress and strain state at the corners of simply supported bending plate should be based on the Mindlin–Reissner plate theory or 3D modelling.

Computations have been performed using two meshes with $NSE = 100$ and $NSE = 256$ using package ROBOT (2006). The vizualization of principal stresses is more clear for the coarse mesh with $NSE = 10 \times 10$. In this and the next example, the global coordinate system used in numerical calculations was adopted with axis z directed upwards and this is the reason for the opposite signs of the check values obtained from analytical and numerical calculations (see Box 8.5).

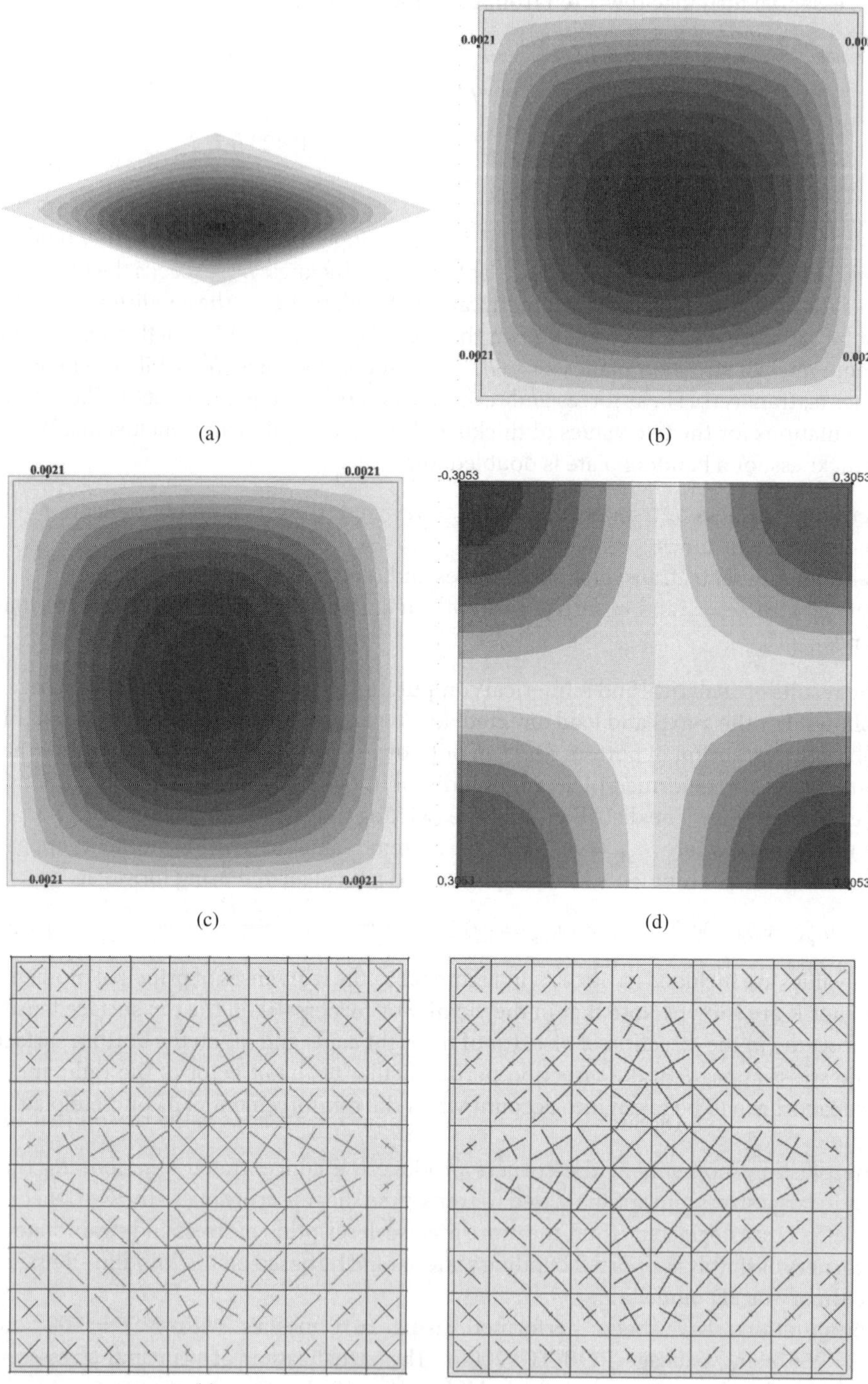

Figure 8.16 Simply supported square plate with uniform load: (a) deflected surface $w(x, y)$, contour maps of moments [kNm/m]: (b) m_x, (c) m_y, (d) m_{xy}; vizualization of directions and magnitudes of principal stresses (for FE mesh with $NSE = 100$) on two surfaces: (e) bottom and (f) top; FEM results from ROBOT. *Subfigures (e) and (f) shown in Plate section for color representation of this figure.*

8.8.2 Results Obtained using STSM and FEM for a Clamped Plate

Using the coefficients and formulae taken from the table relevant for a clamped plate (see Timoshenko and Woinowsky-Krieger 1959) the values of characteristic quantities for points C and S shown in Figure 8.15b have been calculated using the equations:

$$
\begin{aligned}
&w_{\mathrm{C}}^{\mathrm{STSM\,(B)}} = 0.00126\,\hat{p}\,L^4/D^m \\
&m_{x\mathrm{C}}^{\mathrm{STSM\,(B)}} = m_{y\mathrm{C}}^{\mathrm{STSM\,(B)}} = 0.0231\,\hat{p}\,L^2 \\
&m_{\nu\,\mathrm{S}}^{\mathrm{STSM\,(B)}} = -0.0513\,\hat{p}\,L^2
\end{aligned}
\tag{8.70}
$$

The values are given for case B in Box 8.5. In the analysed plate with all edges clamped and with positive deflection (downwards), the signs of extreme bending moments change from positive at the centre C to negative on the clamped edges. Approaching the clamped edges the absolute value of the bending moment rapidly increases and achieves the extremum in the middle of each edge (point S). Thus a finer mesh should be applied in the neighbourhood of the clamped edges.

The results of FEM computations are presented in Figure 8.17 and have been obtained using package ROBOT (2006) for two meshes with $NSE = 100$ and $NSE = 256$ FEs. In Box 8.5 selected values from numerical calculations are compared with the analytical results (note again different directions of axis z and different signs).

The following observations have to be taken into account when the reinforcement of a concrete plate is designed: (i) in the central part of the plate – on its bottom surface – tensile stresses occur in circumferential and radial directions and (ii) in the middle of the clamped boundaries in the direction normal to the edge tensile stresses occur on the top surface and compressive stresses on the bottom one.

We encourage the reader to complete this example by the analysis of the clamped plate with unequal dimensions, for example, $L_y/L_x = 2$. From an engineering point of view, one should observe the changes of the deflection and both bending moments at the centre, as well as the moments at the centres of the clamped edges.

8.9 Rectangular Plate with a Uniform Load and Various Boundary Conditions – Comparison of STSM and FEM Results

In this section, a plate with uniform load and various boundary conditions (two simply supported edges, one clamped and one free) is analysed. The considered configuration is shown in Figure 8.18. Such a diversity of boundary conditions is the reason that the analytical solution of the displacement equation of the bending plate is much more complicated since the following kinematic, static and mixed boundary conditions have to be satisfied:

- edge $x = 0, y \in [0, b]$: $w = 0, \quad \vartheta_x = 0$
- edge $x = a, y \in [0, b]$: $w = 0, \quad m_x = 0$
- edge $y = 0, x \in [0, a]$: $w = 0, \quad m_y = 0$
- edge $y = b, x \in [0, a]$: $m_y = 0, \quad \tilde{t}_y = 0$

Solutions of simple examples, coming from DTSM and STSM, and their superposition enable the authors of tables for engineers to work out more complicated examples, like the considered plate.

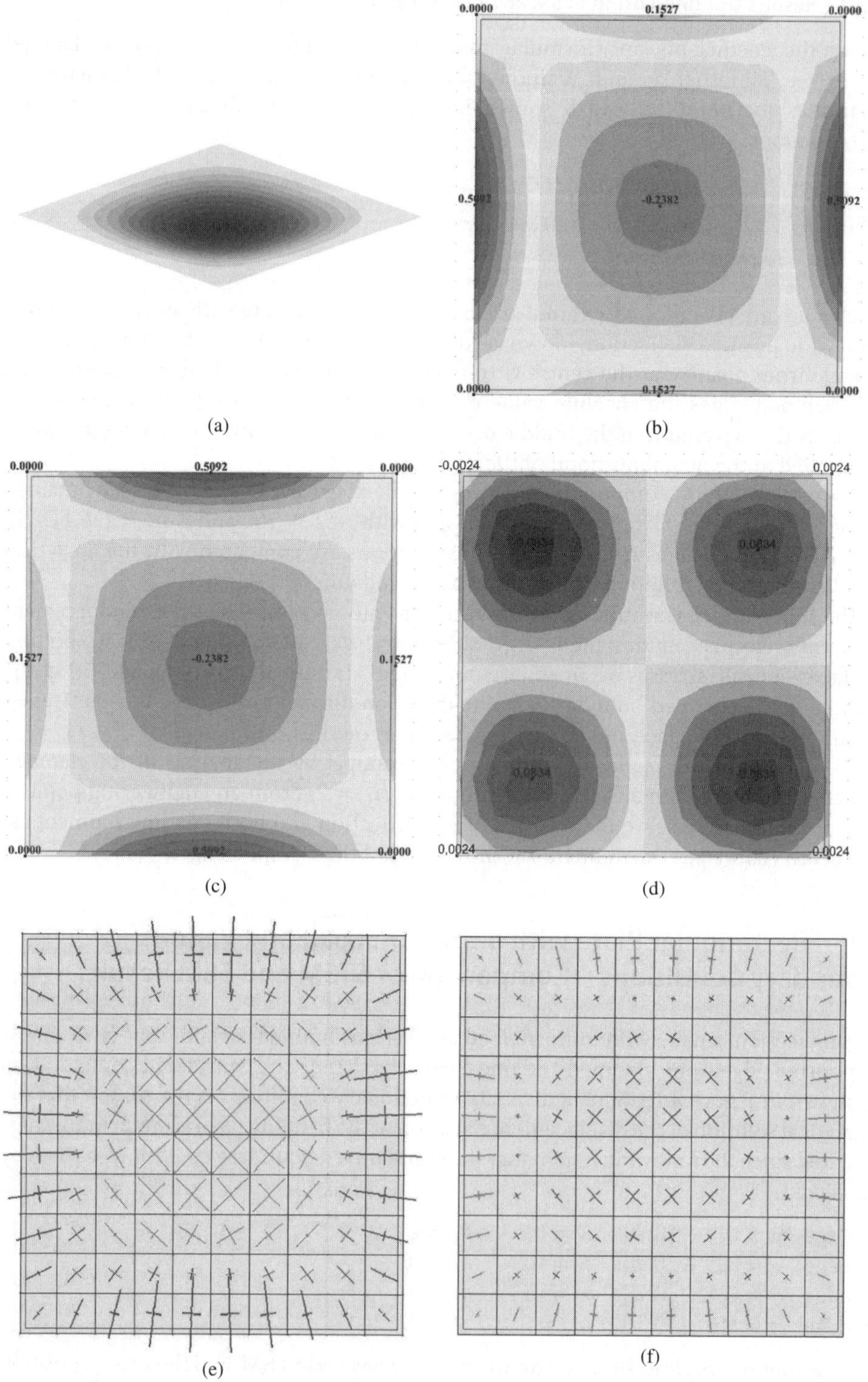

(a) (b) (c) (d) (e) (f)

Figure 8.17 Clamped square plate with a uniform load: (a) deflected surface $w(x, y)$, contour maps of moments [kNm/m]: (b) m_x, (c) m_y, (d) m_{xy}; vizualization of directions and magnitudes of principal stresses (for FEs mesh with $NSE = 100$) on two surfaces: (e) bottom and (f) top; FEM results from ROBOT. *Subfigures (e) and (f) shown in Plate section for color representation of this figure.*

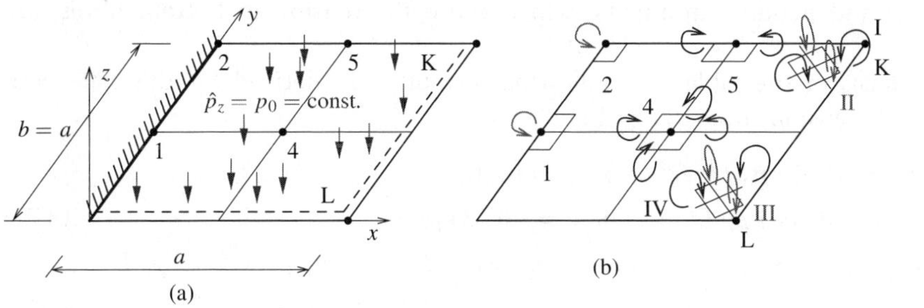

Figure 8.18 Plate with various boundary conditions: (a) configuration and (b) actual senses of bending moments at points (1–5, K, L)

The analysed plate is defined by the following parameters: material constants E, ν, dimensions a, b, h and load p_0. Their values are listed in Box 8.6. The adopted value of Young's modulus indicates that the tables for concrete plates must be used, see for example Starosolski (2009). Note that the z-axis is directed upwards, which affects the signs of load, deflection and bending moments. Thus, load and displacement directed

Box 8.6 Plate with a uniform load and various boundary conditions

Data

$E = 2.0 \times 10^7\ \text{kN/m}^2$, $\nu = 0.0$, $L = a = b = 3.0\ \text{m}$, $h = 0.15\ \text{m}$
$D^m = 5625.0\ \text{kNm}$, $p_0 = -10.0\ \text{kN/m}^2$

Check values from analytical solution

$w_5 = -0.000777\ \text{m}$, $|w_5| \approx 0.005h$
$m_{x1} = 9.82\ \text{kNm/m}$, $m_{x2} = 12.06\ \text{kNm/m}$
$m_{x4} = -4.61\ \text{kNm/m}$, $m_{y4} = -1.18\ \text{kNm/m}$, $m_{x5} = -6.02\ \text{kNm/m}$
$m_{\text{K}}^{\text{I}} = -0.423\ \text{kNm/m}$, $m_{\text{K}}^{\text{II}} = 0.423\ \text{kNm/m}$
$m_{\text{L}}^{\text{III}} = -3.82\ \text{kNm/m}$, $m_{\text{L}}^{\text{IV}} = 3.82\ \text{kNm/m}$

Check values from FEM solution using ROBOT

$NSE = 10 \times 10$, $NEN = 4$
$w_5^{\text{FEM}} = -0.000777\ \text{m} = w_5^{\text{anal}}$
$m_{x2}^{\text{FEM}} = 13.02\ \text{kNm/m} = 1.08\ m_{x2}^{\text{anal}}$, $m_{x5}^{\text{FEM}} = -6.24\ \text{kNm/m} = 1.04\ m_{x5}^{\text{anal}}$
$NSE = 30 \times 30$, $NEN = 4$
$w_5^{\text{FEM}} = -0.000769\ \text{m} = 0.990\ w_5^{\text{anal}}$
$m_{x2}^{\text{FEM}} = 11.51\ \text{kNm/m} = 0.954 m_{x2}^{\text{anal}}$, $m_{x5}^{\text{FEM}} = -6.06\ \text{kNm/m} = 1.007\ m_{x5}^{\text{anal}}$

Plate configuration shown in Figure 8.18a

downwards and bending moments, which cause the tension in bottom fibres, are negative.

On the basis of the tables the following formulae are derived for the plate with $L = a = b$, dimensions ratio $a/b = 1$ and $p_0 < 0$:

$$\begin{aligned} &w_5 = 0.0054\, p_0\, L^4/D^m, \quad w_{max} = |w_5| \\ &m_{x\,1} = -0.1091\, p_0\, L^2, \quad m_{x\,2} = -0.134\, p_0\, L^2 \\ &m_{x\,4} = 0.0513\, p_0\, L^2, \quad m_{y\,4} = 0.0131\, p_0\, L^2, \quad m_{x\,5} = 0.0669\, p_0\, L^2 \end{aligned} \tag{8.71}$$

The values of extremum deflection and bending moments at characteristic points 1–5 are included in Box 8.6.

Attention should be paid to the moments at corners K and L. The bending moments are equal to zero there but the twisting moments are not, see Figure 8.18b. The nonzero twisting moments are the source of principal bending moments at the two corners, with the directions marked as I, II at point K and III, IV at point L (it is a new notation, not related to maximum and minimum principal moments). The bending moments acting in directions I–IV are:

$$\begin{aligned} m^{II}_{K,max} &= -m^{I}_{K,min} = -0.0047\, p_0\, L^2 > 0\,, \\ m^{IV}_{L,max} &= -m^{III}_{L,min} = -0.0424\, p_0\, L^2 > 0, \end{aligned} \tag{8.72}$$

Their values are included in Box 8.6.

The presence of these moments affects the way in which the reinforcement is designed. For instance, at corner L (see Figure 8.18b), the reinforcement is arranged in the diagonal direction on the top surface and perpendicular to it on the bottom surface.

The FEM analysis has been performed using package ROBOT (2006). Two discretizations of the square plate with $NSE = 10 \times 10$ and $NSE = 30 \times 30$ are considered. In this example the z-axis is directed upwards both in the analytical and numerical calculations, which results in identical signs of transverse load, deflections and moments.

In FEM, when the displacement-based FE model is used, only kinematic constraints are imposed at the boundary nodes:

- edge $x = 0, y \in [0, b]$: $w = 0, \quad \vartheta_x = 0, \quad \vartheta_y = 0$
- edge $x = a, y \in [0, b]$: $w = 0, \quad (\vartheta_y = 0)$
- edge $y = 0, x \in [0, a]$: $w = 0, \quad (\vartheta_x = 0)$

Note that for a clamped edge ($x = 0$) not only is zero deflection required but also zero rotations in both directions – normal and tangential to edge. In the case of simply supported boundary ($x = a$ or $y = 0$) apart from the zero value of deflection w, the zero value of rotation in the tangent direction is requested. It is the so-called hard simple support in contrast to soft simple support where only $w = 0$.

In Figure 8.18b the real directions of bending moments for specified points are shown for $\hat{p}_z < 0$. The results of numerical analysis are presented (in a graphical form) in Figures 8.19 and 8.20. In Figure 8.19b the deformed middle surface of the plate is shown. Contour plots of the deflected surface, bending and twisting moments and a vizualization of directions and magnitudes of principal stresses are depicted in Figure 8.20.

Three quantities: extreme deflection w_5 and two bending moments m_{x2}, m_{x5} have been chosen to compare the results of the analytical and numerical calculations.

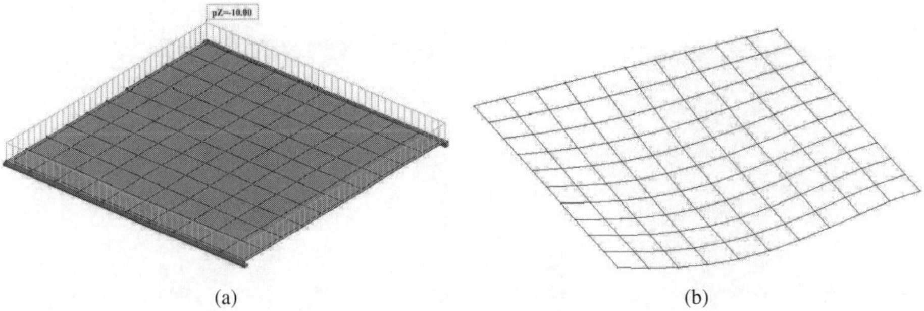

Figure 8.19 Plate with various boundary conditions along edges, (a) numerical model and (b) deformed middle surface; FEM results from ROBOT

In preliminary FEM computations two regular meshes of finite elements have been used. In Box 8.6 a comparison of the analytical and numerical check values is presented for both meshes (10 × 10 and 30 × 30). However, to obtain more accurate values of bending moments, it is recommended to generate a finer mesh in a neighbourhood of the clamped edge.

The behaviour of the considered plate is characterized by the following observations. At the centre of the plate (point 4) both bending moments (m_x and m_y) cause the tensile stress in the fibres at the bottom surface of the plate. In the middle of the free edge ($y = a$, point 5) the tensile stress at the bottom surface results from nonzero bending moment m_x ($m_y = 0$ at point 5). In the middle of the clamped edge (point 1) bending moment m_x results in tension at the top surface. At the corner (point 2) an extreme value of bending moment m_x is computed.

8.10 Uniformly Loaded Rectangular Plate with Clamped and Free Boundary Lines – Comparison of STSM and FEM Results

A balcony slab considered in this section is an example of a real engineering application. Attention is focused on the intensification of bending effects caused by the sudden change of boundary conditions: end of clamped part of the edge at points 4 and 4′, see Figure 8.21 and 8.22.

The considered plate is defined by the following parameters: material constants E, ν, dimensions L, h and surface load p_0. Their values are listed in Box 8.7. Note that for the z-axis directed upwards, the transverse load $\hat{p}_z = p_0 < 0$.

Using the tables for engineers Starosolski (2009), the values of bending moments at four characteristic points can be calculated from the following formulae:

$$\begin{aligned} &m_{y1} = 0.155\, p_0\, L^2, \quad m_{x2} = -0.005\, p_0\, L^2, \quad m_{y2} = 0.134\, p_0\, L^2 \\ &m_{x3} = -0.260\, p_0\, L^2, \quad m_{x4} = -0.500\, p_0\, L^2, \quad m_{y4,\max} = -1.031\, p_0\, L^2 \end{aligned} \tag{8.73}$$

The values of these moments for the adopted data are presented in Box 8.7. In this example the axes of the global coordinate systems in the analytical and numerical calculations have been chosen in such a way that axis z is directed upwards in both cases, but the other axes in ROBOT are perpendicular to the respective axes in the analytical solution. In consequence the following relations between the compared bending moments hold: $m_x^{\mathrm{FEM}} = m_y^{\mathrm{anal}}$, $m_y^{\mathrm{FEM}} = m_x^{\mathrm{anal}}$ (signs are the same).

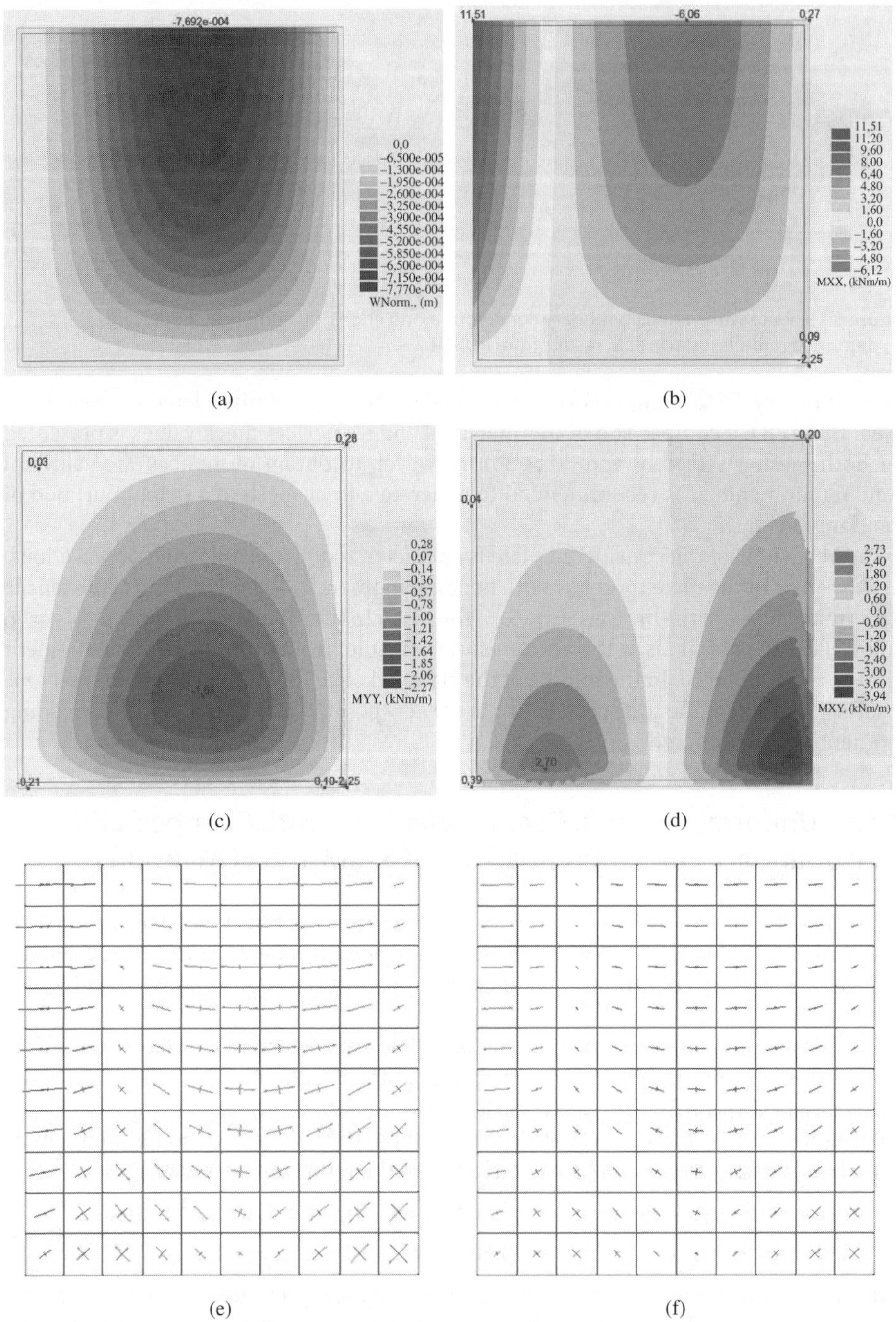

Figure 8.20 Plate with various boundary conditions, contour plots of: (a) deflected surface $w(x, y)$ [m], moments: (b) m_x, (c) m_y, (d) m_{xy} [kNm/m]; vizualization of directions and magnitudes of principal stresses (for FEs mesh with $NSE = 10 \times 10$) on two surfaces: (e) bottom and (f) top; FEM results from ROBOT

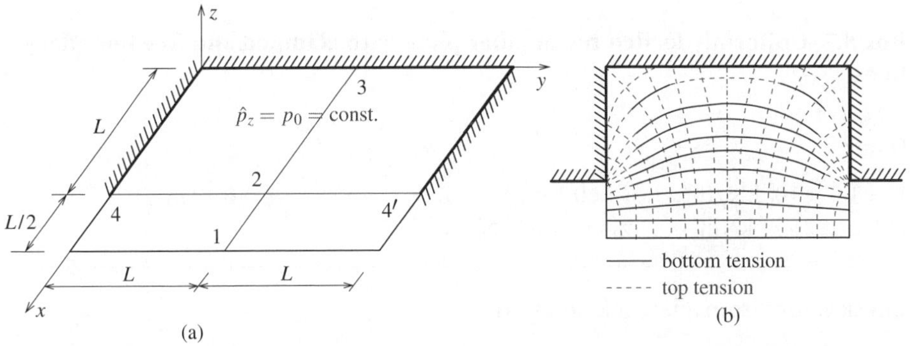

Figure 8.21 Plate with protruding cantilever part: (a) configuration and (b) engineering sketch of trajectory lines of principal moments (with tension lines at bottom and top)

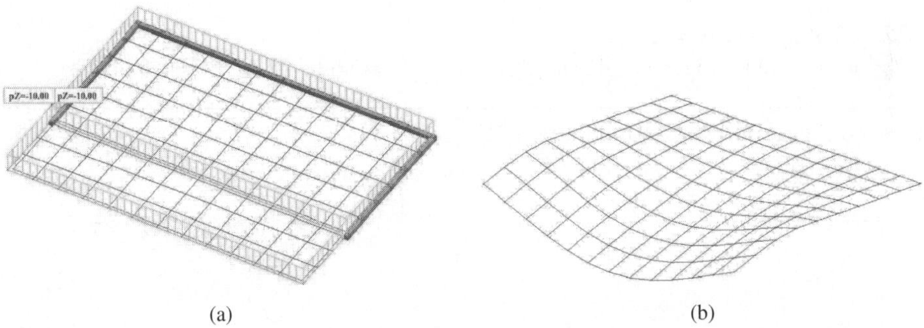

Figure 8.22 Plate with protruding cantilever part: (a) numerical model and (b) deformed middle surface; FEM results from ROBOT

The discontinuity of boundary conditions on two lines $y = 0$ and $y = 2\,L$ causes difficulties (singularity) in the analytical solution. On the contrary, in FEM analysis where the boundary is represented by a finite set of points it is easy to introduce arbitrary kind of constraints. In Figures 8.21b, 8.23 and 8.24 the local concentration of stresses (resulting from bending moments) can be noticed in the neighbourhood of points 4 and 4′. However, the crucial issue is the proper discretization in the regions of discontinuous boundary conditions. In the vicinity of the end points of the clamped lines a significant increase of moments occurs, which requires a local mesh refinement. On the other hand, one has to be very careful since FEs that are too small can lead to overestimation of extreme values.

On the basis of the results shown in Figure 8.24a,b a proper reinforcement of the plate can be designed. Appropriate densification of reinforcing bars in the neighbourhood of points 4 and 4′ is necessary.

The description of these three examples for plates under bending (Sections 8.8–8.10) ends by emphasizing the necessity: (i) to check the orientation of the global coordinate system in the analytical calculations (tables for engineers) and in numerical computations (using some FE code) and, additionally, (ii) to pay attention to the notation

Box 8.7 Uniformly loaded rectangular plate with clamped and free boundary lines

Data

$E = 1.5 \times 10^7\ \text{kN/m}^2,\quad \nu = 0.1667,\quad L = 3.0\ \text{m},\quad h = 0.12\ \text{m}$
$D^m = 2221.71\ \text{kNm},\quad p_0 = -10.0\ \text{kN/m}^2$

Check values from analytical solution

$m_{y1} = -13.95\ \text{kNm/m},\quad m_{x2} = 0.45\ \text{kNm/m},\quad m_{y2} = -12.06\ \text{kNm/m}$
$m_{x3} = 23.40\ \text{kNm/m},\quad m_{x4} = 45.00\ \text{kNm/m}\quad m_{y4} = 92.79\ \text{kNm/m}$

Check values from FEM solution using ROBOT

$NSE = 9 \times 12,\ \ NEN = 4$
$m_{x2}^{\text{FEM}} = -13.52\ \text{kNm/m} = 1.121 m_{y2}^{\text{anal}},\quad m_{x4}^{\text{FEM}} = 86.67\ \text{kNm/m} = 0.934 m_{y4}^{\text{anal}}$

Plate configuration shown in Figure 8.21

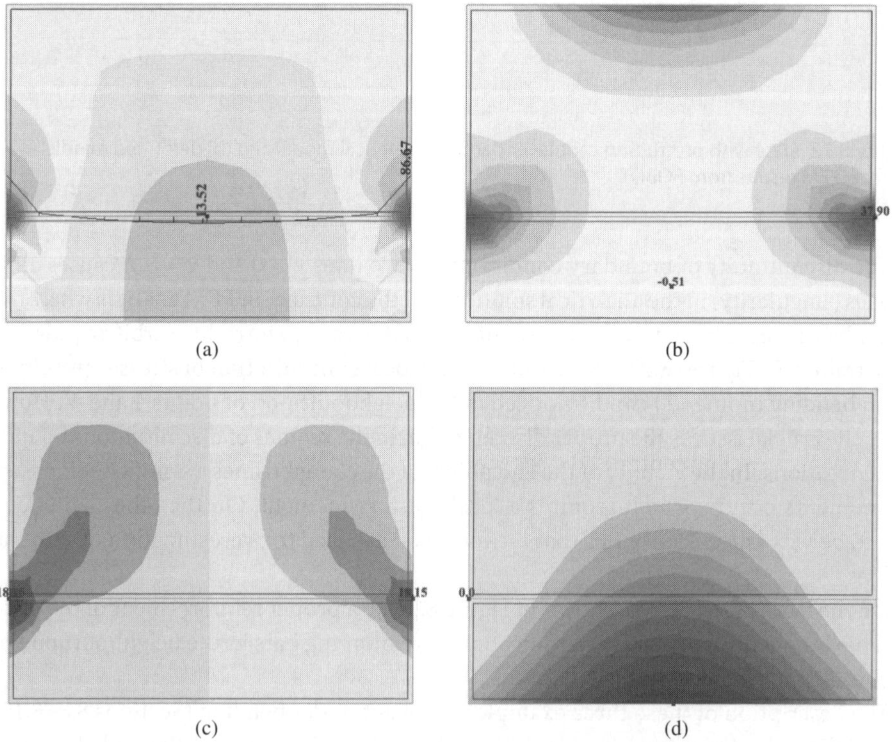

Figure 8.23 Plate with protruding cantilever part – contour plots of: (a) bending moment $m_x^{\text{FEM}} = m_y^{\text{anal}}$, (b) bending moment $m_y^{\text{FEM}} = m_x^{\text{anal}}$, (c) twisting moment m_{xy} [kNm/m] and (d) deflection $w(x, y)$; FEM results from ROBOT

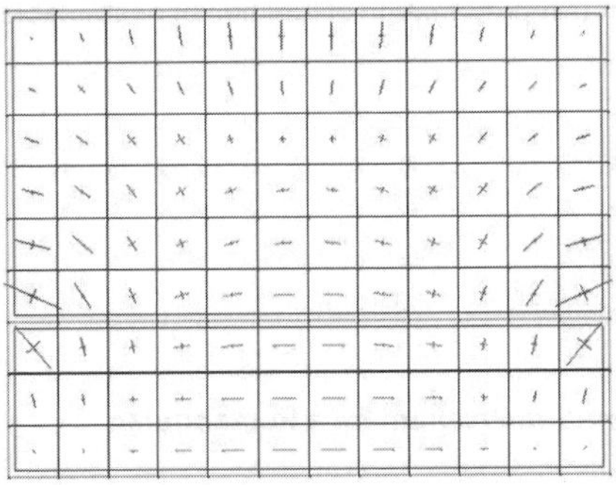

(a)

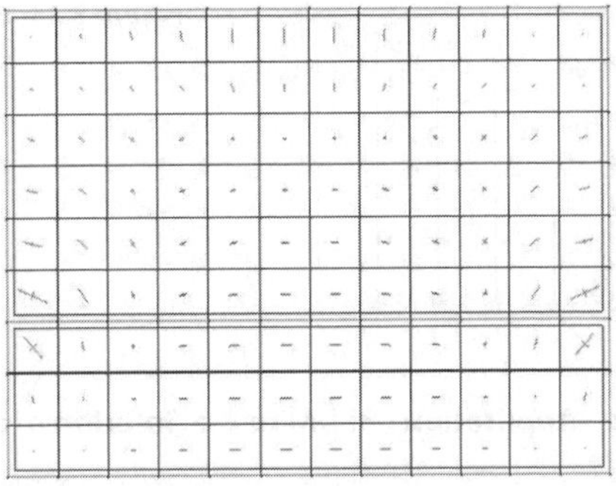

(b)

Figure 8.24 Plate with protruding cantilever part, vizualization of directions and magnitudes of principal stresses (for FE mesh with $NSE = 9 \times 12$) on limiting surfaces: (a) bottom and (b) top; FEM results from ROBOT

and sign convention for forces and moments (related to a local or global coordinate system) defined in books or user manuals of the applied computer codes.

8.11 Approximate Solution to a Plate Bending Problem using FDM

8.11.1 Idea of FDM

The finite difference method (FDM) is a numerical method used to solve differential equations. All derivatives that appear in the problem formulation are approximated by finite difference quotients, which are written as appropriate linear combinations of function values at the grid points.

When the bending plate is considered its domain has to be represented by a finite set of grid points. Then, the derivatives which appear in: (i) displacement differential equations, (ii) relations between generalized resultant forces and deflection function and (iii) relations resulting from boundary conditions, have to be replaced by suitable difference quotients. As a result, the numerical model in the form of a linear system of equations with unknown values of the deflection function at the grid points is obtained.

Before the FDM is applied to the two-dimensional problem of a rectangular bending plate the difference quotients for four derivatives of the function of one variable $f(x)$ are introduced using central difference formulae with constant distance between points. Next, the formulae for approximated (first, second, third and fourth) derivatives of function $f(x)$ calculated at the central point i are listed (the notation shown in Figure 8.25 is employed):

$$
\begin{aligned}
f_i' &\approx \frac{1}{2(\Delta x)}[(-1)f_{i-1} + (+1)f_{i+1}] \\
f_i'' &\approx \frac{1}{(\Delta x)^2}[(+1)f_{i-1} + (-2)f_i + (+1)f_{i+1}] \\
f_i''' &\approx \frac{1}{2(\Delta x)^3}[(-1)f_{i-2} + (+2)f_{i-1} + (-2)f_{i+1} + (+1)f_{i+2}] \\
f_i^{\mathrm{IV}} &\approx \frac{1}{(\Delta x)^4}[(+1)f_{i-2} + (-4)f_{i-1} + (+6)f_i + (-4)f_{i+1} + (+1)f_{i+2}]
\end{aligned}
\tag{8.74}
$$

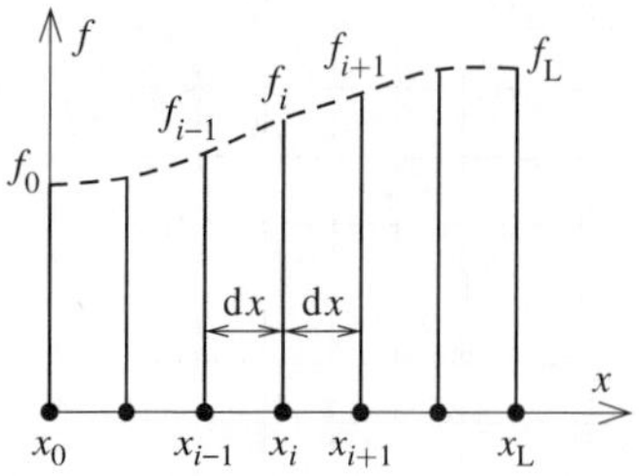

Figure 8.25 Discrete representation of function of one variable

8.11.2 Application of FDM to the Solution of a Bending Problem for a Rectangular Plate

In the local formulation the behaviour of bending plate is described using:

- the fourth-order displacement differential equation with partial derivatives of the deflection function

$$\nabla^2\nabla^2 w(x,y) = \hat{p}_z/D^m \tag{8.75}$$

- coupled constitutive and kinematic equations, which allow one to find the distributions of bending and twisting moments on the basis of deflection values determined earlier:

$$\begin{aligned} m_x &= -D^m(w_{,xx} + \nu w_{,yy}), \qquad m_y = -D^m(w_{,yy} + \nu w_{,xx}) \\ m_{xy} &= -D^m(1-\nu)w_{,xy} \end{aligned} \tag{8.76}$$

The formulation has to be completed by adequate boundary conditions.

An example of a rectangular plate with various boundary conditions is given (Figure 8.26):

- edge $x = 0, y \in [0, b]$: $w = 0, \quad \vartheta_x = -w_{,x} = 0$
- edge $x = a, y \in [0, b]$: $m_x = -D^m(w_{,xx} + \nu\, w_{,yy}) = 0$, $\tilde{t}_x = -D^m\,[w_{,xxx} + (2-\nu)w_{,yyx}] = 0$
- edge $y = 0, x \in [0, a]$: $w = 0, \quad m_y = -D^m(w_{,yy} + \nu\, w_{,xx}) = 0$
- edge $y = b, x \in [0, a]$: $m_y = -D^m(w_{,yy} + \nu\, w_{,xx}) = 0$, $\tilde{t}_y = -D^m\,[w_{,yyy} + (2-\nu)w_{,xxy}] = 0$
- corner $x = a,\ y = b$: $T = -2\,D^m(1-\nu)w_{,xy} = 0$

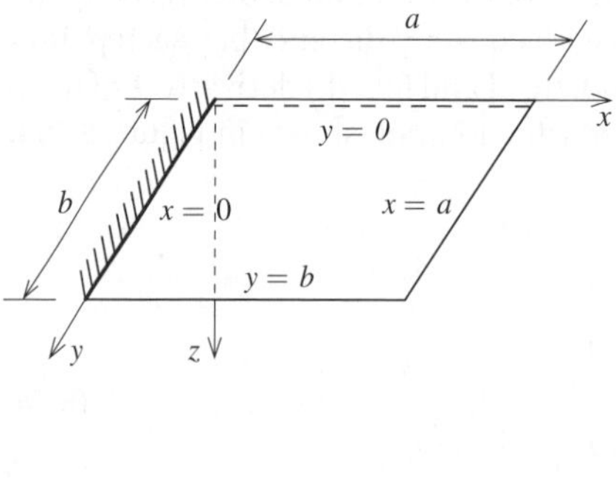

Figure 8.26 Rectangular plate under bending with various boundary constraints

Box 8.8 FDM – solution algorithm for a plate bending problem

1) Replacement of the given plate domain by a grid of points (generally, the grid is formed by a set of lines in x and y directions) with a constant distance between them $\Delta x = \Delta y = \lambda$
2) Formulation of boundary conditions
3) Replacement of all derivatives (in the main displacement differential equation, relations between generalized resultant forces and deflection function and in boundary conditions) by appropriate difference quotients
4) Formulation of the system of linear algebraic equations including: (i) differential equation of bending plate written for all grid points from the plate domain with nonzero deflections and (ii) algebraic equations corresponding to boundary conditions
5) Solution of the system of linear algebraic equations, i.e. computation of values of deflection at the grid points
6) Calculation of values of bending moments m_x, m_y and twisting moments m_{xy} at grid points
7) Plots of approximated distribution of w, m_x, m_y, m_{xy} drawn on the basis of appropriate values known at grid points

In Figure 8.26 three types of boundaries are shown: clamped, simply supported and free edge. The first two are discussed in detail later in this subsection. In the cases of free edges $x = a$ and $y = b$ and not supported corner $x = a, y = b$ the consideration of boundary conditions is more complicated and so-called stars of difference quotient coefficients presented in Figures 8.29 and 8.30 should be used. The thorough discussion of these boundary conditions is omitted.

The algorithm of an approximate solution for the plate bending problem using FDM is presented in Box 8.8.

The replacement of partial derivatives occurring in the displacement differential equation for the bending plate

$$\nabla^2\nabla^2 w(x, y) = w_{,xxxx} + 2\, w_{,xxyy} + w_{,yyyy} = \hat{p}_z / D^m \tag{8.77}$$

requires the use of 12 points around the central point (with indices i, k) from the two-dimensional grid of points. Equation (8.77) can be presented in a graphical form with coefficients by which the deflection values at grid points must be multiplied. Omitting the detailed derivation, the star of difference quotient coefficients is shown in Figure 8.27. Notice that the range of the star for $\nabla^2\nabla^2 w$ is equal to 4λ in both the x and y directions. The star coefficients are related to the numerator of the difference quotient whereas in the denominator, λ^4 appears for a square grid of points.

$$\nabla^2\nabla^2 w|_{i,k} = (w_{,xxxx} + 2w_{,xxyy} + w_{,yyyy})|_{i,k} \approx$$

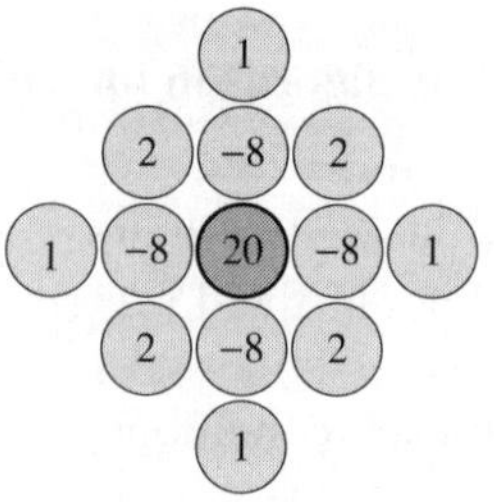

$$\times \frac{1}{\lambda^4} = \frac{p_{i,k}}{D^m}$$

Figure 8.27 Graphical representation of a fourth-order differential equation for a plate under bending, adopted in the finite difference method

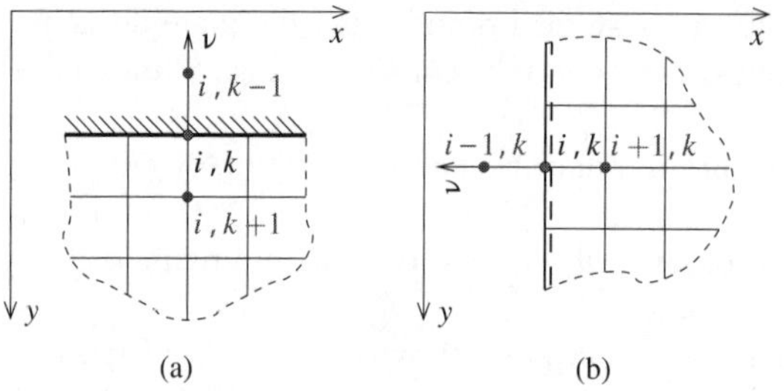

Figure 8.28 Grid points in vicinity of two types of plate edge: (a) clamped and (b) simply supported

The detailed discussion of the notation of boundary conditions in the FDM convention is limited to two types of supports:

Case A – clamped edge, see Figure 8.28a

For a clamped edge (with direction ν parallel to axis y) the following two kinematic conditions must be satisfied:

$$w = 0, \qquad \vartheta_y = -\frac{\partial w}{\partial y} = 0 \tag{8.78}$$

Introducing the FDM grid we obtain at the point with indices i, k on the boundary line:

$$w_{i,k} = 0, \quad \vartheta_y|_{i,k} = -\left.\frac{\partial w}{\partial y}\right|_{i,k} = -\frac{1}{2\Delta y}[(-1)\ w_{i,k-1} + (+1)\ w_{i,k+1}] = 0 \tag{8.79}$$

and finally the following relationships are obtained:

$$w_{i,k} = 0, \qquad w_{i,k+1} = w_{i,k-1} \tag{8.80}$$

Case B – simply supported edge, see Figure 8.28b

For a simply supported edge, not loaded by moment $\hat{m}_\nu$, two homogeneous mixed boundary conditions are postulated:

$$w = 0, \qquad m_\nu = -D^m(w_{,\nu\nu} + \nu\ w_{,ss}) = 0 \tag{8.81}$$

In the case when the direction of the normal to an edge is parallel to axis x, the boundary conditions for FDM are written as:

$$w_{i,k} = 0, \qquad m_\nu|_{i,k} = m_x|_{i,k} = -D^m(w_{,xx} + \nu\ w_{,yy})|_{i,k} = 0 \tag{8.82}$$

We assume $w_{,ss} = w_{,yy} = 0$, hence the relation for moment m_x is reduced to

$$\left.\frac{\partial^2 w}{\partial x^2}\right|_{i,k} \approx \frac{1}{\lambda^2}[(+1)\ w_{i-1,k} + (-2)\ w_{i,k} + (+1)\ w_{i+1,k}] = 0 \tag{8.83}$$

(a)

$$m_x|_{i,k} = -D^m(w_{,xx} + \nu w_{,yy})|_{i,k} \approx \boxed{-1}\,\boxed{2+2\nu}\,\boxed{-1} \times \frac{D^m}{\lambda^2}$$

(b)

$$m_y|_{i,k} = -D^m(w_{,yy} + \nu w_{,xx})|_{i,k} \approx \boxed{-\nu}\,\boxed{2+2\nu}\,\boxed{-\nu} \times \frac{D^m}{\lambda^2}$$

(c)

$$m_{xy}|_{i,k} = -D^m(1-\nu)w_{,xy}|_{i,k} \approx \qquad \times \frac{D^m(1-\nu)}{4\lambda^2}$$

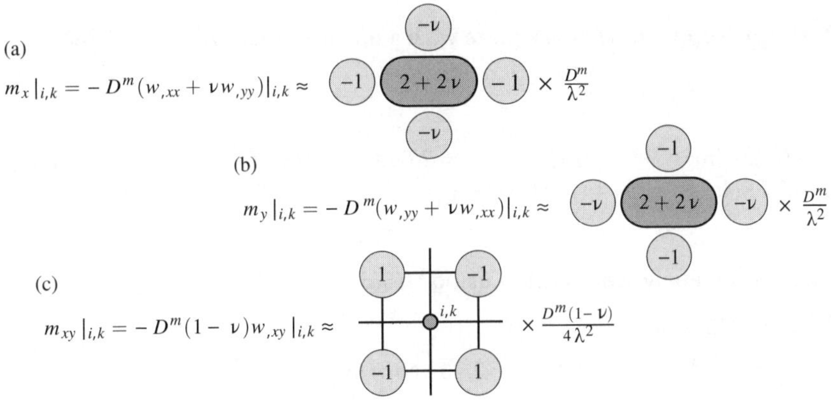

Figure 8.29 Graphical representation of difference formulae for moments: (a) $m_x|_{i,k}$, (b) $m_y|_{i,k}$ and (c) $m_{xy}|_{i,k}$

$$t_x|_{i,k} = D^m(w_{,xxx} + \nu w_{,yyx})|_{i,k} \approx \qquad \times \frac{D^m}{2\lambda^3}$$

(a)

$$\tilde{t}_x|_{i,k} = [t_x + (m_{xy})_{,y}]\,|_{i,k} \approx \qquad \times \frac{D^m}{\lambda^3}$$

(b)

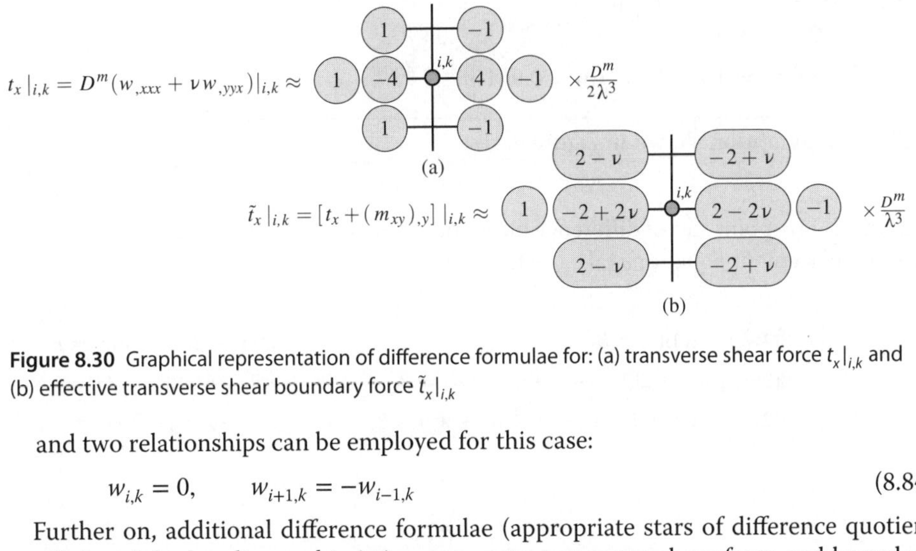

Figure 8.30 Graphical representation of difference formulae for: (a) transverse shear force $t_x|_{i,k}$ and (b) effective transverse shear boundary force $\tilde{t}_x|_{i,k}$

and two relationships can be employed for this case:

$$w_{i,k} = 0, \qquad w_{i+1,k} = -w_{i-1,k} \tag{8.84}$$

Further on, additional difference formulae (appropriate stars of difference quotient coefficients) for bending and twisting moments, transverse shear force and boundary effective transverse force (Figures 8.29 and 8.30) are presented.

8.11.3 Simply Supported Square Plate with a Uniform Load

The considered plate is described by the following parameters: material constants E, ν, dimensions a, h and surface load p_0. Their values are listed in Box 8.9.

The model of the plate is shown in Figure 8.31. The regular grid of points with constant distance between points $\lambda = \Delta x = \Delta y = a/4$ is introduced. The point numbering takes into account: (i) four symmetry planes and (ii) relations between internal and external points resulting from boundary conditions. The boundary points on the supported edges have zero deflections, and the external points in the vicinity of the simply supported lines have the deflections of the opposite sign in relation to the neighbouring inner points. The external (fictitious) points are given the minus sign in front of the numbers on the scheme in Figure 8.31b. These relations result from the mixed boundary conditions ($w = 0$, $m_\nu = 0$) described previously. In fact, the solution of the plate requires the calculation of deflections for three internal points only.

Box 8.9 Simply supported square plate with a uniform load (DTSM, FDM)

Data

$E = 1.5 \times 10^7\ \text{kN/m}^2$, $\nu = 0.15$, $a = 2.0\ \text{m}$, $h = 0.1\ \text{m}$
$D^m = 1278.77\ \text{kNm}$, $p_0 = 100.0\ \text{kN/m}^2$

Check values from analytical solution using DTSM

$w_{\max}^{\text{DTSM,anal}} = 0.005079\ \text{m}$

$m_{\max}^{\text{DTSM,anal}} = m_{x1}^{\text{DTSM,anal}} = m_{y1}^{\text{DTSM,anal}} = 17.20\ \text{kNm/m}$

Check values from FDM solution

$\Delta x = \Delta y = \lambda = a/4 = 0.5\ \text{m}$
$w_{\max}^{\text{FDM}} = 0.005043\ \text{m} = 0.993\ w_{\max}^{\text{DTSM,anal}} \approx 0.05\ h$

$m_{x1}^{\text{FDM}} = m_{y1}^{\text{FDM}} = 16.18\ \text{kNm/m} = 0.941\ m_{\max}^{\text{DTSM,anal}}$

Plate configuration shown in Figure 8.31

Using the 13-point star of difference quotient coefficients, as shown in Figure 8.27, the three displacement equations take the form:

$$
\begin{aligned}
&20w_1 - 8(4w_2) + 2(4w_3) = B_1 && \rightarrow && 20w_1 - 32w_2 + 8w_3 = B \\
&20w_2 - 8(2w_3 + w_1) + 2(2w_2) + 1(-w_2 + w_2) = B_2 && \rightarrow && -8w_1 + 24w_2 - 16w_3 = B \\
&20w_3 - 8(2w_2) + 2w_1 + 1(-w_3 - w_3 + w_3 + w_3) = B_3 && \rightarrow && 2w_1 - 16w_2 + 20w_3 = B
\end{aligned}
\tag{8.85}
$$

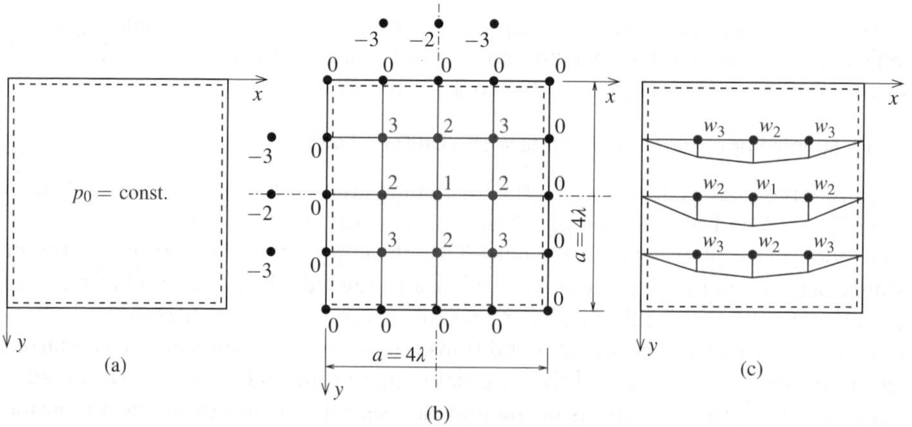

Figure 8.31 Simply supported square plate with uniform load: (a) configuration, (b) set of grid points for FDM analysis and (c) visualization of deflection function on the basis of a finite set of deflection values

These equations can be rewritten in a matrix form:

$$\mathbf{A}\,\mathbf{W} = \mathbf{B} \tag{8.86}$$

$$\begin{bmatrix} 20 & -32 & 8 \\ -8 & 24 & -16 \\ 2 & -16 & 20 \end{bmatrix} \begin{bmatrix} w_1 \\ w_2 \\ w_3 \end{bmatrix} = \begin{bmatrix} B \\ B \\ B \end{bmatrix} \tag{8.87}$$

In each of the three equations the right-hand side is the same

$$B = B_1 = B_2 = B_3 = p_0\ \lambda^4/D^m = 0.00489\ \ \text{m} \tag{8.88}$$

From the solution of the set of equations, the following values of deflections are obtained:

$$w_1 = 0.005043\ \text{m}, \qquad w_2 = 0.003668\ \text{m}, \qquad w_3 = 0.002674\ \text{m} \tag{8.89}$$

In addition, the values of bending moments can be calculated, for example for grid point 1 using relevant difference formula

$$m_{x\,1} = m_{y\,1} = D^m[(2+2\nu)\ w_1 - 1(2\ w_2) - \nu(2\ w_2)]/\lambda^2 = 16.18\ \ \text{kNm/m} \tag{8.90}$$

The values of $w_{\max}$ and $m_{\max}$ obtained using FDM are compared with the ones calculated with DTSM and listed in Box 8.9.

The comparison of the results shows that, even for a coarse grid in FDM, the approximation of the solution is quite satisfactory.

On the basis of the values of deflections calculated at grid points, the approximated surface of the deformed plate is sketched in the form of segment lines along three cross sections, see Figure 8.31c.

8.11.4 Simply Supported Uniformly Loaded Square Plate Resting on a One-Parameter Elastic Foundation

The plate considered in the previous subsection is now analysed with the assumption that it rests on fine-grained sand. For simplicity, according to Winkler's assumption, the subsoil is described by one coefficient k given in Box 8.10. The analysed configuration is presented in Figure 8.32.

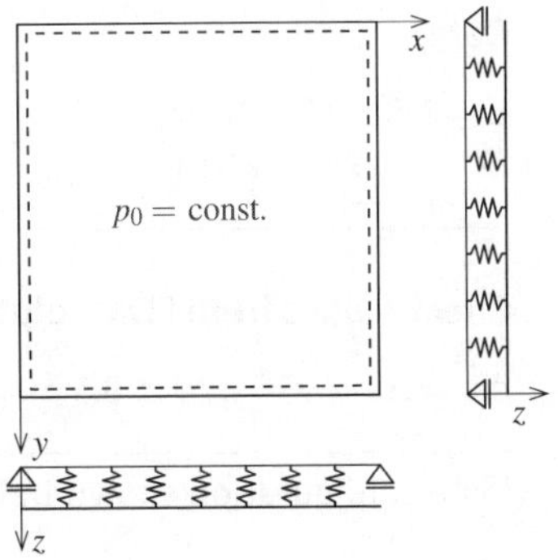

Figure 8.32 Simply supported square plate under bending resting on elastic foundation

The reaction of the elastic subsoil is calculated as a substitute load according to

$$\tilde{p}_z(x, y\ ; k) = -k\ w(x, y) \tag{8.91}$$

The modified displacement equation for that case takes the form

$$\nabla^2\nabla^2 w + \left(\frac{k}{D^m}\right) w = \frac{\hat{p}_z}{D^m} \tag{8.92}$$

The additional component on the left-hand side of the equation takes the interaction with the subsoil into account. The new coefficient K is defined as

$$K = \frac{k}{D^m} = \frac{50\,000}{1278.77} = 39.100\ \text{m}^{-4} \tag{8.93}$$

The grid of points introduced in the previous section is employed to solve the problem again using FDM. The set of three algebraic equations is modified by the addition of component $K\ \lambda^4$ to diagonal elements

$$\tilde{A}_{ii} = A_{ii} + K\ \lambda^4 = A_{ii} + 39.100 \times 0.5^4 = A_{ii} + 2.44 \tag{8.94}$$

Eventually it takes the form:

$$\tilde{\mathbf{A}}\ \mathbf{W} = \mathbf{B} \tag{8.95}$$

$$\begin{bmatrix} 22.44 & -32 & 8 \\ -8 & 26.44 & -16 \\ 2 & -16 & 22.44 \end{bmatrix} \begin{bmatrix} w_1 \\ w_2 \\ w_3 \end{bmatrix} = \begin{bmatrix} 0.00489 \\ 0.00489 \\ 0.00489 \end{bmatrix} \tag{8.96}$$

and new calculated values of deflection are:

$$w_1 = 0.001741\ \text{m}, \qquad w_2 = 0.001319\ \text{m}, \qquad w_3 = 0.001003\ \text{m} \tag{8.97}$$

The value of maximum deflection is presented in Box 8.10. It indicates the significant reduction of deflection of the plate caused by the subsoil resistance.

Box 8.10 Simply supported square plate with a uniform load resting on one-parameter elastic foundation

Data

$E = 1.5 \times 10^7\ \text{kN/m}^2$, $\quad \nu = 0.15$, $\quad a = 2.0\ \text{m}$, $\quad h = 0.1\ \text{m}$
$D^m = 1278.77\ \text{kNm}$, $\quad p_0 = 100.0\ \text{kN/m}^2$, $\quad k = 50\,000\ \text{kN/m}^3$

Check values from FDM solution

$\Delta x = \Delta y = \lambda = a/4 = 0.5\ \text{m}$, $\quad w_{\max}^{\text{FDM}} = 0.001741\ \text{m}$

Plate configuration shown in Figure 8.32

8.12 Approximate Solution to a Bending Plate Problem using the Ritz Method

8.12.1 Idea of the Ritz Method

The Ritz method is based on the principle of minimum total potential energy of a mechanical system, which reads:

Among kinematically admissible displacement fields (satisfying kinematic boundary conditions) the solution fulfilling the conditions of equilibrium of the system is the one for which the total potential energy functional Π^m reaches its minimum.

The necessary condition for minimum of the total potential energy is the zero value of its first variation

$$\delta\, \Pi^m = 0 \tag{8.98}$$

The approximate solution of the bending plate problem is sought assuming the following form of the deflection function

$$w_N(x, y) = \sum_{i=1}^{N} W_i\, \psi_i(x, y) \tag{8.99}$$

where: W_i – unknown scalar multipliers, called degrees of freedom of the solution; they are the coefficients of the linear combination of base functions $\psi_i(x, y)$, which are, for example, polynomials of the appropriate degree in both x and y.

The admissible base functions must be at least of continuity class C^2, linearly independent and should ensure the satisfaction of kinematic boundary conditions.

Analysing the energy functional Π^m expressed using the proposed functions $w_N(x, y)$ dependent on N degrees of freedom W_i, for $i = 1, \ldots, N$, the condition of minimum of the function $\Pi^m(W_1, W_2, \ldots, W_N)$ is considered

$$\delta\Pi^m = \frac{\partial\Pi^m}{\partial W_1}\,\delta W_1 + \frac{\partial\Pi^m}{\partial W_2}\,\delta W_2 + \cdots + \frac{\partial\Pi^m}{\partial W_N}\,\delta W_N = 0 \tag{8.100}$$

Condition $\delta\Pi^m = 0$ has to be satisfied for any variation δW_i, which leads to a system of N linear equations

$$\frac{\partial\Pi^m}{\partial W_i} = 0 \quad \rightarrow \quad \sum_{j=1}^{N} a_{ij}\, W_j = b_i, \quad \text{for} \quad i = 1, 2, \ldots, N \tag{8.101}$$

The total bending energy for rectangular bending plates under surface load (for simplicity without boundary tractions) is written in the following form – Equations (8.22)–(8.24) are rewritten for the completeness of the description of the problem:

$$\begin{aligned}
\Pi^m &= U^m - W^m \\
U^m &= \frac{D^m}{2}\int_A [(w_{,xx})^2 + (w_{,yy})^2 + 2\,\nu\, w_{,xx}\, w_{,yy} + 2(1-\nu)(w_{,xy})^2]\; \mathrm{d}x\,\mathrm{d}y \\
&= \frac{D^m}{2}\int_A \{(\nabla^2 w)^2 - 2(1-\nu)[w_{,xx}\, w_{,yy} - (w_{,xy})^2]\}\; \mathrm{d}x\,\mathrm{d}y \\
W^m &= \int_A \hat{p}_z(x, y)\; w(x, y)\; \mathrm{d}x\,\mathrm{d}y
\end{aligned} \tag{8.102}$$

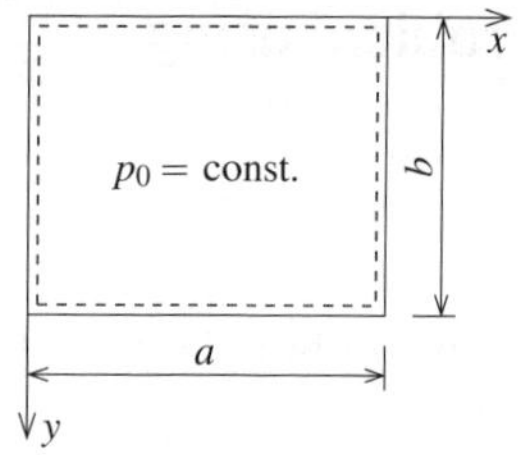

Figure 8.33 Simply supported rectangular plate with uniform load

In special cases of rectangular plate with infinitely stiff support and rectangular plate clamped at the boundaries, the expression for bending energy reduces significantly and takes the form

$$U^m = \frac{D^m}{2} \int_A (\nabla^2 w)^2 \, dx \, dy \tag{8.103}$$

Assuming the simplest base function with one degree of freedom $N = 1$, one algebraic equation is obtained:

$$\begin{aligned} a_{11} \, W_1 &= b_1, \quad \text{where:} \\ a_{11} &= D^m \int_A (\psi_{1,xx} + \psi_{1,yy})(\psi_{1,xx} + \psi_{1,yy}) \, dx \, dy \\ b_1 &= \int_A \hat{p}_z \, \psi_1 \, dx \, dy \end{aligned} \tag{8.104}$$

8.12.2 Simply Supported Rectangular Plate with a Uniform Load

The considered plate, shown in Figure 8.33, is described by the following parameters: material constants E, ν, dimensions a, b, h and surface load p_0. Their values are listed in Box 8.11.

The purpose of the calculation is to find the maximum deflection of the plate. It can be estimated with acceptable accuracy using one kinematically admissible base function

Box 8.11 Simply supported square plate with a uniform load (DTSM, Ritz)

Data

$E = 1.5 \times 10^7$ kN/m^2, $\quad \nu = 0.15, \quad L = a = b = 2.0$ m, $\quad h = 0.1$ m
$D^m = 1278.77$ kNm, $\quad p_0 = 100.0$ kN/m^2

Check values from the analytical – DTSM solution

$w_{max}^{\text{DTSM, anal}} = 0.005079$ m

Check values from solution based on the Ritz method

$w_{max}^{\text{Ritz}} = W_1 = 0.005206 \text{ m} = 1.025 w_{max}^{\text{DTSM, anal}}$

Plate configuration shown in Figure 8.33

$\psi_1(x,y)$ with one degree of freedom W_1

$$w(x,y) = W_1\,\psi_1(x,y) = W_1\,\sin\frac{\pi\,x}{a}\,\sin\frac{\pi\,y}{b} \tag{8.105}$$

The adopted base function ψ_1 and thus the deflection function $w(x,y)$ satisfy kinematic boundary conditions $w = 0$ along all edges of the plate. It can be easily noticed that for four lines $x = 0, x = a, y = 0, y = b$ both functions $\psi_1(x,y)$ and $w(x,y)$ are equal to zero.

Now, the detailed algorithm of calculations is presented. For a start, one has to write an expression for the total energy Π^m. Then, using the principle of minimum total potential energy the derivative of Π^m is calculated with respect to one degree of freedom W_1 and the obtained formula is equated to zero $\partial\Pi^m/\partial W_1 = 0$:

$$\begin{aligned}\Pi^m &= \frac{D^m}{2}\int_A W_1^2\left(\frac{\pi^2}{a^2}+\frac{\pi^2}{b^2}\right)^2\left(\sin\frac{\pi\,x}{a}\right)^2\left(\sin\frac{\pi\,y}{b}\right)^2 \mathrm{d}x\,\mathrm{d}y\\ &\quad - W_1\,p_0\int_A \sin\frac{\pi\,x}{a}\,\sin\frac{\pi\,y}{b}\,\mathrm{d}x\,\mathrm{d}y\end{aligned} \tag{8.106}$$

$$\begin{aligned}&\frac{\partial\Pi^m}{\partial W_1} = \left[D^m\,\pi^4\left(\frac{1}{a^2}+\frac{1}{b^2}\right)^2\frac{a\,b}{4}\right]W_1 - \frac{4}{\pi^2}\,p_0\,a\,b = 0\\ &a_{11}W_1 - b_1 = 0, \qquad W_1 = \frac{b_1}{a_{11}} = \frac{16}{\pi^6\left(1+\frac{a^2}{b^2}\right)^2}\,\frac{p_0\,a^4}{D^m}\\ &\to \quad w(x,y) = W_1\,\sin\frac{\pi\,x}{a}\,\sin\frac{\pi\,y}{b}\end{aligned} \tag{8.107}$$

For a square plate with $L = a = b$ the solution is given as:

$$\begin{aligned}a_{11} &= D^m\int_A(\psi_{1,xx}+\psi_{1,yy})(\psi_{1,xx}+\psi_{1,yy})\,\mathrm{d}x\,\mathrm{d}y\\ &= D^m\,\pi^4\left(\frac{1}{L^2}+\frac{1}{L^2}\right)^2\frac{L^2}{4} = 1278.77\,\pi^4\left(\frac{1}{2.0^2}+\frac{1}{2.0^2}\right)^2\frac{2.0^2}{4}\\ &= 31\,140.96\ \text{kN/m}\\ b_1 &= \int_A \hat{p}_z\,\psi_1\,\mathrm{d}x\,\mathrm{d}y = \frac{4}{\pi^2}\,p_0\,L^2 = \frac{4}{\pi^2}\,100.0\times 2.0^2 = 162.11\ \text{kN}\end{aligned} \tag{8.108}$$

The formula for maximum deflection at the centre of the plate reads

$$w_{\max} = w\left(\frac{L}{2},\frac{L}{2}\right) = W_1 = \frac{b_1}{a_{11}} = \frac{4}{\pi^6}\,\frac{p_0\,L^4}{D^m} = 0.00416\,\frac{p_0\,L^4}{D^m} \tag{8.109}$$

The obtained value and its comparison to the results of the double series method (DTSM) can be found in Box 8.11.

8.13 Plate with Variable Thickness

A plate with two subdomains and a jump of the thickness between them is considered. In practice, in such a plate, the top surface is common for both parts and the change of thickness occurs at the bottom, see Figure 8.34a. When FEM is used the jump of the thickness can be modelled either using 3D solid FEs or assigning different thickness to 2D FEs in two subdomains, see Figure 8.34b.

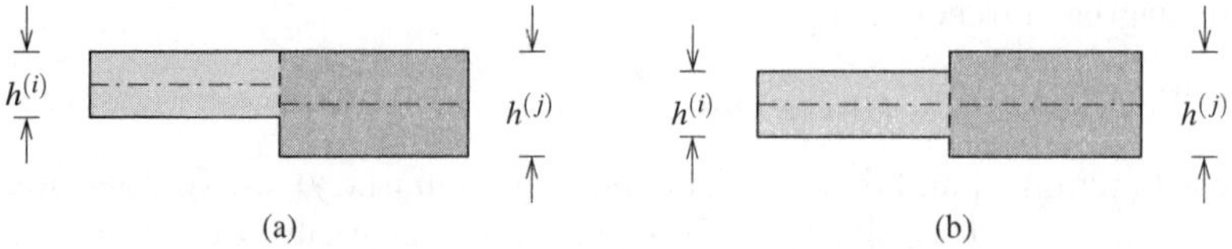

Figure 8.34 Two kinds of thickness jump in a plate

8.13.1 Description of Deformation

In the analysis of deformation of the bending plate two curvatures κ_x, κ_y and two bending moments m_x, m_y are only considered along the boundary of subdomains (i) and (j) with different thicknesses and thus different values of bending stiffness:

$$h^{(i)} \neq h^{(j)}, \qquad D^{m\ (i)} \neq D^{m\ (j)} \tag{8.110}$$

Based on the shape of the deformed surface the equality of tangent curvatures ($s = y$) holds

$$\kappa_s^{(i)} = \kappa_s^{(j)} \quad \rightarrow \quad \kappa_y^{(i)} = \kappa_y^{(j)} \tag{8.111}$$

The balance of normal moments ($\nu = x$) is postulated

$$m_x^{(i)} = m_x^{(j)} \tag{8.112}$$

From the constitutive relations for moments m_x on two sides of the interface of subdomains:

$$m_x^{(i)} = D^{m\ (i)}(\kappa_x^{(i)} + \nu\ \kappa_y^{(i)}), \qquad m_x^{(j)} = D^{m\ (j)}(\kappa_x^{(j)} + \nu\ \kappa_y^{(j)}) \tag{8.113}$$

different curvatures κ_x and different values of moments m_y are obtained:

$$\kappa_x^{(i)} \neq \kappa_x^{(j)}, \qquad m_y^{(i)} \neq m_y^{(j)} \tag{8.114}$$

For $h^{(j)} = 2\ h^{(i)}$, while the simplifying assumption of zero Poisson's constant $\nu = 0$ is adopted, the ratio of bending moments m_y for both sides of subdomain interface can be written as

$$\frac{m_y^{(i)}}{m_y^{(j)}} = \frac{D^{m\ (i)}\ \kappa_y^{(i)}}{D^{m\ (j)}\ \kappa_y^{(j)}} = \frac{D^{m\ (i)}}{D^{m\ (j)}} = \left[\frac{h^{(i)}}{h^{(j)}}\right]^3 = \left[\frac{h^{(i)}}{2\ h^{(i)}}\right]^3 = \frac{1}{8} \tag{8.115}$$

It can be noticed that, for the side with doubled thickness, the bending moment $m_y^{(j)}$ is eight times larger than $m_y^{(i)}$ for the other part, that is $m_y^{(j)} = 8\ m_y^{(i)}$.

8.13.2 FEM Results

In numerical analysis, a simply supported rectangular plate with two subdomains of different thicknesses is considered, see Figure 8.35a. Relationships $h^{(1)} = 2\ h^{(2)}$ and $L_x = 2\ L_y$ are specified in the computations.

The FEM analysis has been performed using package ROBOT (2006). Essential results of computations, that is contour plots of bending moments m_x, m_y and visualization of directions and magnitudes of principal moments are presented in Figure 8.35. Additionally, in Figure 8.35f a distribution of bending moment m_y along a horizontal line with evident jump of value at the interface of the subdomains of different thicknesses

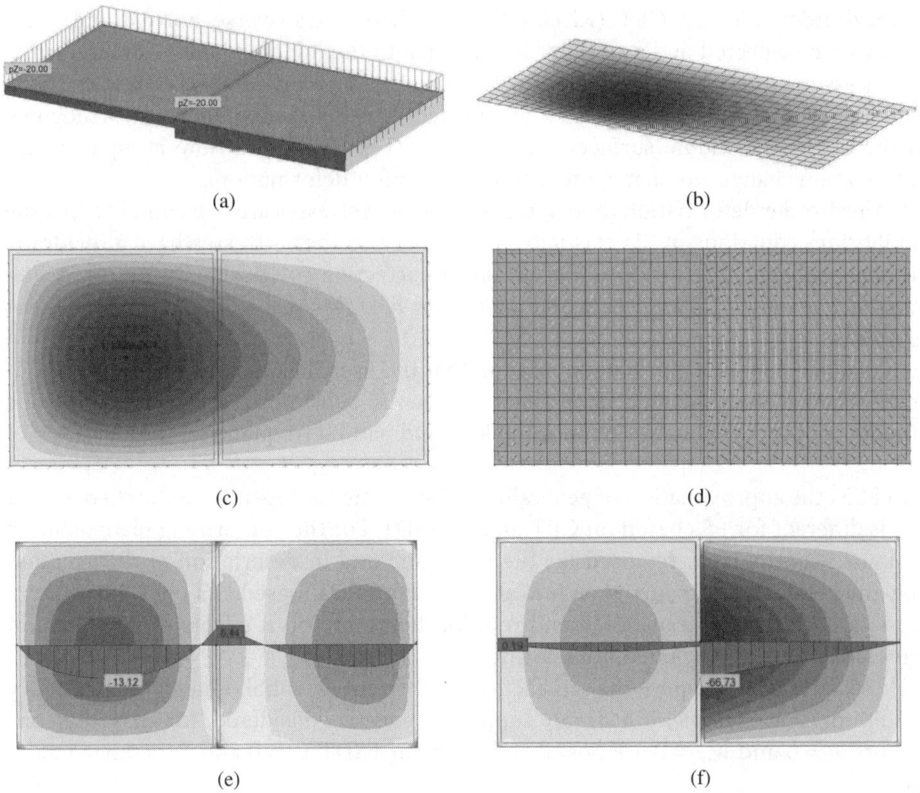

Figure 8.35 Plate under bending with a jump of thickness: (a) configuration of the plate, (b) axonometric view of plate deflection, (c) contour map of plate deflection field, (d) directions of principal moments; contour plots of bending moments: (e) m_x and (f) m_y [kNm/m]; FEM results from ROBOT

is shown. In the right subdomain higher values of moment m_y and thus larger principal moments can be noticed. The maximum deflection occurs in the left subregion with two times smaller thickness and thus eight times lower bending stiffness.

8.14 Analysis of Thin and Moderately Thick Plates in Bending

8.14.1 Preliminary Remarks

The purpose of the section is to complete the information on Mindlin–Reissner theory included in Subsection 1.4.2 and Section 8.5 as well as on different types of FEs incorporated in Subsection 5.2.5. Let us recall the fact that the distinction between thin and moderately thick plates is done on the basis of geometrical parameter, which is a ratio of side to thickness dimensions L/h. The plate is classified as thin if $L/h > 10$, while it is treated as moderately thick if $L/h \leq 10$.

In books by Reddy (1999, 2007) the following abbreviations are used: CPT (classical plate theory) and SDPT (shear deformation plate theory). In addition, two cases of SDPT are distinguished: FSDT (first-order shear deformation theory) and TSDT (a third-order one). In CPT (Kirchhoff–Love theory) transverse normal and shear stresses are neglected and a straight fibre normal to the middle plane before deformation is assumed to remain straight and normal after deformation. According to FSDT (Mindlin–Reissner theory) the straight normal remains straight but not orthogonal to the deformed middle surface. Finally, if TSDT is applied, a fibre straight before deformation changes its shape into a cubic curve after deformation.

In third-order deformation theory, transverse shear stresses are determined using the equilibrium equations of 3D elasticity. The transverse shear stresses have a quadratic distribution along the thickness and the shear correction coefficient k is not used. The difference between the deflection predicted using FSDT (with $k = 5/6$) and TSDT is not significant.

In the literature one can find an opinion that the use CPT in static analysis of moderately thick plates leads to underestimated values of deflection. At the same time, one obtains overestimated values of buckling load and natural frequencies, see Chapters 14 and 15.

In FEM the approximation of generalized displacements depends on the theory used and is different for FEs based on CPT and on FSDT. For the conforming plate bending FE the C^1 continuity is required for the deflection and in description of deformation only the bending effects are taken into account. For FSDT-based FEs the C^0-continuity is guaranteed for the translation and rotation fields and both effects of bending and transverse shear are included in FE formulation.

These approaches require the satisfaction of different essential (kinematic) boundary conditions. For example, if a simply supported edge is considered, for CPT the constraints $w = 0$ and $w_{,s} = 0$ are postulated while for FSDT $w = 0$ and $\vartheta_s = 0$ have to be satisfied.

8.14.2 Simply Supported Square Plate with Uniform Load – Analytical and FEM Results

In Table 8.1 some results of parametric analysis of simply supported plate with uniform load, taken from the book by Reddy (2005) (see Table 12.5.1, p. 657) are presented. They show the dependence of the deflection at the plate centre on: (i) ratio L/h (inverse of slenderness), (ii) type of FE used and mesh density and (iii) order of numerical integration (NI).

Table 8.1 Effect of L/h, mesh density and NI on dimensionless centre deflection

	Analytical solution			FEM solution for $\tilde{w}$	
				Linear FEs	Quadratic FEs
L/h	$\tilde{w}$	$\tilde{\tilde{w}}$	NI	$NSE = 1 \times 1$	$NSE = 2 \times 2$
100	4.572	0.004065	FI	0.011	4.482
			RI	3.579	4.580
10	4.791	0.004259	FI	0.964	4.770
			RI	3.950	4.799

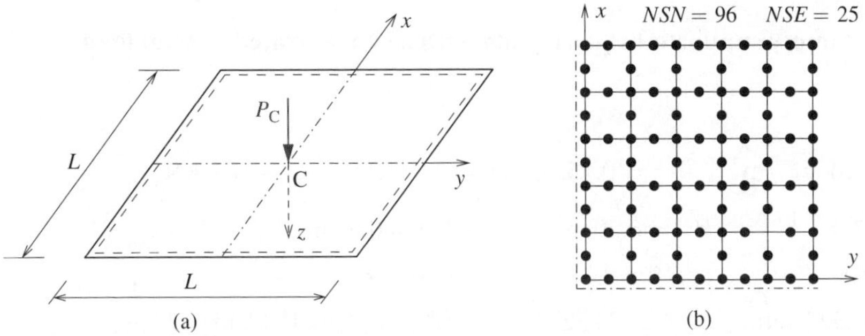

Figure 8.36 Simply supported, square plate with point load: (a) configuration of the plate and (b) discretization of quarter of the domain using eight-node FEs

In computations, Poisson's ratio $\nu = 0.25$ has been adopted. The values of dimensionless deflections $\tilde{w}$ and $\tilde{\tilde{w}}$ are included in Table 8.1 and used in the formula written in (8.116) to find the dimensional maximum deflection $w_{\max}$

$$w_{\max} = \tilde{w}\,\frac{\hat{p}_0\, a^4}{100\, E\, h^3} = \tilde{w}\,\frac{\hat{p}_0\, a^4}{100 \times 12(1-\nu^2)\, D^m} = 0.000899\,\tilde{w}\,\frac{\hat{p}_0\, a^4}{D^m} = \tilde{\tilde{w}}\,\frac{\hat{p}_0\, a^4}{D^m} \tag{8.116}$$

For a thin plate ($L/h = 100$) discretized for testing purposes with just one FE with linear approximation of the three fields of deflection and two rotations, and with full integration, the obtained value of approximation error is unacceptable $\tilde{w}^{\mathrm{FEM,\,FI}}/\tilde{w}^{\mathrm{anal}} = 0.011/4.572 = 0.002$. It indicates the occurrence of locking phenomenon, see also Subsection 17.3.2. To improve the accuracy of numerical results reduced integration (RI) is used and the value of the ratio $\tilde{w}^{\mathrm{FEM,\,FI}}/\tilde{w}^{\mathrm{anal}} = 3.579/4.572 = 0.78$ is calculated. For better accuracy, mesh densification (even with $NSE = 2 \times 2$) and higher approximation (quadratic FEs) are recommended.

8.14.3 Simply Supported Plate with a Concentrated Central Load – Analytical and FEM Solutions

A simply supported square plate with a point load at the centre is considered (Figure 8.36).

The plate is defined by the following parameters: material constants E, ν, dimensions L, $h^{(i)}$ and force P_C. Their values are listed in Box 8.12.

According to the K–L theory (CPT) the maximum deflection at the central point is calculated as

$$w_{\mathrm{C}}^{(i)} = 0.01160\, P_{\mathrm{C}}\, L^2/D^{m(i)} \tag{8.117}$$

The values of deflection for three different thicknesses $h^{(i)}$, $i = 1, 2, 3$ (and thus different values of bending stiffness $D^{m(i)}$) are listed in Box 8.12.

When the M–R theory (FSDT) is adopted the maximum deflection is found using relation

$$w_{\mathrm{C}}^{(i)} = \alpha_{(i)}^{\mathrm{M-R}}\, P_{\mathrm{C}}\, L^2/D^{m(i)} \tag{8.118}$$

in which both coefficients $\alpha_{(i)}^{\mathrm{M-R}}$ and bending stiffness $D^{m(i)}$ depend on ratios $L/h^{(i)}$ (see Box 8.12).

Box 8.12 Simply supported square plate with a concentrated central load

Data

$E = 1.5 \times 10^7\ \text{kN/m}^2, \quad \nu = 0.15, \quad L = 3.0\ \text{m}, \quad P_C = 10.0\ \text{kN}$

$h^{(1)} = 0.03\ \text{m}, \quad h^{(2)} = 0.30\ \text{m}, \quad h^{(3)} = 0.60\ \text{m}$

$L/h^{(1)} = 100, \quad L/h^{(2)} = 10, \quad L/h^{(3)} = 5$

$D^{m(1)} = 34.527\ \text{kNm} \quad D^{m(2)} = 34\,527\ \text{kNm} \quad D^{m(3)} = 276\,215\ \text{kNm}$

Check values from analytical solution

$\alpha^{K-L}_{(1)} = 0.01160 \quad \alpha^{K-L}_{(2)} = 0.01160 \quad \alpha^{K-L}_{(3)} = 0.01160$

$w_C^{(1)\ K-L} = 3.024 \times 10^{-2}\ \text{m} \quad w_C^{(2)\ K-L} = 3.024 \times 10^{-5}\ \text{m} \quad w_C^{(3)\ K-L} = 3.780 \times 10^{-6}\ \text{m}$

$w_C^{(1)\ K-L} = 1.008\ h^{(1)} \quad w_C^{(2)\ K-L} = 1.008 \times 10^{-4}\ h^{(2)} \quad w_C^{(3)\ K-L} = 6.299 \times 10^{-6}\ h^{(3)}$

$\alpha^{M-R}_{(1)} = 0.01170 \quad \alpha^{M-R}_{(2)} = 0.01353 \quad \alpha^{M-R}_{(3)} = 0.01801$

$w_C^{(1)\ M-R} = 3.050 \times 10^{-2}\ \text{m} \quad w_C^{(2)\ M-R} = 3.527 \times 10^{-5}\ \text{m} \quad w_C^{(3)\ M-R} = 5.868 \times 10^{-6}\ \text{m}$

Check values from FEM solution using ANKA

$NSE = 10 \times 10, \quad NEN = 4, \quad NNDOF = 4, \quad NEDOF = 16$

FE PMK3 ANKA (1993) – FEs based on K–L theory

$w_C^{(1)\ FEM} = 3.022 \times 10^{-2}\ \text{m}, \quad w_C^{(2)\ FEM} = 3.022 \times 10^{-5}\ \text{m}, \quad w_C^{(3)\ FEM} = 3.778 \times 10^{-6}\ \text{m}$

$NSE = 5 \times 5, \quad NEN = 8, \quad NNDOF = 5, \quad NEDOF = 40$

FE SQR1 ANKA (1993) – FEs based on M–R theory

$w_C^{(1)\ FEM} = 3.022 \times 10^{-2}\ \text{m} \quad w_C^{(2)\ FEM} = 3.465 \times 10^{-5}\ \text{m} \quad w_C^{(3)\ FEM} = 5.779 \times 10^{-6}\ \text{m}$

$w_C^{(1)\ FEM} = 0.990\ w_C^{(1)\ anal} \quad w_C^{(2)\ FEM} = 0.981\ w_C^{(2)\ anal} \quad w_C^{(3)\ FEM} = 1.022\ w_C^{(3)\ anal}$

Plate configuration shown in Figure 8.36a

For a moderately thick plate ($L/h^{(3)} = 5$) with a point load the inequality is obtained $w_C^{(3)\ K-L} = 3.780 \times 10^{-6}\ \text{m} < w_C^{(3)\ M-R} = 5.868 \times 10^{-6}\ \text{m}$ from analytical solution. It is concluded that if a plate is classified as moderately thick, but calculated according to CPT, neglecting the effects of transverse shear results in an underestimation of deflection.

Numerical analysis has been performed using FEM program ANKA (1993). The plate domain has been discretized with two types of FEs: PMK3 – four-node, conforming, CPT-based FE, and SQR1 – eight-node, FSDT-based FE, suitable for analysis of both thin and moderately thick plates. A quarter of the plate has been discretized using two meshes: $NSE = 10 \times 10 = 100$ with PMK3 FEs or $NSE = 5 \times 5 = 25$ with SQR1 FEs, see

Figure 8.36b. The values of deflection at the centre for the first discretization (CPT) and three different values of thickness are calculated and included in Box 8.12. The obtained numerical results are consistent with the analytical solution. Results computed for the second discretization (FSDT) can also be found in Box 8.12.

After all, it should be pointed out that in the neighbourhood of the point load a more thorough static analysis of stress should be based on 3D formulation and dense meshes.

Please note that deflection $w_C^{(1)}$ and thickness $h^{(1)}$ for the thinnest plate are the same order, compared with the values in Box 8.12. In this case, statics of the plate should be described using von Kármán theory, that is according to the theory of plates with moderately large deflection, see Subsection 1.5.3 and Box 1.1. Fortunately, the value of deflection for linear theory is overestimated, hence it is on the safe side when the serviceability limit condition $w < w_{\text{lim}}$ according to design recommendation is checked.

References

ANKA 1993 ANKA – computer code for nonlinear analysis of structures: User's manual. Technical report, Cracow University of Technology, Cracow (in Polish).

Girkmąnn K 1956 *Flächentragwerke*. Springer-Verlag, Wien.

Kączkowski Z 1980 *Plates. Static calculations*. Arkady, Warsaw (in Polish).

Reddy JN 1999 *Theory and Analysis of Elastic Plates*. Taylor & Francis.

Reddy JN 2005 *An Introduction to the Finite Element Method* 3rd edn. McGraw-Hill, New York.

Reddy JN 2007 *Theory and Analysis of Elastic Plates and Shells* 2nd edn. CRC Press/Taylor & Francis, Boca Raton-London-New York.

ROBOT 2006 ROBOT Millennium: User's Guide. Technical report, RoboBAT, Cracow (in Polish).

Starosolski W 2009 *Reinforced Concrete Structures According to PN-B-03264:2002 and Eurocode 2* 12th edn. PWN, Warsaw (in Polish).

Timoshenko S and Woinowsky-Krieger S 1959 *Theory of Plates and Shells*. McGraw-Hill, New York-Auckland.

Waszczyszyn Z and Radwańska M 1995 Basic equations and calculations methods for elastic shell structures. In Borkowski A, Cichoń C, Radwańska M, Sawczuk A and Waszczyszyn Z, *Structural Mechanics: Computer Approach*, vol. **3**. Arkady, Warsaw, chapter 9, pp. 11–190 (in Polish).

9

Circular and Annular Plates under Bending

9.1 General State

In this chapter we consider thin circular and annular plates under bending. In a plate domain, described in a cylindrical coordinate system (r, θ, z), the following external (known) loads and constraints written in a vector form are taken into account:

- one-component vector of transverse surface load $\hat{\mathbf{p}}^m_{(1\times1)}(r,\theta) = [\hat{p}_z(r,\theta)]$ [N/m^2]
- vector of boundary loads – forces and moments distributed along contour lines $\hat{\mathbf{p}}^m_{b(2\times1)}(s) = [\hat{t}_\nu, \hat{m}_\nu]^\mathrm{T}$ [N/m], [Nm/m]
- vector of kinematic boundary constrains $\hat{\mathbf{u}}^m_{b(2\times1)}(s) = [\hat{w}, \hat{\vartheta}_\nu]^\mathrm{T}$ [m], [–]

The solution of the boundary value problem (BVP) related to the Kirchhoff–Love (K–L) thin plate theory includes the following quantities defined at any point in the middle plane, grouped in vectors:

- one-component deflection vector $\mathbf{u}^m_{(1\times1)}(r,\theta) = [w(r,\theta)]$ [m]
- bending strain vector – two curvatures and warping
  $\mathbf{e}^m_{(3\times1)}(r,\theta) = [\kappa_r, \kappa_\theta, \chi_{r\theta}]^\mathrm{T}$ [1/m]
- vector of generalized resultant forces for bending state $\mathbf{s}^m_{(5\times1)}(r,\theta) = [\mathbf{m}^\mathrm{T}, \mathbf{t}^\mathrm{T}]^\mathrm{T}$:
  – moments $\mathbf{m}_{(3\times1)}(r,\theta) = [m_r, m_\theta, m_{r\theta}]^\mathrm{T}$ [Nm/m]
  – transverse shear forces $\mathbf{t}_{(2\times1)}(r,\theta) = [t_r, t_\theta]^\mathrm{T}$ [N/m]

The positive moments and transverse shear forces are shown on respective edges of an elementary segment in Figure 9.1.

The local formulation of the BVP for circular and annular plates according to the K–L theory is written as a set of nine differential and algebraic equations with nine unknown functions, see Box 9.1. We recall that the transverse shear forces are treated as inactive because of zero transverse shear strains $\gamma_{rz} = \gamma_{\theta z} = 0$. In addition, the global formulation represented by a functional of internal strain energy (expressed using partial derivatives of the deflection function after taking into account appropriate kinematic and constitutive equations) and external work is included in Box 9.1.

Plate and Shell Structures: Selected Analytical and Finite Element Solutions, First Edition.
Maria Radwańska, Anna Stankiewicz, Adam Wosatko and Jerzy Pamin.
© 2017 John Wiley & Sons Ltd. Published 2017 by John Wiley & Sons Ltd.

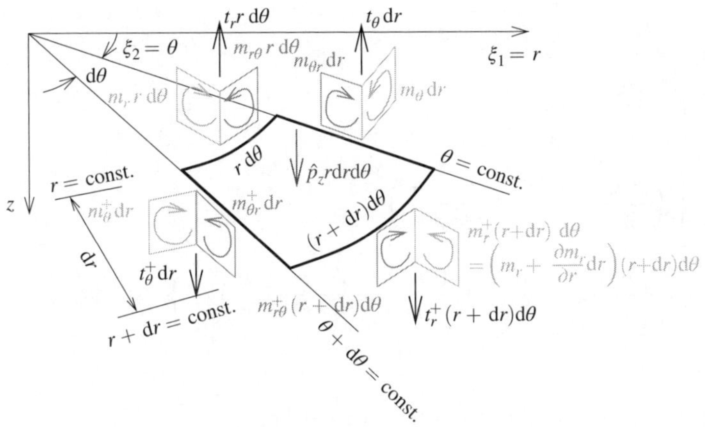

Figure 9.1 Elementary segment of circular plate: surface load and generalized resultant forces for general bending state

Box 9.1 Equations for thin circular and annular plates according to Kirchhoff–Love theory for a general bending state

Local formulation – set of nine equations and boundary conditions

(I) kinematic equations (3):

$$\kappa_r = -\frac{\partial^2 w}{\partial r^2}, \qquad \kappa_\theta = -\frac{1}{r^2}\frac{\partial^2 w}{\partial \theta^2} - \frac{1}{r}\frac{\partial w}{\partial r}, \qquad \chi_{r\theta} = -2\frac{\partial^2}{\partial r\,\partial\theta}\left(\frac{w}{r}\right)$$

(II) equilibrium equations (3):

$$\frac{1}{r}\frac{\partial(r\,t_r)}{\partial r} + \frac{1}{r}\frac{\partial t_\theta}{\partial\theta} + \hat{p}_z = 0$$

$$\frac{1}{r}\frac{\partial(r\,m_r)}{\partial r} + \frac{1}{r}\frac{\partial m_{r\theta}}{\partial\theta} - \frac{m_\theta}{r} - t_r = 0$$

$$\frac{1}{r}\frac{\partial(r\,m_{r\theta})}{\partial r} + \frac{1}{r}\frac{\partial m_\theta}{\partial\theta} + \frac{m_{r\theta}}{r} - t_\theta = 0$$

(III) constitutive equations (3):

$$m_r = D^m(\kappa_r + \nu\kappa_\theta), \quad m_\theta = D^m(\kappa_\theta + \nu\kappa_r), \quad m_{r\theta} = D^m\,(1-\nu)\,\chi_{r\theta}/2$$

where $D^m = E\,h^3/[12(1-\nu^2)]$

(IV) boundary conditions:

a) kinematic: $w = \hat{w},\ \vartheta_\nu = \hat{\vartheta}_\nu$

b) static: $\tilde{t}_\nu = t_\nu + \frac{\partial m_{\nu s}}{\partial s} = \hat{t}_\nu,\ m_\nu = \hat{m}_\nu$

c) mixed (one kinematic and one static)

Global formulation

Total potential energy $\Pi^m = U^m - W^m$

Internal strain energy

$$U^m = \frac{D^m}{2}\int_A \left\{ \left(\frac{\partial^2 w}{\partial r^2} + \frac{1}{r}\frac{\partial w}{\partial r} + \frac{1}{r^2}\frac{\partial^2 w}{\partial \theta^2} \right)^2 - 2(1-\nu)\left[\frac{\partial^2 w}{\partial r^2}\left(\frac{1}{r^2}\frac{\partial^2 w}{\partial \theta^2} + \frac{1}{r}\frac{\partial w}{\partial r} \right) - \left(\frac{\partial^2}{\partial r\,\partial \theta}\frac{w}{r} \right)^2 \right] \right\} r\,\mathrm{d}r\,\mathrm{d}\theta$$

External load work

$$W^m = \int_A \hat{p}_z(r,\theta)\, w(r,\theta)\, r\,\mathrm{d}r\mathrm{d}\theta + \int_{\partial A_\sigma} \left[\hat{t}_\nu\, w + \hat{m}_\nu\, \vartheta_\nu \right] \mathrm{d}s$$

Fourth-order displacement differential equation for thin circular and annular plates in the general bending state has the form

$$\nabla^2\nabla^2 w(r,\theta) = \frac{\hat{p}_z(r,\theta)}{D^m} \tag{9.1}$$

where the bi-Laplacian operator in the polar coordinate system is expressed as

$$\nabla^2\nabla^2 w(r,\theta) = \left(\frac{\partial^2}{\partial r^2} + \frac{1}{r}\frac{\partial}{\partial r} + \frac{1}{r^2}\frac{\partial^2}{\partial \theta^2} \right)\left(\frac{\partial^2 w}{\partial r^2} + \frac{1}{r}\frac{\partial w}{\partial r} + \frac{1}{r^2}\frac{\partial^2 w}{\partial \theta^2} \right) \tag{9.2}$$

9.2 Axisymmetric State

For thin circular or annular plates with $\theta \in [0, 360°]$ and axisymmetric external actions (transverse surface load $\hat{p}_z(r)$ as well as boundary loads $\hat{\mathbf{p}}_b^m(s) = [\hat{t}_\nu, \hat{m}_\nu]^{\mathrm{T}}$ and constraints $\hat{\mathbf{u}}_b^m(s) = [\hat{w}, \hat{\vartheta}_\nu]^{\mathrm{T}}$), the functions of interest do not depend on the circumferential coordinate θ. The solution of BVP for the axisymmetric bending state requires the determination of the following six functions of one coordinate r grouped in vectors:

- one-component deflection vector $\mathbf{u}^m_{(1\times1)}(r) = [w]$ [m]
- bending strain vector – two curvatures in radial and circumferential directions $\mathbf{e}^m_{(2\times1)}(r) = [\kappa_r, \kappa_\theta]^{\mathrm{T}}$ [1/m]
- vector of generalized resultant forces for bending state $\mathbf{s}^m_{(3\times1)}(r) = [\mathbf{m}^{\mathrm{T}}, \mathbf{t}]^{\mathrm{T}}$:
  - two bending moments $\mathbf{m}_{(2\times1)}(r) = [m_r, m_\theta]^{\mathrm{T}}$ [Nm/m]
  - one transverse shear force $\mathbf{t}_{(1\times1)}(r) = [t_r]$ [N/m]

The local formulation of the axisymmetric bending state in the form of a set of six differential and algebraic equations is presented in Box 9.2. To complete the description the global formulation is written as a function of total potential energy and given in Box 9.2.

Box 9.2 Equations for thin circular and annular plates according to Kirchhoff–Love theory for an axisymmetric state

Local formulation – set of six equations and boundary conditions

(I) kinematic equations (2):

$$\kappa_r = -\frac{d^2 w}{dr^2}, \qquad \kappa_\theta = -\frac{1}{r}\frac{dw}{dr}$$

(II) equilibrium equations (2):

$$\frac{1}{r}\frac{d(r\, t_r)}{dr} + \hat{p}_z = 0, \qquad \frac{1}{r}\frac{d(r\, m_r)}{dr} - \frac{m_\theta}{r} - t_r = 0$$

(III) constitutive equations (2):

$$m_r = D^m(\kappa_r + \nu\kappa_\theta), \qquad m_\theta = D^m(\kappa_\theta + \nu\kappa_r)$$

where $D^m = E\,h^3/[12(1-\nu^2)]$

(IV) boundary conditions:

a) kinematic: $w = \hat{w}$, $\vartheta_r = -dw/dr = \hat{\vartheta}_r$

b) static: $t_r = \hat{t}_r$, $m_r = \hat{m}_r$

c) mixed (one kinematic and one static)

Global formulation

Total potential energy $\Pi^m = U^m - W^m$

Internal strain energy

$$U^m = \frac{D^m}{2}\int_A \left[\left(\frac{d^2 w}{dr^2} + \frac{1}{r}\frac{dw}{dr}\right)^2 - \frac{2}{r}(1-\nu)\frac{d^2 w}{dr^2}\frac{dw}{dr}\right] r\, dr\, d\theta$$

External load work

$$W^m = \int_A \hat{p}_z(r)\, w(r)\, r\, dr\, d\theta + \int_{\partial A_\sigma} [\hat{t}_r w + \hat{m}_r \vartheta_r]\, ds$$

In the case of axisymmetric state the displacement differential equation is simplified compared to Equations (9.1) and (9.2), and depends on ordinary derivatives of the deflection function (with respect to radial coordinate r)

$$\nabla^2\nabla^2 w(r) = \frac{\hat{p}_z(r)}{D^m} \tag{9.3}$$

where

$$\nabla^2\nabla^2 w(r) = \frac{d^4 w}{dr^4} + \frac{2}{r}\frac{d^3 w}{dr^3} - \frac{1}{r^2}\frac{d^2 w}{dr^2} + \frac{1}{r^3}\frac{dw}{dr} \tag{9.4}$$

Alternatively, Equation (9.3) can be rewritten in the following form, which is very convenient for direct integration

$$\frac{1}{r}\frac{\mathrm{d}}{\mathrm{d}r}\left\{r\frac{\mathrm{d}}{\mathrm{d}r}\left[\frac{1}{r}\frac{\mathrm{d}}{\mathrm{d}r}\left(r\frac{\mathrm{d}w(r)}{\mathrm{d}r}\right)\right]\right\}=\frac{\hat{p}_z(r)}{D^m} \tag{9.5}$$

9.3 Analytical Solution using a Trigonometric Series Expansion

The analytical solution expressed by a trigonometric series expansion requires the separation of variables. The sought quantities are represented as a product of two functions, one dependent on radial coordinate r and the other being a function of circumferential coordinate θ and written in the form of trigonometric (sine or cosine) series.

The following functions are represented by a cosine series:

$$\begin{aligned}
&\hat{p}_z(r,\theta)=\sum_{j=0}^{J}\hat{p}_z^{(j)}(r)\cos(j\theta), && w(r,\theta)=\sum_{j=0}^{J}w^{(j)}(r)\cos(j\theta)\\
&m_r(r,\theta)=\sum_{j=0}^{J}m_r^{(j)}(r)\cos(j\theta), && m_\theta(r,\theta)=\sum_{j=0}^{J}m_\theta^{(j)}(r)\cos(j\theta)\\
&t_r(r,\theta)=\sum_{j=0}^{J}t_r^{(j)}(r)\cos(j\theta)
\end{aligned} \tag{9.6}$$

and the other two functions by a sine series:

$$m_{r\theta}(r,\theta)=\sum_{j=0}^{J}m_{r\theta}^{(j)}(r)\sin(j\theta),\qquad t_\theta(r,\theta)=\sum_{j=0}^{J}t_\theta^{(j)}(r)\sin(j\theta) \tag{9.7}$$

The Laplacian operator ∇^2 written in polar coordinates r and θ, see Equation (9.2), becomes the product of operator $\nabla^2_{(j)}$ acting on the deflection amplitude function $w^{(j)}(r)$ and function $\cos(j\,\theta)$

$$\nabla^2\,[w(r,\theta)]=\nabla^2_{(j)}\,[w^{(j)}(r)]\cos(j\theta),\quad \text{where}\quad \nabla^2_{(j)}=\frac{\mathrm{d}^2}{\mathrm{d}r^2}+\frac{1}{r}\frac{\mathrm{d}}{\mathrm{d}r}-\frac{j^2}{r^2} \tag{9.8}$$

In the considered approach, the displacement differential equation (9.1) can be transformed into a set of $J+1$ fourth-order differential equations with ordinary derivatives of the unknown function of the deflection amplitude $w^{(j)}(r)$ and the applied transverse load amplitude $\hat{p}_z^{(j)}(r)$

$$\left(\frac{\mathrm{d}^2}{\mathrm{d}r^2}+\frac{1}{r}\frac{\mathrm{d}}{\mathrm{d}r}-\frac{j^2}{r^2}\right)\left(\frac{\mathrm{d}^2w^{(j)}(r)}{\mathrm{d}r^2}+\frac{1}{r}\frac{\mathrm{d}w^{(j)}(r)}{\mathrm{d}r}-\frac{j^2\,w^{(j)}(r)}{r^2}\right)=\frac{\hat{p}_z^{(j)}(r)}{D^m} \tag{9.9}$$

The general integrals of these equations for $j=0,1$ and $j>1$ are given as:

$$\begin{aligned}
w^{o(j=0)}&=C_{10}+C_{20}r^2+C_{30}\,\log r+C_{40}r^2\,\log r\\
w^{o(j=1)}&=C_{11}r+C_{21}r^3+C_{31}r^{-1}+C_{41}r\,\log r\\
w^{o(j>1)}&=C_{1j}r^j+C_{2j}r^{j+2}+C_{3j}r^{-j}+C_{4j}r^{-j+2}
\end{aligned} \tag{9.10}$$

Equation (9.9) is solved separately for each j and must be completed by boundary conditions in order to calculate four integration constants $C_{1j}, \dots, C_{4j}$.

For $j = 0$ the differential equation describing a special case of axisymmetric state is obtained in the following form, see Equation (9.3)

$$\nabla^2\nabla^2 w(r) = \frac{d^4 w}{dr^4} + \frac{2}{r}\frac{d^3 w}{dr^3} - \frac{1}{r^2}\frac{d^2 w}{dr^2} + \frac{1}{r^3}\frac{dw}{dr} = \frac{\hat{p}_z(r)}{D^m} \tag{9.11}$$

Its solution is the sum of general and particular integrals

$$w(r) = w^o(r) + \overline{w}(r) \tag{9.12}$$

The general integral contains the following four components with unknown constants

$$w^o(r) = C_{10} + C_{20}\, r^2 + C_{30}\, \log r + C_{40}\, r^2 \log r \tag{9.13}$$

The second index of integration constants indicates axisymmetry characterized by the number of circumferential waves $j = 0$ with $\cos(j\theta) = 1$. The particular integral for uniformly distributed load $\hat{p}_z(r) = p_0 = \text{const.}$ has the form

$$\overline{w}(r) = \frac{p_0\, r^4}{64\, D^m} \tag{9.14}$$

Four integration constants C_{i0}, where $i = 1, \dots, 4$, are calculated from boundary conditions. For example, in the case of annular plate, four boundary conditions are formulated for internal and external contours, two for each of them.

The solution of the problem includes five functions (dependent on coordinate r, surface load p_0 and integration constants) describing the deflection, rotation, two bending moments and transverse shear force as follows:

$$\begin{aligned}
w(r) &= C_{10} + C_{20}\, r^2 + C_{30}\, \log r + C_{40}\, r^2 \log r + \frac{p_0\, r^4}{64\, D^m}\\
\vartheta_r(r) &= -\frac{dw}{dr} = -C_{20}\, 2r - C_{30}\, \frac{1}{r} - C_{40}(2r \log r + r) - \frac{p_0\, r^3}{16\, D^m}\\
\frac{d^2 w}{dr^2} &= C_{20}\, 2 - C_{30}\, \frac{1}{r^2} + C_{40}(2 \log r + 3) + \frac{3\, p_0\, r^2}{16\, D^m}\\
\frac{d^3 w}{dr^3} &= C_{30}\, \frac{2}{r^3} + C_{40}\, \frac{2}{r} + \frac{3\, p_0\, r}{8\, D^m}\\
m_r(r) &= -D^m \left(\frac{d^2 w}{dr^2} + \frac{\nu}{r}\frac{dw}{dr} \right) = -\frac{(3+\nu)\, p_0\, r^2}{16}\\
&\quad - D^m \left[C_{20}\, (2 + 2\nu) + C_{30}\, \frac{(\nu - 1)}{r^2} + C_{40}(2 \log r + 2\nu \log r + 3 + \nu) \right]\\
m_\theta(r) &= -D^m \left(\frac{1}{r}\frac{dw}{dr} + \nu \frac{d^2 w}{dr^2} \right) = -\frac{(1+3\nu)\, p_0\, r^2}{16}\\
&\quad - D^m \left[C_{20}\, (2 + 2\nu) + C_{30}\, \frac{(1 - \nu)}{r^2} + C_{40}(2 \log r + 2\nu \log r + 1 + 3\nu) \right]\\
t_r(r) &= D^m \left(\frac{1}{r^2}\frac{dw}{dr} - \frac{1}{r}\frac{d^2 w}{dr^2} - \frac{d^3 w}{dr^3} \right) = -C_{40}\, D^m\, \frac{4}{r} - \frac{p_0\, r}{2}
\end{aligned} \tag{9.15}$$

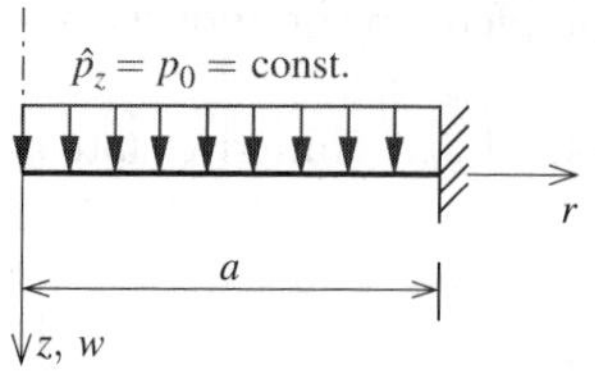

Figure 9.2 Configuration of clamped circular plate: dimensions, load and boundary constraints

9.4 Clamped Circular Plate with a Uniformly Distributed Load

In this section the analytical solution for a circular plate shown in Figure 9.2 is discussed. The considered structure is loaded by uniform surface load $\hat{p}_z(r) = p_0 = \text{const.}$ The external edge of the plate is clamped. Solution of the problem requires the specification of material constants E, ν, dimensions r_{ext}, h and surface load p_0. The input values are listed in Box 9.3.

For a complete description of the solution, the differential Equation (9.3) and its integral are recalled:

$$\begin{aligned} &\nabla^2\nabla^2 w(r) = \hat{p}_z/D^m \\ &w(r) = C_{10} + C_{20}\, r^2 + C_{30} \log r + C_{40}\, r^2 \log r + p_0\, r^4/(64\, D^m) \end{aligned} \tag{9.16}$$

To solve the problem four integration constants C_{i0}, $i = 1, \ldots, 4$ have to be determined. In the circular plate a point on the axis of symmetry with $r \to 0$ has to be considered. It should be noted that $\lim_{r\to 0} \log r = -\infty$. The general integral $w^o(r)$ includes component $C_{30} \log r$ and element $C_{40} \log r$ appears in the formulae for bending moments $m_r(r)$ and $m_\theta(r)$, see Equations $(9.15)_{3,4}$. To ensure the finite values of $w(0)$, $m_r(0)$ and $m_\theta(0)$, the components including log 0 have to be eliminated, thus

$$C_{30} = C_{40} = 0 \tag{9.17}$$

The other two constants C_{10} and C_{20} are determined from kinematic boundary conditions for external edge, that is for $r = a$. On a clamped edge deflection $w(a)$ and rotation

Box 9.3 Clamped circular plate with a uniformly distributed load

Data

$E = 2.0 \times 10^7\ \text{kN/m}^2$, $\nu = 0.1$, $r_{ext} = a = 3.0\ \text{m}$, $h = 0.1\ \text{m}$
$D^m = 1683.50\ \text{kNm}$, $\hat{p}_z(r) = p_0 = 6.0\ \text{kN/m}^2 = \text{const.}$

Check values from analytical solution

$w^{anal}_{max} = w(0.0) = 0.00451\ \text{m}$
$m^{anal}_r(0.0) = m^{anal}_\theta(0.0) = 3.713\ \text{kNm/m}$
$m^{anal}_r(3.0) = -6.75\ \text{kNm/m}$, $m^{anal}_\theta(3.0) = -0.675\ \text{kNm/m}$
$t_r(0.0) = 0.0$, $t_r(3.0) = -9.0\ \text{kN/m}$

Plate configuration shown in Figure 9.2

$\vartheta_r(a)$ (in a plane normal to the middle plane, with the rotation vector tangent to the contour) are equal to zero:

$$\begin{aligned} w(a) &= 0 & &\rightarrow & C_{10} + a^2\,C_{20} + \frac{p_0\,a^4}{64\,D^m} &= 0 \\ \vartheta_r(a) &= -\frac{\mathrm{d}w}{\mathrm{d}r} = 0 & &\rightarrow & -2a\,C_{20} - \frac{p_0\,a^3}{16\,D^m} &= 0 \end{aligned} \tag{9.18}$$

From this system of two algebraic equations, two integration constants are calculated:

$$C_{10} = \frac{p_0\,a^4}{64\,D^m}, \qquad C_{20} = -\frac{p_0\,a^2}{32\,D^m} \tag{9.19}$$

Now, all functions describing behaviour of the plate are determined:

- deflection

$$w(r) = \frac{p_0\,a^4}{64\,D^m}\left[1 - \left(\frac{r}{a}\right)^2\right]^2 \tag{9.20}$$

with maximum value

$$w_{\max} = w(0) = \frac{p_0\,a^4}{64\,D^m} \tag{9.21}$$

- two bending moments:

$$m_r(r) = \frac{p_0\,a^2}{16}\left[(1+\nu) - (3+\nu)\,\frac{r^2}{a^2}\right], \quad m_\theta(r) = \frac{p_0\,a^2}{16}\left[(1+\nu) - (1+3\,\nu)\,\frac{r^2}{a^2}\right] \tag{9.22}$$

with extreme ordinates:

$$\begin{aligned} m_r(0) &= m_\theta(0) = (1+\nu)\,\frac{p_0\,a^2}{16} \\ m_r(a) &= -\frac{p_0\,a^4}{8}, \qquad m_\theta(a) = -\nu\,\frac{p_0\,a^4}{8} \end{aligned} \tag{9.23}$$

- transverse shear force:

$$t_r(r) = -\frac{p_0\,r}{2}, \qquad t_r(0) = 0, \qquad t_r(a) = -\frac{p_0\,a}{2} \tag{9.24}$$

The characteristic feature of the axially symmetric state of a circular plate is the equality of moments $m_r(0)$ and $m_\theta(0)$ at the point on symmetry axis ($r = 0$), see Equation $(9.23)_1$.

The data used in calculations are listed in Box 9.3. Formulae (9.20)–(9.24) are applied to plot the diagrams of four functions describing the behaviour of the clamped circular plate under uniformly distributed load, shown in Figure 9.3, and to calculate the characteristic values given in Box 9.3.

The analysis of circular plates using FEM is a common student assignment. The crucial issue is the choice of a proper FE type and discretization. It is not recommended to generate the finite element mesh on the basis of radial (θ = const.) and circumferential (r = const.) lines since it leads to excessive mesh density (too small FEs) in the central part of the domain, see Figure 9.4a. Then, one of the possibilities is to discretize this subdomain with arbitrary FEs (3-, 4-, 8/9-noded) and introduce gradually elements based on radial and circumferential lines of division in the area near the boundary, see

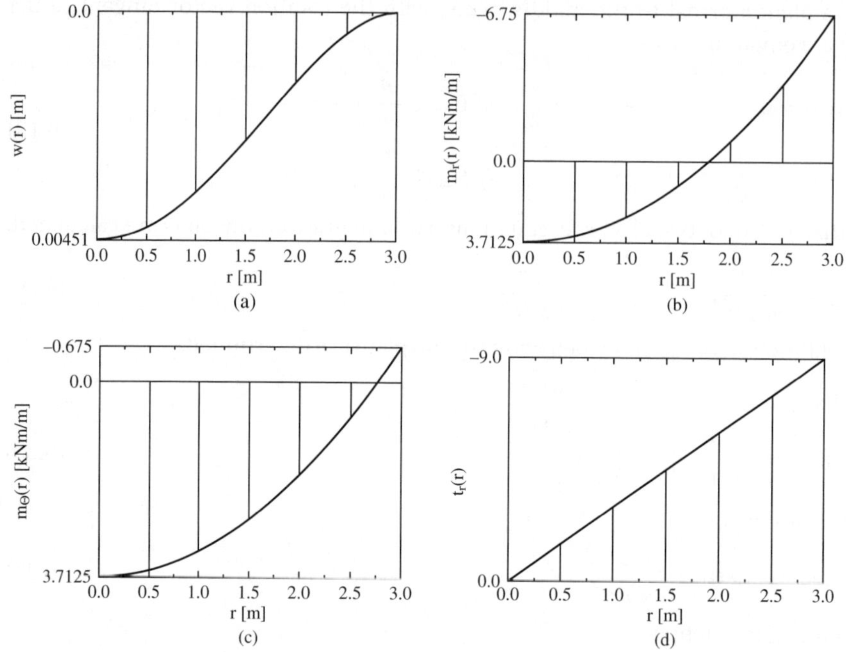

Figure 9.3 Clamped circular plate with uniformly distributed load – functions obtained from analytical solution: (a) $w(r)$, (b) $m_r(r)$, (c) $m_\theta(r)$, (d) $t_r(r)$

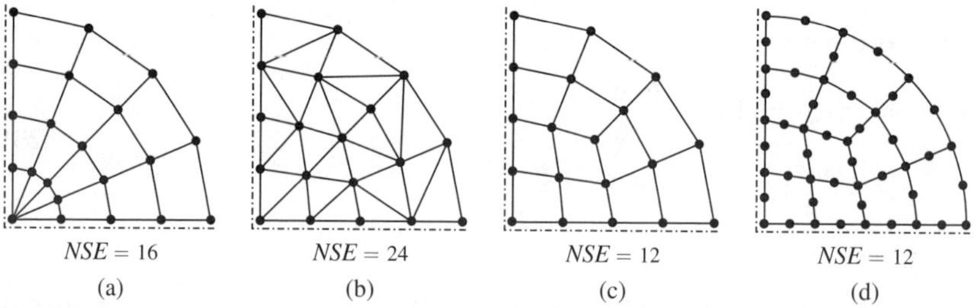

Figure 9.4 Different meshes for a quarter of circular plate: (a) discretization with improper FEs around the centre, (b) three-node triangular FEs, (c) four-node quadrilateral FEs and (d) eight-node isoparametric FEs

Figure 9.4b,c. Additionally, it should be noted that the eight-noded isoparametric elements are able to reconstruct the curved external boundary line (see Figure 9.4d).

In the work of Katili (1993) the results of a broad parametric analysis of circular plates discretized with triangular DKMT and quadrilateral DKMQ elements (see Subsection 18.3.5) employed for modelling of thin and moderately thick plate structures are presented. The different discretizations of a quarter of a circular plate are used. Denser and denser meshes with $NSE = 6, 24, 54, 96$ for DKMT FEs and $NSE = 3, 12, 27, 48$ for DKMQ FEs are considered (see Figure 9.4b,c). The convergence of results is shown on the basis of monitored values of deflection and radial bending moment at the central point. The results for clamped and a simply supported plate under uniform load for three values of R/h ratio: 50, 5, 2 are presented in Katili (1993). For thin plates the limit $R/h > 10$ is assumed to hold. A comparison of numerical and analytical results with check values taken from the book by Batoz and Dhatt (1990) is worth attention. Example results refer to a thin ($R/h = 50$) clamped plate discretized with DKMT FEs and characterized by the following unitless data: $E = 10.92$, $\nu = 0.3$, $R = 5$, $h = 0.1$ and $p_0 = 1.0$. The observed convergence of the central deflection (from above) to exact solution $v_C^{exact} = 9783.5$ is as follows: $v_C^{NSE=6} = 10\,306.0$, $v_C^{NSE=24} = 9995.6$, $v_C^{NSE=54} = 9883.5$, $v_C^{NSE=96} = 9847.5$.

9.5 Simply Supported Circular Plate with a Concentrated Central Force

A simply supported circular plate under point load presented in Figure 9.5 is considered. Analysis of this example is one of the most important and commonly discussed academic assignments referring to plate bending problems. The analytical solution can be found in the book of Timoshenko and Woinowsky-Krieger (1959). The complete definition of the problem requires the specification of material constants E, ν, dimensions a, h and concentrated force P. The input values are given in Box 9.4.

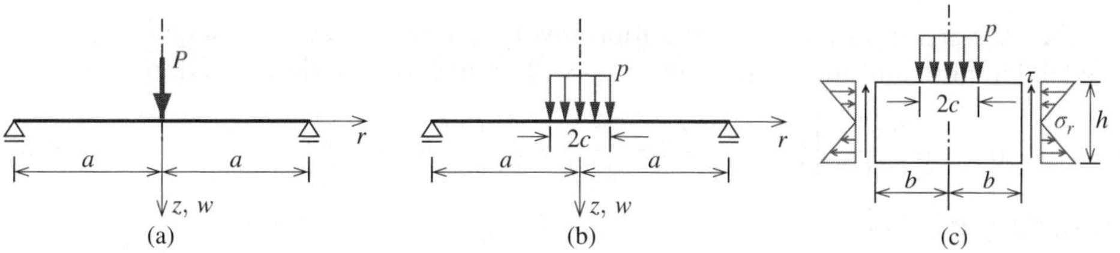

Figure 9.5 Simply supported circular plate with concentrated central force: (a), (b) configurations for two different approaches and (c) 3D model of the loaded central part of the domain

Box 9.4 Simply supported circular plate with a concentrated central force

Data

$E = 2.0 \times 10^7$ kN/m^2, $\nu = 0.1$, $r_{ext} = a = 3.0$ m, $h = 0.1$ m
$D^m = 1683.50$ kNm, $P = 1.0$ kN

Check values from analytical solution

$w_{max}^{anal} = w(0.0) = 0.00030$ m $= 0.003\,h$
$m_r^{anal}(0.12) = 0.298$ kNm/m, $m_r^{anal}(3.0) = 0.0$
$m_\theta^{anal}(0.12) = 0.369$ kNm/m, $m_\theta^{anal}(3.0) = 0.072$ kNm/m
$t_r(0.12) = -1.592$ kN/m, $t_r(3.0) = -0.053$ kN/m

Plate configuration shown in Figure 9.5a

Two approaches can be used to solve the considered problem. In the first one (see Figure 9.5a), the third-order differential equation (presented in Timoshenko and Woinowsky-Krieger 1959) is used as a start point

$$D^m \frac{\mathrm{d}}{\mathrm{d}r}\left[\frac{1}{r}\frac{\mathrm{d}}{\mathrm{d}r}\left(r\,\frac{\mathrm{d}w(r)}{\mathrm{d}r}\right)\right] = -t_r(r) \tag{9.25}$$

The function describing the distribution of transverse shear force on the right-hand side of the Equation (9.25) is given as

$$t_r = -\frac{P}{2\pi r} \tag{9.26}$$

It has been obtained from the equilibrium equation in direction z for a central circular subdomain with running radius r.

Triple integration of Equation (9.25) leads to the following form of the general integral

$$w^o(r) = C_1\,\frac{r^2}{4} + C_2\,\log r + C_3 \tag{9.27}$$

with three integration constants, determined from:

- two mixed boundary conditions: $w(a) = 0$, $m_r(a) = 0$
- condition resulting from axial symmetry of deflection function: $\mathrm{d}w/\mathrm{d}r|_{r=0} = 0$

The satisfaction of the third condition involves the elimination of singularity in the deflection function since $\lim_{r\to 0} \log r = -\infty$. The final form of the deflection function is

$$w(r) = \frac{P\,a^2}{16\pi D^m}\left[\left(\frac{3+\nu}{1+\nu}\right)\left(1-\frac{r^2}{a^2}\right) + 2\left(\frac{r}{a}\right)^2 \log\left(\frac{r}{a}\right)\right] \tag{9.28}$$

with the maximum

$$w_{\max} = w(0) = \frac{P\,a^2}{16\pi D^m}\left(\frac{3+\nu}{1+\nu}\right) \tag{9.29}$$

From coupled constitutive and kinematic equations the following functions of two bending moments are obtained:

$$m_r(r) = -\frac{P(1+\nu)}{4\pi}\,\log\left(\frac{r}{a}\right), \qquad m_\theta(r) = -\frac{P}{4\pi}\left[(1+\nu)\,\log\left(\frac{r}{a}\right) - 1 + \nu\right] \tag{9.30}$$

with known distribution of the transverse shear force

$$t_r(r) = -\frac{P}{2\pi r} \tag{9.31}$$

In formulae describing the distribution of bending moments (9.30) and transverse force (9.31) singularity occurs for $r \to 0$. However, this effect is physically justified and describes the fact that the concentrated force causes a sharp increase of values of bending moments m_r, m_θ and transverse shear force t_r for $r \to 0$, see Figure 9.6b,c,d. These formulae can be used in the calculations for $r > 0$.

In the second approach the modified configuration of the plate presented in Figure 9.5b is adopted. The concentrated force P is replaced by uniform pressure $p = P/(\pi c^2)$ acting in the circular domain of radius $r = c$. Then, the plate with two intervals [0,c] and [c,a] is considered. The solution has to satisfy boundary conditions for $r = a$ and continuity conditions at the interface of the intervals $r = c$. Finally, it is

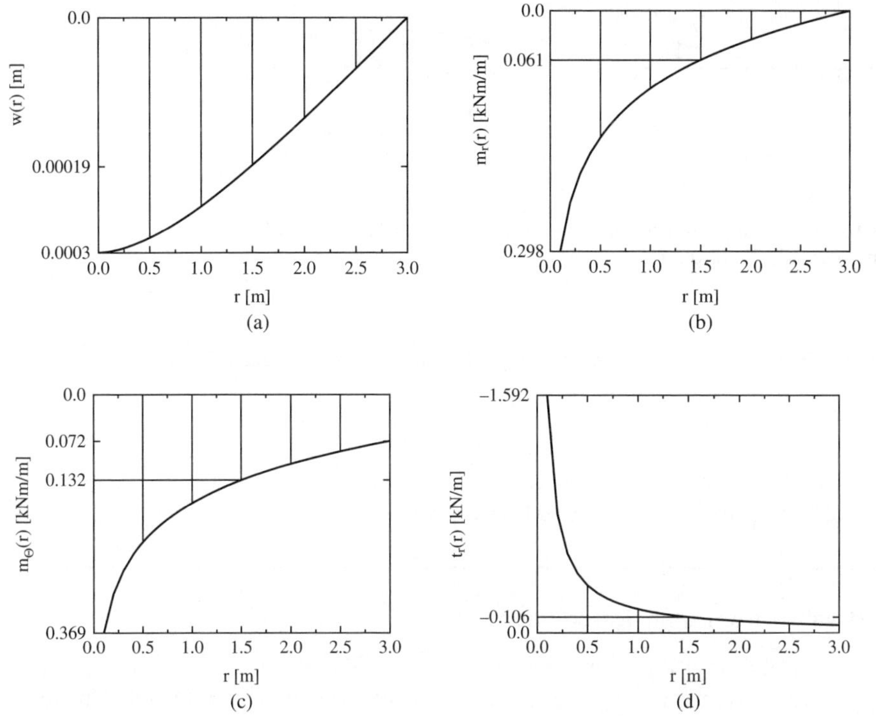

Figure 9.6 Simply supported circular plate with concentrated central force – functions obtained from analytical solution: (a) $w(r)$, (b) $m_r(r)$, (c) $m_\theta(r)$ and (d) $t_r(r)$

assumed that radius $c \to 0$ and the value of resultant load remains finite and is equal to P. For $r \in [c, a]$ the behaviour of the structure is described by the classical theory of thin plates. The solution in the interval [0,c] requires a 3D analysis of axisymmetric stress state in a cylinder of height h with radius c, loaded by pressure $p = P/(\pi c^2)$ and tractions σ_r and $\tau_{rz} = \tau$ acting on the lateral surface and resulting from the interaction of two plate parts (see Figure 9.5c). The problem is discussed in the book by Timoshenko and Woinowsky-Krieger (1959).

Attention should be paid to the fact that the concentrated transverse force induces shear stress τ_{rz} of considerable value which, as a rule, is neglected in the thin plate theory. The detailed analysis related to point loads and/or supports is however beyond our consideration.

9.6 Simply Supported Circular Plate with an Asymmetric Distributed Load

Next, a simply supported circular plate under asymmetric load is considered (see Figure 9.7). The transverse load is a product of functions of two coordinates r and θ. It is defined by a linear function of r (visible in a radial section of the circular domain) and by a trigonometric function $\cos\theta$ in the circumferential direction

$$p(r,\theta) = \hat{p}^{(1)} \frac{r}{a} \cos\theta \tag{9.32}$$

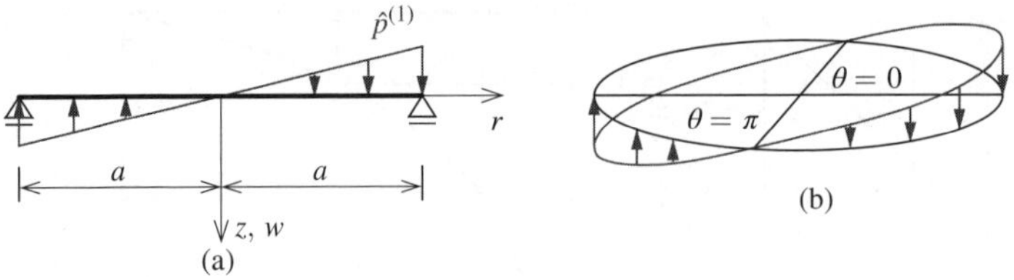

Figure 9.7 Simply supported circular plate with asymmetric load: (a) plate configuration and (b) transverse load distribution

Box 9.5 Simply supported circular plate with an asymmetric load

Data

$E = 1.0 \times 10^7$ kN/m^2, $\nu = 0.1667$, $r_{ext} = a = 2.0$ m, $h = 0.1$ m
$D^m = 857.10$ kNm, $\hat{p}^{(1)} = 10.0$ kN/m^2, $j = 1$

Check values from analytical solution

$w_{max}^{anal} = w(1.0) = 7.34 \times 10^{-4}$ m
$m_{r\,max}^{anal} = 1.615$ kNm/m, $m_{\theta}^{anal}(2.0) = 0.5171$ kNm/m, $m_{\theta\,max}^{anal} = 0.8288$ kNm/m

---

Check values from FEM solution using ANKA

$NSE = 20$, $NEN = 2$, $NEDOF = 4$, $j = 1$, SRK FEs
$w_{max}^{FEM}(1.0) = 7.34 \times 10^{-4}$ m $= w_{max}^{anal}(1.0)$
$m_{r\,max}^{FEM} = 1.615$ kNm/m $= m_{r\,max}^{anal}$, $m_{\theta}^{FEM}(2.0) = 0.5125$ kNm/m $= 0.991\ m_{\theta}^{anal}(2.0)$

Plate configuration shown in Figure 9.7

The input data such as material constants E, ν, dimensions $r_{ext} = a$, h, amplitude of surface load $\hat{p}^{(1)}$ and bending stiffness D^m are given in Box 9.5.

9.6.1 Analytical Solution

The applied surface load and the resulting deflection are expressed using one cosine function with $j = 1$. The solution of the displacement differential equation (9.9) has the form

$$w(r, \theta) = w^{(j=1)}(r)\ \cos\theta \tag{9.33}$$

where the amplitude of deflection function is

$$w^{(j=1)}(r) = w^{(1)}(r) = C_{11}\, r + C_{21}\, r^3 + C_{31}\, r^{-1} + C_{41}\, r \log r + \frac{\hat{p}^{(1)}\, r^5}{192\, D^m\, a} \tag{9.34}$$

The upper index (1) is used in the functions of amplitudes of deflection and bending moments to emphasize their distribution in circumferential direction described by function $\cos(j\theta) = \cos(1\ \theta)$. Similarly, the integration constants are denoted by two lower indices, where the second one indicates $j = 1$. Bending moments are expressed as:

$$\begin{aligned} m_r^{(1)}(r,\theta) &= -D^m \left[\frac{\mathrm{d}^2 w^{(1)}}{\mathrm{d}r^2} + \frac{\nu}{r} \left(\frac{\mathrm{d}w^{(1)}}{\mathrm{d}r} - \frac{w^{(1)}}{r} \right) \right] \cos\theta \\ m_\theta^{(1)}(r,\theta) &= -D^m \left[\frac{1}{r} \left(\frac{\mathrm{d}w^{(1)}}{\mathrm{d}r} - \frac{w^{(1)}}{r} \right) + \nu \frac{\mathrm{d}^2 w^{(1)}}{\mathrm{d}r^2} \right] \cos\theta \end{aligned} \tag{9.35}$$

The components with r^{-1} and $\log r$ have to be eliminated to avoid a singularity of the solution for $r \to 0$. To obtain finite values of deflection $w^{(1)}(0)$ and bending moments $m_r^{(1)}(0)$ and $m_\theta^{(1)}(0)$ two constants have to be equal to zero

$$C_{31} = C_{41} = 0 \tag{9.36}$$

The other two constants C_{11} and C_{21} are calculated from the mixed boundary conditions related to the simply supported edge $r = a$:

$$\begin{aligned} w^{(1)}(a) = 0 \quad &\to \quad C_{11}\ a + C_{21}\ a^3 + \frac{\hat{p}^{(1)}\ a^4}{192\ D^m} = 0 \\ m_r^{(1)}(a) = 0 \quad &\to \quad -2\ D^m(3+\nu)\ C_{21}\ a - \frac{\hat{p}^{(1)}\ a^2 (5+\nu)}{48} = 0 \end{aligned} \tag{9.37}$$

and read:

$$C_{11} = \frac{\hat{p}^{(1)}\ a^3 (7+\nu)}{192\ D^m (3+\nu)}, \qquad C_{21} = -\frac{\hat{p}^{(1)}\ a (5+\nu)}{96\ D^m (3+\nu)} \tag{9.38}$$

Finally, the solution is obtained as four functions describing the distributions of the deflection, two bending moments and a twisting moment for $\rho \in [0,1]$, where $\rho = r/a$:

$$\begin{aligned} w(\rho,\theta) &= \frac{\hat{p}^{(1)}\ a^4}{192\ D^m}\ \rho \left(\frac{7+\nu}{3+\nu} - \rho^2 \right) (1-\rho^2)\ \cos\theta \\ m_r(\rho,\theta) &= \frac{5+\nu}{48}\ \hat{p}^{(1)}\ a^2\ \rho(1-\rho^2)\ \cos\theta \\ m_\theta(\rho,\theta) &= \frac{\hat{p}^{(1)}\ a^2}{48}\ \rho \left[\frac{(5+\nu)(1+3\nu)}{3+\nu} - (1+5\nu)\ \rho^2 \right] \cos\theta \\ m_{r\theta}(\rho,\theta) &= \frac{\hat{p}^{(1)}\ a^2}{48}\ (1-\nu) \left[\frac{5+\nu}{3+\nu} - \rho^2 \right] \sin\theta \end{aligned} \tag{9.39}$$

In Figure 9.8 the diagrams of the deflection and bending moments are presented. To show the sign change in the two parts of the diagram for the radial cross-sectional line with $\theta = \pi$ and $\theta = 0$ the whole range $r \in [-2.0, 2.0]$ is considered. Three check values obtained from the analytical and numerical solutions are compared in Box 9.5.

Finally, we discuss two additional functions of transverse shear forces $t_r(r,\theta)$ and $t_\theta(r,\theta)$, which are nonzero in the solution even though they are treated as inactive since the transverse shear strains are zero. These forces can be determined from two equilibrium equations (see Box 9.1) on the basis of earlier derived functions of bending and twisting moments (and their derivatives). Note that force $t_r(r,\theta)$ is described by a cosine function in the circumferential direction and force $t_\theta(r,\theta)$ by a sine function.

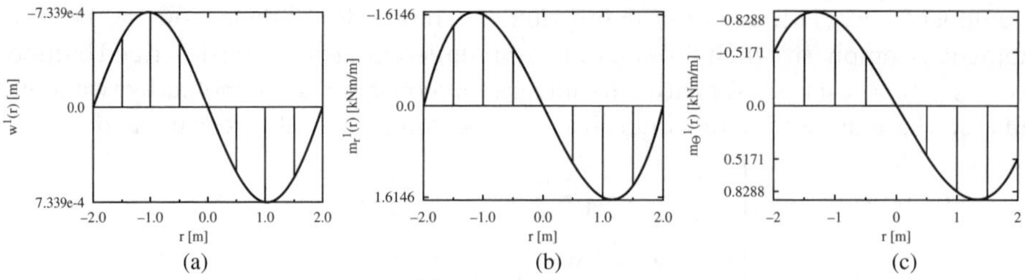

Figure 9.8 Simply supported circular plate with asymmetric load – functions: (a) $w(r)$, (b) $m_r(r)$ and (c) $m_\theta(r)$ for $r \in [-2.0, 0.0]$, $\theta = \pi$ and for $r \in [0.0, 2.0]$, $\theta = 0$

9.6.2 FEM Solution

The numerical solution is obtained using FEM program ANKA (1993). The interval $[0, a]$ related to a radial section of the circular plate is discretized with two-node one-dimensional finite elements SRK designed for analysis of circular/annular membranes/plates as well as axisymmetric shells (see Subsection 5.2.8 for details). Asymmetric behaviour of the structure is forced by setting (in the input file) the number of circumferential waves $j = 1$. As shown in Box 9.5 the values of $w_{\max}$ and $m_{r\,\max}$ obtained from the analytical and numerical calculations are the same.

9.7 Uniformly Loaded Annular Plate with Static and Kinematic Boundary Conditions

The annular plate shown in Figure 9.9a is considered. The functions of the deflection, two bending moments and transverse shear force are derived analytically and compared with the numerical FEM solution. The input data including material constants E, ν, dimensions r_{int}, r_{ext}, h, surface load p_0, uniform boundary traction $\hat{t}_r$ along the internal edge and bending stiffness D^m are listed in Box 9.6.

9.7.1 Analytical Solution

For a start the displacement fourth-order differential equation for the axisymmetric state

$$\nabla^2\nabla^2 w(r) = \frac{\hat{p}_z(r)}{D^m} \tag{9.40}$$

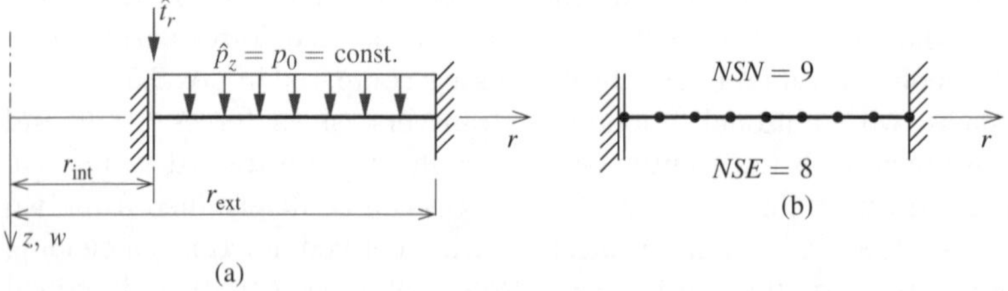

Figure 9.9 Annular plate: (a) configuration with loads and boundary constraints and (b) FE model discretized with SRK FEs

Box 9.6 Annular plate with a uniformly distributed load and different boundary conditions

Data

$E = 2.0 \times 10^7$ kN/m^2, $\nu = 0.1$, $r_{\text{int}} = 1.0$ m, $r_{\text{ext}} = 3.0$ m, $h = 0.1$ m
$D^m = 1683.5$ kNm, $\hat{t}_r = 3.0$ kN/m, $\hat{p}_z(r) = p_0 = 6.0$ kN/m^2 = const.

Check values from analytical and FEM solutions using ANKA

$NSE = 8$, $NEN = 2$, $NEDOF = 4$, $j = 0$, SRK FEs
$w_{\max}^{\text{anal}}(1.0) = w_{\max}^{\text{FEM}}(1.0) = 2.27 \times 10^{-3}$ m
$m_r^{\text{anal}}(1.0) = m_r^{\text{FEM}}(1.0) = 6.00$ kNm/m, $m_r^{\text{anal}}(3.0) = m_r^{\text{FEM}}(3.0) = -6.00$ kNm/m
$m_\theta^{\text{anal}}(1.0) = m_\theta^{\text{FEM}}(1.0) = 0.60$ kNm/m, $m_\theta^{\text{anal}}(3.0) = m_\theta^{\text{FEM}}(3.0) = -0.60$ kNm/m
$t_r^{\text{anal}}(1.0) = t_r^{\text{FEM}}(1.0) = -3.00$ kNm/m, $t_r^{\text{anal}}(3.0) = t_r^{\text{FEM}}(3.0) = -9.00$ kNm/m

---

Plate configuration shown in Figure 9.9

and its integral

$$w(r) = C_{10} + C_{20}\, r^2 + C_{30}\, \log r + C_{40}\, r^2 \log r + \frac{p_0\, r^4}{64\, D^m} \tag{9.41}$$

are recalled. Four integration constants C_{i0}, for $i = 1, \ldots, 4$, are derived from four boundary conditions formulated for the internal and external (two for each) contours of the annular plate:

- mixed conditions for the clamped edge with vertical translation allowed for r_{int}:

$$\vartheta_r(r_{\text{int}}) = -\frac{\mathrm{d}w}{\mathrm{d}r} = 0, \quad t_r(r_{\text{int}}) = -\hat{t}_r \tag{9.42}$$

- kinematic conditions for the clamped edge with radius r_{ext}:

$$w(r_{\text{ext}}) = 0, \quad \vartheta_r(r_{\text{ext}}) = -\frac{\mathrm{d}w}{\mathrm{d}r} = 0 \tag{9.43}$$

Using the formulae presented in Section 9.3 the following set of four algebraic equations, resulting from these boundary conditions, is written:

$$\begin{aligned}
&-2\, r_{\text{int}}\, C_{20} - \frac{1}{r_{\text{int}}}\, C_{30} - (2\, r_{\text{int}} \log r_{\text{int}} + r_{\text{int}})\, C_{40} - \frac{p_0\, r_{\text{int}}^3}{16\, D^m} = 0 \\
&-D^m \frac{4}{r_{\text{int}}}\, C_{40} - \frac{p_0\, r_{\text{int}}}{2} = -\hat{t}_r \\
&C_{10} + r_{\text{ext}}^2\, C_{20} + \log r_{\text{ext}}\, C_{30} + r_{\text{ext}}^2 \log r_{\text{ext}}\, C_{40} + \frac{p_0\, r_{\text{ext}}^4}{64\, D^m} = 0 \\
&-2\, r_{\text{ext}}\, C_{20} - \frac{1}{r_{\text{ext}}}\, C_{30} - (2\, r_{\text{ext}} \log r_{\text{ext}}) + r_{\text{ext}})\, C_{40} - \frac{p_0\, r_{\text{ext}}^3}{16\, D^m} = 0
\end{aligned} \tag{9.44}$$

Taking the numerical data included in Box 9.6 into account and dividing the second equation by the factor $-4\ D^m$ the system of equations in a matrix form can be written as follows

$$\begin{bmatrix} 0.0 & -2.0 & -1.0 & -1.0 \\ 0.0 & 0.0 & 0.0 & 1.0 \\ 1.0 & 9.0 & 1.098 & 9.88 \\ 0.0 & -6.0 & -0.333 & -9.592 \end{bmatrix} \begin{bmatrix} C_{10} \\ C_{20} \\ C_{30} \\ C_{40} \end{bmatrix} = \begin{bmatrix} 0.0002227 \\ 0.0 \\ -0.004511 \\ 0.006014 \end{bmatrix} \tag{9.45}$$

The values of calculated integration constants are:

$$\begin{aligned} C_{10} &= 0.333 \times 10^{-2}\ \text{m}, \qquad C_{20} = -0.111 \times 10^{-2}\ \text{m}^{-1} \\ C_{30} &= 0.200 \times 10^{-2}\ \text{m}, \qquad C_{40} = 0 \end{aligned} \tag{9.46}$$

Introducing the constants into the formulae describing functions $w(r)$, $m_r(r)$, $m_\theta(r)$, and $t_r(r)$ the diagrams for interval $[r_{\text{int}}, r_{\text{ext}}] = [1.0, 3.0]$ are obtained and presented in Figure 9.10. The characteristic values of the deflection and bending moments are listed in Box 9.6.

It is worthwhile to show the actual generalized forces (transverse shear forces and two bending moments) acting along the radial and circumferential cross-sectional lines limiting two elementary segments located in the vicinity of the internal and external contours of the annular plate, see Figure 9.11. The signs of generalized forces depend on the axes of cylindrical coordinate system (here axis z is directed downwards) and result from the adopted sign convention presented in Figure 9.1.

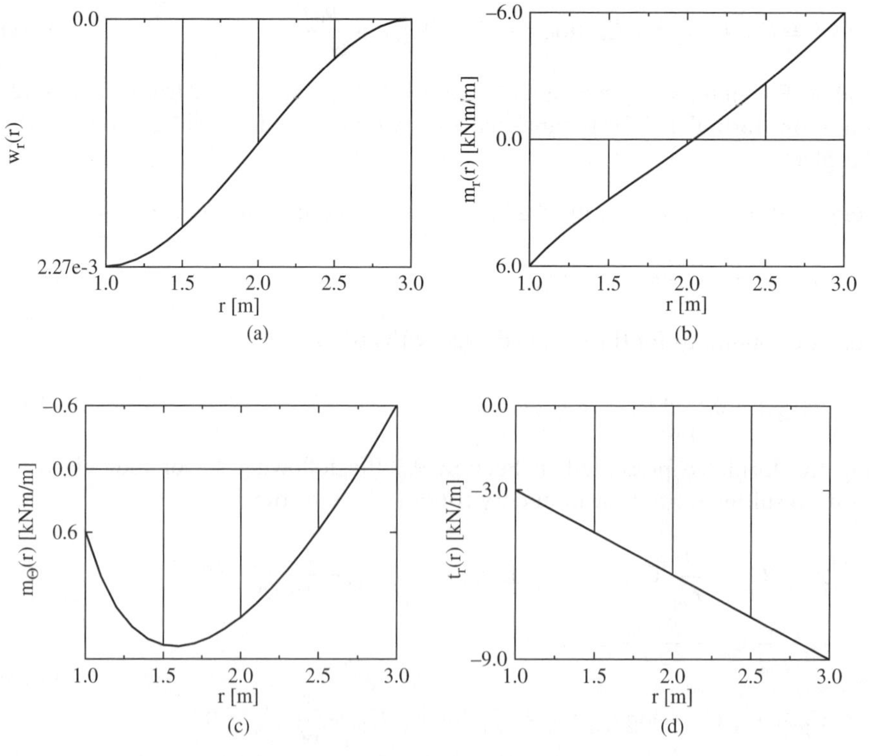

Figure 9.10 Annular plate – results of analytical solution: (a) deflection $w(r)$, (b) bending moment $m_r(r)$, (c) bending moment $m_\theta(r)$ and (d) transverse shear force $t_r(r)$

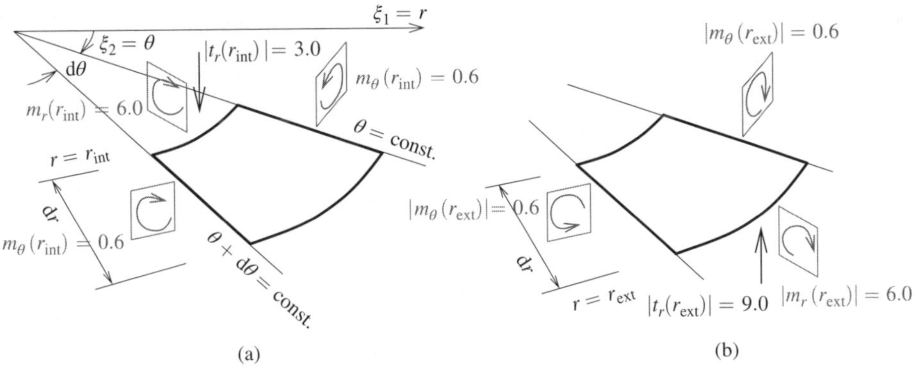

Figure 9.11 Annular plate – two elementary segments with bending moments and transverse shear forces in the vicinity of the: (a) internal edge and (b) external edge

9.7.2 Numerical Solution using FEM

The numerical solution is obtained using FEM program ANKA (1993). The annular plate is discretized (as in the previous example) with two-node one-dimensional SRK FEs (Figure 9.9b). The axisymmetric behaviour of the structure is forced by setting $j = 0$ in the input file. Each node of an SRK element represents the whole circle $r =$ const. The introduction of kinematic boundary conditions (zero deflection w for $r = r_{\text{ext}}$ and zero rotations ϑ_r for $r = r_{\text{int}}$ and $r = r_{\text{ext}}$) is straightforward, while the static boundary condition for $r = r_{\text{int}}$ requires the calculation of the resultant value for the whole internal perimeter, which is equal to $2\pi\, r_{\text{int}}\, \hat{t}_r$. Conversely, to find the intensity of the uniformly distributed vertical reaction at the external supported edge the calculated nodal value of the resultant reaction force has to be divided by the length of the external contour $2\pi\, r_{\text{ext}}$. When the SRK FEs are used, the deflection field in the bending plate is approximated by cubic Hermitean polynomials of coordinate r, which guarantees high solution quality. In Box 9.6 the results of analytical and numerical calculations are compared and the check values are equal.

References

ANKA 1993 ANKA – computer code for nonlinear analysis of structures: User's manual. Technical report, Cracow University of Technology, Cracow (in Polish).

Batoz JL and Dhatt G 1990 *Modélisation des Structures par Élément Finis*. Hermes, Paris.

Katili I 1993 A new discrete Kirchhoff–Mindlin element based on Mindlin–Reissner plate theory and assumed shear strain fields. Part I: An extended DKT element for thick-plate bending analysis, Part II: An extended DKQ element for thick-plate bending analysis. *International Journal for Numerical Methods in Engineering* **36**(11), 1859–1883, 1885–1908.

Timoshenko S and Woinowsky-Krieger S 1959 *Theory of Plates and Shells*. McGraw-Hill, New York-Auckland.

Waszczyszyn Z, Cichoń C and Radwańska M 1994 *Stability of Structures by Finite Element Methods*. Elsevier, Amsterdam.

Part 3

Shells

10

Shells in the Membrane State

10.1 Introduction

The membrane state of shells is the most advantageous as regards to the strength of materials. It guarantees the optimal use of the material in the cross section and thus in the whole structure. The stress distribution along the thickness is uniform (constant) and as a result of integration in the thickness direction only membrane forces occur, that is there are no bending or twisting moments and transverse shear forces.

The following assumptions have to be satisfied to assure the membrane state:

- middle surface is smooth (without sharp bends, i.e. internal edges)
- surface loads are continuous and smooth, boundary loads and reactions act tangentially to the middle surface
- displacement w normal to the middle surface and rotation ϑ_ν in the plane normal to the boundary are not constrained

The attention is focused on the description of shells with axisymmetric geometry. Depending on the shape of the shell different coordinate systems: spherical ($\xi_1 = \varphi$, $\xi_2 = \theta$) or cylindrical ($\xi_1 = x$, $\xi_2 = \theta$) are used. As previously stated in Section 1.3, the geometry of an axisymmetric shell is defined by means of:

- meridian equation $r = r(\xi_1)$
- two Lame parameters $A_\alpha(\xi_1)$, $\alpha = 1, 2$
- two radii of principal curvature $R_\alpha(\xi_1)$, $\alpha = 1, 2$

which are all independent of $\xi_2 = \theta$. Note that the second Lame parameter for both coordinate systems is $A_2 = r(\xi_1)$.

The assumption of axisymmetric geometry results in the cancellation of some components in the generally formulated Sanders kinematic and equilibrium equations due to the conditions $\partial A_\alpha / \partial \xi_2 = 0$ (see Section 3.1).

Numerous analytical solutions of BVPs for shells in membrane state can be found in the books of Girkmann (1956); Kolkunov (1972); Timoshenko and Woinowsky-Krieger (1959). A selection of typical examples is considered in this chapter.

Plate and Shell Structures: Selected Analytical and Finite Element Solutions, First Edition.
Maria Radwańska, Anna Stankiewicz, Adam Wosatko and Jerzy Pamin.
© 2017 John Wiley & Sons Ltd. Published 2017 by John Wiley & Sons Ltd.

10.2 General Membrane State in Shells of Revolution

For the surface loads $\hat{\mathbf{p}}_{(3\times1)} = [\hat{p}_1, \hat{p}_2, \hat{p}_n]^{\mathrm{T}}$ [N/m^2], with boundary actions $\hat{\mathbf{p}}_{b(2\times1)} = [\hat{n}_v, \hat{n}_{vs}]^{\mathrm{T}}$ [N/m] and kinematic constraints $\hat{\mathbf{u}}_{b(2\times1)} = [\hat{u}_v, \hat{u}_s]^{\mathrm{T}}$ [m] the following fields are generated:

- displacements (translations) $\mathbf{u}_{(3\times1)}(\xi_1, \theta) = [u_1, u_2, w]^{\mathrm{T}} = [u, v, w]^{\mathrm{T}}$ [m]
- membrane strains $\mathbf{e}^n_{(3\times1)}(\xi_1, \theta) = [\varepsilon_{11}, \varepsilon_{22}, \gamma_{12}]^{\mathrm{T}}$ [−]
- membrane forces $\mathbf{s}^n_{(3\times1)}(\xi_1, \theta) = \mathbf{n} = [n_{11}, n_{22}, n_{12}]^{\mathrm{T}}$ [N/m]

Coordinate ξ_1 depends on the coordinate system ($\xi_1 = \varphi$ or $\xi_1 = x$) and remains a variable whereas $\xi_2 = \theta$. Hereafter, the notation $r' = \partial r/\partial \xi_1$ and $A = A_1$ is introduced in the presented equations.

The local and global formulations of the mathematical model for the considered problem are presented in Box 10.1.

Box 10.1 Equations for shells of revolution in a general membrane state

Local formulation – set of nine equations and boundary conditions

(I) kinematic equations (3):

$$\varepsilon_{11} = \frac{1}{A}\frac{\partial u_1}{\partial \xi_1} + \frac{w}{R_1}$$

$$\varepsilon_{22} = \frac{1}{r}\frac{\partial u_2}{\partial \theta} + \frac{r'}{rA}u_1 + \frac{w}{R_2}$$

$$\gamma_{12} = \frac{r}{A}\frac{\partial}{\partial \xi_1}\left(\frac{u_2}{r}\right) + \frac{1}{r}\frac{\partial u_1}{\partial \theta}$$

(II) equilibrium equations (3):

$$\frac{\partial(r\, n_{11})}{\partial \xi_1} + A\frac{\partial n_{12}}{\partial \theta} - r'\, n_{22} + A\, r\, \hat{p}_1 = 0$$

$$\frac{\partial(r\, n_{12})}{\partial \xi_1} + A\frac{\partial n_{22}}{\partial \theta} + r'\, n_{12} + A\, r\, \hat{p}_2 = 0$$

$$n_{11}/R_1 + n_{22}/R_2 - \hat{p}_n = 0$$

(III) constitutive equations (3):

$$n_{11} = D^n(\varepsilon_{11} + \nu\varepsilon_{22}), \qquad n_{22} = D^n(\varepsilon_{22} + \nu\varepsilon_{11}), \qquad n_{12} = D^n\,(1-\nu)\,\gamma_{12}/2$$

where $D^n = E\,h/(1-\nu^2)$

(IV) boundary conditions:

a) kinematic: $u_v = \hat{u}_v$, $u_s = \hat{u}_s$

b) static: $n_v = \hat{n}_v$, $n_{vs} = \hat{n}_{vs}$

c) mixed (one kinematic and one static)

Global formulation

Total potential energy $\Pi^n = U^n - W^n$

Internal strain energy

$$U^n = \frac{D^n}{2}\int_{\Omega}(n_{11}\,\varepsilon_{11} + n_{12}\,\gamma_{12} + n_{22}\,\varepsilon_{22})\,\mathrm{d}\Omega$$

External load work

$$W^n = \int_{\Omega}(\hat{p}_1\,u_1 + \hat{p}_2\,u_2 + \hat{p}_n\,w)\,\mathrm{d}\Omega + \int_{\partial\Omega_\sigma}(\hat{n}_\nu\,u_\nu + \hat{n}_{\nu s}\,u_s)\,\mathrm{d}\partial\Omega$$

One can notice in Box 10.1 that there are three unknown membrane forces in three equilibrium equations (II). These equations can be solved when suitable static boundary conditions are postulated. This means that the shells in membrane state are statically determinate structures. Having the membrane forces calculated, one can compute membrane strains from algebraic constitutive equations (III) and then determine membrane displacements integrating kinematic differential equations (I) and using kinematic boundary conditions.

The set of nine equations can be reduced to three displacement differential equations with unknowns u_1, u_2, w. The boundary conditions can be formulated in terms of displacements u_1 and u_2 since in the kinematic equations only derivatives $\partial u_1/\partial \xi_\alpha$ and $\partial u_2/\partial \xi_\alpha$ occur. Normal displacement w cannot be used in the formulation of boundary conditions due to the lack of its derivatives in the kinematic equations. In fact, according to the assumptions of membrane state, the normal displacement w cannot be constrained.

10.3 Axisymmetric Membrane State

In axisymmetric shells with zero surface load $\hat{p}_2 = 0$ the displacement u_2 tangent to the parallel, shear force n_{12} and strain γ_{12} equal zero.

To describe the behaviour of a shell in axisymmetric membrane state six functions (of one coordinate ξ_1) have to be determined:

- displacements (translations) $\mathbf{u}_{(2\times1)}(\xi_1) = [u, w]^{\mathrm{T}}$ [m]
- membrane strains $\mathbf{e}^n_{(2\times1)}(\xi_1) = [\varepsilon_{11}, \varepsilon_{22}]^{\mathrm{T}}$ [−]
- membrane forces $\mathbf{s}^n_{(2\times1)}(\xi_1) = \mathbf{n} = [n_{11}, n_{22}]^{\mathrm{T}}$ [N/m]

The set of six equations (local formulation) and the formulae for the global formulation related to the considered case are presented in Box 10.2. In the equations of axisymmetric state all derivatives with respect to circumferential coordinate are zero $\partial(\ )/\partial\theta = 0$ and derivatives $\mathrm{d}(\ )/\mathrm{d}\xi_1$ are denoted by $(\ )'$.

Box 10.2 Equations for shells of revolution in an axisymmetric membrane state

Local formulation – set of six equations and boundary conditions

(I) kinematic equations (2):

$$\varepsilon_{11} = \frac{1}{A}\,u' + \frac{w}{R_1}, \qquad \varepsilon_{22} = \frac{r'}{r\,A}\,u + \frac{w}{R_2}$$

(II) equilibrium equations (2):

$$(r\,n_{11})' - r'\,n_{22} + A\,r\,\hat{p}_1 = 0, \qquad n_{11}/R_1 + n_{22}/R_2 - \hat{p}_n = 0$$

(III) constitutive equations (2):

$$n_{11} = D^n(\varepsilon_{11} + \nu\varepsilon_{22}), \qquad n_{22} = D^n(\varepsilon_{22} + \nu\varepsilon_{11})$$

where $D^n = E\,h/(1-\nu^2)$

(IV) boundary conditions – one kinematic or one static:

$$u_\nu = \hat{u}_\nu \quad \text{or} \quad n_\nu = \hat{n}_\nu$$

Global formulation

Total potential energy $\Pi^n = U^n - W^n$

Internal strain energy

$$U^n = \frac{D^n}{2}\int_\Omega (n_{11}\,\varepsilon_{11} + n_{22}\,\varepsilon_{22})\,\mathrm{d}\Omega$$

External load work

$$W^n = \int_\Omega (\hat{p}_1\,u + \hat{p}_n\,w)\,\mathrm{d}\Omega + \int_{\partial\Omega_\sigma} \hat{n}_\nu\,u_\nu\,\mathrm{d}\partial\Omega$$

In the analytical solution for the axisymmetric membrane state, two formulae describing the membrane forces, meridional $n_{11}(\xi_1)$ and circumferential $n_{22}(\xi_1)$, can be found using equilibrium equations only as shown in the following subsections.

10.3.1 Membrane Forces for Shells Described in a Spherical Coordinate System

The relations necessary to find the distributions of meridional force $n_\varphi(\varphi)$ and circumferential force $n_\theta(\varphi)$ for a shell defined in the spherical coordinate system ($\xi_1 = \varphi, \xi_2 = \theta$) are presented in this section.

Let us consider a segment of a shell (Figure 10.1a) bounded by the initial angle φ_0 and running angle φ.

To formulate the equations in the spherical coordinate system it is necessary to define the following geometric parameters of the shell:

$$r = r(\varphi), \qquad A_1 = R_1(\varphi), \qquad A_2 = r(\varphi), \qquad R_2 = r(\varphi)/\sin\varphi \tag{10.1}$$

The condition $\Sigma P = 0$ includes the projections of all loads onto the vertical axis: (i) the resultant of the meridional force $n_\varphi(\varphi)$ occurring in the horizontal intersection of the

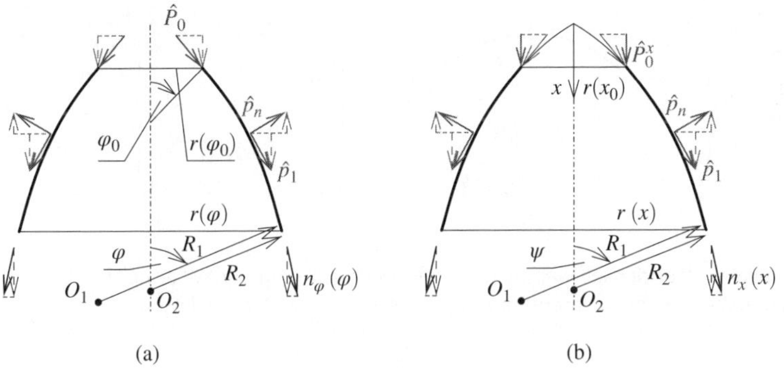

Figure 10.1 Shells of revolution in axisymmetric membrane state – analysis of equilibrium of shell segment described in: (a) spherical and (b) cylindrical coordinate system

shell with running coordinate φ, (ii) the resultant of the boundary load $\hat{P}_0$ acting at the top contour with radius $r(\varphi_0)$ and (iii) the resultant of surface loads $\hat{p}_1(\varphi)$ and $\hat{p}_n(\varphi)$

$$2\pi\ r(\varphi)\ [n_\varphi(\varphi)\ \sin\varphi] + 2\pi\ r(\varphi_0)\ [\hat{P}_0\ \sin\varphi_0]$$
$$+2\pi \int_{\varphi_0}^{\varphi} [\hat{p}_1(\varphi)\ \sin\varphi - \hat{p}_n(\varphi)\ \cos\varphi]\ r(\varphi)\ R_1(\varphi)\ \mathrm{d}\varphi = 0 \tag{10.2}$$

From this equation the general formula for the meridional force is derived

$$n_\varphi(\varphi) = -\frac{1}{r(\varphi)\ \sin\varphi}\left[\hat{P}_0\ \sin\varphi_0\ r(\varphi_0) + \int_{\varphi_0}^{\varphi} [\hat{p}_1(\varphi)\ \sin\varphi - \hat{p}_n(\varphi)\ \cos\varphi]\ r(\varphi)\ R_1(\varphi)\ \mathrm{d}\varphi\right] \tag{10.3}$$

The circumferential force is determined from the second algebraic equation of equilibrium (see Box 10.2)

$$n_\theta(\varphi) = -\frac{R_2(\varphi)}{R_1(\varphi)}\ n_\varphi(\varphi) + R_2(\varphi)\ \hat{p}_n(\varphi) \tag{10.4}$$

10.3.2 Membrane Forces for Shells Described in a Cylindrical Coordinate System

To describe a segment of a shell defined in the cylindrical coordinate system with $\xi_1 = x$ and $\xi_2 = \theta$ (Figure 10.1b) the following information about the shell geometry is used:

$$\begin{aligned} &r = r(x), \qquad r' = \mathrm{d}r/\mathrm{d}x \\ &A_1 = \sqrt{1+(r')^2}, \qquad A_2 = r(x), \qquad R_1 = R_1(x), \qquad R_2(x) = r(x)/\sin\psi \end{aligned} \tag{10.5}$$

where angle ψ is marked in Figure 10.1b. Now, the equilibrium equation $\Sigma P_x = 0$ is written

$$2\pi\ \frac{r(x)}{\sqrt{1+(r')^2}}\ n_x(x) + 2\pi\ r(x_0)\ \hat{P}_0^x + 2\pi \int_{x_0}^{x} [\hat{p}_1(x) - \hat{p}_n(x)\ r']\ r(x)\ \mathrm{d}x = 0 \tag{10.6}$$

from which the general formula for the meridional force is determined

$$n_x(x) = -\frac{\sqrt{1+(r')^2}}{r(x)}\left[\hat{P}_0^x\ r(x_0) + \int_{x_0}^{x} [\hat{p}_1(x) - \hat{p}_n(x)\ r']\ r(x)\ \mathrm{d}x\right] \tag{10.7}$$

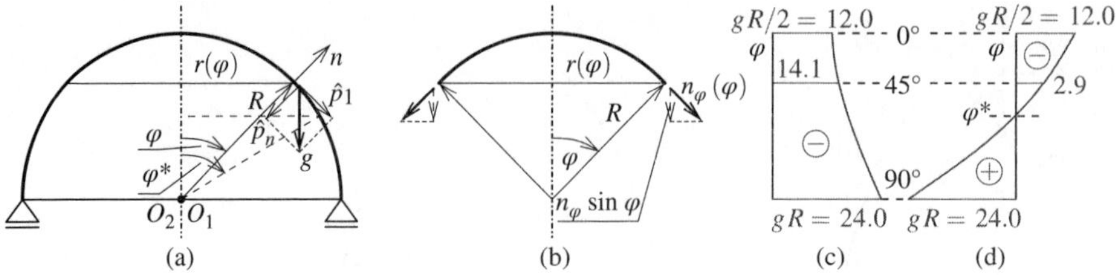

Figure 10.2 Hemispherical shell under self weight: (a) configuration, (b) segment of a shell in equilibrium; diagrams of forces [kN/m]: (c) meridional $n_\varphi(\varphi)$ and (d) circumferential $n_\theta(\varphi)$

The circumferential force $n_\theta(x)$ is obtained from the second algebraic equilibrium equation

$$n_\theta(x) = -\frac{R_2(x)}{R_1(x)}\, n_x(x) + R_2(x)\, \hat{p}_n(x) \tag{10.8}$$

10.4 Hemispherical Shell

10.4.1 Shell under Self Weight – Analytical Solution

In the example, only meridional force $n_\varphi(\varphi)$ and circumferential force $n_\theta(\varphi)$ (independent of coordinate θ) are determined (Figure 10.2).

The geometry of an axisymmetric hemispherical shell is defined in the spherical coordinate system by:

$$\begin{aligned}&\xi_1 = \varphi, \quad r(\varphi) = R\ \sin\varphi, \qquad A_1 = R_1 = R_2 = R = \text{const.}, \quad A_2 = r(\varphi) = R\ \sin\varphi\\ &r_0 = 0, \qquad \varphi_0 = 0^\circ, \qquad \varphi_L = 90^\circ\end{aligned} \tag{10.9}$$

The self weight has two nonzero components ($\hat{p}_2 = 0$):

$$\hat{p}_1(\varphi) = g\ \sin\varphi, \qquad \hat{p}_n(\varphi) = -g\ \cos\varphi \tag{10.10}$$

The input data $(R, h, r_0, \varphi_0, \varphi_L, \gamma_c, \hat{P}_0)$ adopted in the calculations and selected check results are included in Box 10.3. If the only purpose of the calculations is to find the functions of membrane forces, it is not necessary to adopt particular values of material parameters E, ν.

The formula describing the distribution of the meridional force, derived in Subsection 10.3.1 and modified for the considered case takes the form

$$\begin{aligned}n_\varphi(\varphi) &= -\frac{1}{(R\ \sin\varphi)\ \sin\varphi}\int_{0^\circ}^{\varphi} [\hat{p}_1(\varphi)\ \sin\varphi - \hat{p}_n(\varphi)\ \cos\varphi](R\ \sin\varphi)\, R\, \mathrm{d}\varphi\\ &= -\frac{1}{R\ \sin^2\varphi}\int_{0^\circ}^{\varphi} [g\ \sin^2\varphi + g\ \cos^2\varphi](R^2\ \sin\varphi)\ \mathrm{d}\varphi = -\frac{g\,R}{1+\cos\varphi}\end{aligned} \tag{10.11}$$

and the formulae for characteristic values are obtained as follows:

$$n_\varphi(0^\circ) = -g\,R/2, \qquad n_\varphi(45^\circ) = -g\,R/(1+0.707), \qquad n_\varphi(90^\circ) = -g\,R \tag{10.12}$$

Box 10.3 Hemispherical shell under self weight in an axisymmetric membrane state

Data

$R = 5.0$ m, $h = 0.2$ m, $r_0 = 0$ m, $\varphi_0 = 0°$, $\varphi_L = 90°$
$\gamma_c = 24.0$ kN/m^3, $g = \gamma_c\, h = 4.8$ kN/m^3, $\hat{P}_0 = 0$

Check values from analytical solution

$n_\varphi(0°) = -12.0$ kN/m, $n_\varphi(45°) = -14.06$ kN/m, $n_\varphi(90°) = -24.0$ kN/m
$n_\theta(0°) = -12.0$ kN/m, $n_\theta(45°) = -2.91$ kN/m, $n_\theta(90°) = 24.0$ kN/m

---

Shell configuration shown in Figure 10.2

Then, the expression for the circumferential force is written and the characteristic values are calculated and presented in Box 10.3:

$$n_\theta(\varphi) = -\frac{R_2}{R_1}\, n_\varphi(\varphi) + R_2\, \hat{p}_n(\varphi) = \frac{g\,R}{1+\cos\varphi} - R\,g\cos\varphi = -\frac{g\,R(\cos\varphi - \sin^2\varphi)}{1+\cos\varphi} \tag{10.13}$$

$$n_\theta(0°) = -g\,R/2 = n_\varphi(0°), \qquad n_\theta(45°) = -g\,R\,\frac{0.707 - 0.707^2}{1+0.707}, \qquad n_\theta(90°) = g\,R \tag{10.14}$$

In a closed spherical shell in an axisymmetric state, at the point of symmetry axis $\varphi = 0°$, meridional and circumferential forces are equal (analogy to equality of two membrane forces in circular membrane and two bending moments in circular plate).

Note that the circumferential force changes the sign at the level defined by angle φ^*, for which the expression in brackets in the numerator of formula (10.13) is equal to zero

$$(\cos\varphi^* - \sin^2\varphi^*) = 0 \quad \rightarrow \quad \varphi^* = 51°49' \tag{10.15}$$

In the case of a hemisphere the meridional force at the bottom edge is taken over by the continuous vertical support. When $\varphi_L \neq 90°$ then the vertical component of the meridional force becomes a distributed reaction $r_V = n_\varphi(\varphi_L)\sin\varphi_L$ [kN/m], while the horizontal component of the compressive meridional force $r_H = n_\varphi(\varphi_L)\cos\varphi_L$ acts as a uniform load of the supporting ring, inducing tensile axial force in it.

10.4.2 Shell under Uniform Pressure – Analytical Solution

Now the behaviour of a hemispherical shell with geometry characterized in the previous subsection and loaded by uniform pressure $\hat{p}_n = p =$ const. ($\hat{p}_1 = \hat{p}_2 = 0$) is considered, see Figure 10.3a. In this example strains and displacements are found next to membrane forces.

The input data $(E, \nu, R, h, \varphi_0, \varphi_L, p)$ adopted in the calculations and selected check results are included in Box 10.4.

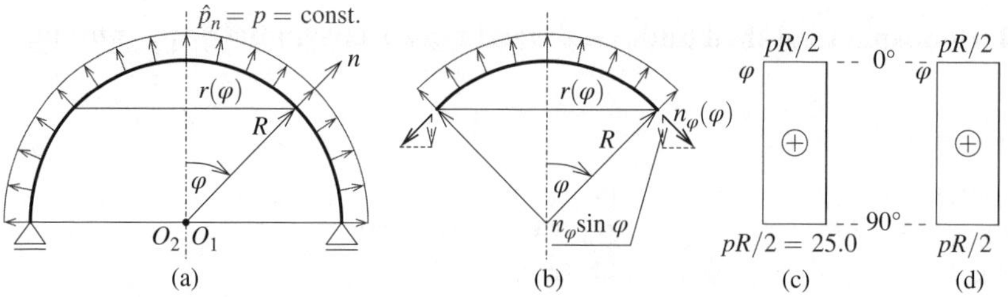

Figure 10.3 Hemispherical shell under normal pressure: (a) configuration, (b) segment of shell limited by running angle φ, analysed in equilibrium state; diagrams of membrane forces [kN/m]: (c) meridional $n_\varphi(\varphi)$ and (d) circumferential $n_\theta(\varphi)$

Box 10.4 Hemispherical shell under normal pressure in an axisymmetric membrane state

Data

$E = 2.0 \times 10^7$ kN/m^2, $\nu = 0.1$
$R = 5.0$ m, $h = 0.2$ m, $\varphi_0 = 0°$, $\varphi_L = 90°$
$\hat{p}_n = p = 10.0$ kN/m^2

Check values from analytical solution

$n_\varphi = n_\theta = 25.0$ kN/m, $\varepsilon_\varphi = \varepsilon_\theta = 0.437 \times 10^{-5}$, $w = 2.18 \times 10^{-5}$ m

---

Shell configuration shown in Figure 10.3

The general formulae for the membrane forces related to the spherical coordinate system are rewritten, see Equations (10.3) and (10.4):

$$n_\varphi(\varphi) = -\frac{1}{r(\varphi)\ \sin\varphi}\left[\hat{P}_0\ \sin\varphi_0\ r(\varphi_0) + \int_{\varphi_0}^{\varphi} [\hat{p}_1(\varphi)\ \sin\varphi - \hat{p}_n(\varphi)\ \cos\varphi]\ r(\varphi)\ R_1(\varphi)\ \mathrm{d}\varphi\right]$$
$$n_\theta(\varphi) = -\frac{R_2(\varphi)}{R_1(\varphi)}\ n_\varphi(\varphi) + R_2(\varphi)\ \hat{p}_n(\varphi) \tag{10.16}$$

Substituting the following equations:

$$r(\varphi) = R\ \sin\varphi, \qquad R_1 = R_2 = R, \qquad \varphi_0 = 0°, \qquad \hat{P}_0 = 0, \qquad \hat{p}_n = p \tag{10.17}$$

new formulae for membrane forces are derived:

$$n_\varphi(\varphi) = \frac{1}{R\ \sin^2\varphi}\int_{0°}^{\varphi} [p\ \cos\varphi](R\ \sin\varphi)\ R\ \mathrm{d}\varphi = \frac{p\,R}{\sin^2\varphi}\ \frac{\sin^2\varphi}{2} = \frac{p\,R}{2} = \text{const.}$$
$$n_\theta(\varphi) = -n_\varphi(\varphi) + R\,\hat{p}_n = -\frac{p\,R}{2} + R\,p = \frac{p\,R}{2} = \text{const.} \tag{10.18}$$

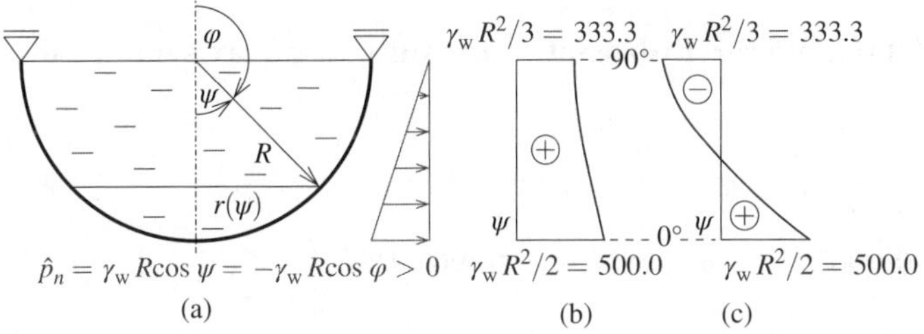

Figure 10.4 Hemispherical tank: (a) configuration; diagrams of membrane forces [kN/m]: (b) meridional $n_\varphi(\psi)$ and (c) circumferential $n_\theta(\psi)$

The calculated constant membrane forces result in constant strain fields obtained from the constitutive equations adopting material parameters E and ν

$$\varepsilon_\varphi = \varepsilon_\theta = (1-\nu)p\,R/(2\,E\,h) = \text{const.} \tag{10.19}$$

The only nonzero displacement w is determined from the algebraic kinematic equation

$$\varepsilon_\theta = w/R \quad \rightarrow \quad w = \varepsilon_\theta\,R \quad \rightarrow \quad w = (1-\nu)p\,R^2/(2\,E\,h) = \text{const.} \tag{10.20}$$

Summarizing, in the considered case of spherical shell under uniform pressure, constant fields of membrane forces, strains and normal displacement occur.

10.4.3 Suspended Tank under Hydrostatic Pressure – Analytical Solution

The considered shell is suspended at the top contour (described by coordinate $\psi = 90°$ or $\varphi = 90°$) in such a way that only the vertical displacement u (tangent to the meridian) is prevented. Moreover, due to axially symmetric state the rigid movement of the top boundary is not allowed in the horizontal plane. In the solution two alternative angular coordinates $\psi \in [0°, 90°]$ and $\varphi = 180° - \psi \in [90°, 180°]$ are introduced, see Figure 10.4a.

The hydrostatic pressure in a hemispherical shell, fully filled with water is described by the function

$$\hat{p}_n(\psi) = \gamma_w\,R\,\cos\psi = -\gamma_w\,R\,\cos\varphi > 0 \tag{10.21}$$

The input data R, γ_w and the range of angular coordinates ψ and φ are required in calculations of membrane forces and are listed in Box 10.5.

The distribution of meridional force is described by the following indefinite integral as a modification of Equation (10.3) using angular coordinate ψ

$$n_\varphi(\psi) = -\frac{1}{(R\,\sin\psi)\,\sin\psi}\int[-\hat{p}_n(\psi)\,\cos\psi]\,r(\psi)\,R_1(\psi)\,\mathrm{d}\psi$$
$$= -\frac{1}{R\,\sin^2\psi}\int[-(\gamma_w\,R\,\cos\psi)\,\cos\psi]\,R^2\,\sin\psi\,\mathrm{d}\psi = -\frac{\gamma_w}{R\,\sin^2\psi}\left[\frac{1}{3}\,R^3\,\cos^3\psi + C_1\right] \tag{10.22}$$

Box 10.5 Suspended tank under hydrostatic pressure in an axisymmetric membrane state

Data

$R = 10.0$ m, $\quad \psi \in [0°, 90°] \quad \varphi = 180° - \psi \in [90°, 180°]$
$\gamma_w = 10.0$ kN/m^3

Check values from analytical solution

$n_\varphi(0°) = 500.0$ kN/m, $\quad n_\varphi(90°) = 333.0$ kN/m
$n_\theta(0°) = 500.0$ kN/m, $\quad n_\theta(90°) = -333.0$ kN/m

---

Shell configuration shown in Figure 10.4

The integration constant C_1 is determined assuming a finite value of the meridional force at the point on the symmetry axis with $\psi = 0°$. Since $\sin\psi \to 0$ occurs in the denominator the expression in brackets in Equation (10.22) has to be equal zero, giving the value of C_1

$$\frac{1}{3} R^3 \cos^3(0°) + C_1 = 0 \quad \rightarrow \quad C_1 = -R^3/3 \tag{10.23}$$

Substituting constant C_1 into Equation (10.22) and eliminating $\sin^2\psi$ from the denominator (using trigonometric relation $\sin^2\psi = 1 - \cos^2\psi$) the final formula for the meridional force is obtained

$$n_\varphi(\psi) = \frac{\gamma_w R^2}{\sin^2\psi}\left[\frac{\cos^3\psi}{3} - \frac{1}{3}\right] = \frac{\gamma_w R^2}{3}\left[\frac{1}{1+\cos\psi} + \cos\psi\right] \tag{10.24}$$

The circumferential force is described by the following formula

$$n_\theta(\psi) = -n_\varphi(\psi) + R\,\hat{p}_n = \frac{\gamma_w R^2}{3}\left[-\frac{1}{1+\cos\psi} + 2\cos\psi\right] \tag{10.25}$$

The characteristic ordinates shown in the diagrams in Figure 10.4b,c are calculated using the formulae:

$$\begin{aligned} &n_\varphi(90°) = 0.333\,\gamma_w R^2 = -n_\theta(90°) \\ &n_\varphi(0°) = n_\theta(0°) = 0.333\,\gamma_w R^2 \times 1.5 = 0.5\,\gamma_w R^2 \end{aligned} \tag{10.26}$$

The computed values are listed in Box 10.5.

The problem can also be solved for a shell described with the alternative angular coordinate $\varphi \in [90°, 180°]$. Taking into account relations: $\varphi = 180° - \psi$ and $\cos\varphi = -\cos\psi$ the following formulae for the two membrane forces are derived:

$$n_\varphi(\varphi) = \frac{\gamma_w R^2}{3}\left[\frac{1}{1-\cos\varphi} - \cos\varphi\right], \qquad n_\theta(\varphi) = \frac{\gamma_w R^2}{3}\left[-\frac{1}{1-\cos\varphi} - 2\cos\varphi\right] \tag{10.27}$$

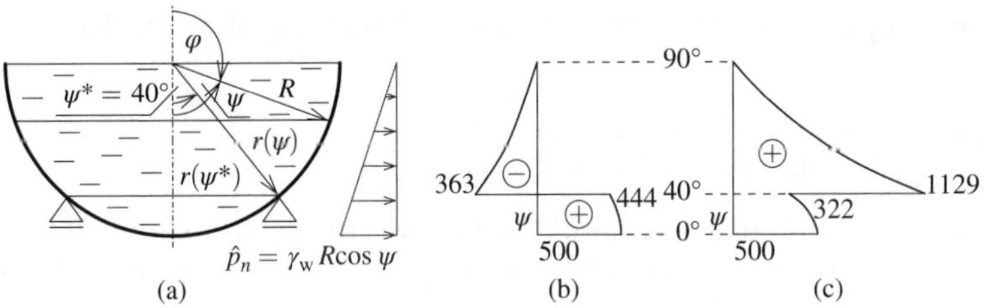

Figure 10.5 Hemispherical tank supported along the intermediate parallel: (a) configuration; diagrams of membrane forces [kN/m]: (b) meridional $n_\varphi(\psi)$ and (c) circumferential $n_\theta(\psi)$

10.4.4 Supported Tank under Hydrostatic Pressure – Analytical and FEM Solutions

A hemispherical tank under hydrostatic pressure

$$\hat{p}_n = \gamma_w R \cos\psi \tag{10.28}$$

supported along the parallel described by the angular coordinate $\psi^* = 40°$ as shown in Figure 10.5 is analysed. The input data R, γ_w and ψ^* are listed in Box 10.6.

Box 10.6 Supported tank under hydrostatic pressure in an axisymmetric membrane state

Data

$R = 10.0$ m, $\gamma_w = 10.0$ kN/m^3, $\psi^* = 40°$, $r(\psi^*) = 6.428$ m

Case I – upper shell part: $\psi \in [40°, 90°]$, index u
Case II – lower shell part: $\psi \in [0°, 40°]$, index l

Check values from analytical solution

Case I $n_\varphi^{u}(90°) = n_\theta^{u}(90°) = 0$
$n_\varphi^{u}(40°) = -363$ kN/m, $n_\theta^{u}(40°) = 1129$ kN/m
Case II $n_\varphi^{l}(40°) = 444$ kN/m, $n_\theta^{l}(40°) = 322$ kN/m
$n_\varphi^{l}(0°) = n_\theta^{l}(0°) = 500$ kN/m

$\Delta n_\varphi(\psi^*) = 807.0$ kN/m
$r_H = 618.2$ kN/m, $r_V = 518.7$ kN/m
$N = -618.2 \times 6.428 = -3973.8$ kN

Check values from FEM solution using ANSYS

SHELL93 FEs
$n_\varphi^{u,FEM}(40°) = -328.61 \text{ kN/m} = 0.9053\, n_\varphi^{u,anal}(40°)$
$n_\varphi^{l,FEM}(0°) = n_\theta^{l,FEM}(0°) = 500 \text{ kN/m} = n_\varphi^{l,anal}(0°) = n_\theta^{l,anal}(0°)$

Shell configuration shown in Figure 10.5

The location of the support along the intermediate parallel imposes the consideration of two intervals:

- case I – the upper part of the shell with $\psi \in [40°, 90°]$
- case II – the lower part of the shell with $\psi \in [0°, 40°]$

The analytical solution of the considered problem involves the determination of two sets of formulae for two membrane forces $n_\varphi(\psi)$ and $n_\theta(\psi)$, related to each interval.

10.4.4.1 Case I – upper shell part $\psi \in [40°, 90°]$

The relation for the meridional force $n^{\mathrm{u}}_\varphi(\psi)$ in the upper part of the shell (index u) is derived from Equation (10.22) taking into account the particular load and the shape of the meridian

$$n^{\mathrm{u}}_\varphi(\psi) = -\frac{\gamma_{\mathrm{w}}}{R\,\sin^2\psi}\left[\frac{1}{3}\,R^3\,\cos^3\psi + C_1\right] \tag{10.29}$$

The integration constant C_1 is calculated from the static boundary condition written for the meridional force. At the top unloaded and unsupported contour $n^{\mathrm{u}}_\varphi(90°) = 0$ and thus $C_1 = 0$. Finally, two membrane forces are expressed as follows:

$$\begin{aligned} n^{\mathrm{u}}_\varphi(\psi) &= -\frac{\gamma_{\mathrm{w}}\,R^2\,\cos^3\psi}{3\,\sin^2\psi} \\ n^{\mathrm{u}}_\theta(\psi) &= -n_\varphi(\psi) + R\,\hat{p}_n = \frac{\gamma_{\mathrm{w}}\,R^2}{3}\,\cos\psi\,[3 + \cot^2\psi] \end{aligned} \tag{10.30}$$

To find the characteristic values of membrane forces for the two contours, supported ($\psi^* = 40°$) and free ($\psi = 90°$), the formulae presented next are used:

$$n^{\mathrm{u}}_\varphi(40°) = -0.363\,\gamma_{\mathrm{w}}\,R^2, \quad n^{\mathrm{u}}_\theta(40°) = 1.129\,\gamma_{\mathrm{w}}\,R^2, \quad n^{\mathrm{u}}_\varphi(90°) = n^{\mathrm{u}}_\theta(90°) = 0 \tag{10.31}$$

The computed values are included in Box 10.6.

10.4.4.2 Case II – lower shell part $\psi \in [0°, 40°]$

Now, the distributions of the meridional and circumferential forces in the lower part of the shell (index l) are expressed by the same formulae as in the previous Subsection 10.4.3. However, this time they are valid for $\psi \in [0°, 40°]$ and read:

$$n^{\mathrm{l}}_\varphi(\psi) = \frac{\gamma_{\mathrm{w}}\,R^2}{3}\left[\frac{1}{1+\cos\psi} + \cos\psi\right], \qquad n^{\mathrm{l}}_\theta(\psi) = \frac{\gamma_{\mathrm{w}}\,R^2}{3}\left[-\frac{1}{1+\cos\psi} + 2\cos\psi\right] \tag{10.32}$$

At the supported contour with $\psi^* = 40°$, in the considered lower part, the membrane forces are determined using the formulae:

$$n^{\mathrm{l}}_\varphi(40°) = 0.444\,\gamma_{\mathrm{w}}\,R^2, \qquad n^{\mathrm{l}}_\theta(40°) = 0.322\,\gamma_{\mathrm{w}}\,R^2 \tag{10.33}$$

In the axisymmetric state at point $\psi = 0°$ the meridional and circumferential forces are equal to each other

$$n^{\mathrm{l}}_\varphi(0°) = n^{\mathrm{l}}_\theta(0°) = 0.5\,\gamma_{\mathrm{w}}\,R^2 \tag{10.34}$$

The computed check values are listed in Box 10.6.

It should be noted that for the supported contour ($\psi^* = 40°$, at the interface between the two parts) a discontinuity of the meridional force denoted by $\Delta n_\varphi(\psi^*)$ occurs. Its horizontal r_H and vertical r_V components represent the uniform reaction of supporting rim:

$$r_H = \Delta n_\varphi(\psi^*) \cos \psi^*, \qquad r_V = \Delta n_\varphi(\psi^*) \sin \psi^* \tag{10.35}$$

When the radial displacement is allowed at each point of the rim axis, the horizontal action of the shell on the rim causes the compressive axial force N in it

$$N = -r_H \, r(\psi^*) \tag{10.36}$$

The computed characteristic values of membrane forces for the upper and lower parts of the tank, the action on the rim and the force in it are listed in Box 10.6. The results of analytical calculations are compared with the numerical ones obtained using FEM package ANSYS (2013) and also included in Box 10.6.

In FEM calculations the considered shell is discretized by two-dimensional elements SHELL93. For the parallel defined by $\psi^* = 40°$, separating the upper and lower part of the shell, the kinematic constraints are introduced, which prevent the vertical displacement and the rigid movement in the horizontal plane.

From the numerical analysis two characteristic check values of meridional force close to the analytical solution are obtained. They are shown in Figure 10.6c and presented in Box 10.6.

Referring to the assumptions of membrane state (see Section 10.1), local bending effects should be considered in the neighbourhood of the supported contour $\psi^* = 40°$. The source of bending is the support which also constrains the displacement w in the direction normal to the middle surface. This results in the distributed reaction along the supporting rim and in the discontinuity of the meridional force.

10.5 Open Conical Shell under Self Weight

An open axisymmetric conical shell under self weight, supported along the bottom edge as shown in Figure 10.7a, is analysed. In the example only the functions of membrane forces and their characteristic values are determined. The shell is described in the cylindrical coordinate system by following relations:

$$\begin{aligned} &x \in [x_0, x_L], \qquad r(x) = x \ \tan\alpha, \qquad r'(x) = \tan\alpha, \qquad A_1 = \sqrt{1 + (r')^2} = 1/\cos\alpha \\ &R_1 = \infty, \quad R_2 = \frac{r(x)}{\cos\alpha} = \frac{x \ \sin\alpha}{\cos^2\alpha} \end{aligned} \tag{10.37}$$

where α defines the direction of meridian and $(\ )' = \mathrm{d}(\ )/\mathrm{d}\,x$.

The distributions of membrane forces are determined using the general Equations (10.7) and (10.8), see Subsection 10.3.2:

$$\begin{aligned} n_x(x) &= -\frac{\sqrt{1 + [r'(x)]^2}}{r(x)} \left[\hat{P}_0^x \, r(x_0) + \int_{x_0}^{x} [\hat{p}_1(x) - \hat{p}_n(x) r'(x)] \, r(x) \, \mathrm{d}x \right] \\ n_\theta(x) &= -\frac{R_2(x)}{R_1(x)} \, n_x(x) + R_2(x) \, \hat{p}_n(x) \end{aligned} \tag{10.38}$$

The two components of self weight (tangent and normal to the cone meridian) are:

$$\hat{p}_1 = g \ \cos\alpha, \qquad \hat{p}_n = -g \ \sin\alpha \tag{10.39}$$

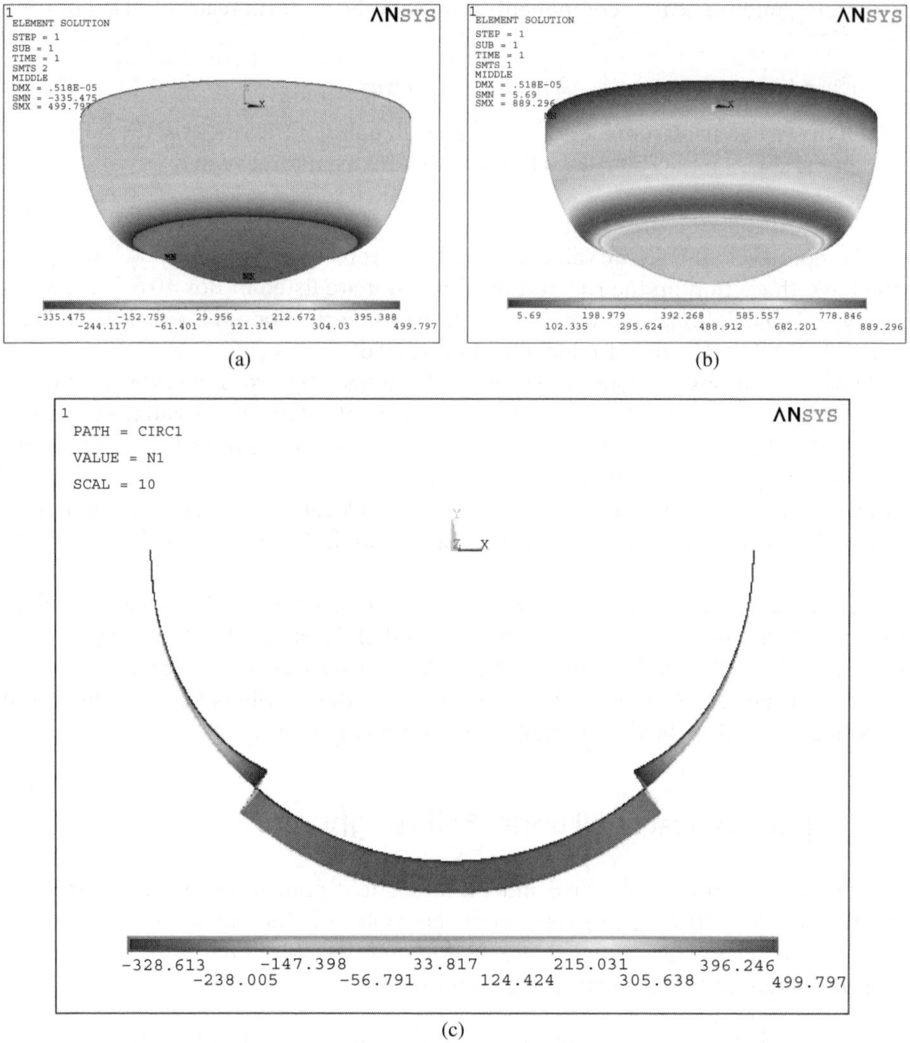

Figure 10.6 Hemispherical shell supported at the intermediate parallel, subjected to hydrostatic pressure – contour maps of membrane forces: (a) meridional, (b) circumferential and (c) diagram of meridional force along the section described by θ = const., with extreme ordinates $n_\varphi^{\mathrm{u,FEM}}(40°) = -328.61$ kN/m and $n_\varphi^{\mathrm{l,FEM}}(0°) = n_\theta^{\mathrm{l,FEM}}(0°) = 500$ kN/m; FEM results from ANSYS. *See Plate section for color representation of this figure.*

The boundary load $\hat{P}_0^x$ is not applied.

Taking into account the shell geometry and its load, two formulae are derived:

$$
\begin{aligned}
n_x(x) &= -\frac{1}{x\ \sin\alpha}\int_{x_0}^{x}[g\ \cos\alpha + g\ \sin\alpha\tan\alpha]\,x\ \tan\alpha\ \mathrm{d}x \\
&= -\frac{g}{x\ \cos^2\alpha}\int_{x_0}^{x} x\ \mathrm{d}x = -\frac{g\ (x^2 - x_0^2)}{2\ x\ \cos^2\alpha} \\
n_\theta(x) &= R_2(x)\,\hat{p}_n = \frac{x\ \sin\alpha}{\cos^2\alpha}(-g\ \sin\alpha) = -g\ x\tan^2\alpha
\end{aligned}
\tag{10.40}
$$

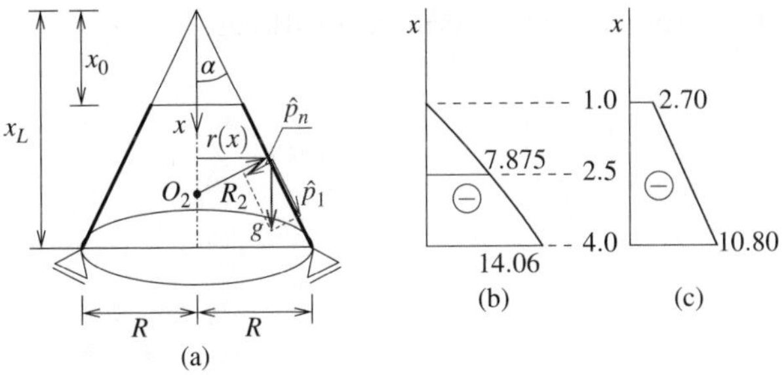

Figure 10.7 Conical shell: (a) configuration, diagrams of membrane forces [kN/m]: (b) meridional $n_x(x)$ and (c) circumferential $n_\theta(x)$

Box 10.7 Open conical shell under self weight in an axisymmetric membrane state

Data

$R = 3.0$ m, $x_0 = 1.0$ m, $x_L = 4.0$ m, $h = 0.2$ m
$\sin\alpha = 0.6$, $\cos\alpha = 0.8$, $\tan\alpha = 0.75$
$g = \gamma_c\, h = 24.0 \times 0.2 = 4.8$ kN/m^2, $\hat{p}_1 = 3.84$ kN/m^2, $\hat{p}_n = -2.88$ kN/m^2

Check values from analytical solution

$n_x(1.0) = 0$, $n_x(2.5) = -7.875$ kN/m, $n_x(4.0) = -14.06$ kN/m
$n_\theta(1.0) = -2.70$ kN/m, $n_\theta(4.0) = -10.80$ kN/m

Shell configuration shown in Figure 10.7

The input data and characteristic ordinates from the diagrams shown in Figure 10.7 are included in Box 10.7.

10.6 Cylindrical Shell

10.6.1 Equations of the Axisymmetric Membrane State

The axisymmetric membrane state in a circular cylindrical shell (see Figure 10.8) characterized by:

$$r(x) = R = \text{const.}, \qquad A_1 = 1, \qquad R_1 = \infty, \qquad A_2 = R_2 = R \tag{10.41}$$

is described by six unknown functions:

- two displacements $\mathbf{u}_{(2\times1)}(x) = [u, w]^{\mathrm{T}}$ [m]
- two membrane strains $\mathbf{e}^n_{(3\times1)}(x) = [\varepsilon_x, \varepsilon_\theta]^{\mathrm{T}}$ [−]
- two membrane forces $\mathbf{s}^n_{(2\times1)}(x) = \mathbf{n} = [n_x, n_\theta]^{\mathrm{T}}$ [N/m]

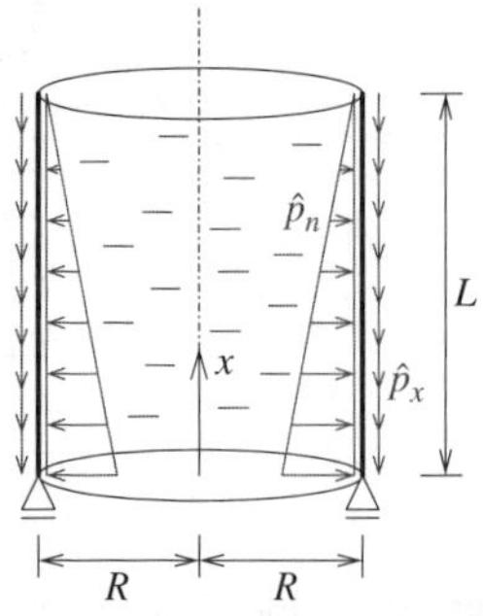

Figure 10.8 Cylindrical shell – configuration and loads

Box 10.8 Equations for circular cylindrical shells in an axisymmetric membrane state

Local formulation – set of six equations and boundary conditions

(I) kinematic equations (2):

$$\varepsilon_x = u', \qquad \varepsilon_\theta = w/R, \qquad ()' = \mathrm{d}()/\mathrm{d}x$$

(II) equilibrium equations (2):

$$n_x' + \hat{p}_x = 0, \qquad n_\theta/R - \hat{p}_n = 0$$

(III) constitutive equations (2):

$$\varepsilon_x = (n_x - \nu n_\theta)/(E\,h), \qquad \varepsilon_\theta = (n_\theta - \nu n_x)/(E\,h)$$

(IV) boundary conditions – one kinematic or one static:

$$u = \hat{u} \qquad \text{or} \qquad n_x = \hat{n}_x$$

Global formulation

Total potential energy $\Pi^n = U^n - W^n$
Internal strain energy

$$U^n = \frac{D^n}{2} \int_A (n_x\,\varepsilon_x + n_\theta\,\varepsilon_\theta)\,R\,\mathrm{d}x\,\mathrm{d}\theta$$

External load work

$$W^n = \int_A (\hat{p}_x\,u + \hat{p}_n\,w)\,R\,\mathrm{d}x\,\mathrm{d}\theta + \int_{\partial A_\sigma} \hat{n}_x\,u\,R\,\mathrm{d}\theta$$

They can be determined solving the boundary value problem described by six equations, that is the mathematical model in local formulation. Both the local and global formulations related to the considered case are presented in Box 10.8.

10.6.2 Circular Cylindrical Shell under Self Weight and Hydrostatic Pressure – Analytical Solution

The aim of the analysis is to determine six functions describing the behaviour of a cylindrical tank in membrane state, loaded by self weight and hydrostatic pressure. The considered shell is simply supported at the bottom edge in such a way that the vertical translation u equals zero and the horizontal translation w in the direction normal to the middle surface is not prevented (Figure 10.8).

Two nonzero components of surface load are taken into account:

$$\hat{p}_x = -\gamma_c\, h, \qquad \hat{p}_n = \gamma_w\, (L - x) \tag{10.42}$$

The input data $(E, \nu, R, L, h, \gamma_c, \gamma_w)$ adopted for the calculations are listed in Box 10.9.

In the case of statically determinate shell the unknown functions are determined in three steps: (i) membrane forces from equilibrium equations, (ii) membrane strains using constitutive equations and (iii) displacements taking into account kinematic relations, see Box 10.8.

The solution starts with the integration of the first equilibrium equation

$$n_x' + \hat{p}_x = 0 \quad \rightarrow \quad n_x(x) = -\hat{p}_x\, x + C_1 \tag{10.43}$$

Constant C_1 is calculated from the static boundary condition for the free unloaded top contour $(x = L)$

$$n_x(L) = 0 \quad \rightarrow \quad n_x(L) = -\hat{p}_x\, L + C_1 = 0 \quad \rightarrow \quad C_1 = \hat{p}_x\, L \tag{10.44}$$

The distribution of the meridional force is described as follows

$$n_x(x) = \hat{p}_x(L - x) = -\gamma_c\, h\, (L - x) = -2.88\,(20.0 - x) \tag{10.45}$$

The function of the circumferential force is derived from the second algebraic equilibrium equation

$$n_\theta(x) = R\, \hat{p}_n(x) = \gamma_w\, R\, (L - x) = 100.0\,(20.0 - x) \tag{10.46}$$

Box 10.9 Cylindrical shell under self weight and hydrostatic pressure in an axisymmetric membrane state

Data

$E = 1.5 \times 10^7$ kN/m^2, $\nu = 0.1667$ $R = 10.0$ m, $L = 20.0$ m, $h = 0.12$ m
$\gamma_c = 24.0$ kN/m^3, $\hat{p}_x = -2.88$ kN/m^2
$\gamma_w = 10.0$ kN/m^3, $\hat{p}_n = 10.0\,(20.0 - x)$ kN/m^2

Check values from analytical solution

$n_x(0) = -57.6$ kN/m, $n_x(20.0) = 0$, $n_\theta(0) = 2000.0$ kN/m, $n_\theta(20.0) = 0$
$\varepsilon_x(0.0) = -0.000217$, $\varepsilon_x(20.0) = 0$, $\epsilon_\theta(0.0) = 0.00112$, $\epsilon_\theta(20.0) = 0$
$u(0.0) = 0$, $u(10.0) = -0.00162$ m, $u(20.0) = -0.00217$ m
$w(0.0) = 0.0112$ m, $w(20.0) = 0$

Shell configuration shown in Figure 10.8

From the algebraic constitutive relations the two functions of membrane strains are determined:

$$\begin{aligned} \varepsilon_x(x) &= \frac{1}{E\,h}[n_x(x) - \nu n_\theta(x)] = \frac{1}{E\,h}(-\gamma_c\, h - \nu\, \gamma_w\, R)(L - x) \\ &= -\,1.086 \times 10^{-5}\ (20.0 - x) \\ \varepsilon_\theta(x) &= \frac{1}{E\,h}[n_\theta(x) - \nu n_x(x)] = \frac{1}{E\,h}(\gamma_w\, R + \nu\, \gamma_c\, h)(L - x) \\ &= 5.58 \times 10^{-5}\ (20.0 - x) \end{aligned} \tag{10.47}$$

Finally, two displacements are derived using the kinematic equations. From the first-order differential kinematic equation function $u(x)$ is obtained

$$u' = \varepsilon_x = \frac{1}{E\,h}(-\gamma_c\, h - \nu\, \gamma_w\, R)(L - x) = -1.086 \times 10^{-5}\ (20.0 - x) \tag{10.48}$$

After integration the following expression with unknown constant C_2 is obtained

$$u(x) = -0.543 \times 10^{-5}(40.0\, x - x^2) + C_2 \tag{10.49}$$

At the bottom edge ($x = 0$) the meridional vertical displacement is not admitted and thus zero value of C_2 and the final formula for function $u(x)$ is obtained

$$u(0) = 0 \quad \rightarrow \quad C_2 = 0 \quad \rightarrow \quad u(x) = -0.543 \times 10^{-5}\ (40.0 - x)\, x \tag{10.50}$$

From the second algebraic kinematic equation the displacement function $w(x)$ is derived

$$w(x) = R\, \varepsilon_\theta(x) = \frac{R}{E\,h}(\gamma_w\, R + \nu \gamma_c\, h)\ (L - x) = 0.558 \times 10^{-3}\ (20.0 - x) \tag{10.51}$$

The characteristic values of the determined functions are presented in the diagrams in Figure 10.9 and in Box 10.9.

In practice, the construction of a cylindrical tank with a circular bottom plate implies a rigid connection between the shell and the plate. It involves a conflict with the boundary conditions assumed in the previous consideration and becomes a source of bending effects. The static analysis of such a real structure is going to be continued in Subsections 11.1.5 and 11.1.6. The solution of the membrane-bending state requires the consideration of two stages. Stage I (membrane state) provides the particular integral $\overline{w}(x)$ of the membrane-bending displacement differential equation that is solved in stage II. Four functions determined for a membrane state: $n_\theta(x)$, $\varepsilon_x(x)$, $\varepsilon_\theta(x)$, $w(x)$ are going to be modified due to bending effects. Analysing the bending state, the distributions of bending moments, transverse shear forces and curvature change are derived. Then, the extreme values of the moments and forces, which occur in the vicinity of the clamped bottom contour, are determined. They play a crucial role in engineering design.

10.6.3 FEM Results

The rotationally symmetric cylindrical shell has been discretized with two different kinds of elements: (i) one-dimensional SRK (ANKA 1993) and (ii) two-dimensional SHELL93 (ANSYS 2013). In the case of axisymmetric membrane state, contour maps presented for the whole periphery of a cylindrical shell have the form of horizontal stripes. According to the analytical solution, the distributions of membrane forces, strains and of the normal displacement w are linear functions of coordinate x. Therefore, the width of stripes in the contour maps is constant. A gradual change of values is presented by different colours or shades of gray. Comparing the analytical solution

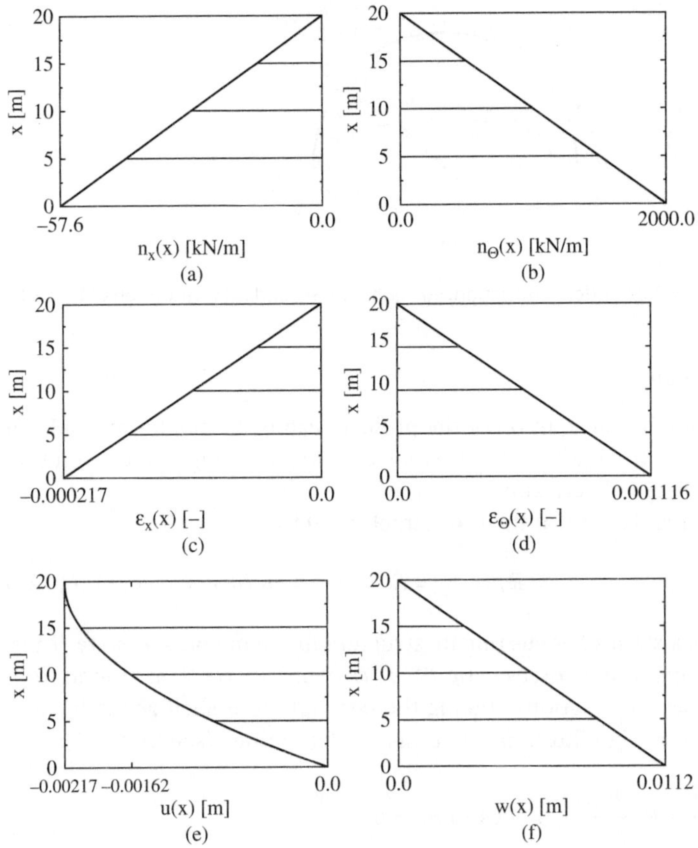

Figure 10.9 Cylindrical shell – diagrams of characteristic functions: (a) $n_x(x)$, (b) $n_\theta(x)$, (c) $\varepsilon_x(x)$, (d) $\varepsilon_\theta(x)$, (e) $u(x)$ and (f) $w(x)$

(represented by diagrams of six functions shown in Figure 10.9) and the numerical one, a satisfactory accuracy of the FEM results is stated.

In Subsections 11.1.5–11.1.7 the analytical and numerical analyses of long and short cylindrical shells in membrane-bending state, clamped at the bottom edge, are presented.

10.7 Hemispherical Shell with an Asymmetric Wind Action

Wind action is considered as a normal pressure, variable in both meridional and circumferential directions (see Figure 10.10). Such a load is described by the formula

$$\hat{p}_n(\varphi, \theta) = -p \ \sin\varphi \ \cos\theta \tag{10.52}$$

The geometry of the shell, boundary constraints and loading fulfil the conditions of the membrane state. In this example the membrane forces, strains and displacements vary in the meridional and circumferential direction according to appropriate trigonometric functions of coordinates φ and θ, respectively.

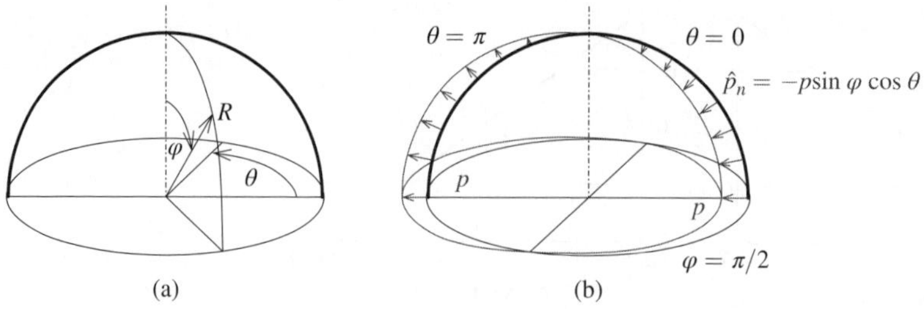

Figure 10.10 Hemispherical shell under wind action: (a) configuration and (b) distribution of normal wind pressure

10.7.1 Analytical Solution

The discussion of the analytical solution of the problem can be found, for example, in the work by Waszczyszyn and Radwańska (1985). Here, the main stages of the solution are presented without detailed derivations.

The geometry of the hemispherical shell is characterized by:

$$r(\varphi) = R \ \sin\varphi = A_2, \qquad R_1 = R_2 = A_1 = R, \qquad r' = \partial r/\partial\varphi = R \ \cos\varphi \quad (10.53)$$

The considered shell is a statically determinate structure in the membrane state. Thus, the functions of three membrane forces $n_\varphi(\varphi,\theta)$, $n_\theta(\varphi,\theta)$ and $n_{\varphi\theta}(\varphi,\theta)$ are found from the set of three equilibrium equations. Taking the description of shell geometry into account, the following three equilibrium equations are determined (see Box 10.1):

$$\begin{aligned}
&\frac{\partial[(R\ \sin\varphi)\ n_\varphi]}{\partial\varphi} + R\frac{\partial n_{\varphi\theta}}{\partial\theta} - (R\ \cos\varphi)\ n_\theta = 0\\
&\frac{\partial[(R\ \sin\varphi)\ n_{\varphi\theta}]}{\partial\varphi} + R\frac{\partial n_\theta}{\partial\theta} + (R\ \cos\varphi)\ n_{\varphi\theta} = 0\\
&n_\varphi/R + n_\theta/R - \hat{p}_n = 0
\end{aligned} \qquad (10.54)$$

According to Equation (10.52), the function describing the distribution of normal load $\hat{p}_n(\varphi,\theta)$ is a product of two functions of separated coordinates φ and θ. This corresponds to one component of cosine series expansion with respect to circumferential variable θ for $j = 1$. As a consequence, meridional and circumferential forces are expressed as cosine functions while tangent force as a sine function. Additionally, unknown amplitude functions dependent only on angle φ are introduced:

$$\begin{aligned}
&n_\varphi(\varphi,\theta) = n_\varphi^{(1)}(\varphi)\ \cos\theta, \qquad n_\theta(\varphi,\theta) = n_\theta^{(1)}(\varphi)\ \cos\theta\\
&n_{\varphi\theta}(\varphi,\theta) = n_{\varphi\theta}^{(1)}(\varphi)\ \sin\theta
\end{aligned} \qquad (10.55)$$

Using the third equilibrium equation $(10.54)_3$ the formula for the amplitude of circumferential force is obtained

$$n_\theta^{(1)} = R\ p_n^{(1)} - n_\varphi^{(1)} \qquad (10.56)$$

which is substituted into the first and second equilibrium equations. Next, after some mathematical manipulations two first-order ordinary differential equations with unknown amplitude functions $n_{\varphi}^{(1)}(\varphi)$, $n_{\varphi\theta}^{(1)}(\varphi)$ are written as:

$$\begin{aligned} &[n_{\varphi}^{(1)}]' + 2\, \cot\varphi\, n_{\varphi}^{(1)} + \frac{1}{\sin\varphi}\, n_{\varphi\theta}^{(1)} = R\, \cot\varphi\, p_n^{(1)} \\ &[n_{\varphi\theta}^{(1)}]' + 2\, \cot\varphi\, n_{\varphi\theta}^{(1)} + \frac{1}{\sin\varphi}\, n_{\varphi}^{(1)} = \frac{R}{\sin\varphi}\, p_n^{(1)} \end{aligned} \tag{10.57}$$

where $(\ )' = \partial(\ )/\partial\varphi$. To simplify the notation, the upper index (1) resulting from $j = 1$ is omitted in further considerations.

Equations (10.57) can be uncoupled by the substitutions:

$$U = n_{\varphi} + n_{\varphi\theta}, \qquad V = n_{\varphi} - n_{\varphi\theta} \tag{10.58}$$

After integration of two independent equations and using the inverse substitution

$$n_{\varphi} = \frac{1}{2}(U + V), \qquad n_{\varphi\theta} = \frac{1}{2}(U - V) \tag{10.59}$$

the amplitudes of membrane forces are calculated as:

$$\begin{aligned} n_{\varphi}(\varphi) &= \frac{1}{2\,\sin^3\varphi}\left[(C_1 + C_2) + (C_1 - C_2)\, \cos\varphi + 2\, p\, R\, \cos\varphi\, (\cos\varphi - \frac{1}{3}\, \cos^3\varphi)\right] \\ n_{\varphi\theta}(\varphi) &= \frac{1}{2\,\sin^3\varphi}\left[(C_1 - C_2) + (C_1 + C_2)\, \cos\varphi + 2\, p\, R\, (\cos\varphi - \frac{1}{3}\, \cos^3\varphi)\right] \end{aligned} \tag{10.60}$$

Two integration constants are calculated using the assumption that the functions should have finite values for $\varphi \to 0°$. Omitting the detailed derivation C_1 and C_2 are calculated as:

$$C_1 = -C_2 = -2\, p\, R/3 \tag{10.61}$$

Finally, three formulae for membrane forces are expressed as:

$$\begin{aligned} n_{\varphi}(\varphi, \theta) &= p\, R\, \left[\frac{\cos\varphi}{\sin^3\varphi}\left(-\frac{2}{3} + \cos\varphi - \frac{1}{3}\, \cos^3\varphi\right)\right] \cos\theta \\ n_{\theta}(\varphi, \theta) &= p\, R\, \left[\frac{1}{\sin^3\varphi}\left(\frac{2}{3}\, \cos\varphi - \sin^2\varphi - \frac{2}{3}\, \cos^4\varphi\right)\right] \cos\theta \\ n_{\varphi\theta}(\varphi, \theta) &= p\, R\, \left[\frac{1}{\sin^3\varphi}\left(-\frac{2}{3} + \cos\varphi - \frac{1}{3}\, \cos^3\varphi\right)\right] \sin\theta \end{aligned} \tag{10.62}$$

The input data (E, ν, R, h, p) adopted for the analytical and FEM solutions are listed in Box 10.10. In Figure 10.11 the diagrams of three membrane forces are shown along the meridian lines with coordinates $\theta = 0, \pi/2, \pi, 3\pi/2$ and along the bottom contour with coordinate $\varphi = \pi/2$.

Below, the formulae used to calculate the characteristic values of load and membrane (meridional, circumferential and tangent) forces are written:

$$\begin{aligned} &p_n(90°, 0°) = -p, && p_n(90°, 180°) = p \\ &n_{\varphi}(30°, 0°) = -0.114\, p\, R, && n_{\varphi}(30°, 180°) = 0.114\, p\, R \\ &n_{\varphi}(60°, 0°) = -0.160\, p\, R, && n_{\varphi}(60°, 180°) = 0.160\, p\, R \\ &n_{\theta}(30°, 0°) = -0.380\, p\, R, && n_{\theta}(30°, 180°) = 0.380\, p\, R \\ &n_{\theta}(90°, 0°) = -p\, R, && n_{\theta}(90°, 180°) = p\, R \\ &n_{\varphi\theta}(90°, 90°) = -0.667\, p\, R, && n_{\varphi\theta}(90°, 270°) = 0.667\, p\, R \end{aligned} \tag{10.63}$$

Box 10.10 Hemispherical shell under asymmetric wind action

Data

$E = 2.1 \times 10^7$ kN/m^2, $\nu = 0.1667$, $R = 10.0$ m, $h = 0.08$ m
$p = 0.010$ kN/m^2, $p_n(90°, 0°) = -0.010$ kN/m^2, $p_n(90°, 180°) = 0.010$ kN/m^2

Check values from analytical solution

$n_\varphi(30°, 0°) = -0.0114$ kN/m, $n_\varphi(30°, 180°) = 0.0114$ kN/m
$n_\varphi(60°, 0°) = -0.0160$ kN/m, $n_\varphi(60°, 180°) = 0.0160$ kN/m
$n_\theta(30°, 0°) = -0.0380$ kN/m, $n_\theta(30°, 180°) = 0.0380$ kN/m
$n_\theta(90°, 0°) = -0.1000$ kN/m, $n_\theta(90°, 180°) = 0.1000$ kN/m
$n_{\varphi\theta}(90°, 90°) = -0.0667$ kN/m, $n_{\varphi\theta}(90°, 270°) = 0.0667$ kN/m

---

Check values from FEM solution using ANSYS

SHELL93 FEs
$n^{\mathrm{FEM}}_{\varphi,\mathrm{extr}} = \pm 0.01625$ kN/m
$n^{\mathrm{FEM}}_{\theta,\mathrm{extr}} = \pm 0.09865$ kN/m $= 0.987\, n^{\mathrm{anal}}_\theta(90°, 180°$ or $0°)$
$n^{\mathrm{FEM}}_{\varphi\theta,\mathrm{extr}} = \pm 0.0640$ kN/m $= 0.960\, n^{\mathrm{anal}}_{\varphi\theta}(90°, 270°$ or $90°)$

Shell configuration shown in Figure 10.10

The check values from the analytical solution are listed in Box 10.10 and two of them are compared with the results of FEM analysis.

10.7.2 FEM Results

The computations have been carried out using the computer package ANSYS. The considered shell has been discretized with two-dimensional elements SHELL93. The input data (loading) and results of numerical analysis are presented in Figures 10.12 and 10.13.

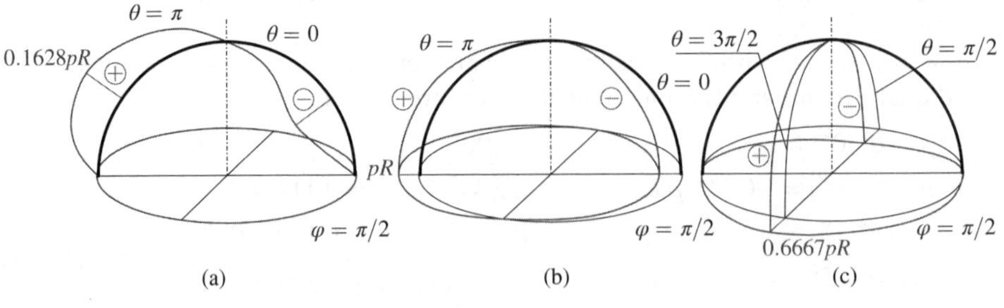

Figure 10.11 Hemispherical shell under wind action – analytical solution, diagrams of three membrane forces: (a) $n_\varphi(\varphi, \theta)$, (b) $n_\theta(\varphi, \theta)$ and (c) $n_{\varphi\theta}(\varphi, \theta)$

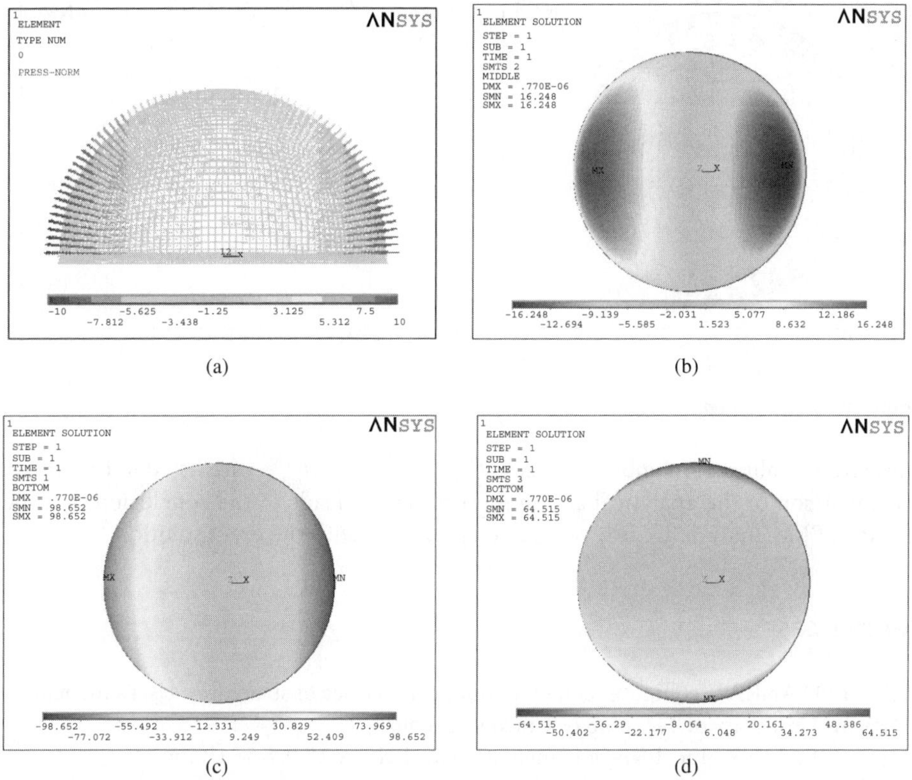

Figure 10.12 Hemispherical shell under wind action: (a) vector vizualization of wind load, contour maps of membrane forces [N/m]: (b) meridional n_{φ}, (c) circumferential n_{θ} and (d) tangent $n_{\varphi\theta}$; FEM results from ANSYS. *See Plate section for color representation of this figure.*

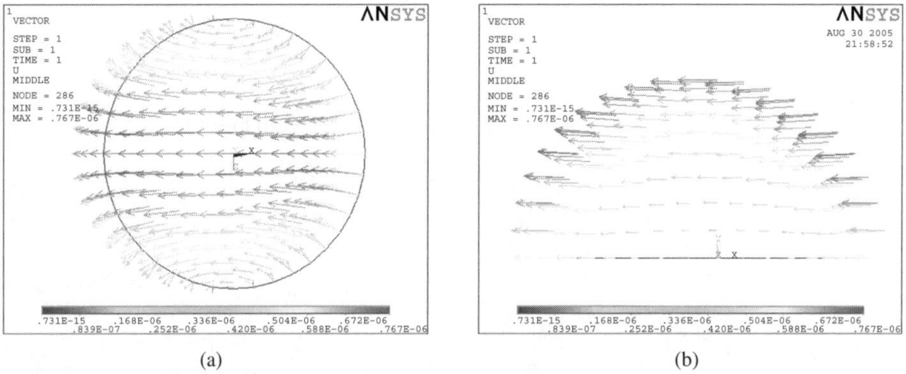

Figure 10.13 Hemispherical shell under wind action – vector visualization of displacement field: (a) top view, (b) side view; the figure is continued on next page and presents directions of principal stress at bottom limiting surface: (c) top view and (d) side view; FEM results from ANSYS. *Subfigures (a) and (b) shown in Plate section for color representation of this figure.*

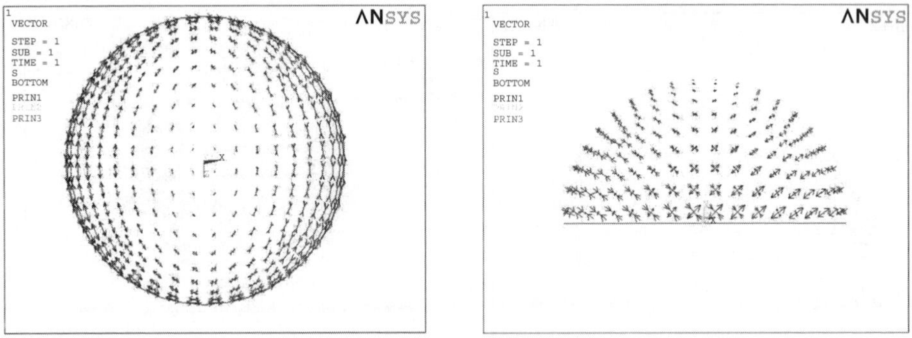

(c) (d)

Figure 10.13 *(Continued)*

The extreme values of membrane forces $n^{\text{FEM}}_{\varphi,\text{extr}}$, $n^{\text{FEM}}_{\theta,\text{extr}}$ and $n^{\text{FEM}}_{\varphi\theta,\text{extr}}$ are listed in Box 10.10. A comparison of the analytical and numerical results leads to the conclusion that the results of FEM analysis are in a very good agreement with the exact solution.

References

ANKA 1993 ANKA – computer code for nonlinear analysis of structures: User's manual. Cracow University of Technology, Cracow (in Polish).

ANSYS 2013 ANSYS Inc. PDF Documentation for Release 15.0. SAS IP, Inc.

Girkmann K 1956 *Flächentragwerke*. Springer Verlag, Wien.

Kolkunov NV 1972 *Foundations of the Analysis of Elastic Shells*. Izdatel'stvo Vishaya Shkola, Moskow (in Russian).

Timoshenko S and Woinowsky-Krieger S 1959 *Theory of Plates and Shells*. McGraw-Hill, New York-Auckland.

Waszczyszyn Z and Radwańska M 1985 *Plates and Shells*. Cracow University of Technology, Cracow (in Polish).

11

Shells in the Membrane-Bending State

11.1 Cylindrical Shells

11.1.1 Description of Geometry

Let us adopt a cylindrical coordinate system in which $\xi_1 = x$, $\xi_2 = \theta$. The geometry of a cylindrical shell is described by meridian equation $r(x) = R = \text{const.}$, two Lame parameters $A_1 = 1$, $A_2 = r = R$ and two radii of principal curvature $R_1 = \infty$, $R_2 = R$. The origin of axis x is set in a plane of one of the curved contours of the shell. The outward normal determines the third versor $\mathbf{n}$ in local base $(\mathbf{e}_1, \mathbf{e}_2, \mathbf{n})$, see Figure 11.1. The second coordinate θ varies in the range $[-\alpha, \alpha]$ in the case of cylindrical segment (Figure 11.1a) or takes values from 0° to 360° for shells of revolution (Figure 11.1b).

11.1.2 Equations of the Membrane-Bending State

Introducing the description of geometry of circular cylindrical shell into the general Sanders shell equations (see Section 3.1) the local formulation in the form of the set of 17 equations with 17 unknowns describing this particular type of the structure is obtained. The local formulation is presented in Box 11.1 and the global formulation in Box 11.2.

Box 11.1 Equations for circular cylindrical shells in a general membrane-bending state – local formulation

Set of 17 equations and boundary conditions

(I) kinematic equations (6):

$$\varepsilon_x = \frac{\partial u_x}{\partial x}, \qquad \varepsilon_\theta = \frac{1}{R}\frac{\partial u_\theta}{\partial \theta} + \frac{w}{R}, \qquad \gamma_{x\theta} = \frac{1}{R}\frac{\partial u_x}{\partial \theta} + \frac{\partial u_\theta}{\partial x}$$

$$\kappa_x = -\frac{\partial^2 w}{\partial x^2}, \qquad \kappa_\theta = \frac{1}{R^2}\frac{\partial u_\theta}{\partial \theta} - \frac{1}{R^2}\frac{\partial^2 w}{\partial \theta^2}$$

$$\chi_{x\theta} = -\frac{1}{2R^2}\frac{\partial u_x}{\partial \theta} + \frac{3}{2R}\frac{\partial u_\theta}{\partial x} - \frac{2}{R}\frac{\partial^2 w}{\partial x \partial \theta}$$

Plate and Shell Structures: Selected Analytical and Finite Element Solutions, First Edition.
Maria Radwańska, Anna Stankiewicz, Adam Wosatko and Jerzy Pamin.
© 2017 John Wiley & Sons Ltd. Published 2017 by John Wiley & Sons Ltd.

(II) equilibrium equations (5):

$$\frac{\partial n_x}{\partial x} + \frac{1}{R}\frac{\partial n_{x\theta}}{\partial \theta} - \frac{1}{2R^2}\frac{\partial m_{x\theta}}{\partial \theta} + \hat{p}_x = 0$$

$$\frac{\partial n_{x\theta}}{\partial x} + \frac{1}{R}\frac{\partial n_\theta}{\partial \theta} + \frac{1}{2R}\frac{\partial m_{x\theta}}{\partial x} + \frac{1}{R}t_\theta + \hat{p}_\theta = 0$$

$$\frac{\partial t_x}{\partial x} + \frac{1}{R}\frac{\partial t_\theta}{\partial \theta} - \frac{1}{R}n_\theta + \hat{p}_n = 0$$

$$\frac{\partial m_x}{\partial x} + \frac{1}{R}\frac{\partial m_{x\theta}}{\partial \theta} - t_x = 0$$

$$\frac{\partial m_{x\theta}}{\partial x} + \frac{1}{R}\frac{\partial m_\theta}{\partial \theta} - t_\theta = 0$$

(III) constitutive equations (6):

$$n_x = D^n(\varepsilon_x + \nu\varepsilon_\theta), \quad n_\theta = D^n(\varepsilon_\theta + \nu\varepsilon_x), \quad n_{x\theta} = D^n(1-\nu)\gamma_{x\theta}/2$$

$$m_x = D^m(\kappa_x + \nu\kappa_\theta), \quad m_\theta = D^m(\kappa_\theta + \nu\kappa_x), \quad m_{x\theta} = D^m(1-\nu)\chi_{x\theta}/2$$

where: $D^n = Eh/(1-\nu^2)$, $\quad D^m = Eh^3/[12(1-\nu^2)]$

(IV) boundary conditions:

a) kinematic: $u_\nu = \hat{u}_\nu,\ u_s = \hat{u}_s,\ w = \hat{w},\ \vartheta_\nu = \hat{\vartheta}_\nu$

b) static: $n_\nu = \hat{n}_\nu,\ \tilde{n}_{\nu s} = \hat{n}_{\nu s},\ \tilde{t}_\nu = \hat{t}_\nu,\ m_\nu = \hat{m}_\nu$

c) mixed

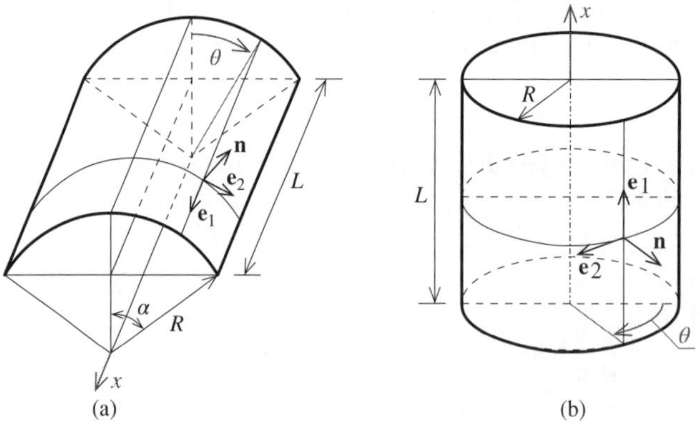

Figure 11.1 Configuration of circular cylindrical shell: (a) cylindrical segment and (b) shell of revolution

Box 11.2 Equations for circular cylindrical shells in a general membrane-bending state – global formulation
Total potential energy $\Pi = U^n + U^m - W$
Internal strain energy

$$U^n = \frac{D^n}{2} \int_{\Omega} (n_x \varepsilon_x + n_{x\theta} \gamma_{x\theta} + n_\theta \varepsilon_\theta) \, \mathrm{d}\Omega$$

$$U^m = \frac{D^m}{2} \int_{\Omega} (m_x \kappa_x + m_{x\theta} \chi_{x\theta} + m_\theta \kappa_\theta) \, \mathrm{d}\Omega$$

External load work

$$W = \int_{\Omega} (\hat{p}_x u_x + \hat{p}_\theta u_\theta + \hat{p}_n w) \, \mathrm{d}\Omega + \int_{\partial\Omega_\sigma} (\hat{n}_\nu u_\nu + \hat{n}_{\nu s} u_s + \hat{t}_\nu w + \hat{m}_\nu \vartheta_\nu) \, \mathrm{d}\partial\Omega$$

where: $\mathrm{d}\Omega = R \, \mathrm{d}x \, \mathrm{d}\theta$

11.1.3 Equations of the Axisymmetric Membrane-Bending State

The axisymmetric membrane-bending state is associated with a zero value of the surface load $\hat{p}_\theta = 0$ and entails vanishing of the following functions: displacement u_θ tangent to the parallel, shear force $n_{x\theta}$ and strain $\gamma_{x\theta}$, twisting moment $m_{x\theta}$, warping $\chi_{x\theta}$ and transverse shear force t_θ. The following external actions must be known: surface loads $\hat{\mathbf{p}}_{(2\times1)} = [\hat{p}_x, \hat{p}_n]^{\mathrm{T}}$ [N/m^2], distributed boundary loads $\hat{\mathbf{p}}_{b(3\times1)} = [\hat{n}_x, \hat{t}_x, \hat{m}_x]^{\mathrm{T}}$ [N/m], [Nm/m] and kinematic constraints $\hat{\mathbf{u}}_{b(3\times1)} = [\hat{u}_x, \hat{w}, \hat{\vartheta}_x]^{\mathrm{T}}$ [m], [−].

The behaviour of a cylindrical shell in axisymmetric membrane-bending state is then described by 11 unknown functions grouped in the following vectors:

- displacements (transitions) $\mathbf{u}_{(2\times1)} = [u(x), w(x)]^{\mathrm{T}}$ [m]
- membrane strains $\mathbf{e}^n_{(2\times1)} = [\varepsilon_x(x), \varepsilon_\theta(x)]^{\mathrm{T}}$ [−]
- changes of curvature $\mathbf{e}^m_{(2\times1)} = [\kappa_x(x), \kappa_\theta(x)]^{\mathrm{T}}$ [1/m]
- membrane forces $\mathbf{s}^n_{(2\times1)} = \mathbf{n} = [n_x(x), n_\theta(x)]^{\mathrm{T}}$ [N/m]
- generalized resultant forces $\mathbf{s}^m_{(3\times1)}(x) = [\mathbf{m}^{\mathrm{T}}, \mathbf{t}]^{\mathrm{T}}$:
  - bending moments $\mathbf{m}_{(2\times1)} = [m_x(x), m_\theta(x)]^{\mathrm{T}}$ [Nm/m]
  - transverse shear force $\mathbf{t}_{(1\times1)} = [t_x(x)]$ [N/m]

Note that the membrane force vector is sometimes denoted by **n** and should not be confused with the normal direction.

The set of 17 equations from Box 11.1 is now reduced to 11 equations describing the special case of axisymmetric membrane-bending state and is presented in Box 11.3.

Box 11.3 Equations for circular cylindrical shells of revolution in an axisymmetric membrane-bending state

Local formulation – set of 11 equations and boundary conditions

(I) kinematic equations (4):

$$\varepsilon_x = u', \qquad \varepsilon_\theta = \frac{w}{R}, \qquad \kappa_x = -w'', \qquad \kappa_\theta = 0, \qquad (\,)' = \frac{\mathrm{d}(\,)}{\mathrm{d}x}$$

(II) equilibrium equations (3):

$$n_x' + \hat{p}_x = 0, \qquad t_x' - \frac{n_\theta}{R} + \hat{p}_n = 0, \qquad m_x' - t_x = 0$$

(III) constitutive equations (4):

$$n_x = D^n(\varepsilon_x + \nu\varepsilon_\theta), \qquad n_\theta = D^n(\varepsilon_\theta + \nu\varepsilon_x)$$

$$m_x = D^m\kappa_x, \qquad m_\theta = D^m\nu\kappa_x$$

where: $D^n = Eh/(1-\nu^2)$, $D^m = Eh^3/[12(1-\nu^2)]$

(IV) boundary conditions:

a) kinematic: $u_x = \hat{u}_x$, $w = \hat{w}$, $\vartheta_x = \hat{\vartheta}_x$

b) static: $n_x = \hat{n}_x$, $t_x = \hat{t}_x$, $m_x = \hat{m}_x$

c) mixed

Global formulation

Total potential energy $\Pi = U^n + U^m - W$

Internal strain energy

$$U^n = \frac{D^n}{2}\int_\Omega (n_x\varepsilon_x + n_\theta\varepsilon_\theta)\,\mathrm{d}\Omega, \qquad U^m = \frac{D^m}{2}\int_\Omega m_x\kappa_x\,\mathrm{d}\Omega$$

External load work

$$W = \int_{\partial\Omega_\sigma} (\hat{n}_x u_x + \hat{t}_x w + \hat{m}_x\vartheta_x)\,\mathrm{d}\partial\Omega$$

where: $\mathrm{d}\Omega = R\,\mathrm{d}x\,\mathrm{d}\theta$, $\mathrm{d}\partial\Omega = R\,\mathrm{d}\theta$

In the equations of axisymmetric state all derivatives with respect to the circumferential coordinate are zero $\partial(\,)/\partial\theta = 0$ and derivatives $\mathrm{d}(\,)/\mathrm{d}x$ are denoted by $(\,)'$.

Note that the mathematical model for the cylindrical shell has already been presented in Subsection 3.3.3 but is now recalled for the completeness of the chapter.

It should be mentioned that the zero value of curvature change $\kappa_\theta = 0$ results from the fact that in a general case (see Box 11.1) it is expressed by two components that vanish in the case of axial symmetry. Moreover, for materials with zero Poisson's ratio ν, circumferential bending moment m_θ also equals zero.

11.1.4 Equations of the Axisymmetric Membrane State

The analysis of membrane-bending state is preceded by a solution of the boundary value problem with an assumed membrane state. The function of the normal displacement

$\overline{w}(x)$ derived for the membrane state will further be used to represent the shell deformation in the membrane-bending state. The behaviour of the cylindrical shell in axisymmetric membrane state (see Section 10.6) is described by six unknown functions:

- displacements (translations) $\mathbf{u}_{(2\times1)} = [u(x), w(x)]^{\mathrm{T}}$ [m]
- membrane strains $\mathbf{e}^n_{(2\times1)} = [\varepsilon_x(x), \varepsilon_\theta(x)]^{\mathrm{T}}$ [−]
- membrane forces $\mathbf{s}^n_{(2\times1)} = \mathbf{n} = [n_x(x), n_\theta(x)]^{\mathrm{T}}$ [N/m]

The unknown quantities are derived from the set of six equations specified in Box 10.8 (Chapter 10) and recalled next:

(I) kinematic equations (2):

$$\varepsilon_x = u', \qquad \varepsilon_\theta = w/R \tag{11.1}$$

(II) equilibrium equations (2):

$$n_x' + \hat{p}_x = 0, \qquad n_\theta/R - \hat{p}_n = 0 \tag{11.2}$$

(III) constitutive equations (2) (these equations are now written with unknown strains on the left-hand sides):

$$\varepsilon_x = (n_x - \nu n_\theta)/(Eh), \qquad \varepsilon_\theta = (n_\theta - \nu n_x)/(Eh) \tag{11.3}$$

11.1.5 Analytical Solution for the Axisymmetric Membrane-Bending State

The combination of the kinematic (I), equilibrium (II) and constitutive (III) relations shown in Box 11.3 leads to two displacement differential equations in terms of two unknown displacement functions $u(x)$ and $w(x)$:

$$u'' + \frac{\nu}{R}w' = -\frac{\hat{p}_x}{D^n}, \qquad \frac{\nu u'}{R} + \frac{h^2}{12}w^{\mathrm{IV}} + \frac{w}{R^2} = \frac{\hat{p}_n}{D^n} \tag{11.4}$$

where: $D^n = (Eh)/(1-\nu^2)$ and $h^2/12 = D^m/D^n$.

The derivative $u'(x)$ is determined from the first equation and introduced into the second one. The final differential equation of the fourth order for the function of displacement $w(x)$ normal to the middle surface of a cylindrical shell is obtained in books by Girkmann (1956); Timoshenko and Woinowsky-Krieger (1959)

$$w^{\mathrm{IV}} + 4\beta^4 w = \frac{1}{D^m}\left[\hat{p}_n + \frac{\nu}{R}\left(\int \hat{p}_x \mathrm{d}x + \overline{C}_1\right)\right] \tag{11.5}$$

where

$$\beta = \sqrt{\frac{1}{Rh}}\sqrt[4]{3(1-\nu^2)} \tag{11.6}$$

The solution of Equation (11.5) is a sum of general integral $w^{\mathrm{o}}(x)$ and particular integral $\overline{w}(x)$

$$w(x) = w^{\mathrm{o}}(x) + \overline{w}(x) \tag{11.7}$$

The general integral is given as a sum of four components

$$w^{\mathrm{o}}(x) = e^{-\beta x}(C_1 \cos\beta x + C_2 \sin\beta x) + e^{\beta x}(C_3 \cos\beta x + C_4 \sin\beta x) \tag{11.8}$$

Below, the calculation algorithm referring to one interval with $x \in [0, L]$ is presented. The solution of a shell in membrane-bending state is carried out in two stages. The membrane state is considered in the preliminary stage I. The boundary conditions (external load and supports) enabling the occurrence of the membrane state are taken into account. The function $\overline{w}(x)$ is derived to be used as a particular integral in the further stage. In stage II the original static and kinematic boundary conditions are examined. Four constants of integration $C_1, \dots, C_4$ appearing in the general integral are calculated from boundary conditions written for quantities describing bending effects in membrane-bending state.

The list of three derivatives of the general integral, which occur in the derivation of all essential formulae, is as follows:

$$\begin{aligned} w^{o\prime}(x) &= \beta e^{-\beta x}[C_1(-\cos\beta x - \sin\beta x) + C_2(\cos\beta x - \sin\beta x)] \\ &\quad + \beta e^{\beta x}[C_3(-\cos\beta x - \sin\beta x) + C_4(\cos\beta x + \sin\beta x)] \\ w^{o\prime\prime}(x) &= 2\beta^2 e^{-\beta x}(C_1 \sin\beta x - C_2\cos\beta x) + 2\beta^2 e^{\beta x}(-C_3\sin\beta x + C_4\cos\beta x) \\ w^{o\prime\prime\prime}(x) &= 2\beta^3 e^{-\beta x}[C_1(\cos\beta x - \sin\beta x) + C_2(\cos\beta x + \sin\beta x)] \\ &\quad + 2\beta^3 e^{\beta x}[C_3(-\cos\beta x - \sin\beta x) + C_4(\cos\beta x - \sin\beta x)] \end{aligned} \tag{11.9}$$

It should be noted that the first two components of the general integral

$$w_0^o(x) = e^{-\beta x}(C_1\cos\beta x + C_2\sin\beta x) \tag{11.10}$$

which are products of exponential function $e^{-\beta x}$ and trigonometric functions $\cos\beta x$ and $\sin\beta x$, respectively, quickly vanish with the increase of x and take nonzero values only in the vicinity of the contour $x = x_0 = 0$ (see diagrams of $F_1(x)$, $F_2(x)$ in Figure 11.2a,b). In contrast, the two remaining components

$$w_L^o(x) = e^{\beta x}(C_3\cos\beta x + C_4\sin\beta x) \tag{11.11}$$

grow fast when approaching the other end of the interval $x = x_L = L$ (see distributions of $F_3(x)$, $F_4(x)$ in Figure 11.2c,d).

In the description of the local bending state the attention should be paid to the rapid drop (or growth) of two succeeding amplitudes of the exponential-trigonometric functions occurring in the solution of the considered problem, computed as the ratio

$$\frac{e^{-\beta(x+\lambda)}}{e^{-\beta x}} = \frac{1}{e^{\pi}} \approx \frac{1}{23} \qquad \text{or} \qquad \frac{e^{\beta(x+\lambda)}}{e^{\beta x}} = e^{\pi} \approx \frac{23}{1} \tag{11.12}$$

The parameter $\lambda = \pi/\beta$ is introduced to distinguish between:

- long shells for $3\lambda < L$
- short shells for $3\lambda \geq L$

In the former case of a long shell if only one source of bending occurs, the general integral has two components:

- the source of bending state in the vicinity of $x = x_0 = 0$, see Equation (11.10)
- the source of bending state in the vicinity of $x = x_L = L$, see Equation (11.11)

The solution is expressed using two integration constants C_1, C_2 or C_3, C_4 calculated from the boundary conditions written for the contour with a disturbance of the membrane state.

For a long shell with sources of bending state in the vicinity of two edges $x = x_0 = 0$ and $x = x_L = L$ the solution is formed using both Equation (11.10) and Equation (11.11). Then, two approaches are possible. One option is to formulate two independent sets of two algebraic equations with two unknown constants C_1, C_2 and C_3, C_4. Otherwise, the full set of four equations is solved and four integration constants C_1, C_2, C_3, C_4 are derived simultaneously. Regardless of the used approach, finally the character of the exponential-trigonometric functions results in such a distribution of the solution

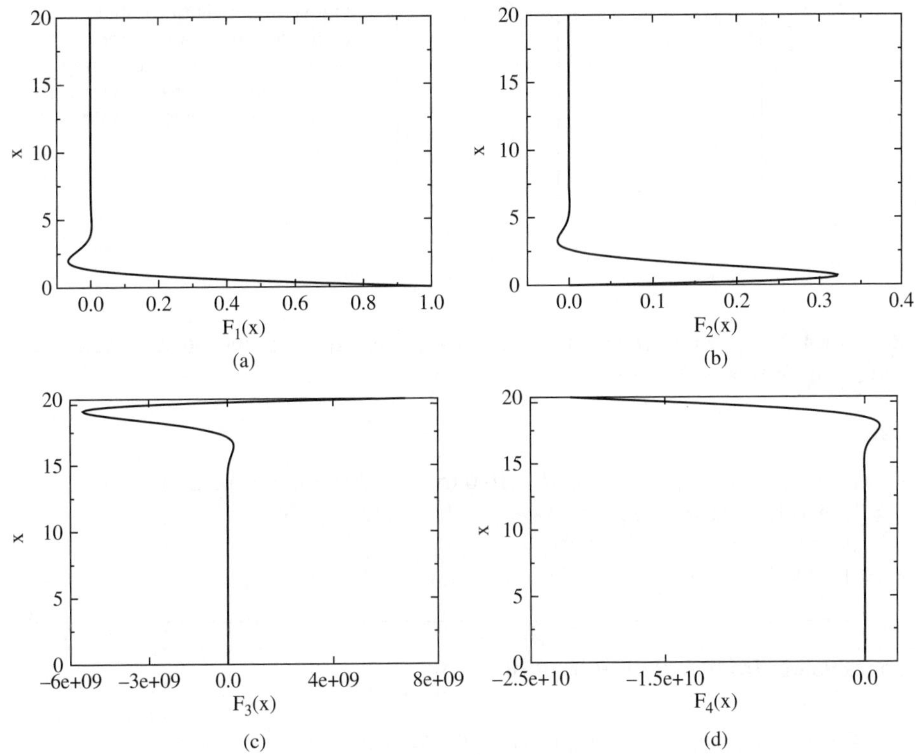

Figure 11.2 Cylindrical shell – distributions of exponential-trigonometric functions in interval $x \in [0, L] = [0, 20]$: (a) $F_1(x) = e^{-\beta x} \cos \beta x$, (b) $F_2(x) = e^{-\beta x} \sin \beta x$, (c) $F_3(x) = e^{\beta x} \cos \beta x$ and (d) $F_4(x) = e^{\beta x} \sin \beta x$, for $\beta = 1.193$ 1/m

functions that the local bending effects occur in a vicinity of two boundaries, while a membrane state takes place in the remaining part of the shell.

In the latter case of a short shell all four components of the general integral, see Equation (11.8), have to be used to derive the solution function, irrespectively of the number and location of the bending sources.

11.1.6 Long Cylindrical Shell Clamped on the Bottom Edge under Self Weight and Hydrostatic Pressure – Analytical and FEM Solutions

A long cylindrical shell, clamped along the bottom contour and loaded by self weight and hydrostatic pressure is examined (see Figure 11.3a). The dimensions and material parameters describing the shell are included in Box 11.4.

Two nonzero components of surface load are taken into account:

$$\hat{p}_x = -\gamma_c h, \qquad \hat{p}_n = \gamma_w (L - x) \tag{11.13}$$

As indicated in the previous subsection, the solution of the membrane-bending problem is carried out in two stages.

11.1.6.1 Stage I – analysis of the membrane state

In stage I the membrane state with changed kinematic boundary conditions at the bottom contour is examined. To fulfil the assumptions of the membrane state the shell is simply supported at the bottom edge in such a way that the vertical translation $\overline{u}(0)$

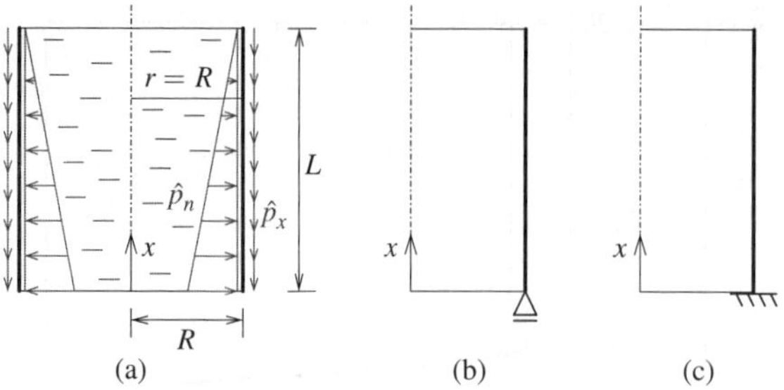

Figure 11.3 Cylindrical shell: (a) description of geometry and loads; boundary conditions: (b) in stage I for membrane state and (c) in stage II for the membrane-bending state

Box 11.4 Long cylindrical shell in an axisymmetric membrane-bending state – analytical solution

Data

$E = 1.5 \times 10^7\ \text{kN/m}^2,\ \nu = 1/6,\ R = 10.0\ \text{m},\ L = 20.0\ \text{m},\ h = 0.12\ \text{m}$
$\beta = 1.1930\ 1/\text{m},\ \lambda = 2.6334\ \text{m},\ 3\lambda < L,\ D^m = 2221.74\ \text{kNm}$
$\gamma_c = 24.0\ \text{kN/m}^3,\ \hat{p}_x = -2.88\ \text{kN/m}^2$
$\gamma_w = 10.0\ \text{kN/m}^3,\ \hat{p}_n = 10.0\ (20.0 - x)\ \text{kN/m}^2$

Check values from analytical solution

$w(0) = 0,$	$w_{max}(2.323) = 0.01028$ m,	$w(20) = 0$
$\kappa_{x,min}(0) = -0.03045$ 1/m,	$\kappa_{x,max}(1.299) = 0.00661$ 1/m,	$\kappa_x(20) = 0$
$\kappa_\theta(0) = 0,$		$\kappa_\theta(20) = 0$
$m_{x,min}(0) = -67.643$ kNm/m,	$m_{x,max}(1.299) = 14.683$ kNm/m,	$m_x(20) = 0$
$m_{\theta,min}(0) = -11.274$ kNm/m,	$m_{\theta,max}(1.299) = 2.447$ kNm/m,	$m_\theta(20) = 0$
$t_{x,max}(0) = 164.923$ kN/m,	$t_{x,min}(1.957) = -11.295$ kN/m,	$t_x(20) = 0$
$u(0) = 0,$	$u(10) = -0.001629$ m,	$u(20) = -0.002172$ m
$n_{x,min}(0) = -57.60$ kN/m,		$n_x(20) = 0$
$n_{\theta,min}(0) = -9.60$ kN/m,	$n_{\theta,max}(2.324) = 1841.33$ kN/m,	$n_\theta(20) = 0$
$\varepsilon_x(0) = -0.311 \times 10^{-4},$	$\varepsilon_{x,min}(2.287) = -1.988 \times 10^{-4},$	$\varepsilon_x(20) = 0$
$\varepsilon_\theta(0) = 0,$	$\varepsilon_{\theta,max}(2.323) = 0.001028,$	$\varepsilon_\theta(20) = 0$

Shell configuration shown in Figure 11.3a

equals zero and horizontal translation $\overline{w}(0)$ in the direction normal to the middle surface is not prevented (see Figure 11.3b).

The aim of the analysis in stage I is to determine six functions describing the behaviour of a cylindrical tank in membrane state loaded by self weight and hydrostatic pressure. The analysis is focused on kinematic, equilibrium and constitutive equations describing the membrane state of a cylindrical shell, which have been discussed in Subsection 10.6.2 and recalled as Equations (11.1)–(11.3). For the two first-order differential equations the following boundary conditions are postulated for $x = 0$ and $x = L$, respectively:

$$\overline{u}(0) = 0, \qquad \overline{n}_x(L) = 0 \tag{11.14}$$

Notice that the functions derived in Stage I are marked by overbar like the particular integral.

For completeness of the solution the derived functions are rewritten here from Subsection 10.6.2:

$$\begin{aligned}
\overline{u}(x) &= \frac{1}{Eh}(-\gamma_c h - \nu\gamma_w R)(Lx - 0.5x^2) = -0.5430 \times 10^{-5}\ (40.0x - x^2)\\
\overline{w}(x) &= \frac{R}{Eh}(\nu\gamma_c h + \gamma_w R)(L - x) = 0.5582 \times 10^{-3}\ (20.0 - x)\\
\overline{n}_x(x) &= -\gamma_c h\ (L - x) = -2.88\ (20.0 - x)\\
\overline{n}_\theta(x) &= \gamma_w\ R\ (L - x) = 100.0\ (20.0 - x)\\
\overline{\varepsilon}_x(x) &= \frac{1}{Eh}\ [\overline{n}_x(x) - \nu\overline{n}_\theta(x)]\\
&= \frac{1}{Eh}\ (-\gamma_c h - \nu\gamma_w R)(L - x) = -1.0859 \times 10^{-5}\ (20.0 - x)\\
\overline{\varepsilon}_\theta(x) &= \frac{1}{Eh}\ [\overline{n}_\theta(x) - \nu\overline{n}_x(x)]\\
&= \frac{1}{Eh}(\nu\gamma_c h + \gamma_w\ R)(L - x) = 5.582 \times 10^{-5}\ (20.0 - x)
\end{aligned} \tag{11.15}$$

The distribution of vertical displacement $\overline{u}(x)$ and meridional force $\overline{n}_x(x)$ do not change in the final solution of the problem, that is $u(x) = \overline{u}(x)$ and $n_x(x) = \overline{n}_x(x)$.

11.1.6.2 Stage II – analysis of the membrane-bending state

The consideration of membrane-bending effects influences the functions describing displacement $w(x)$ normal to the middle surface, circumferential force $n_\theta(x)$ and both strains $\varepsilon_x(x)$ and $\varepsilon_\theta(x)$. In addition, four new quantities, two bending moments $m_x(x)$, $m_\theta(x)$, curvature $\kappa_x(x)$ and transverse shear force $t_x(x)$, need to be determined.

Now, the shell clamped at the bottom contour is considered. The constraints imposed on displacement $w(x)$ and rotation $\vartheta_x(x)$ become the local sources of bending effects in the vicinity of the edge $x = 0$.

For a start the parameter β is calculated and used to classify the shell as short or long, which affects the form of general integral $w^o(x)$. For data presented in Box 11.4:

$$\beta = 1.1930\ 1/\text{m}, \qquad \lambda = \pi/\beta = 2.6334\ \text{m} \tag{11.16}$$

are calculated. The shell is regarded as long, because $3\lambda < L\,(L = 20.0\ \text{m})$.

For the long shell with only one source of the bending state for $x = 0$ the displacement normal to the middle surface, expressed as in Equation (11.7) with the general integral $w^o(x)$, is reduced to $w^o_0(x)$ containing only two constants C_1 and C_2, see Equation (11.10).

Thus, the bending state is described by the following seven functions:

$$\begin{aligned}
w(x) &= w^o_0(x) + \overline{w}(x)\\
&= e^{-\beta x}(C_1 \cos\beta x + C_2 \sin\beta x) + \frac{R}{Eh}(\nu\gamma_c h + \gamma_w R)(L - x)\\
w'(x) &= w^{o\prime}_0(x) + \overline{w}'(x)\\
&= \beta e^{-\beta x}\left[C_1(-\cos\beta x - \sin\beta x) + C_2(\cos\beta x - \sin\beta x)\right]\\
&\quad - \frac{R}{Eh}(\nu\gamma_c h + \gamma_w R)
\end{aligned}$$

$$
\begin{aligned}
\kappa_x(x) &= -w_0^{o\prime\prime}(x) = 2\beta^2 e^{-\beta x}(-C_1 \sin\beta x + C_2 \cos\beta x) \\
\kappa_\theta(x) &\equiv 0 \\
m_x(x) &= D^m\left[-w_0^{o\prime\prime}(x)\right] = D^m 2\beta^2 e^{-\beta x}(-C_1 \sin\beta x + C_2 \cos\beta x) \\
m_\theta(x) &= \nu D^m\left[-w_0^{o\prime\prime}(x)\right] = \nu D^m 2\beta^2 e^{-\beta x}(-C_1 \sin\beta x + C_2 \cos\beta x) \\
t_x(x) &= m_x'(x) = D^m\left[-w_0^{o\prime\prime\prime}(x)\right] \\
&= D^m 2\beta^3 e^{-\beta x}\left[C_1(-\cos\beta x + \sin\beta x) + C_2(-\cos\beta x - \sin\beta x)\right]
\end{aligned}
\tag{11.17}
$$

Analysing the ratio of two factors appearing in function $w(x)$ and resulting from hydrostatic pressure and self weight the dominant effect of the pressure on the solution can be stated

$$
\frac{\gamma_w R}{\nu\gamma_c h} = \frac{10.0\times 10.0}{0.1667\times 24.0\times 0.12} = \frac{100.0}{0.48} \approx 208.33 \tag{11.18}
$$

The modified distributions of the circumferential force $n_\theta(x)$, two strains $\varepsilon_x(x)$ and $\varepsilon_\theta(x)$ are given as:

$$
\begin{aligned}
n_\theta(x) &= n_\theta^o(x) + \overline{n}_\theta(x) = \frac{Eh}{R}e^{-\beta x}(C_1\cos\beta x + C_2 \sin\beta x) + \gamma_w R(L-x) \\
\varepsilon_x(x) &= \frac{1}{Eh}[\overline{n}_x(x) - \nu\overline{n}_\theta(x) - \nu n_\theta^o(x)] \\
&= -\frac{\nu}{R}e^{-\beta x}(C_1\cos\beta x + C_2\sin\beta x) - \frac{1}{Eh}(\gamma_c h + \nu\gamma_w R)(L-x) \\
\varepsilon_\theta(x) &= \frac{1}{Eh}[-\nu\overline{n}_x(x) + \overline{n}_\theta(x) + n_\theta^o(x)] \\
&= \frac{1}{R}e^{-\beta x}(C_1\cos\beta x + C_2\sin\beta x) + \frac{1}{Eh}(\nu\gamma_c h + \gamma_w R)(L-x)
\end{aligned}
\tag{11.19}
$$

Two integration constants C_1 and C_2 are calculated from the kinematic boundary conditions written for the clamped edge $x = 0$ (the second equation is divided by β):

$$
w(0) = 0, \qquad w'(0) = 0 \tag{11.20}
$$

$$
\begin{bmatrix} 1.0 & 0.0 \\ -1.0 & 1.0 \end{bmatrix}\begin{bmatrix} C_1 \\ C_2 \end{bmatrix} = \begin{bmatrix} -0.01116 \\ 0.000468 \end{bmatrix} \tag{11.21}
$$

and read:

$$
C_1 = -0.01116\ \text{m}, \qquad C_2 = -0.01070\ \text{m} \tag{11.22}
$$

Now, the final formulae of the eleven functions are written:

$$
\begin{aligned}
w(x) &= e^{-1.1930x}\,[-0.01116\cos(1.1930x) - 0.01070\sin(1.1930x)] \\
&\quad + 0.5582\times 10^{-3}(20.0 - x) \\
\kappa_x(x) &= e^{-1.1930x}\,[-0.03045\cos(1.1930x) + 0.03178\sin(1.1930x)] \\
\kappa_\theta(x) &= 0 \\
m_x(x) &= e^{-1.1930x}\,[-67.643\cos(1.1930x) + 70.602\sin(1.1930x)] \\
m_\theta(x) &= e^{-1.1930x}\,[-11.274\cos(1.1930x) + 11.767\sin(1.1930x)] \\
t_x(x) &= e^{-1.1930x}\,[164.923\cos(1.1930x) - 3.5301\sin(1.1930x)] \\
\\
u(x) &= \overline{u}(x) = -0.5430\times 10^{-5}(40.0x - x^2) \\
n_x(x) &= \overline{n}_x(x) = -2.88(20.0 - x)
\end{aligned}
\tag{11.23}
$$

$$n_\theta(x) = e^{-1.1930x}\,[-2009.6\cos(1.1930x) - 1925.4\sin(1.1930x)] + 100.0(20.0 - x) \tag{11.24}$$

$$\varepsilon_x(x) = e^{-1.1930x} \times 10^{-5}\,[-18.607\cos(1.1930x) + 17.827\sin(1.1930x)] - 1.0859 \times 10^{-5}(20.0 - x)$$

$$\varepsilon_\theta(x) = e^{-1.1930x} \times 10^{-5}\,[-111.6\cos(1.1930x) + 107.0\sin(1.1930x)] - 5.582 \times 10^{-5}(20.0 - x)$$

Four of them are presented in Figure 11.4. As can be seen in diagrams $m_x(x)$ and $t_x(x)$ in Figure 11.4c,b nonzero ordinates are in interval of length $2\lambda \approx 0.25L$.

The characteristic values of functions obtained in analytical solution, calculated for $x_0 = 0.0$ and $x_L = 20.0$ m as well as extreme values in interval [0, 20], are included in Box 11.4. We emphasize that the three values $m_{x,\min}(0)$, $t_{x,\max}(0)$, $n_{\theta,\max}(2.324)$ have essential significance for design.

In numerical analysis, two types of FEs have been used: (i) one-dimensional, two-node elements SRK (ANKA 1993) and (ii) two-dimensional, eight-node SHELL93 elements (ANSYS 2013).

In the case of the SRK elements the axially symmetric state has been ensured by $j = 0$ set in the input data file. The meridian of the shell has been discretized with $NSE = 26$ and an increased mesh density has been introduced in the vicinity of the bottom edge $x = 0$ for a better representation of bending effects. The obtained values of meridional

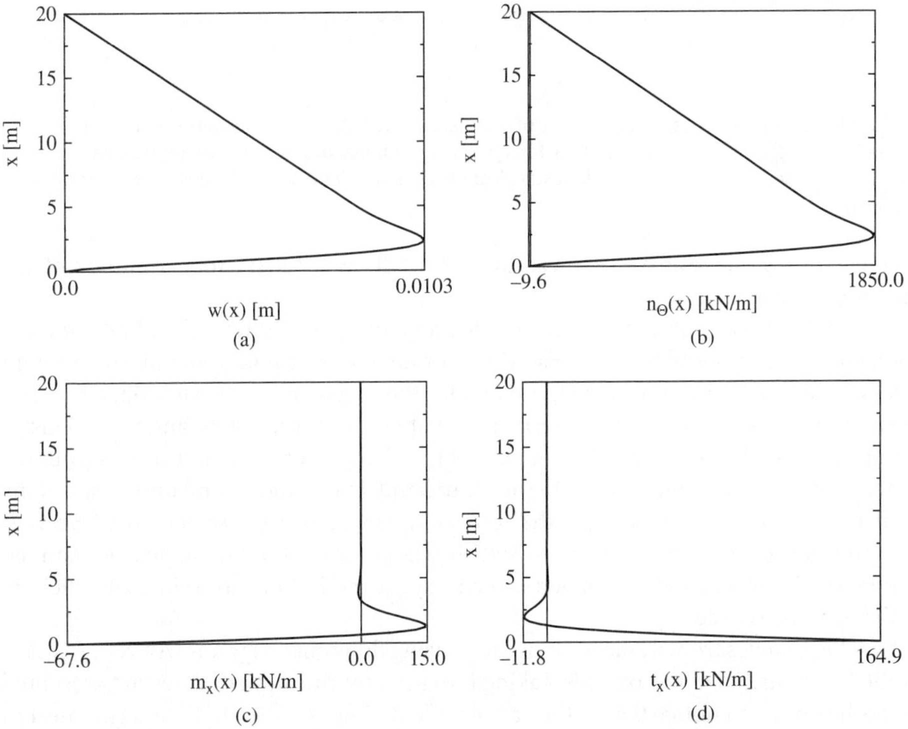

Figure 11.4 Long cylindrical shell (analytical solution) – characteristic functions: (a) normal displacement $w(x)$, (b) circumferential force $n_\theta(x)$, (c) meridional moment $m_x(x)$ and (d) meridional transverse shear force $t_x(x)$

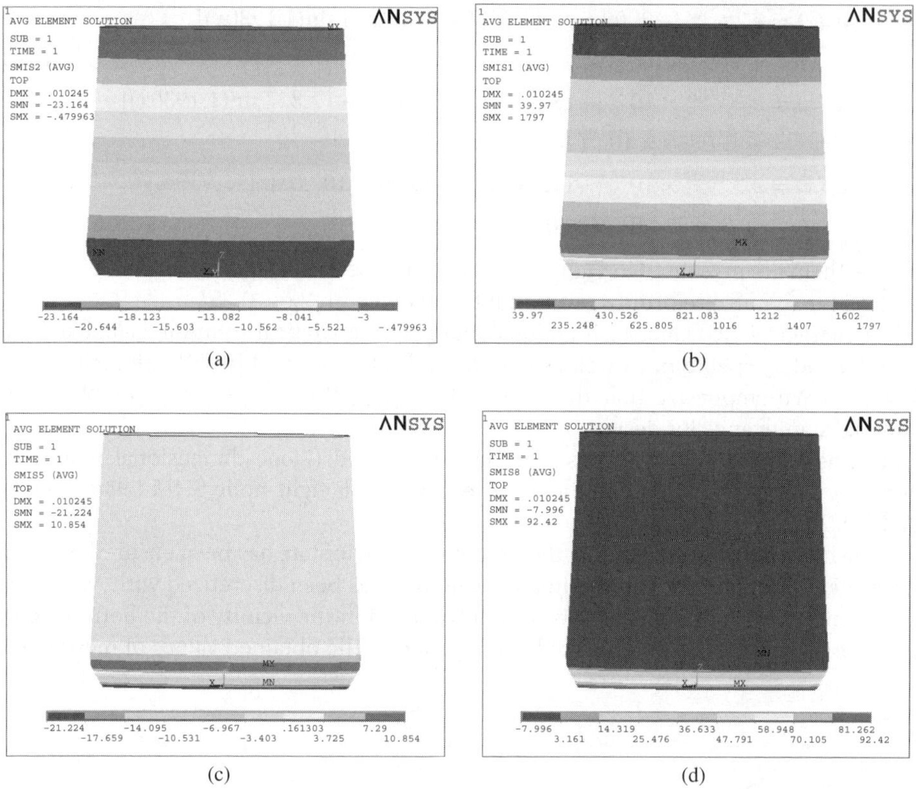

Figure 11.5 Long cylindrical shell – results of calculations with 2D FEs – distributions of functions: (a) meridional force n_x, (b) circumferential force n_θ, (c) meridional moment m_x, (d) meridional transverse shear force t_x, [kN], [m]; FEM results from ANSYS. *See Plate section for color representation of this figure.*

moment $m_x(0)$ and transverse shear force $t_x(0)$ show a satisfactory accuracy of the numerical solution.

Using the ANSYS computer code, the side area of the cylindrical shell has been discretized with surface SHELL93 FEs. The contour maps of four generalized resultant forces: n_x, n_θ, m_x, t_x are shown in Figure 11.5 for the coarse mesh. Due to axial symmetry the presented plots have forms of horizontal stripes. In the case of m_x and t_x the densification of stripes is observed in the vicinity of the clamped contour and results from the concentration of bending effects. On the other hand, in the middle and upper part of the shell, one colour corresponding to the zero value can be noticed (see Figure 11.5c,d). It confirms the occurrence of membrane state in a large part of the shell above the clamped boundary. To obtain a solution of better accuracy the mesh should be refined in a vicinity of the bending source.

The check values from numerical solutions using different programs (ANKA, ANSYS, ROBOT) are included in Box 11.5. Taking into account the numerical results presented in this box one can notice that in the case of calculations with ROBOT and two meshes ($NSE = 800, 2400$) an essential improvement of accuracy is obtained with respect to meridional bending moment at the bottom clamped edge. The values of meridional transverse shear force are still biased by significant errors. Both the extreme values of

Box 11.5 Long cylindrical shell in an axisymmetric membrane-bending state – FEM solution

Data

$E = 1.5 \times 10^7$ kN/m^2, $\nu = 1/6 \approx 0.1667$, $R = 10.0$ m, $L = 20.0$ m, $h = 0.12$ m
$\beta = 1.193$ 1/m, $\lambda = 2.633$ m, $3\lambda < L$, $D^m = 2221.74$ kNm
$\gamma_c = 24.0$ kN/m^3, $\hat{p}_x = -2.88$ kN/m^2
$\gamma_w = 10.0$ kN/m^3, $\hat{p}_n = 10.0\ (20.0 - x)$ kN/m^2

Check values from FEM solution

ANKA, $NSE = 26$, SRK FEs
$m_x^{\mathrm{FEM}}(0) = -62.46$ kNm/m $= 0.923 m_x^{\mathrm{anal}}(0)$
$t_x^{\mathrm{FEM}}(0) = 161.1$ kN/m $= 0.977 t_x^{\mathrm{anal}}(0)$
ANSYS, $NSE = 2800$, SHELL 93 FEs
$w^{\mathrm{FEM}}(2.32) = 0.01020$ m $= 0.992\ w^{\mathrm{anal}}(2.32)$
$n_\theta^{\mathrm{FEM}}(2.32) = 1847.8$ kN/m $= 1.004\ n_\theta^{\mathrm{anal}}(2.32)$
$m_x^{\mathrm{FEM}}(0) = -58.19$ kNm/m $= 0.860\ m_x^{\mathrm{anal}}(0)$
$m_x^{\mathrm{FEM}}(1.30) = 13.72$ kNm/m $= 0.934\ m_x^{\mathrm{anal}}(1.30)$

ROBOT, $NSE = 800$
$w^{\mathrm{FEM}}(2.32) = 0.01005$ m $= 0.978\ w^{\mathrm{anal}}(2.32)$
$n_\theta^{\mathrm{FEM}}(2.32) = 1877.45$ kN/m $= 1.020\ n_\theta^{\mathrm{anal}}(2.32)$
$m_x^{\mathrm{FEM}}(0) = -52.46$ kNm/m $= 0.776\ m_x^{\mathrm{anal}}(0)$
$m_x^{\mathrm{FEM}}(1.30) = 18.68$ kNm/m $= 1.272\ m_x^{\mathrm{anal}}(1.299)$
$t_x^{\mathrm{FEM}}(0) = 74.65$ kNm/m $= 0.453\ t_x^{\mathrm{anal}}(0)$
$t_x^{\mathrm{FEM}}(1.96) = -6.37$ kNm/m $= 0.564\ t_x^{\mathrm{anal}}(1.96)$
ROBOT, $NSE = 2400$
$w^{\mathrm{FEM}}(2.32) = 0.01020$ m $= 0.992\ w^{\mathrm{anal}}(2.32)$
$n_\theta^{\mathrm{FEM}}(2.32) = 1841.07$ kN/m $= 0.999\ n_\theta^{\mathrm{anal}}(2.32)$
$m_x^{\mathrm{FEM}}(0) = -63.86$ kNm/m $= 0.944\ m_x^{\mathrm{anal}}(0)$
$m_x^{\mathrm{FEM}}(1.30) = 14.92$ kNm/m $= 1.016\ m_x^{\mathrm{anal}}(1.30)$
$t_x^{\mathrm{FEM}}(0) = 123.34$ kNm/m $= 0.748\ t_x^{\mathrm{anal}}(0)$
$t_x^{\mathrm{FEM}}(1.96) = -10.33$ kNm/m $= 1.076\ t_x^{\mathrm{anal}}(1.96)$

the meridional bending moment and meridional transverse shear force appear in the narrow zone with local bending effects, and the FEM approximation of these fields is not easy. On the contrary the smooth distribution of circumferential membrane force is observed and the maximum ordinate is reproduced by the FE approximation exactly (see Box 11.5).

Summarizing, the numerical calculations carried out with the use of two previously mentioned FE approaches shows high utility of the one-dimensional elements in the analysis of axisymmetric shells with deformation represented by an arbitrary number of circumferential waves (specified in the input data).

11.1.7 Short Cylindrical Shell Clamped on the Bottom Edge under Self Weight and Hydrostatic Pressure – Analytical and FEM Solutions

Now, a cylindrical shell classified as short ($3\lambda > L$) is examined. The shell is clamped at the bottom contour and loaded by self weight and hydrostatic pressure as in the previous example. The dimensions and material parameters describing the shell are included in Box 11.6.

Box 11.6 Short cylindrical shell in an axisymmetric membrane-bending state

Data

$E = 1.5 \times 10^7$ kN/m^2, $\nu = 1/6$, $R = 10.0$ m, $L = 5.0$ m, $h = 0.12$ m
$\beta = 1.1930$ 1/m, $\lambda = 2.6334$ m, $3\lambda > L$, $D^m = 2221.74$ kNm
$\gamma_c = 24.0$ kN/m^3, $\hat{p}_x = -2.88$ kN/m^2
$\gamma_w = 10.0$ kN/m^3, $\hat{p}_n = 10.0(5.0 - x)$ kN/m^2

Check values from analytical solution

$w(0) = 0$, $w_{max}(1.803) = 0.001736$ m, $w(5) = 0$
$\kappa_{x,min}(0) = -0.00661$ 1/m, $\kappa_{x,max}(1.240) = 0.00166$ 1/m, $\kappa_x(5) = 0$
$\kappa_\theta(0) = 0$, $\kappa_\theta(5) = 0$
$m_{x,min}(0) = -14.691$ kNm/m, $m_{x,max}(1.240) = 3.698$ kNm/m, $m_x(5) = 0$
$m_{\theta,min}(0) = -2.449$ kNm/m, $m_{\theta,max}(1.240) = 0.6164$ kNm/m, $m_\theta(20) = 0$
$t_{x,max}(0) = 38.583$ kN/m, $t_{x,min}(1.899) = -2.847$ kN/m, $t_x(20) = 0$

$u(0) = 0$, $u(2.5) = -10.18 \times 10^{-5}$ m, $u(5) = -13.57 \times 10^{-5}$ m
$n_{x,min}(0) = -14.40$ kN/m, $n_x(5) = 0$
$n_{\theta,min}(0) = -2.40$ kN/m, $n_{\theta,max}(1.805) = 311.01$ kN/m, $n_\theta(5) = 0$
$\varepsilon_x(0) = -0.778 \times 10^{-5}$, $\varepsilon_{x,min}(1.728) = -3.397 \times 10^{-5}$, $\varepsilon_x(5) = 0$
$\varepsilon_\theta(0) = 0$, $\varepsilon_{\theta,max}(1.803) = 17.36 \times 10^{-5}$, $\varepsilon_\theta(20) = 0$

Shell configuration shown in Figure 11.3a

Despite the fact that the source of the bending state is located only at the bottom edge, for a short shell all four components of the general integral have to be taken into account. The solution functions have the following form:

$$
\begin{aligned}
w(x) &= w^{o}(x) + \overline{w}(x) \\
&= e^{-\beta x}(C_1 \cos \beta x + C_2 \sin \beta x) + e^{\beta x}(C_3 \cos \beta x + C_4 \sin \beta x) \\
&\quad + \frac{R}{Eh}(\nu \gamma_c h + \gamma_w R)(L - x)
\end{aligned}
$$

$$
w'(x) = w^{o\prime}(x) + \overline{w}'(x)
$$

$$= \beta e^{-\beta x}\left[C_1(-\cos\beta x - \sin\beta x) + C_2(\cos\beta x - \sin\beta x)\right]$$
$$+ \beta e^{\beta x}\left[C_3(-\cos\beta x - \sin\beta x) + C_4(\cos\beta x + \sin\beta x)\right]$$
$$- \frac{R}{Eh}(\nu\gamma_c h + \gamma_w R)$$

$$\kappa_x(x) = -w^{o\prime\prime}(x) \tag{11.25}$$
$$= 2\beta^2 e^{-\beta x}(-C_1\sin\beta x + C_2\cos\beta x) + 2\beta^2 e^{\beta x}(C_3\sin\beta x - C_4\cos\beta x)$$
$$\kappa_\theta(x) \equiv 0$$
$$m_x(x) = D^m\left[-w^{o\prime\prime}(x)\right]$$
$$= D^m\left[2\beta^2 e^{-\beta x}(-C_1\sin\beta x + C_2\cos\beta x) + 2\beta^2 e^{\beta x}(C_3\sin\beta x - C_4\cos\beta x)\right]$$
$$m_\theta(x) = \nu D^m\left[-w^{o\prime\prime}(x)\right]$$
$$= \nu D^m\left[2\beta^2 e^{-\beta x}(-C_1\sin\beta x + C_2\cos\beta x) + 2\beta^2 e^{\beta x}(C_3\sin\beta x - C_4\cos\beta x)\right]$$
$$t_x(x) = D^m\left[-w^{o\prime\prime\prime}(x)\right]$$
$$= D^m 2\beta^3 e^{-\beta x}\left[C_1(-\cos\beta x + \sin\beta x) + C_2(-\cos\beta x - \sin\beta x)\right]$$
$$+ D^m 2\beta^3 e^{\beta x}\left[C_3(\cos\beta x + \sin\beta x) + C_4(-\cos\beta x + \sin\beta x)\right]$$
$$u(x) = \overline{u}(x) = \frac{1}{Eh}(-\gamma_c h - \nu\gamma_w R)(Lx - 0.5x^2)$$
$$n_x(x) = \overline{n}_x(x) = -\gamma_c h(L - x)$$
$$n_\theta(x) = n_\theta^o(x) + \overline{n}_\theta(x)$$
$$= \frac{Eh}{R}\left[e^{-\beta x}(C_1\cos\beta x + C_2\sin\beta x) + e^{\beta x}(C_3\cos\beta x + C_4\sin\beta x)\right]$$
$$+ \gamma_w R(L - x)$$

$$\varepsilon_x(x) = \frac{1}{Eh}\left[\overline{n}_x(x) - \nu\overline{n}_\theta(x) - \nu n_\theta^o(x)\right] \tag{11.26}$$
$$= -\frac{\nu}{R}\left[e^{-\beta x}(C_1\cos\beta x + C_2\sin\beta x) + e^{\beta x}(C_3\cos\beta x + C_4\sin\beta x)\right]$$
$$- \frac{1}{Eh}\left(\gamma_c h + \nu\gamma_w R\right)(L - x)$$
$$\varepsilon_\theta(x) = \frac{1}{Eh}\left[-\nu\overline{n}_x(x) + \overline{n}_\theta(x) + n_\theta^o(x)\right]$$
$$= \frac{1}{R}\left[e^{-\beta x}(C_1\cos\beta x + C_2\sin\beta x) + e^{\beta x}(C_3\cos\beta x + C_4\sin\beta x)\right]$$
$$+ \frac{1}{Eh}\left(\nu\gamma_c h + \gamma_w R\right)(L - x)$$

The integration constants are calculated from the following four boundary conditions:

- kinematic – for clamped bottom contour $x = 0$:

$$w(0) = 0, \qquad w'(0) = 0 \tag{11.27}$$

- static – for free unloaded top edge $x = L$:

$$m_x(L) = 0, \qquad t_x(L) = 0 \tag{11.28}$$

After detailed calculations the final set of equations in matrix notation has the form

$$\begin{bmatrix} 1.0 & 0.0 & 1.0 & 0.0 \\ -1.0 & 1.0 & -1.0 & 1.0 \\ 0.0008 & 0.00244 & -121.901 & -369.936 \\ 0.00324 & 0.00163 & 248.035 & -491.837 \end{bmatrix} \begin{bmatrix} C_1 \\ C_2 \\ C_3 \\ C_4 \end{bmatrix} = \begin{bmatrix} -0.00279 \\ 0.000468 \\ 0.0 \\ 0.0 \end{bmatrix} \tag{11.29}$$

and the constants are determined:

$$\begin{aligned} &C_1 = -0.002791 \text{ m}, \qquad C_2 = -0.002323 \text{ m} \\ &C_3 = 0.5690 \times 10^{-8} \text{ m}, \qquad C_4 = -2.325 \times 10^{-8} \text{ m} \end{aligned} \tag{11.30}$$

Substituting these values and data from Box 11.6 into Equations $(11.25)_{1,5,7}$ and $(11.26)_3$ the final diagrams of four functions: $w(x)$, $n_\theta(x)$, $m_x(x)$ and $t_x(x)$ are obtained and presented in Figure 11.6. The check values from analytical solution are included in Box 11.6.

It should be noted that (in contrast to the long shell) for a short shell the bending effects resulting from the clamped bottom edge influence the solution in a large part of the shell domain, up to the height of about $0.6L$ (see Figure 11.6c,d). Comparing diagrams $m_x(x)$ and $t_x(x)$ with the diagrams of functions $F_1(x)$ and $F_2(x)$ in Figure 11.2a,b and taking into account the homogeneous static boundary conditions for $x = L$ (Equation 11.28) as

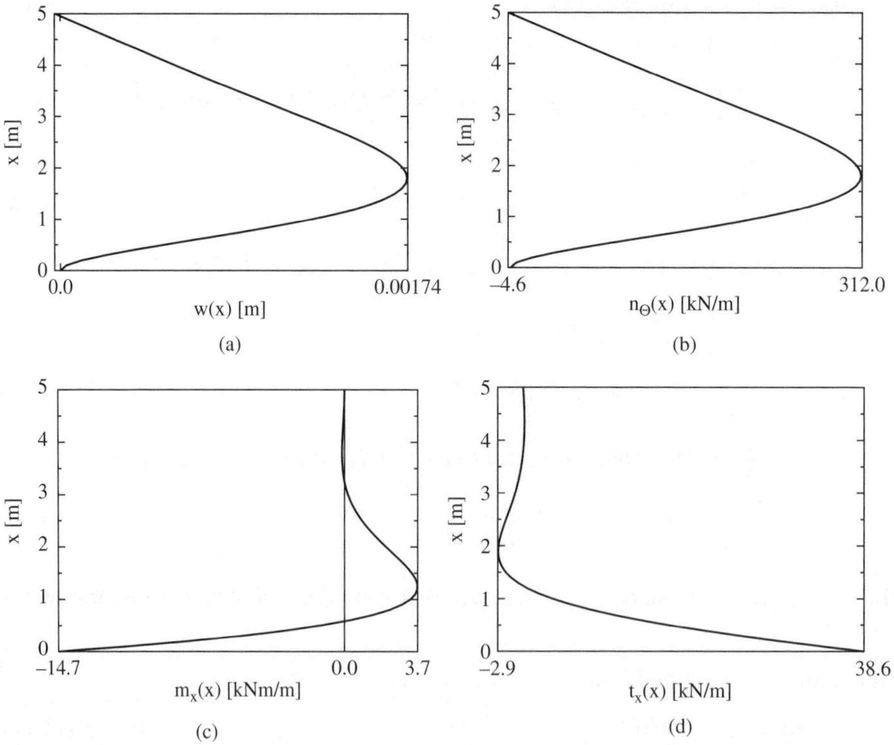

Figure 11.6 Short cylindrical shell (analytical solution) – graphs of four characteristic functions: (a) normal displacement $w(x)$, (b) circumferential force $n_\theta(x)$, (c) meridional moment $m_x(x)$ and (d) meridional transverse shear force $t_x(x)$

well as almost zero values of integration constants C_3 and C_4, one can conclude that the membrane state is not disturbed only in the vicinity of the top shell edge.

11.2 Spherical Shells

11.2.1 Description of Geometry

A spherical shell is described in a spherical coordinate system in which $\xi_1 = \varphi$, $\xi_2 = \theta$. The geometry of the spherical shell is specified by the meridian equation $r(\varphi) = R \sin \varphi$, two Lame parameters $A_1 = R$, $A_2(\varphi) = r(\varphi) = R \sin \varphi$ and two radii of principal curvature $R_1 = R_2 = R$. Next to the angular coordinate φ two auxiliary coordinates φ_1 and φ_2 are introduced. The coordinate φ_1 starts at the bottom and coordinate φ_2 starts from the symmetry axis, see Figure 11.7 (note that for a shell open at the top it would start from the top contour). Additionally, the limiting angle $\varphi = \alpha$ is used and equals for instance 90° for the hemispherical shell.

Introducing the geometry of the spherical shell into the general Sanders shell equations (see Section 3.1) the local formulation describing this particular type of structure can be obtained.

11.2.2 Equations for the Axisymmetric Membrane-Bending State

The axisymmetric membrane-bending state of spherical shells is a typical BVP, described in many books, for example by Girkmann (1956); Kolkunov (1972); Timoshenko and Woinowsky-Krieger (1959). Different notations of quantities and different sign conventions for loads, displacements and generalized resultant forces are used. Both in the works of Girkmann (1956) and Timoshenko and Woinowsky-Krieger (1959) the normal surface load and displacement are positive when their sense is opposite to the external normal. Moreover, positive bending moments induce tensile stress at the internal shell surface.

Throughout this book, notation and sign convention introduced in Section 3.1 in Figures 3.1–3.3 are used. In particular, the positive normal surface load $\hat{p}_n$ and translation w coincide with external normal axis and positive bending moments cause tension at the external surface ($z = h/2$).

The following external actions must be known for the considered case: surface loads $\hat{\mathbf{p}}_{(2\times1)} = [\hat{p}_\varphi, \hat{p}_n]^{\mathrm{T}}$ [N/m²], distributed boundary loads $\hat{\mathbf{p}}_{b(3\times1)} = [\hat{n}_\varphi, \hat{t}_\varphi, \hat{m}_\varphi]^{\mathrm{T}}$ [N/m], [Nm/m] and kinematic constraints $\hat{\mathbf{u}}_{b(3\times1)} = [\hat{u}_\varphi, \hat{w}, \hat{\vartheta}_\varphi]^{\mathrm{T}}$ [m], [–].

The behaviour of a thin spherical shell in axisymmetric membrane-bending state is then described by 11 unknown functions grouped in the following vectors:

- displacements (transitions) $\mathbf{u}_{(2\times1)} = [u(\varphi), w(\varphi)]^{\mathrm{T}}$ [m]
- membrane strains $\mathbf{e}^n_{(2\times1)} = [\varepsilon_\varphi(\varphi), \varepsilon_\theta(\varphi)]^{\mathrm{T}}$ [–]

Figure 11.7 Angular coordinates used in a spherical shell

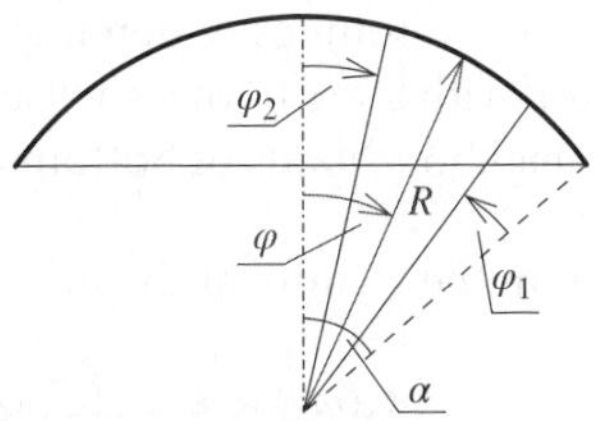

- membrane forces $\mathbf{s}^{n}_{(2\times1)} = \mathbf{n} = [n_{\varphi}(\varphi), n_{\theta}(\varphi)]^{\mathrm{T}}$ [N/m]
- changes of curvature $\mathbf{e}^{m}_{(2\times1)} = [\kappa_{\varphi}(\varphi), \kappa_{\theta}(\varphi)]^{\mathrm{T}}$ [1/m]
- vector of generalized resultant forces $\mathbf{s}^{m}_{(3\times1)}(x) = [\mathbf{m}^{\mathrm{T}}, \mathbf{t}]^{\mathrm{T}}$:
  - bending moments $\mathbf{m}_{(2\times1)} = [m_{\varphi}(\varphi), m_{\theta}(\varphi)]^{\mathrm{T}}$ [Nm/m]
  - transverse shear force $\mathbf{t}_{(1\times1)} = [t_{\varphi}(\varphi)]$ [N/m]

The corresponding system of equations describing the boundary value problem of a spherical shell in axisymmetric membrane-bending state is included in Box 11.7.

The analytical solution presented next is based on Timoshenko and Woinowsky-Krieger (1959). One of the possible approaches to describe the problem of membrane-bending state of spherical shell is to transform the system of 11 equations to two differential equations with unknown transverse shear force t_{φ} and rotation ϑ of the normal to the meridian:

$$\begin{aligned} &\frac{\mathrm{d}^2 t_{\varphi}}{\mathrm{d}\varphi^2} + \cot\varphi \frac{\mathrm{d}t_{\varphi}}{\mathrm{d}\varphi} - (\cot^2\varphi - \nu)t_{\varphi} = Eh\vartheta \\ &\frac{\mathrm{d}^2 \vartheta}{\mathrm{d}\varphi^2} + \cot\varphi \frac{\mathrm{d}\vartheta}{\mathrm{d}\varphi} - (\cot^2\varphi + \nu)\vartheta = -\frac{R^2 t_{\varphi}}{D^m} \end{aligned} \tag{11.31}$$

Taking into account the relationships between functions of transverse shear force t_{φ} and rotation ϑ and their first and second derivatives Equations (11.31) can be reduced to the following form:

$$\frac{\mathrm{d}^2 t_{\varphi}}{\mathrm{d}\varphi^2} = Eh\vartheta, \qquad \frac{\mathrm{d}^2 \vartheta}{\mathrm{d}\varphi^2} = -\frac{R^2 t_{\varphi}}{D^m} \tag{11.32}$$

to eventually obtain one fourth-order differential equation

$$\frac{\mathrm{d}^4 t_{\varphi}}{\mathrm{d}\varphi^4} + 4\varsigma^4 t_{\varphi} = 0, \qquad \text{where} \qquad \varsigma = \sqrt{\frac{R}{h}\sqrt[4]{3(1-\nu^2)}} \tag{11.33}$$

The solution function of the transverse shear force is expressed as

$$t_{\varphi}(\varphi) = e^{-\varsigma\varphi}(C_1 \cos\varsigma\varphi + C_2 \sin\varsigma\varphi) + e^{\varsigma\varphi}(C_3 \cos\varsigma\varphi + C_4 \sin\varsigma\varphi) \tag{11.34}$$

The first two terms in Equation (11.34) (with constants C_1, C_2 and multiplier $e^{-\varsigma\varphi}$) are related to the subdomain in the vicinity of the bottom edge for $\varphi = \alpha$. The third and the fourth component include constants C_3, C_4 and are multiplied by $e^{\varsigma\varphi}$. They describe the distribution of transverse shear force in the neighbourhood of the symmetry axis (or top contour for an open shell) for $\varphi = 0°$.

If two sources of bending state situated at two boundaries: bottom with $\varphi_1 = 0°$ and top with $\varphi_2 = 0°$ of a shell are considered the solution can be defined separately for two neighbourhoods of bottom and top contour:

- in the vicinity of the bottom contour with $\varphi_1 = \alpha - \varphi = 0°$, $\varphi = \alpha$

$$t_{\varphi}(\varphi_1) = e^{-\varsigma\varphi_1}(C_1 \cos\varsigma\varphi_1 + C_2 \sin\varsigma\varphi_1) \tag{11.35}$$

- in the vicinity of the top of a shell where $\varphi_2 = \varphi = 0°$

$$t_\varphi(\varphi_2) = e^{-\varsigma\varphi_2}(C_3 \cos\varsigma\varphi_2 + C_4 \sin\varsigma\varphi_2) \tag{11.36}$$

Notice that using coordinates φ_1 and φ_2, starting from zero on respective contours, both the above functions for t_φ contain the factor $e^{-\varsigma\varphi_1}$ or $e^{-\varsigma\varphi_2}$.

Box 11.7 Equations for spherical shells in an axisymmetric membrane-bending state

Local formulation – set of 11 equations and boundary conditions

(I) kinematic equations (4):

$$\varepsilon_\varphi = \frac{1}{R}\frac{\mathrm{d}u}{\mathrm{d}\varphi} + \frac{w}{R}, \qquad \varepsilon_\theta = \frac{u}{R}\cot\varphi + \frac{w}{R}$$

$$\kappa_\varphi = \frac{1}{R^2}\frac{\mathrm{d}u}{\mathrm{d}\varphi} - \frac{1}{R^2}\frac{\mathrm{d}^2 w}{\mathrm{d}\varphi^2}, \qquad \kappa_\theta = \frac{\cot\varphi}{R^2}\left(u - \frac{\mathrm{d}w}{\mathrm{d}\varphi}\right)$$

(II) equilibrium equations (3):

$$\frac{\mathrm{d}}{\mathrm{d}\varphi}(n_\varphi R\sin\varphi) - n_\theta R\cos\varphi + t_\varphi R\sin\varphi + \hat{p}_\varphi R^2\sin\varphi = 0$$

$$-n_\varphi R\sin\varphi - n_\theta R\sin\varphi + \frac{\mathrm{d}}{\mathrm{d}\varphi}(t_\varphi R\sin\varphi) + \hat{p}_n R^2\sin\varphi = 0$$

$$\frac{\mathrm{d}}{\mathrm{d}\varphi}(m_\varphi R\sin\varphi) - m_\theta R\cos\varphi - t_\varphi R^2\sin\varphi = 0$$

(III) constitutive equations (6):

$$n_\varphi = D^n(\varepsilon_\varphi + \nu\varepsilon_\theta), \qquad n_\theta = D^n(\varepsilon_\theta + \nu\varepsilon_\varphi)$$

$$m_\varphi = D^m(\kappa_\varphi + \nu\kappa_\theta), \qquad m_\theta = D^m(\kappa_\theta + \nu\kappa_\varphi)$$

where: $D^n = Eh/(1-\nu^2)$, $\quad D^m = Eh^3/[12(1-\nu^2)]$

(IV) boundary conditions:

a) kinematic: $u = \hat{u}$, $w = \hat{w}$, $\vartheta_\varphi = \hat{\vartheta}_\varphi$

b) static: $n_\varphi = \hat{n}_\varphi$, $t_\varphi = \hat{t}_\varphi$, $m_\varphi = \hat{m}_\varphi$

c) mixed

Total potential energy $\quad \Pi = U^n + U^m - W$

Internal strain energy

$$U^n = \frac{D^n}{2}\int_\Omega (n_\varphi\varepsilon_\varphi + n_\theta\varepsilon_\theta)\,\mathrm{d}\Omega$$

$$U^m = \frac{D^m}{2}\int_\Omega (m_\varphi\kappa_\varphi + m_\theta\kappa_\theta)\,\mathrm{d}\Omega$$

External load work

$$W = \int_\Omega (\hat{p}_\varphi u + \hat{p}_n w)\,\mathrm{d}\Omega + \int_{\partial\Omega_\sigma} (\hat{n}_\varphi u + \hat{t}_\varphi w + \hat{m}_\varphi\vartheta_\varphi)\,\mathrm{d}\partial\Omega$$

where: $\mathrm{d}\Omega = R^2\sin\varphi\,\mathrm{d}\varphi\,\mathrm{d}\theta$, $\quad \mathrm{d}\partial\Omega = R\sin\varphi\,\mathrm{d}\theta$

Two or four constants appearing in the solution functions are determined from boundary conditions written for the corresponding contour. When the values of constants C_1, C_2, C_3 and C_4 are calculated, the final solution includes two sets of functions describing the transverse shear force t_φ, rotation ϑ, membrane forces n_φ, n_θ and bending moments m_φ, m_θ in the vicinity of the two contours. Next, the equations describing the local membrane-bending state, taken from the book by Kolkunov (1972), are presented:

- at the bottom edge ($\varphi_1 = 0°$):

$$\begin{aligned}
t_\varphi(\varphi_1) &= e^{-\varsigma\varphi_1}\left(C_1 \cos\varsigma\varphi_1 + C_2 \sin\varsigma\varphi_1\right)\\
\vartheta(\varphi_1) &= -\frac{2\varsigma^2}{Eh} e^{-\varsigma\varphi_1}\left(C_1 \sin\varsigma\varphi_1 - C_2 \cos\varsigma\varphi_1\right)\\
n_\varphi(\varphi_1) &= \cot\alpha\, e^{-\varsigma\varphi_1}\left(C_1 \cos\varsigma\varphi_1 + C_2 \sin\varsigma\varphi_1\right)\\
n_\theta(\varphi_1) &= -\varsigma e^{-\kappa\varphi_1}\left[(-C_1 + C_2)\cos\varsigma\varphi_1 - (C_1 + C_2)\sin\varsigma\varphi_1\right]\\
m_\varphi(\varphi_1) &= \frac{h^2\varsigma^3}{6(1-\nu^2)R} e^{-\varsigma\varphi_1}\left[(C_1 + C_2)\cos\varsigma\varphi_1 + (-C_1 + C_2)\sin\varsigma\varphi_1\right]\\
m_\theta(\varphi_1) &= -\frac{h^2\varsigma^2}{6(1-\nu^2)R}\cot\alpha\, e^{-\varsigma\varphi_1}\left(C_1 \sin\varsigma\varphi_1 - C_2 \cos\varsigma\varphi_1\right) + \nu m_\varphi(\varphi_1)
\end{aligned} \tag{11.37}$$

- at the top edge ($\varphi_2 = 0°$):

$$\begin{aligned}
t_\varphi(\varphi_2) &= e^{-\varsigma\varphi_2}\left(C_3 \cos\varsigma\varphi_2 + C_4 \sin\varsigma\varphi_2\right)\\
\vartheta(\varphi_2) &= -\frac{2\varsigma^2}{Eh} e^{-\varsigma\varphi_2}\left(C_3 \sin\varsigma\varphi_2 - C_4 \cos\varsigma\varphi_2\right)\\
n_\varphi(\varphi_2) &= \cot\varphi_2\, e^{-\varsigma\varphi_2}\left(C_3 \cos\varsigma\varphi_2 + C_4 \sin\varsigma\varphi_2\right)\\
n_\theta(\varphi_2) &= -\varsigma e^{-\varsigma\varphi_2}\left[(C_3 - C_4)\cos\varsigma\varphi_2 + (C_3 + C_4)\sin\varsigma\varphi_2\right]\\
m_\varphi(\varphi_2) &= -\frac{h^2\varsigma^3}{6(1-\nu^2)R} e^{-\varsigma\varphi_2}\left[(C_3 + C_4)\cos\varsigma\varphi_2 + (-C_3 + C_4)\sin\varsigma\varphi_2\right]\\
m_\theta(\varphi_2) &= \frac{h^2\varsigma^2}{6(1-\nu^2)R}\cot\varphi_2\, e^{-\varsigma\varphi_2}\left(C_3 \sin\varsigma\varphi_2 - C_4 \cos\varsigma\varphi_2\right) + \nu m_\varphi(\varphi_2)
\end{aligned} \tag{11.38}$$

11.2.3 Clamped or Simply Supported Spherical Shell under Self Weight – Analytical and FEM Solutions

The shell configuration shown in Figure 11.8a–c is analysed. Two types of boundary conditions are considered: clamped (case A) and hinged (case B). The data describing geometry, material and loading of the shell are presented in Box 11.8. For the adopted data, the parameter ς introduced in Subsection 11.2.2 and appearing in Equation (11.33) is computed as

$$\varsigma = \sqrt{\frac{R}{h}}\sqrt[4]{3(1-\nu^2)} = 25.955 \tag{11.39}$$

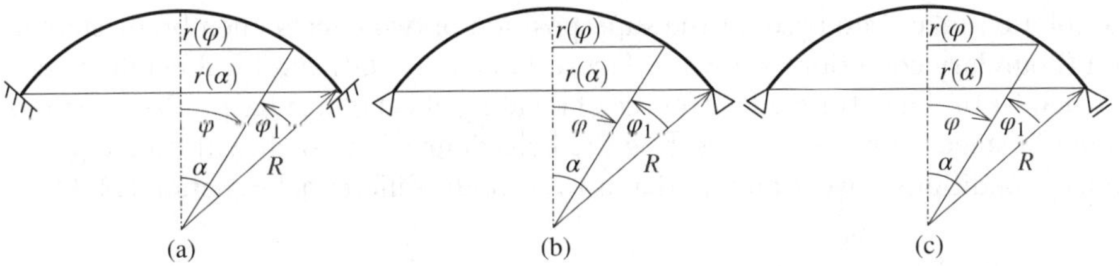

Figure 11.8 Spherical shell – description of geometry and boundary conditions: (a) clamped support (case A), (b) hinged support (case B) and (c) support used for membrane state analysis in stage I

Box 11.8 Spherical shell under self weight: case A – clamped, case B – simply supported

Data

$E = 2.1 \times 10^7\ \text{kN/m}^2$, $\nu = 0.0$, $g = 1.962\ \text{kN/m}^2$
$R = 23.33\ \text{m}$, $h = 0.06\ \text{m}$, $\alpha = \varphi_b = 40°$, $\varphi_1 = 0°$, $r(\alpha) = r_b = 15.0\ \text{m}$
$\sin\alpha = 0.643$, $\cos\alpha = 0.766$, $\cot\alpha = 1.192$

Check values from analytical solution

$\overline{n}_{\varphi,b} = \overline{n}_\varphi(\alpha) = -25.92\ \text{kN/m}$, $\overline{n}_{\theta,b} = \overline{n}_\theta(\alpha) = -9.14\ \text{kN/m}$
$\overline{\sigma}_{\theta,b} = \overline{\sigma}_\theta(\alpha) = -152.3\ \text{kN/m}^2$, $\mathrm{d}\overline{n}_\theta/\mathrm{d}\varphi = 38.87\ \text{kN/m}$
$E\overline{\vartheta}_b = E\overline{\vartheta}(\alpha) = -981.20\ \text{kN/m}^2$

A $m^{\text{anal}}_{\varphi(A)}(\alpha) = 0.1976\ \text{kNm/m}$, $t^{\text{anal}}_{\varphi(A)}(\alpha) = 0.3958\ \text{kN/m}$
$n^{\text{anal}}_{\varphi(A)}(\alpha) = \overline{n}^{\text{anal}}_{\varphi(A)}(\alpha) + n^{\text{anal}}_{\varphi(A)}(\alpha) = -25.92 + 0.47 = -25.45\ \text{kN/m}$

B $t^{\text{anal}}_{\varphi(B)}(\alpha) = 0.176\ \text{kN/m}$, $n^{\text{anal}}_{\varphi(B)}(\alpha) = -25.71\ \text{kN/m}$

Check values from FEM solution using ANSYS

A $m^{\text{FEM}}_{\varphi(A)}(\alpha) = -0.198\ \text{kNm/m} = -m^{\text{anal}}_{\varphi(A)}(\alpha)$
$t^{\text{FEM}}_{\varphi(A)}(\alpha) = -0.352\ \text{kN/m} = -0.89\, t^{\text{anal}}_{\varphi(A)}(\alpha)$
$n^{\text{FEM}}_{\varphi(A)}(\alpha) = -25.47\ \text{kN/m} \approx n^{\text{anal}}_{\varphi(A)}(\alpha)$

B $t^{\text{FEM}}_{\varphi(B)}(\alpha) = -0.128\ \text{kN/m} = -0.73\, t^{\text{anal}}_{\varphi(B)}(\alpha)$
$n^{\text{FEM}}_{\varphi(B)}(\alpha) = -25.72\ \text{kN/m} \approx n^{\text{anal}}_{\varphi(B)}(\alpha)$

Shell configuration shown in Figure 11.8

The solution takes into account the superposition of two effects: membrane state in stage I (boundary conditions shown in Figure 11.8c are valid) and local bending state in the vicinity of the bottom edge (source of bending effects) in stage II. The functions obtained in stage I are used to describe the behaviour of the shell with two types of boundary conditions (case A and B), that is to solve the differential Equation (11.33).

11.2.3.1 Stage I – analysis of the membrane state

From the solution of the membrane state described in Subsection 10.4.1 three functions of angle $\varphi \in [0°, 40°]$ are known:

$$
\begin{aligned}
\overline{n}_\varphi(\varphi) &= \overline{n}_\varphi(\alpha - \varphi_1) = -\frac{gR}{1+\cos\varphi} \\
\overline{n}_\theta(\varphi) &= -\overline{n}_\varphi(\varphi) - gR\cos\varphi = \frac{gR}{1+\cos\varphi} - gR\cos\varphi \\
\overline{\vartheta}(\varphi) &= \frac{1}{Eh}\left[-\frac{\mathrm{d}\overline{n}_\theta}{\mathrm{d}\varphi} + (\overline{n}_\varphi - \overline{n}_\theta)\cot\varphi\right]
\end{aligned}
\tag{11.40}
$$

The characteristic values of selected quantities are calculated at the bottom edge (for $\varphi = \alpha$ and $\varphi_1 = 0°$) according to the formulae:

$$
\begin{aligned}
\overline{n}_{\varphi\mathrm{b}} &= \overline{n}_\varphi(\alpha) = \overline{n}_\varphi|_{\varphi_1=0°} = -\frac{gR}{1+\cos\alpha} \\
\overline{n}_{\theta\mathrm{b}} &= \overline{n}_\theta(\alpha) = \overline{n}_\theta|_{\varphi_1=0°} = -\overline{n}_\varphi(\alpha) - gR\cos\alpha \\
\overline{\sigma}_{\theta\mathrm{b}} &= \overline{\sigma}_\theta(\alpha) = \overline{n}_\theta(\alpha)/h \\
\frac{\mathrm{d}\overline{n}_\theta}{\mathrm{d}\varphi} &= gR\sin\alpha\left[1 + \frac{1}{(1+\cos\alpha)^2}\right] \\
E\overline{\vartheta}_\mathrm{b} &= E\overline{\vartheta}(\alpha)
\end{aligned}
\tag{11.41}
$$

and included in Box 11.8. The lower index b is used to emphasize that these quantities are related to the boundary line. Two values $E\overline{\vartheta}_\mathrm{b}$ and $\overline{\sigma}_{\theta\mathrm{b}}$ are used in stage II of the calculation.

11.2.3.2 Stage II – analysis of the membrane-bending state

Now, the disturbance of the membrane state in the vicinity of the bottom contour is analysed. Separate discussion is provided for two types of the support (case A and B).

In case A two kinematic boundary conditions are formulated: the total rotation is zero $\vartheta(\alpha) = 0$ and the total normal displacement is zero $w(\alpha) = 0$. The second condition requires special clarification. For the clamped edge both displacements, tangent to the meridian u and normal to the middle surface w, are equal to zero. From the second kinematic equation the zero value of circumferential strain $\varepsilon_\theta(\alpha) = 0$ is concluded. Since the zero value of Poisson's ratio $\nu = 0$ has been adopted, the circumferential strain is directly related to circumferential force $n_\theta(\alpha)$ or stress $\sigma_\theta(\alpha)$. The second boundary condition is related to the circumferential stress. The two boundary conditions taking into account the membrane-bending state can be written as:

$$
E\overline{\vartheta}_\mathrm{b} + E\vartheta_\mathrm{b} = 0, \qquad \overline{\sigma}_{\theta\mathrm{b}} + \sigma_{\theta\mathrm{b}} = 0 \tag{11.42}
$$

Using the relations describing the bending state and data from Box 11.8 the following equations are obtained:

$$E\vartheta_b = -\frac{2}{h}\varsigma^2(-C_2) = \frac{2\varsigma^2}{h}C_2 = 22\,455\,C_2$$
$$n_{\theta b} = -\varsigma(C_2 - C_1) \quad \rightarrow \quad \sigma_{\theta b} = -\frac{\varsigma}{h}(C_2 - C_1) = 432.58\,C_1 - 432.58\,C_2 \tag{11.43}$$

These formulae are related to Equation $(11.38)_{2,4}$ after introducing $\varphi_1 = 0$ and in consequence $\cos\varphi_1 = 1$ and $e^{-\varsigma\varphi_1} = 1$.

From the set of two equations

$$\begin{bmatrix} 0.0 & 22\,455.0 \\ 432.58 & -432.58 \end{bmatrix} \begin{bmatrix} C_1 \\ C_2 \end{bmatrix} = \begin{bmatrix} 981.20 \\ 152.3 \end{bmatrix} \tag{11.44}$$

constants C_1 and C_2 are derived

$$C_1 = 0.3958\ \text{kN/m}, \qquad C_2 = 0.0437\ \text{kN/m} \tag{11.45}$$

Finally, the behaviour of the shell in the vicinity of the bottom edge is described by three nonzero functions: meridional bending moment m_φ, transverse shear force t_φ and modified (in stage II) meridional force n_φ:

$$\begin{aligned}
m_\varphi(\varphi_1) &= \frac{h^2\varsigma^3}{6(1-\nu^2)R}e^{-\varsigma\varphi_1}\left[(C_1 + C_2)\cos\varsigma\varphi_1 + (-C_1 + C_2)\sin\varsigma\varphi_1\right] \\
&= e^{-25.955\varphi_1}\left[0.1976\cos(25.955\varphi_1) - 0.15583\sin(25.955\varphi_1)\right] \\
t_\varphi(\varphi_1) &= e^{-\varsigma\varphi_1}\left(C_1\cos\varsigma\varphi_1 + C_2\sin\varsigma\varphi_1\right) \\
&= e^{(-25.955\varphi_1)}\left[0.3958\cos(25.955\varphi_1) + 0.0437\sin(25.955\varphi_1)\right] \\
n_\varphi(\varphi_1) &= \overline{n}_\varphi(\alpha - \varphi_1) + n_\varphi(\varphi_1) \\
&= -\frac{gR}{1+\cos(\alpha - \varphi_1)} + \cot\alpha\ e^{-\varsigma\varphi_1}\left(C_1\cos\varsigma\varphi_1 + C_2\sin\varphi_1\right) \\
&= -\frac{45.77}{1+\cos(\alpha - \varphi_1)} \\
&\quad + e^{(-25.955\varphi_1)}\left[0.4718\cos(25.955\,\varphi_1) + 0.0521\sin(25.955\,\varphi_1)\right]
\end{aligned} \tag{11.46}$$

The characteristic values computed for $\varphi_1 = 0°$ (or $\varphi = \alpha$) according to equations

$$m_\varphi|_{\varphi_1=0°} = \frac{h^2\varsigma^3}{6R}(C_1 + C_2), \qquad t_\varphi|_{\varphi_1=0°} = C_1$$
$$n_\varphi|_{\varphi_1=0°} = \overline{n}_\varphi|_{\varphi_1=0°} + n_\varphi|_{\varphi_1=0°} \tag{11.47}$$

are included in Box 11.8.

In case B (hinged contour) two mixed boundary conditions: $w(\alpha) = 0$ and $m_\varphi(\alpha) = 0$ are formulated. The algorithm of analytical calculations is the same as for the clamped structure, thus the detailed solution is not presented.

Numerical computations for the considered problem have been performed using ANSYS. Selected results are shown in Figure 11.9, where the diagrams in the left column are related to case A and in the right column to case B. Note that opposite sign convention for bending moments and transverse shear forces is applied in the analytical solution and in ANSYS calculations. For case A three and for case B two check values have been listed in Box 11.8 and compared with the analytical results.

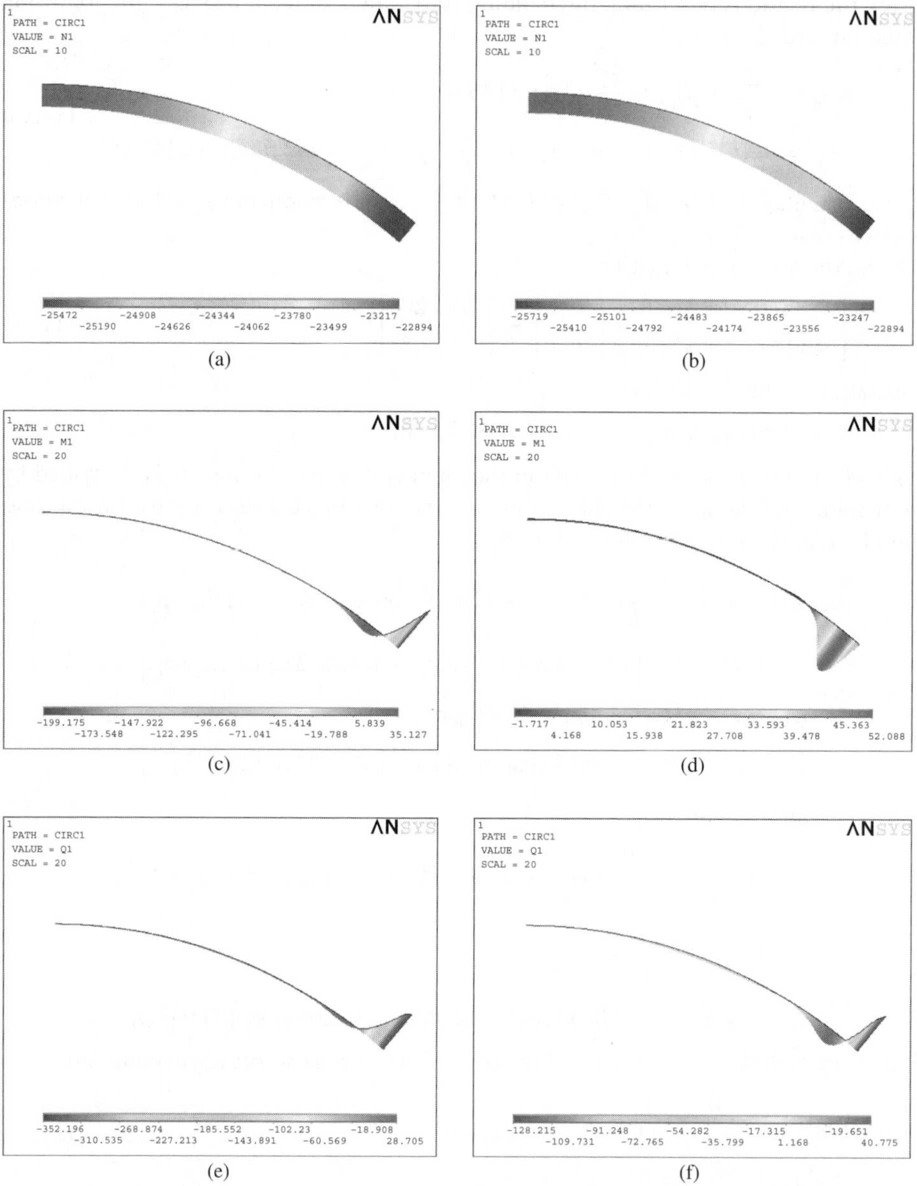

Figure 11.9 Spherical shells under self weight: left column – results for case A (shell clamped along bottom edge), right column – case B (simply supported shell); graphs of following functions: (a), (b) meridional membrane forces $n_\varphi(\varphi)$, (c), (d) meridional moment $m_\varphi(\varphi)$ and (e), (f) meridional transverse shear forces $t_\varphi(\varphi)$, [N],[m]; FEM results from ANSYS. *See Plate section for color representation of this figure.*

11.3 Cylindrical and Spherical Shells Loaded by a Uniformly Distributed Boundary Moment and Horizontal Force

11.3.1 Cylindrical Shell

Long cylindrical shell loaded at the bottom contour by boundary normal moment M [kNm/m] and horizontal force H [kN/m] is considered. The applied loads induce the local axisymmetric bending state in the vicinity of the bottom edge with $x = 0$.

The geometry of the analysed shell with static and kinematic boundary conditions is presented in Figure 11.10. According to the adopted sign convention (see Figure 3.2) boundary moment M corresponds to negative meridional bending moment $m_x(0) < 0$ and horizontal load H to positive meridional transverse shear force $t_x(0) > 0$, that is $m_x(0) = -M$ and $t_x(0) = H$.

The input data adopted in this example are the same as in the example from Subsection 11.1.6 and are presented in Box 11.4. For the adopted values, parameter β is equal to 1.1930 1/m.

The bending state in the vicinity of the loaded contour ($x = 0$) triggered by moment M (case A) and force H (case B) is described by the function of normal displacement $w(x)$ dependent on two integration constants (see Subsection 11.1.5)

$$w(x) = e^{-\beta x}(C_1 \cos \beta x + C_2 \sin \beta x) \tag{11.48}$$

Integration constants C_1 and C_2 are derived from appropriate boundary conditions and functions: $w(x)$, $m_x(x)$, $t_x(x)$, $n_\theta(x)$ and boundary values $w(0)$ and $w'(0)$ are determined separately for the two boundary loads.

11.3.1.1 Case A – distributed boundary moment M

Two integration constants are calculated from the static boundary conditions for $x = 0$:

$$m_x(0) = -D^m w'' = -M, \qquad t_x(0) = -D^m w''' = 0 \tag{11.49}$$

Then, the solution functions are as follows:

$$\begin{aligned} w_M(x) &= \frac{1}{2D^m \beta^2} e^{-\beta x}(\cos \beta x - \sin \beta x)M \\ &= 1.581 \times 10^{-4}\, e^{-1.1930x}\, [\cos(1.1930x) - \sin(1.1930x)]\, M \\ m_x^M(x) &= -D^m w''(x) = -e^{-\beta x}(\cos \beta x + \sin \beta x)M \\ &= -e^{-1.1930x}\, [\cos(1.1930x) + \sin(1.1930x)]\, M \end{aligned} \tag{11.50}$$

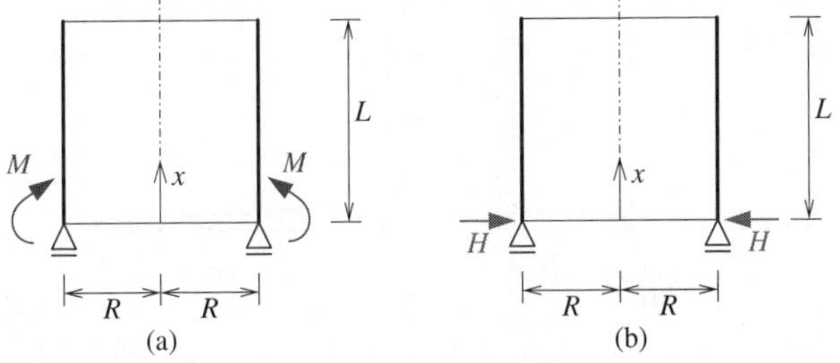

Figure 11.10 Cylindrical shell with two uniformly distributed boundary loads on contour $x = 0$: (a) normal moment M and (b) horizontal force H

$$t_x^M(x) = -D^m w'''(x) = 2\beta e^{-\beta x} \sin \beta x\, M$$
$$= 2.386 e^{-1.1930x} \sin(1.1930x) M$$
$$n_\theta^M(x) = \frac{Eh}{2RD^m \beta^2} e^{-\beta x}(\cos \beta x - \sin \beta x) M$$
$$= 28.86 e^{-1.1930x)} [\cos(1.1930x) - \sin(1.1930x)]\, M$$

and the boundary values at the bottom contour read:

$$w_M(0) = \frac{1}{2D^m \beta^2} M = 1.581 \times 10^{-4} M, \qquad w_M'(0) = -\frac{1}{D^m \beta} M = -3.7725 \times 10^{-4} M \tag{11.51}$$

In Figure 11.11 the distributions of normal displacement $w_M(x)$, meridional moment $m_x^M(x)$ and meridional transverse shear force $t_x^M(x)$ are presented for interval $[0, 20]$ – note that $m_x^M(0) = -1.0$ kNm/m and $w_M(0)$ is given by Equation (11.51).

11.3.1.2 Case B – horizontal boundary traction *H*

For case B, two integration constants are also calculated from the static boundary conditions for $x = 0$:

$$m_x(0) = -D^m w'' = 0, \qquad t_x(0) = -D^m w''' = H \tag{11.52}$$

The final solution functions are as follows:

$$w_H(x) = -\frac{1}{2D^m \beta^3} e^{-\beta x} \cos \beta x\, H = -1.325 \times 10^{-4}\, e^{-1.193x} \cos(1.193x) H$$
$$m_x^H(x) = -D^m w''(x) = \frac{1}{\beta} e^{-\beta x} \sin \beta x\, H = 0.8382 e^{-1.193x} \sin(1.193x) H$$
$$t_x^H(x) = -D^m w'''(x) = e^{-\beta x}(\cos \beta x - \sin \beta x) H \tag{11.53}$$
$$= e^{-1.193x} [\cos(1.193x) - \sin(1.193x)]\, H$$
$$n_\theta^H(x) = -\frac{Eh}{2RD^m \beta^3} e^{-\beta x} \cos \beta x\, H = -23.86 e^{-1.193x} \cos(1.193x) H$$

and the boundary values at the bottom contour are:

$$w_H(0) = -\frac{1}{2D^m \beta^3} H = -1.325 \times 10^{-4} H, \qquad w_H'(0) = \frac{1}{2D^m \beta^2} H = 1.5811 \times 10^{-4} H \tag{11.54}$$

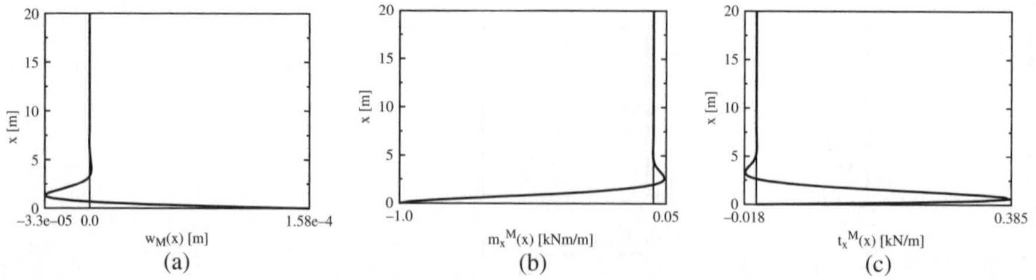

Figure 11.11 Cylindrical shell with boundary load $M = 1$ kNm/m (see Figure 11.10a) – diagrams of: (a) normal displacement $w_M(x)$, (b) meridional moment $m_x^M(x)$ – see ordinate $m_x^M(0) = -1.0$ kNm/m and (c) meridional transverse shear force $t_x^M(x)$

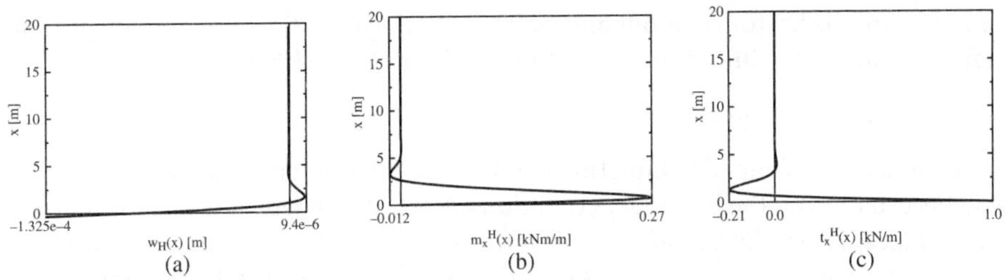

Figure 11.12 Cylindrical shell with boundary load $H = 1$ kN/m (see Figure 11.10b) – diagrams of: (a) normal displacement $w_H(x)$, (b) meridional moment $m_x^H(x)$ and (c) meridional transverse shear force $t_x^H(x)$ with ordinate $t_x^H(0) = 1.0$ kN/m

In Figure 11.12 the distributions of normal displacement $w_H(x)$, meridional moment $m_x^H(x)$ and meridional transverse shear force $t_x^H(x)$ are presented for interval $[0, 20]$ – note that $t_x^H(0) = 1.0$ kN/m and $w_H(0)$ is given by Equation $(11.54)_1$.

The solutions obtained for cases A and B are going to be used in future considerations in Section 11.4 and Subsection 13.3.3.

11.3.2 Spherical shell

A spherical shell loaded at the bottom contour by boundary normal moment M [kNm/m] and horizontal force H [kN/m] is considered. The shell is limited by angle α and described using angular coordinate φ_1 measured from the bottom contour. The geometry of the analysed shell and adopted loads are presented in Figure 11.13. It is assumed that the rigid-body motions of the spherical segment are prevented.

Moment M and horizontal force H induce on edge $\varphi_1 = 0°$ translations denoted by $\delta_M(0)$ and $\delta_H(0)$ and rotations of the normal to the edge $\vartheta_M(0)$ and $\vartheta_H(0)$. The goal of the consideration is to determine the formulae to compute the magnitudes of these four boundary kinematic quantities (Figure 11.13b,c). The knowledge of formulae defining the boundary magnitudes of horizontal translations $\delta_M(0)$, $\delta_H(0)$ and rotations $\vartheta_M(0)$, $\vartheta_H(0)$ caused by loads M and H at each point of bottom contour in axisymmetric state is necessary to build the solution of some problems using superposition. Within this approach a set of two algebraic equations (assuming for example zero translation and zero rotation at the bottom edge) with unknowns M and H is formulated. Then, as a solution, the distributions of meridional bending moment $m_\varphi(\varphi_1)$ and transverse shear force $t_\varphi(\varphi_1)$, which are the main functions describing bending effects are obtained.

The starting point of the analysis is Equation (11.33) from Subsection 11.2.2 which is a differential relation with unknown transverse shear force $t_\varphi(\varphi_1)$ and describes axisymmetric bending state in a spherical shell. Its integral is rewritten for the shell subdomain

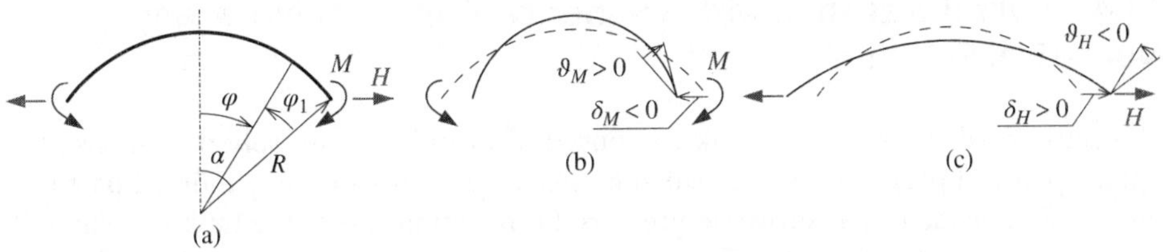

Figure 11.13 (a) Spherical shell with uniformly distributed boundary loads M and H; shell deformation caused by: (b) moment M and (c) horizontal force H

in the vicinity of the bottom contour and expressed using coordinate φ_1 and integration constants C and γ, see Timoshenko and Woinowsky-Krieger (1959)

$$t_\varphi(\varphi_1) = Ce^{-\varsigma\varphi_1}\sin(\varsigma\varphi_1 + \gamma) \tag{11.55}$$

where parameter ς is defined in Equation (11.33). The integration constants C and γ are derived (as usual) from the boundary conditions. Next, the effects of two loads: normal moment M (case A) and horizontal force H (case B) are considered separately. Due to the complexity of mathematical operations only the final formulae taken from Timoshenko and Woinowsky-Krieger (1959) are presented.

11.3.2.1 Case A – distributed boundary moment *M*

For the bottom edge two static boundary conditions are formulated. They are related to the case of pure bending state without membrane effects:

$$m_\varphi|_{\varphi_1=0^\circ} = M, \qquad n_\varphi|_{\varphi_1=0^\circ} = t_\varphi \tan\alpha = 0 \tag{11.56}$$

The final formulae for the boundary values of horizontal translation δ_M (interpreted as the change of the boundary circle radius Δr), rotation ϑ_M and function of meridional moment $m_\varphi^M(\varphi_1)$ take the form:

$$\delta_M(0^\circ) = \Delta r_M(0^\circ) = -\frac{2\varsigma^2\sin\alpha}{Eh}M, \qquad \vartheta_M(0^\circ) = \frac{4\varsigma^3}{ERh}M \tag{11.57}$$

$$m_\varphi^M(\varphi_1) = e^{-\varsigma\varphi_1}\left[\sin(\varsigma\varphi_1) + \cos(\varsigma\varphi_1)\right] M \tag{11.58}$$

11.3.2.2 Case B – horizontal boundary traction *H*

In this case, two static boundary conditions at the bottom edge are also written:

$$m_\varphi|_{\varphi_1=0^\circ} = 0, \qquad n_\varphi|_{\varphi_1=0^\circ} = H\cos\alpha \tag{11.59}$$

As a solution of the boundary value problem the following three formulae for boundary magnitudes of translation δ_H, rotation ϑ_H and function of meridional moment $m_\varphi^H(\varphi_1)$ are obtained:

$$\delta_H(0^\circ) = \Delta r_H(0^\circ) = -\frac{2R\varsigma\sin\alpha}{Eh}H, \qquad \vartheta_H(0^\circ) = -\frac{2\varsigma^2\sin\alpha}{Eh}H \tag{11.60}$$

$$m_\varphi^H(\varphi_1) = -\frac{R}{\varsigma}e^{-\varsigma\varphi_1}\sin\alpha\sin(\varsigma\varphi_1)H \tag{11.61}$$

These formulae are going to be used in Section 11.4 and Subsection 13.2.4 to form the solution of more complex states resulting from static or thermal load, respectively.

11.4 Cylindrical Shell with a Spherical Cap – Analytical and Numerical Solution

The considered structure is a tank composed of cylindrical and spherical shells. The upper spherical part (the cap) is loaded by self weight. The lower cylindrical part (the tank wall) is loaded by hydrostatic pressure (a maximum fill of the tank is assumed), the cap weight and self weight. The configuration is shown in Figure 11.14. The bottom edge of the spherical cap is described by $\varphi = \alpha$ and the top edge of cylindrical shell by coordinate $x = 0$.

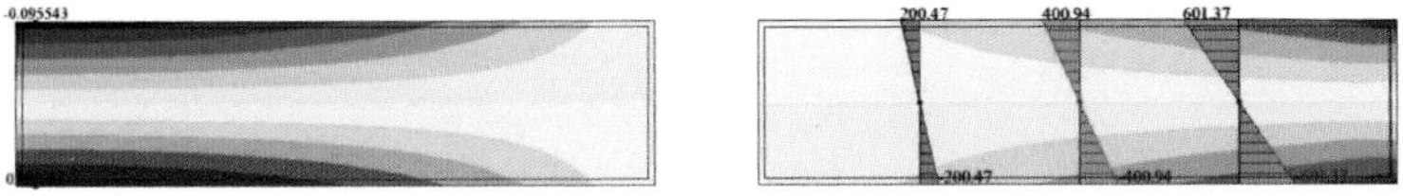

(a) (b)

Figure 6.12 Cantilever beam – contour plots of: (a) translation u_x [cm] and (b) stress σ_x [MPa]

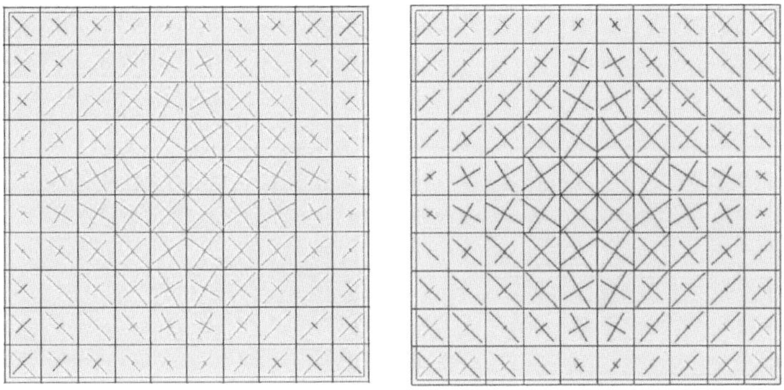

(a) (b)

Figure 8.16 Simply supported square plate with uniform load: vizualization of directions and magnitudes of principal stresses on two surfaces: (a) bottom and (b) top

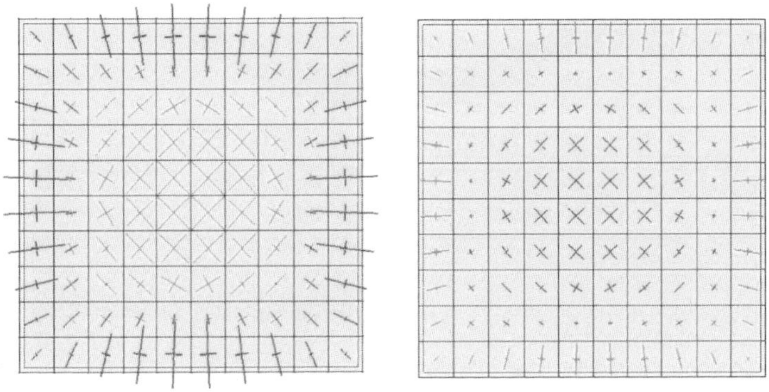

(a) (b)

Figure 8.17 Clamped square plate with uniform load: vizualization of directions and magnitudes of principal stresses on two surfaces: (a) bottom and (b) top

Plate and Shell Structures: Selected Analytical and Finite Element Solutions, First Edition.
Maria Radwańska, Anna Stankiewicz, Adam Wosatko and Jerzy Pamin.
© 2017 John Wiley & Sons Ltd. Published 2017 by John Wiley & Sons Ltd.

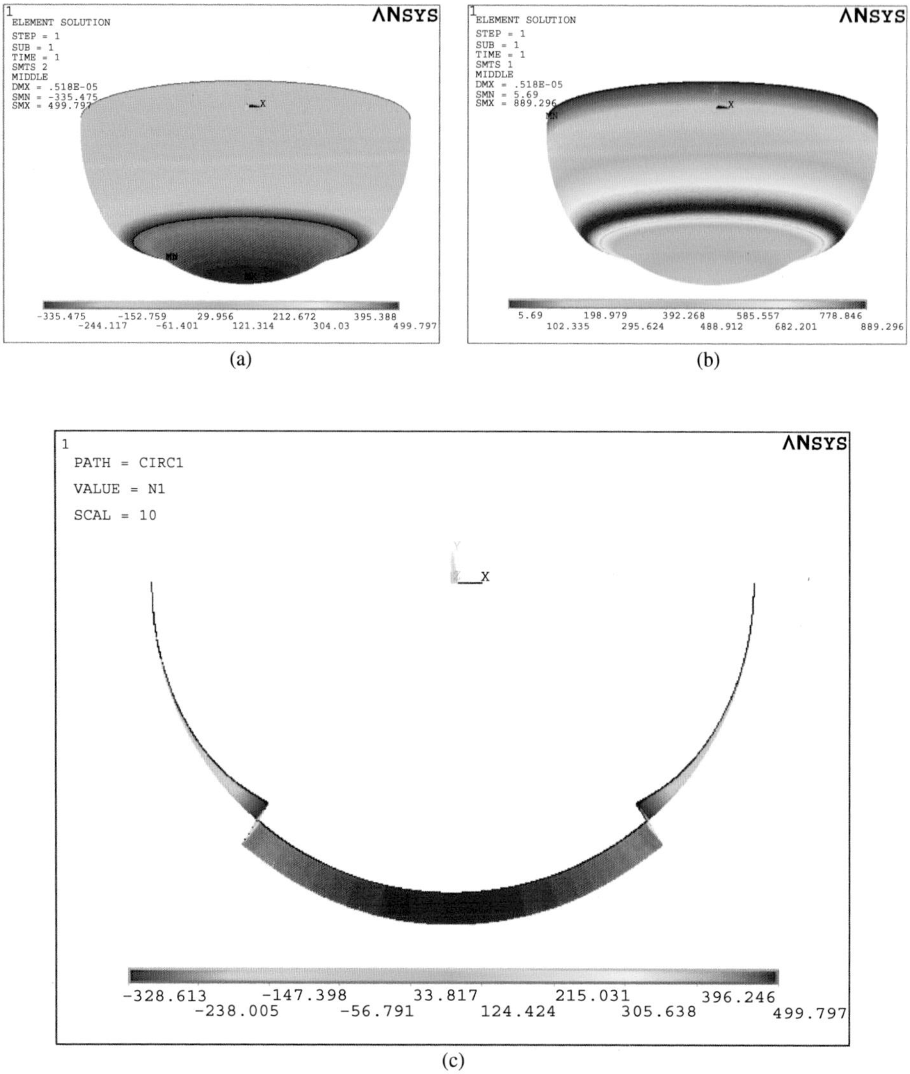

Figure 10.6 Hemispherical shell supported at the intermediate parallel, subjected to hydrostatic pressure – contour maps of membrane forces: (a) meridional, (b) circumferential and (c) diagram of meridional force along the section described by $\theta = \text{const.}$ with extreme ordinates $n_{\varphi}^{\mathrm{u,FEM}}(40°) = -328.61$ kN/m and $n_{\varphi}^{\mathrm{l,FEM}}(0°) = n_{\theta}^{\mathrm{l,FEM}}(0°) = 500$ kN/m

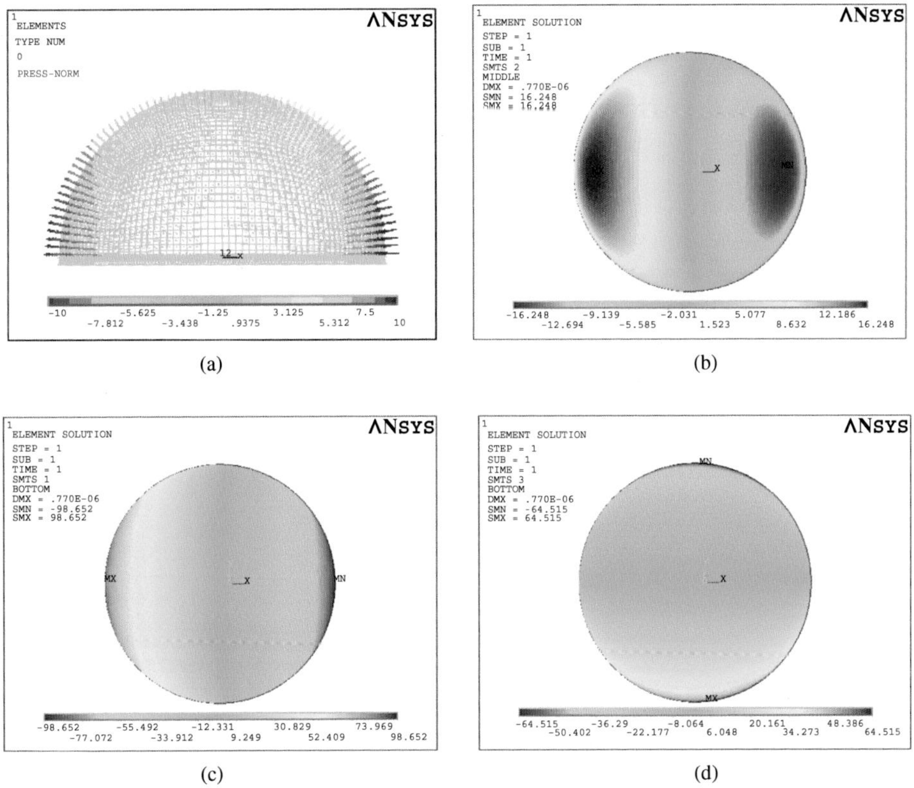

Figure 10.12 Hemisperical shell under wind action: (a) vector vizualization of wind load, contour maps of membrane forces [N/m]: (b) meridional n_{φ}, (c) circumferential n_{θ}, (d) tangent $n_{\varphi\theta}$

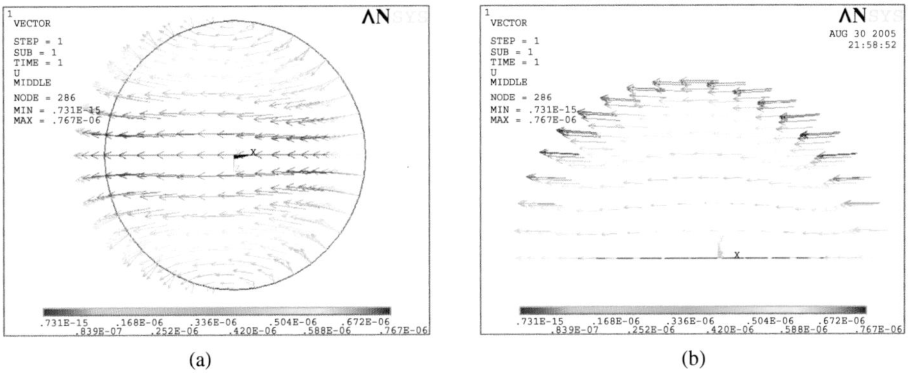

Figure 10.13 Hemisperical shell under wind action: vector vizualization of displacement field: (a) top view, (b) side view

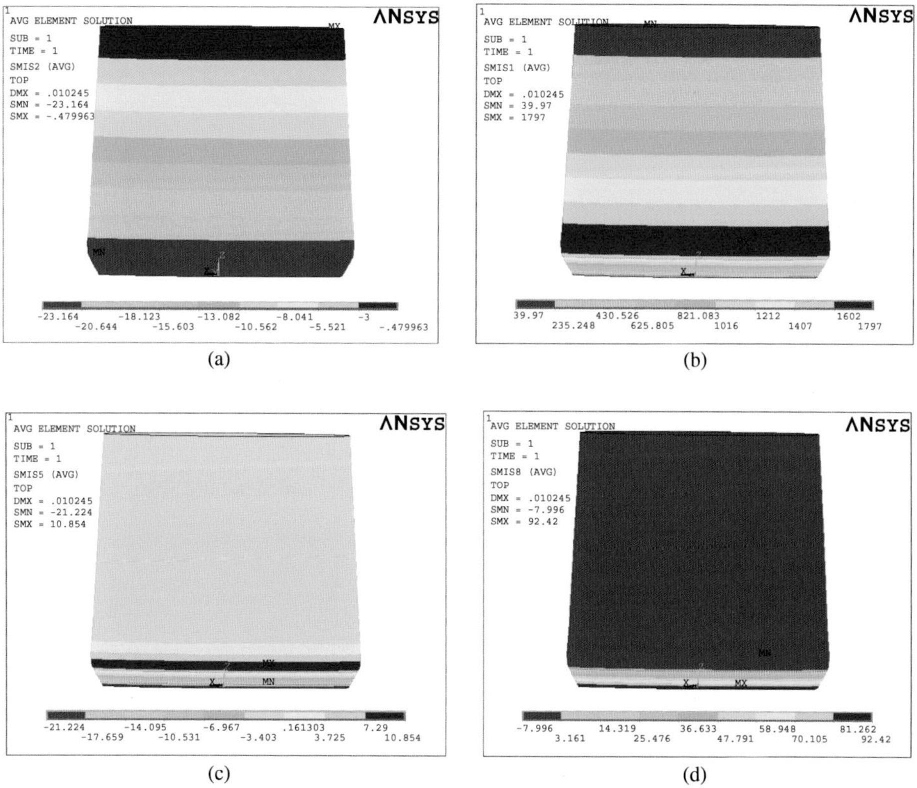

Figure 11.5 Long cylindrical shell – results of calculations with 2D FEs – distributions of functions: (a) meridional force n_x, (b) circumferential force n_θ, (c) meridional moment m_x, (d) meridional transverse shear force t_x, [kN], [m]

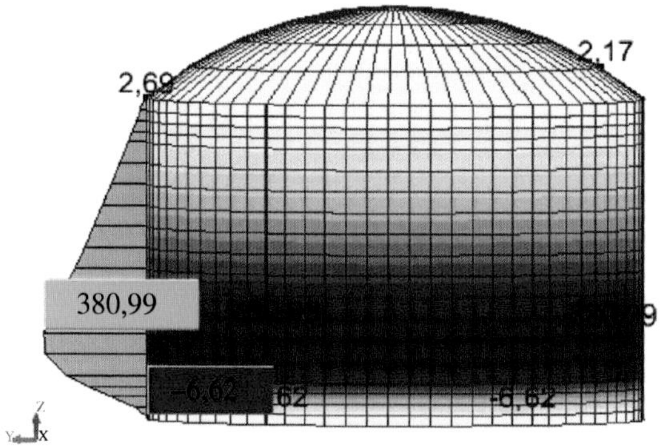

Figure 11.17 Cylindrical shell with spherical cap – distribution of circumferential force caused by hydrostatic pressure

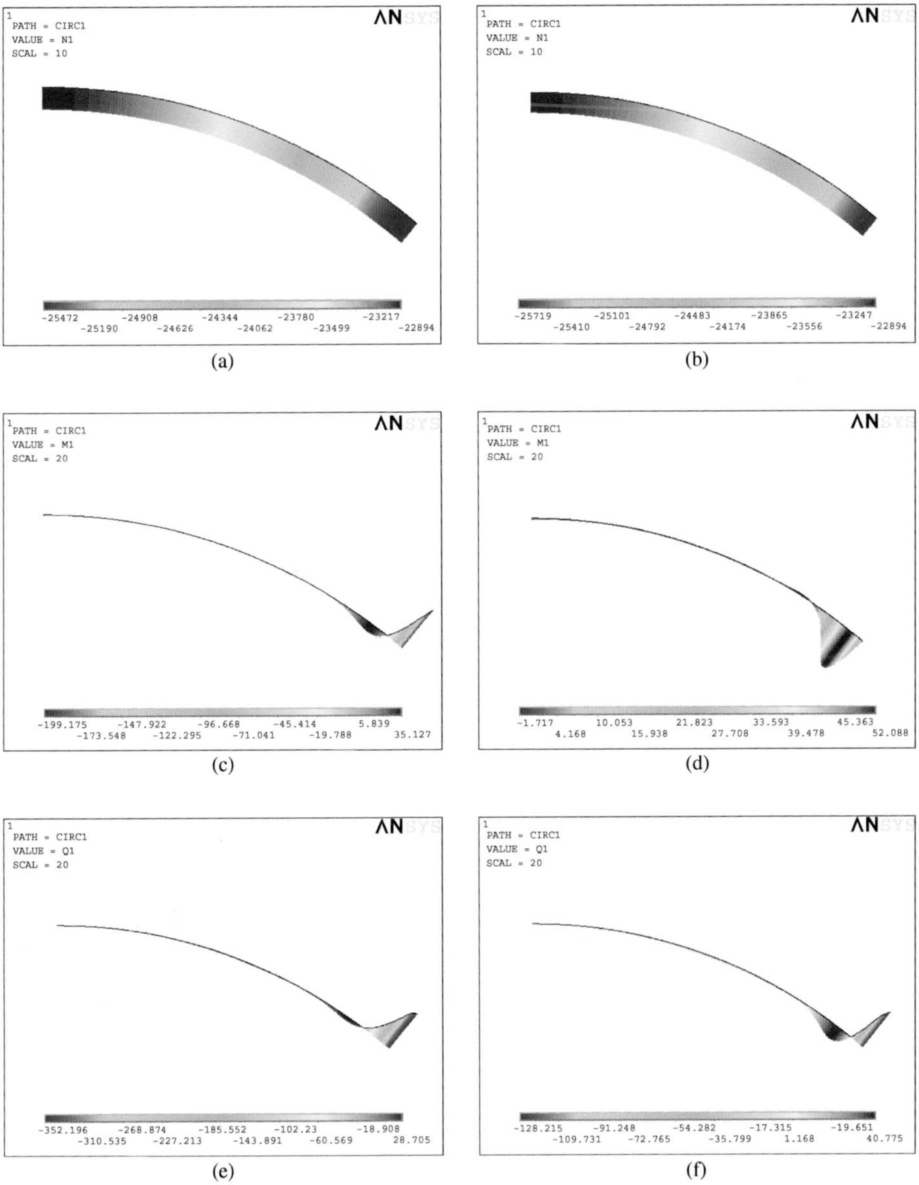

Figure 11.9 Spherical shells under self weight: left column – case A (shell clamped along bottom edge), right column – case B (simply supported shell); graphs of following functions, respectively: (a),(b) meridional membrane forces $n_\varphi(\varphi)$, (c),(d) meridional moment $m_\varphi(\varphi)$ and (e),(f) meridional transverse shear forces $t_\varphi(\varphi)$, [N],[m]

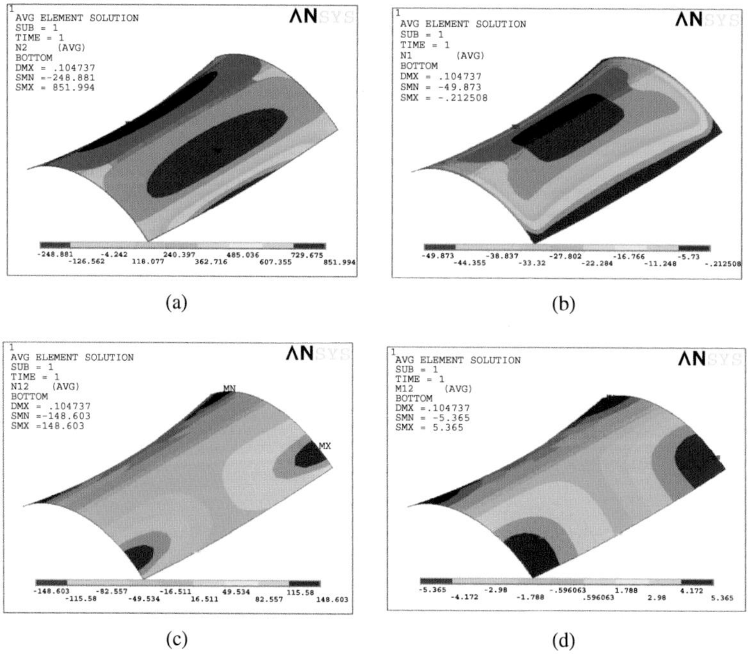

Figure 11.23 Scordelis-Lo roof – contour maps for: membrane forces [kN/m]: (a) n_x, (b) n_θ, (c) $n_{x\theta}$ and (d) twisting moment [kNm/m] $m_{x\theta}$

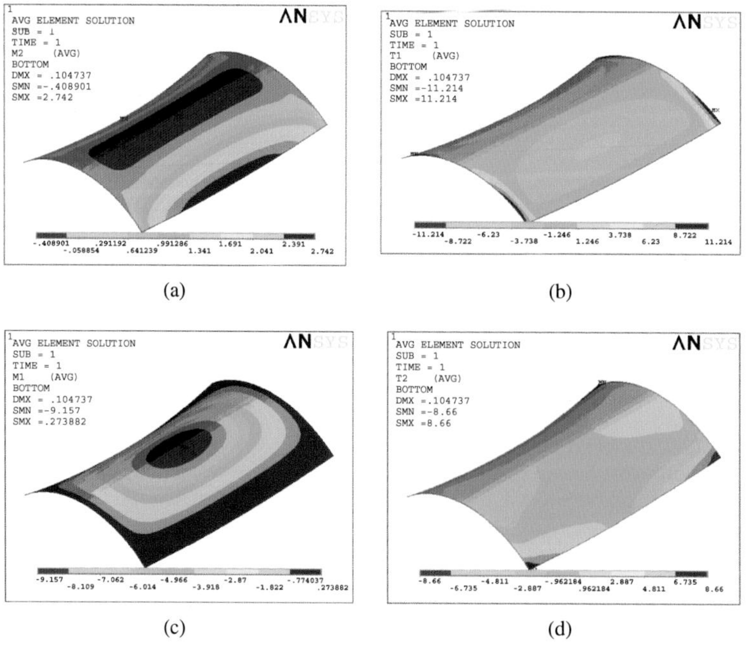

Figure 11.24 Scordelis-Lo roof – contour maps for bending moments [kNm/m]: (a) m_x, (c) m_θ, transverse shear forces [kN/m]: (b) t_x, (d) t_θ

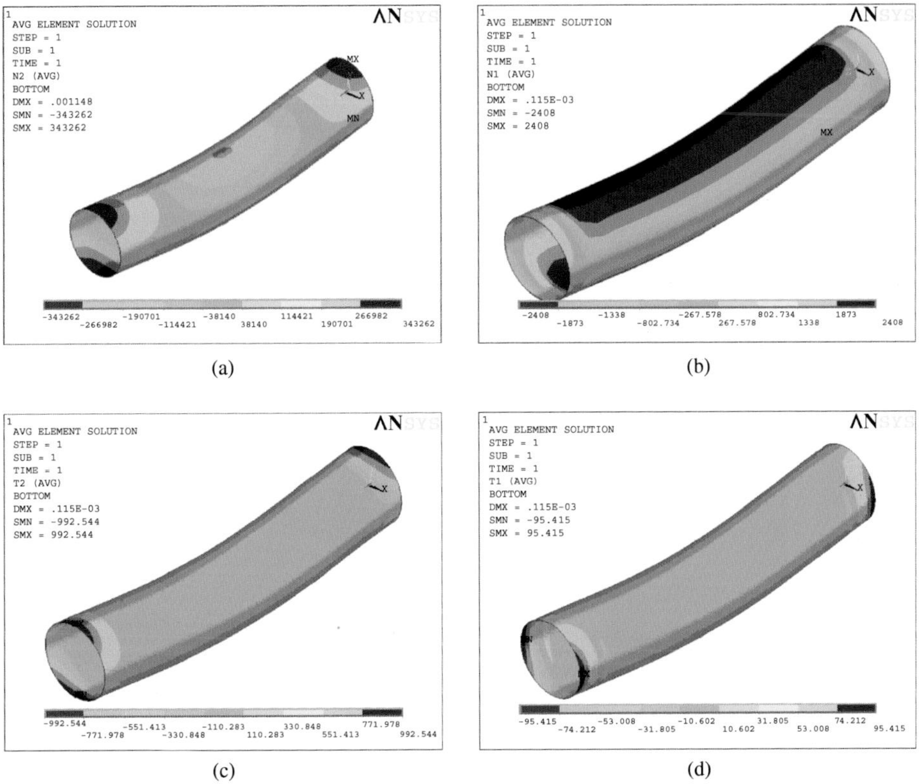

Figure 11.29 Horizontal tube under self weight – contour maps for membrane forces: (a) n_x, (b) n_θ and for transverse shear forces: (c) t_x, (d) t_θ [N,m]

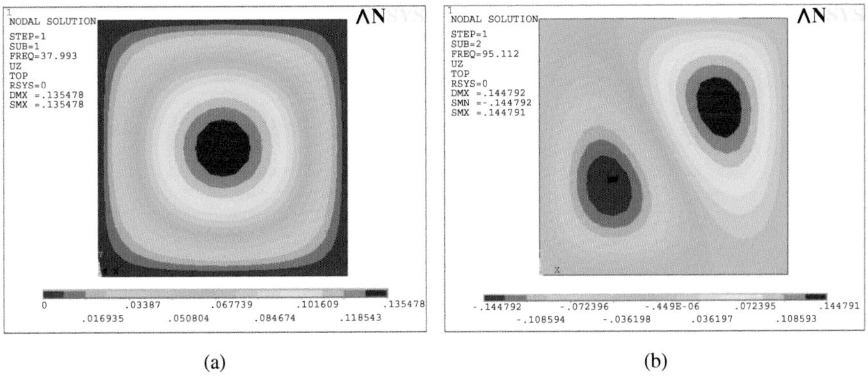

Figure 15.2 Simply supported square plate – two first modes of transverse natural vibrations: (a) for frequency $f^{(1)} = f_{(1,1)} = 37.99$ Hz, (b) for frequency $f^{(2)} = f_{(1,2)} = 95.11$ Hz

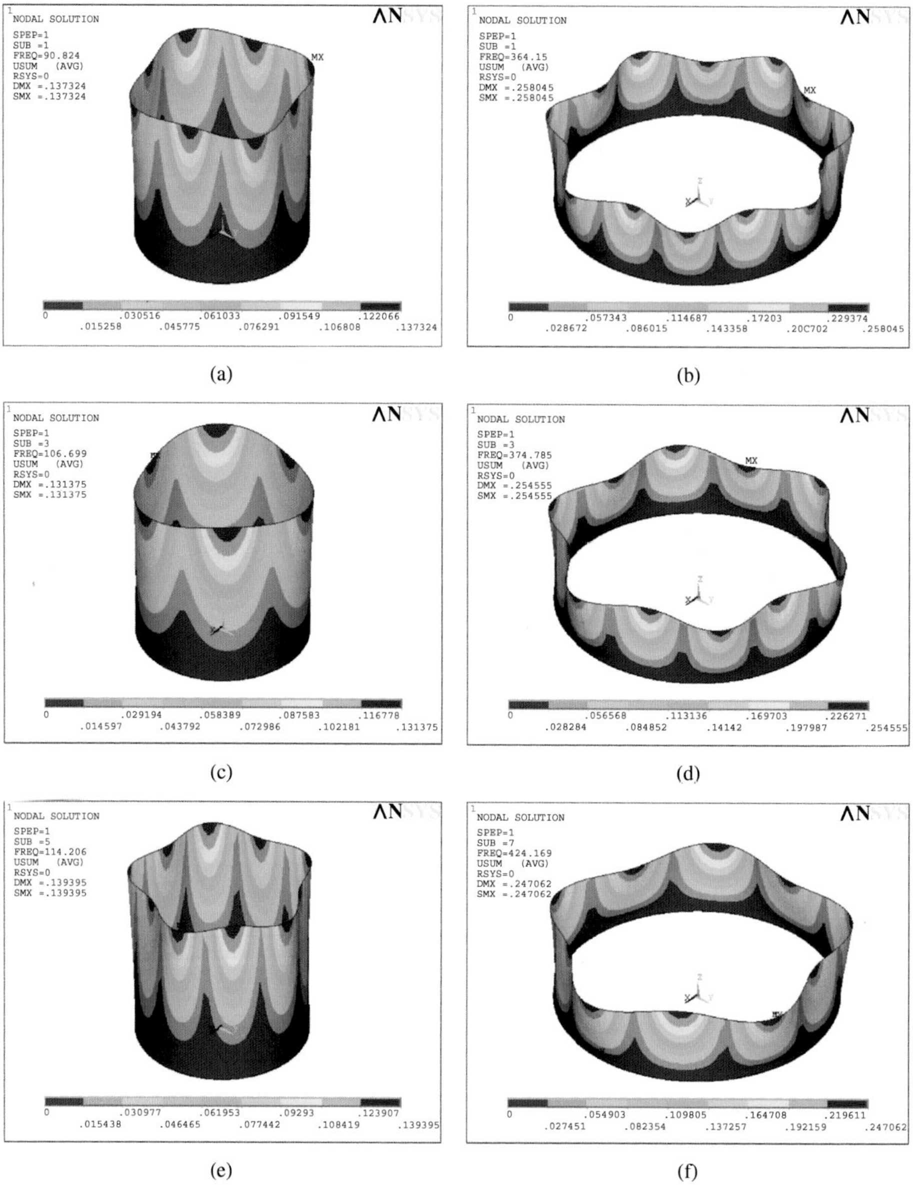

Figure 15.3 Three natural vibration modes related to three initial frequencies for a long shell with $j = 4, 3, 5$ (first column) and for a short shell with $j = 7, 6, 5$ (second column)

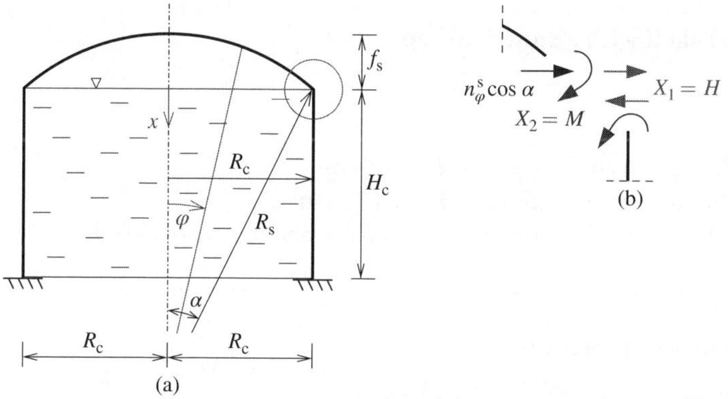

Figure 11.14 Cylindrical shell with spherical cap: (a) configuration with two coordinate systems: spherical with $\varphi \in [0, \alpha]$ and cylindrical with $x \in [0, H_c]$ and (b) interactions H and M introduced in interface of two parts of reservoir

In the tank the bending effects occur, resulting from the clamped bottom edge and rigid connection of cylindrical and spherical parts. The main goal of the following subsection is an analytical description of local bending state in the vicinity of the rigid connection of two parts of the tank. Functions of meridional bending moment, meridional transverse shear force and circumferential membrane force are derived and extreme values are calculated. The numerical solution is also discussed and a comparison of analytical and numerical results is presented in Box 11.9.

11.4.1 Analytical Solution

The problem solution is based on the example discussed in Kobiak and Stachurski (1991), the book dedicated to students and engineers (designers). The essential formulae are quoted and complicated mathematical operations are omitted.

The data characterizing analysed tank are listed in Box 11.9. The material constants for concrete are marked by index 'con'. The quantities associated with cylindrical and spherical parts have indices 'c' and 's', res pectively. The radius, thickness and geometrical parameters for cylindrical and spherical shells are denoted here by: $R_c, R_s, h_c, h_s, H_c, \beta, \varsigma$, respectively:

$$\beta = \sqrt{\frac{1}{R_c h_c}} \sqrt[4]{3(1-\nu^2)} = 1.193\ 1/\mathrm{m}, \qquad \varsigma = \sqrt{\frac{R_s}{h_s}} \sqrt[4]{3(1-\nu^2)} = 14.081 \qquad (11.62)$$

The calculations involve the determination of the interaction of the two shells. It is a computation of the horizontal force $X_1 = H$ and normal moment $X_2 = M$ at the contour of the connection zone of two parts, that is at the bottom edge of the spherical cap with $\varphi = \alpha$ and at the top edge of cylindrical shell with coordinate $x = 0$. The notation X_1 and X_2 comes from the force method used in structural mechanics.

The rigid connection of the two parts of the tank implies:

- continuity of the radial horizontal displacement $\Delta_1 = \Delta r$ (associated energetically with unknown horizontal force X_1), that is the uniform change of radius Δr
- continuity of rotation of normal to meridian $\Delta_2 = \vartheta$ (rotation associated with unknown moment X_2)

Box 11.9 Cylindrical shell with a spherical cap

Data

$E = 1.5 \times 10^7$ kN/m^2, $\nu = 1/6$, $\gamma_{\text{con}} = 24.0$ kN/m^3
$H_c = 7.6$ m, $R_c = 6.0$ m, $h_c = 0.15$ m, $\beta = 1.193$ 1/m
$R_s = 9.282$ m, $f_s = 2.2$ m, $h_s = 0.08$ m, $r_{0,c} = R_c = 6.0$ m, $\varsigma = 14.081$

Check values from analytical solution

$n_\theta^{\text{s,anal}}|_{\varphi=\alpha} = 29.42$ kN/m, $n_\theta^{\text{c,anal}}|_{x=0} = 55.22$ kN/m
$m_\varphi^{\text{s,anal}}|_{\varphi=\alpha} = m_x^{\text{c,anal}}|_{x=0} = 0.70$ kNm/m $= M$
$t_x^{\text{c,anal}}|_{x=0} = 4.31$ kN/m $= -H$

Check values from FEM solution

ANKA, $NSE = 25$, SRK FEs
$m_\varphi^{\text{s,FEM}}|_{\varphi=\alpha} = m_x^{\text{c,FEM}}|_{x=0} = 0.75$ kNm/m $= 1.071 m_\varphi^{\text{s,anal}}|_{\varphi=\alpha}$
$t_x^{\text{c,anal}}|_{x=0} = 4.13$ kN/m $= 0.958 t_x^{\text{c,anal}}|_{x=0}$

ROBOT
$n_\theta^{\text{s,FEM}}|_{\varphi=\alpha} = 28.26$ kN/m $= 0.961 n_\theta^{\text{s,anal}}|_{\varphi=\alpha}$
$n_\theta^{\text{c,FEM}}|_{x=0} = 47.7$ kN/m $= 0.864 n_\theta^{\text{c,anal}}|_{x=0}$
$n_{\theta,\max}^{\text{c,FEM}} = 380.99$ kN/m
$m_\varphi^{\text{c,FEM}}|_{\varphi=\alpha} = 0.70$ kNm/m $= m_\varphi^{\text{c,anal}}|_{\varphi=\alpha} = M$
$m_{x,\min}^{\text{c,FEM}}|_{x=7.6} = -17.50$ kNm/m

Shell configuration shown in Figure 11.14

Two continuity conditions, including the components denoted as in the force method, take into account the contribution of both the spherical and cylindrical shell:

$$\begin{aligned} \Delta r = \Delta_1 &= (\delta_{11}^{\text{s}} + \delta_{11}^{\text{c}})X_1 + (\delta_{12}^{\text{s}} + \delta_{12}^{\text{c}})X_2 + (\Delta_{1p}^{\text{s}} + \Delta_{1p}^{\text{c}}) = 0 \\ \vartheta = \Delta_2 &= (\delta_{21}^{\text{s}} + \delta_{21}^{\text{c}})X_1 + (\delta_{22}^{\text{s}} + \delta_{22}^{\text{c}})X_2 + (\Delta_{2p}^{\text{s}} + \Delta_{2p}^{\text{c}}) = 0 \end{aligned} \tag{11.63}$$

Next, the formulae for components of the previous set of equations are listed:

(I) radius change Δr:
- caused by $X_1 = H = 1$ [kN/m]

$$\delta_{11}^{\text{s}} + \delta_{11}^{\text{c}} = \Delta r_H = \frac{2\varsigma R_s}{E_{\text{con}} h_s} \sin^2\alpha + \frac{2\beta R_c^2}{E_{\text{con}} h_c} \tag{11.64}$$

- caused by $X_2 = M = 1$ [kNm/m]

$$\delta^{\mathrm{s}}_{12} + \delta^{\mathrm{c}}_{12} = \Delta r_M = -\frac{2\varsigma^2}{E_{\mathrm{con}} h_{\mathrm{s}}} \sin\alpha + \frac{2\beta^2 R_{\mathrm{c}}^2}{E_{\mathrm{con}} h_{\mathrm{c}}} \tag{11.65}$$

- caused by external load in membrane state

$$\Delta^{\mathrm{s}}_{1p} + \Delta^{\mathrm{c}}_{1p} = \Delta r_p = \frac{R_{\mathrm{s}}}{E_{\mathrm{con}} h_{\mathrm{s}}} (\overline{n}^{\mathrm{s}}_{\theta} - \nu \overline{n}^{\mathrm{s}}_{\varphi}) \sin\alpha$$
$$+ \overline{n}^{\mathrm{s}}_{\varphi} \cos\alpha \frac{2\varsigma R_{\mathrm{s}}}{E_{\mathrm{con}} h_{\mathrm{s}}} \sin^2\alpha \Big|_{\varphi=\alpha} + \overline{n}^{\mathrm{c}}_{\theta} \frac{R_{\mathrm{c}}}{E_{\mathrm{con}} h_{c}} \Big|_{x=0} \tag{11.66}$$

(II) rotation of a normal to a meridian ϑ:

- caused by $X_1 = H = 1$ [kN/m]

$$\delta^{\mathrm{s}}_{21} + \delta^{\mathrm{c}}_{21} = \vartheta_H = -\frac{2\varsigma^2}{E_{\mathrm{con}} h_{\mathrm{s}}} \sin\alpha + \frac{2\beta^2 R_{\mathrm{c}}^2}{E_{\mathrm{con}} h_{\mathrm{c}}} \tag{11.67}$$

- caused by $X_2 = M = 1$ [kNm/m]

$$\delta^{\mathrm{s}}_{22} + \delta^{\mathrm{c}}_{22} = \vartheta_M = \frac{4\varsigma^3}{E_{\mathrm{con}} h_{\mathrm{s}} R_{\mathrm{s}}} + \frac{4\beta^3 R_{\mathrm{c}}^2}{E_{\mathrm{con}} h_{\mathrm{c}}} \tag{11.68}$$

- caused by external load in membrane state

$$\Delta^{\mathrm{s}}_{2p} + \Delta^{\mathrm{c}}_{2p} = \vartheta_p = \frac{1+\nu}{E_{\mathrm{con}} h_{\mathrm{s}}} (\overline{n}^{\mathrm{s}}_{\varphi} - \overline{n}^{\mathrm{s}}_{\theta}) \cot\alpha - \frac{1}{E_{\mathrm{con}} h_{\mathrm{s}}} \frac{\mathrm{d}(\overline{n}^{\mathrm{s}}_{\theta} - \nu \overline{n}^{\mathrm{s}}_{\varphi})}{\mathrm{d}\varphi} \Big|_{\varphi=\alpha}$$
$$- \overline{n}^{\mathrm{s}}_{\varphi} \cos\alpha \frac{2\varsigma^2}{E_{\mathrm{con}} h_{\mathrm{s}}} \sin\alpha + \frac{R_{\mathrm{c}}}{E_{\mathrm{con}} h_{\mathrm{c}}} \frac{\mathrm{d}(\overline{n}^{\mathrm{c}}_{\theta} - \nu \overline{n}^{\mathrm{c}}_{x})}{\mathrm{d}x} \Big|_{x=0} \tag{11.69}$$

These components are computed at the bottom edge of the spherical cap with $\varphi = \alpha$ and at the top edge of the cylindrical shell with coordinate $x = 0$.

Equations (11.63) are multiplied by the value of Young's modulus for concrete E_{con} in order to avoid too small values of coefficients in the set of equations. Eventually, two algebraic equations with unknowns X_1 and X_2 are:

$$(1363.58 + 677.79)X_1 + (-3202.13 + 933.59)X_2 = 142.41 - 10\,530.42 \tag{11.70}$$
$$(-3202.13 + 933.59)X_1 + (15\,039.35 + 2571.90)X_2$$
$$= -122.02 - 2460.37 + 24\,688.45$$

The solution to the set of equations leads to the following values of X_1 and X_2

$$X_1 = H = -4.31 \text{ kN/m} = -t^{\mathrm{c}}_{x}|_{x=0}, \qquad X_2 = M = 0.70 \text{ kNm/m} = m^{\mathrm{s}}_{\varphi}|_{\varphi=\alpha} = m^{\mathrm{c}}_{x}|_{x=0} \tag{11.71}$$

Now, the description of the local bending state can be completed. In Figures 11.15 and 11.16 the functions of circumferential force and meridional bending moment are plotted. The diagrams are analogous to those in the work of Kobiak and Stachurski (1991), but with the present rules of notation and sign convention.

The main purpose of the presented example was an analytical description of the rigid connection of two parts of the tank with different shapes. It is worth noticing that the value of the meridional bending moment at the interface of the two parts has the same

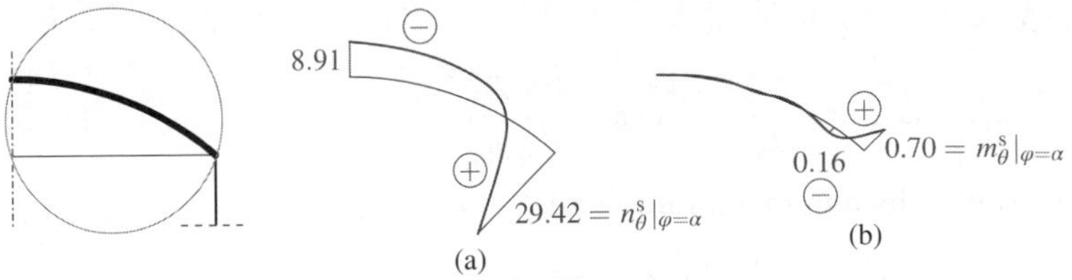

Figure 11.15 Cylindrical shell with spherical cap – distribution of: (a) circumferential force $n_2 = n^{s}_{\theta}$ and (b) meridional moment $m_1 = m^{s}_{\varphi}$ in spherical shell in vicinity of connection of two shells

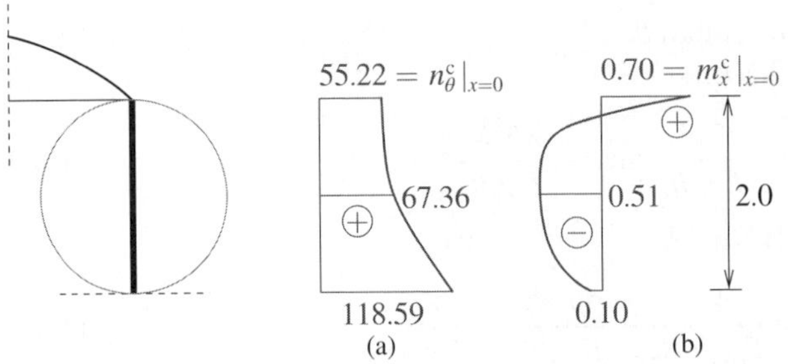

Figure 11.16 Cylindrical shell with spherical cap – distribution of: (a) circumferential force $n_2 = n^{c}_{\theta}$ and (b) meridional moment $m_1 = m^{c}_{x}$ in cylindrical shell in the vicinity of connection of two shells

value for the spherical and cylindrical part, that is $m^{s}_{\varphi}|_{\varphi=\alpha} = m^{c}_{x}|_{x=0}$. The meridional force computed at the bottom edge of the spherical part is compressive for the membrane state and becomes tensile when the bending effects are taken into account. In the analysis of the whole tank a hydrostatic pressure plays a dominant role. It causes substantial tensile circumferential forces on the parallel with $x = 0.75H_c$ and the highest meridional bending moment at the bottom clamped edge $x = H_c$.

11.4.2 FEM Solution

The numerical analysis has been performed using two kinds of FEs: (i) one-dimensional SRK (ANKA) and (ii) two-dimensional (ROBOT).

In the first case the meridian of spherical-cylindrical shell has been discretized with $NSE = 25$ FEs with densification in the zone of cylinder-cap connection and in the vicinity of the clamped edge. The characteristic values of the meridian bending moment computed at the interface of the two parts and of the transverse shear force in the upper part of cylindrical shell are included in Box 11.9.

The results obtained for the second discretization with two-dimensional FEs in ROBOT are presented in Figure 11.17.

The following check values have been chosen for result comparison: circumferential forces in spherical and cylindrical part, the meridional moment at the interface of two tank parts and at the bottom clamped edge. It is worth emphasizing that the absolute value of the meridional moment $m^{c}_{x}|_{x=H_c}$ at the clamped edge is 25 times higher than the positive moment $m^{c}_{x}|_{x=0}$ at the interface of the two shell parts (see also Figure 11.17b).

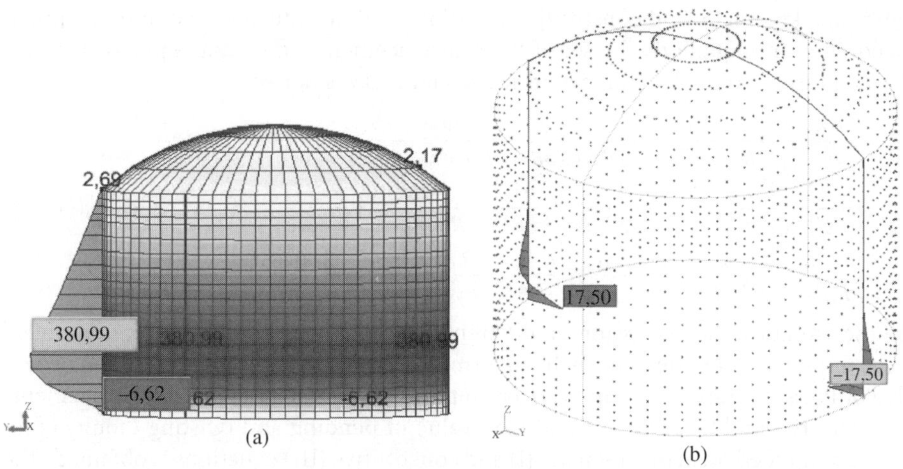

Figure 11.17 Cylindrical shell with spherical cap – distribution of (a) circumferential force caused by hydrostatic pressure and (b) meridional moment in section θ = const. (extreme moment $m_{x,\min}^{c,\mathrm{FEM}}|_{x=H_c} = -17.50$ kNm/m); FEM results from ROBOT. *Subfigure (a) shown in Plate section for color representation of this figure.*

11.5 General Case of Deformation of Cylindrical Shells

In this section the membrane-bending state of a segment of cylindrical shell is considered. This type of shell is very often used as a cylindrical roof or a part of multispan covering. The configuration of the roof is shown in Figure 11.18.

The mathematical model for thin cylindrical shells in the form of a set of 17 equations has been included in Box 11.1. Here, three equilibrium equations obtained after elimination of transverse shear forces t_x and t_θ are written:

$$\begin{aligned}
&R\frac{\partial n_x}{\partial x} + \frac{\partial n_{\theta x}}{\partial \theta} - \frac{1}{2R}\frac{\partial m_{x\theta}}{\partial \theta} + R\hat{p}_x = 0\\
&\frac{\partial n_\theta}{\partial \theta} + R\frac{\partial n_{x\theta}}{\partial x} + \frac{3}{2}\frac{\partial m_{x\theta}}{\partial x} + \frac{1}{R}\frac{\partial m_\theta}{\partial \theta} + R\hat{p}_\theta = 0\\
&-n_\theta + R\frac{\partial^2 m_x}{\partial x^2} + \frac{1}{R}\frac{\partial^2 m_\theta}{\partial \theta^2} + 2\frac{\partial^2 m_{\theta x}}{\partial x \partial \theta} + R\hat{p}_n = 0
\end{aligned} \tag{11.72}$$

Figure 11.18 Configuration of cylindrical segment

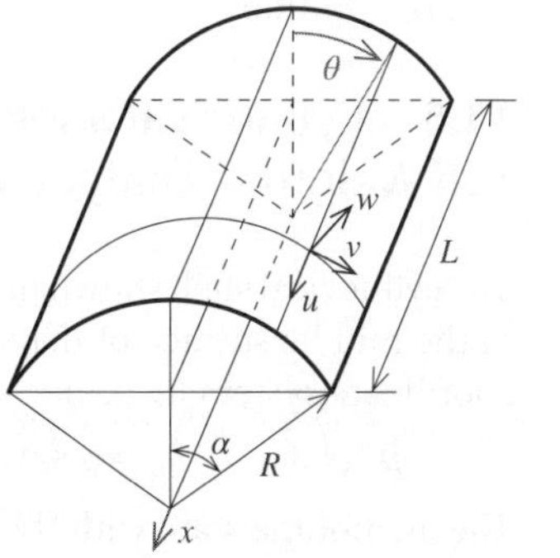

Combining kinematic and constitutive equations and introducing them into the previous equilibrium equations the set of three displacement differential equations for the so-called three-parameter thin cylindrical shell theory is derived:

$$\begin{aligned}
&\frac{\partial^2 u}{\partial x^2} + \frac{1-\nu}{2R^2}\frac{\partial^2 u}{\partial \theta^2} + \frac{1+\nu}{2R}\frac{\partial^2 v}{\partial x \partial \theta} + \frac{\nu}{R}\frac{\partial w}{\partial x} = -\frac{\hat{p}_x}{D^n} \\
&\frac{1+\nu}{2R}\frac{\partial^2 u}{\partial x \partial \theta} + \frac{1-\nu}{2}\frac{\partial^2 v}{\partial x^2} + \frac{1}{R^2}\frac{\partial^2 v}{\partial \theta^2} + \frac{1}{R^2}\frac{\partial w}{\partial \theta} = -\frac{\hat{p}_\theta}{D^n} \\
&\frac{\nu}{R}\frac{\partial u}{\partial x} + \frac{1}{R^2}\frac{\partial v}{\partial \theta} + \frac{w}{R^2} + \frac{h^2}{12}\left(\frac{\partial^4 w}{\partial x^4} + \frac{2}{R^2}\frac{\partial^4 w}{\partial x^2 \partial \theta^2} + \frac{1}{R^4}\frac{\partial^4 w}{\partial \theta^4}\right) = -\frac{\hat{p}_n}{D^n}
\end{aligned} \tag{11.73}$$

In the literature the discussion on different simplification of the considered problem can be found: (i) omission of underlined terms in Equations (11.72), (ii) adoption of zero value of Poisson's ratio $\nu = 0$ and (iii) assumption of a small influence of displacements tangent to the middle surface u and v on value of bending and twisting moments. In effect, a simplified form of kinematic (I) and constitutive (III) equations is obtained. The combination of these two groups of equations leads to the following relations between the generalized forces (membrane forces and moments) and the translations:

$$\begin{aligned}
&n_x = Eh\frac{\partial u}{\partial x}, \qquad n_\theta = Eh\left(\frac{1}{R}\frac{\partial v}{\partial \theta} + \frac{w}{R}\right), \qquad n_{x\theta} = \frac{Eh}{2}\left(\frac{1}{R}\frac{\partial u}{\partial \theta} + \frac{\partial v}{\partial x}\right) \\
&m_x = -D^m\frac{\partial^2 w}{\partial x^2}, \qquad m_\theta = -D^m\frac{1}{R^2}\frac{\partial^2 w}{\partial \theta^2}, \qquad m_{x\theta} = m_{\theta x} = -D^m\frac{\partial^2 w}{\partial x \partial \theta}
\end{aligned} \tag{11.74}$$

For zero Poisson's ratio the set of three displacement differential equations takes the simplified form:

$$\begin{aligned}
&\frac{\partial^2 u}{\partial x^2} + \frac{1}{2R^2}\frac{\partial^2 u}{\partial \theta^2} + \frac{1}{2R}\frac{\partial^2 v}{\partial x \partial \theta} = -\frac{\hat{p}_x}{Eh} \\
&\frac{1}{2R}\frac{\partial^2 u}{\partial x \partial \theta} + \frac{1}{2}\frac{\partial^2 v}{\partial x^2} + \frac{1}{R^2}\frac{\partial^2 v}{\partial \theta^2} + \frac{1}{R^2}\frac{\partial w}{\partial \theta} = -\frac{\hat{p}_\theta}{Eh} \\
&\frac{1}{R^2}\frac{\partial v}{\partial \theta} + \frac{w}{R^2} + \frac{h^2}{12}\left(\frac{\partial^4 w}{\partial x^4} + \frac{2}{R^2}\frac{\partial^4 w}{\partial x^2 \partial \theta^2} + \frac{1}{R^4}\frac{\partial^4 w}{\partial \theta^4}\right) = -\frac{\hat{p}_n}{Eh}
\end{aligned} \tag{11.75}$$

The unique solution of the problem requires appropriate (kinematic or static) boundary conditions. In the case of static boundary conditions the relations between generalized forces and combination of displacements and its derivatives given in Equations (11.74) have to be used.

In the work of Timoshenko and Woinowsky-Krieger (1959) the version of Equations (11.75) can be found, in which the components associated with the displacement w normal to the middle surface have the opposite sign. It is the consequence of a different sign convention.

11.6 Cylindrical Shell with a Semicircular Cross Section under Self Weight – Analytical Solution of Membrane State

The cylindrical shell shown in Figure 11.19 is loaded by self weight. The load per unit area in the middle surface of the shell has two nonzero components in the local cylindrical coordinate system (x, θ, z):

$$\hat{p}_x = 0, \qquad \hat{p}_\theta = g\sin\theta, \qquad \hat{p}_n = -g\cos\theta \tag{11.76}$$

The membrane state with the rigid motions prevented is assumed.

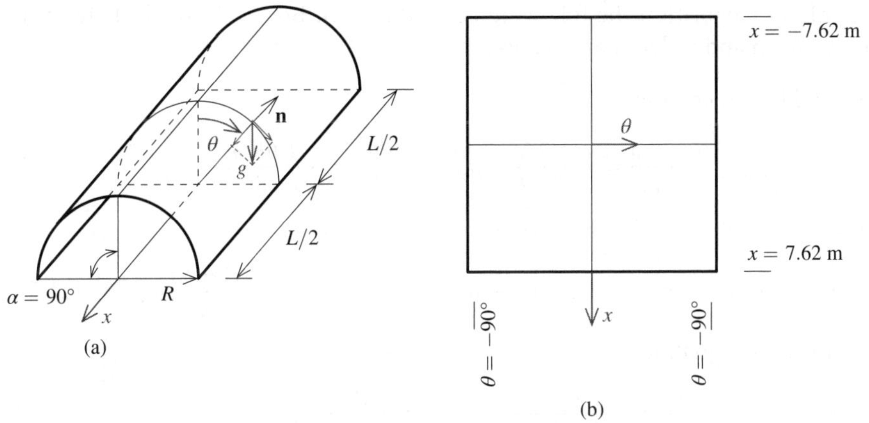

Figure 11.19 Cylindrical shell with semicircular cross section: (a) configuration and (b) peripheral of cylindrical shell as domain with $x \in [-L/2, +L/2]$ and $\theta \in [-90°, 90°]$

Box 11.10 Cylindrical shell of a semicircular cross section under self weight

Data

$E = 2.07 \times 10^7$ kN/m^2, $\quad \nu = 0.0$
$R = 7.62$ m, $\quad L = 15.24$ m, $\quad \alpha = 90°$, $\quad h = 0.0762$ m, $\quad g = 4.3$ kN/m^2

Check values from analytical solution

$n_x(0, 0°) = -32.766$ kN/m, $\quad n_x(x, 90°) = 0.0$
$n_\theta(0, 0°) = -32.766$ kN/m, $\quad n_\theta(x, 90°) = 0.0$
$n_{x\theta}(x, 0°) = 0.0$, $\quad n_{x\theta}(\pm L/2, 90°) = \pm 65.532$ kN/m
$u(0, 0°) = 0.0$, $\quad u(\pm L/2, 90°) = -1.056 \times 10^{-4}$ m, $\quad u(x, 90°) = 0.0$
$v(x, 0°) = 0.0$, $\quad v(\pm L/2, 90°) = 0.0$, $\quad v(0, 90°) = 3.828 \times 10^{-4}$ m
$w(0, 0°) = -5.409 \times 10^{-4}$ m $\quad w(L/2, 0°) = -1.583 \times 10^{-4}$ m, $\quad w(x, 90°) = 0.0$

Shell configuration shown in Figure 11.19

The input data (material and geometrical parameters, and loads) are listed in Box 11.10.

For the cylindrical shell with $\alpha = 90°$ and $\theta \in [-90°, 90°]$ the formulae for all necessary functions can be easily derived as products of two functions: polynomial dependent on coordinate x and trigonometric $\sin\theta$ or $\cos\theta$. The dependence of the load functions $\hat{p}_\theta$ and $\hat{p}_n$ on coordinate θ affects the distributions of three membrane forces, three strains and three displacements in the circumferential direction.

In the solution algorithm the following equations from Box 11.1 are used. They are listed in the form and order of application:

- three equilibrium equations:

$$n_\theta = R\hat{p}_n, \qquad \frac{\partial n_{x\theta}}{\partial x} = -\frac{1}{R}\frac{\partial n_\theta}{\partial \theta} - \hat{p}_\theta, \qquad \frac{\partial n_x}{\partial x} = -\frac{1}{R}\frac{\partial n_{x\theta}}{\partial \theta} - \hat{p}_x \tag{11.77}$$

- three constitutive equations (strains expressed by forces):

$$\varepsilon_x = \frac{1}{Eh}(n_x - \nu n_\theta), \qquad \varepsilon_\theta = \frac{1}{Eh}(n_\theta - \nu n_x), \qquad \gamma_{x\theta} = \frac{2(1+\nu)}{Eh}n_{x\theta} \tag{11.78}$$

- three kinematic equations:

$$\frac{\partial u}{\partial x} = \varepsilon_x, \qquad \frac{\partial v}{\partial x} = -\frac{1}{R}\frac{\partial u}{\partial \theta} + \gamma_{x\theta} \qquad w = -\frac{\partial v}{\partial \theta} + \varepsilon_\theta \tag{11.79}$$

The process of calculations presented next is usual for the analysis of statically determinate structures (shell in membrane state is statically determinate, see Chapter 10).

In the first step three membrane forces are found using Equations (11.77). After integration with respect to x the following formulae are derived:

$$n_x = \frac{g}{R}x^2\cos\theta - \frac{x}{R}\frac{\partial C_1}{\partial \theta} + C_2(\theta), \quad n_\theta = -gR\cos\theta, \quad n_{x\theta} = -2gx\sin\theta + C_1(\theta) \tag{11.80}$$

in which two integration functions $C_1(\theta)$ and $C_2(\theta)$ appear.

The unique solution of two differential equilibrium equations requires the information on the character of membrane force fields $n_x(x,\theta)$ and $n_{x\theta}(x,\theta)$. The following relations can be written for two curved edges ($x = \pm L/2$) on the basis of: (i) symmetry of distribution of meridional force $n_x(x,\theta)$ and (ii) antisymmetry of membrane shear force $n_{x\theta}(x,\theta)$ with respect to plane $x = 0$:

$$n_x(L/2,\theta) = n_x(-L/2,\theta), \qquad n_{x\theta}(L/2,\theta) = -n_{x\theta}(-L/2,\theta) \tag{11.81}$$

Taking into account these conditions, two unknown functions $C_1(\theta)$ and $C_2(\theta)$ are derived:

$$C_1(\theta) = 0, \qquad C_2(\theta) = -gL^2\cos\theta/(4R) \tag{11.82}$$

The final formulae for the membrane forces can be written as:

$$\begin{aligned} n_x(x,\theta) &= -g(L^2 - 4x^2)\cos\theta/(4R) = -g(L^2 - 4x^2)y/(4R^2) \\ n_\theta(x,\theta) &= -gR\cos\theta = -gy \\ n_{x\theta}(x,\theta) &= -2gx\sin\theta = -2gx\sqrt{R^2 - y^2}/R \end{aligned} \tag{11.83}$$

where $y = R\cos\theta$. The characteristic ordinates are:

$$\begin{aligned} &n_x(-L/2,0) = 0, \quad n_x(0,0) = -gL^2/(4R), \quad n_x(L/2,0) = 0, \quad n_x(x,\pi/2) = 0 \\ &n_\theta(x,0) = -gR, \quad n_\theta(x,\pi/2) = 0 \\ &n_{x\theta}(x,0) = 0, \quad n_{x\theta}(-L/2,\pi/2) = gL, \quad n_{x\theta}(0,\pi/2) = 0, \quad n_{x\theta}(L/2,\pi/2) = -gL \end{aligned} \tag{11.84}$$

In the second solution step, distributions of three membrane strains are obtained directly from algebraic constitutive relations, see Equations (11.78):

$$\begin{aligned} \varepsilon_x(x,\theta) &= g(4x^2 - L^2 + 4\nu R^2)\cos\theta/(4EhR) \\ \varepsilon_\theta(x,\theta) &= g\left[\nu(L^2 - 4x^2) - 4R^2\right]\cos\theta/(4EhR) \\ \gamma_{x\theta}(x,\theta) &= -4(1+\nu)gx\sin\theta/(Eh) \end{aligned} \tag{11.85}$$

Eventually, the kinematic Equations (11.79) (two differential and one algebraic) are used to determine three translation fields. Integrating the first equation with derivative $u_{,x}$ and taking into account the kinematic condition (written for the middle curved line)

$$u(0,\theta) = 0 \tag{11.86}$$

the zero function $C_3(\theta) = 0$ is obtained and function $u(x,\theta)$ reads

$$u(x,\theta) = g(4x^3 - 3L^2x + 12\nu R^2x)\cos\theta/(12EhR) \tag{11.87}$$

Then, the derivative of displacement u with respect to θ is calculated

$$\frac{\partial u}{\partial\theta} = g(4x^3 - 3L^2x + 12\nu R^2x)(-\sin\theta)/(12EhR) \tag{11.88}$$

and from the second kinematic Equation $(11.79)_2$ the derivative $v_{,x}$ is determined

$$\frac{\partial v}{\partial x} = -\frac{1}{R}\frac{\partial u}{\partial\theta} + \gamma_{x\theta} = g\left[4x^3 - 3L^2x - 12\nu R^2x(4-3\nu)\right]\sin\theta/(12EhR^2) \tag{11.89}$$

Assuming that the shell is supported on diaphragms (rigid in their planes) along two curved edges, the boundary condition

$$v(\pm L/2,\theta) = 0 \tag{11.90}$$

can be considered. Integrating Equation (11.89) and taking into account the assumed kinematic constraints, function $C_4(\theta)$ is obtained and the final formula for displacement $v(x,\theta)$ is written next as Equation $(11.91)_2$.

Finally, the third kinematic Equation $(11.79)_3$ is used to determine function $w(x,\theta)$. Here, the final forms of the three displacement functions are listed:

$$\begin{aligned} u(x,\theta) &= g(4x^3 - 3L^2x + 12\ \nu R^2x)\cos\theta/(12\ EhR) \\ v(x,\theta) &= g(L^2 - 4x^2)\left[5L^2 - 4x^2 + 24(4+3\nu)R^2\right]\sin\theta/(192\ EhR^2) \\ w(x,\theta) &= -g\left[(L^2 - 4x^2)[5L^2 - 4x^2 + 24(4+\nu)R^2] + 192\ R^4\right] \\ &\quad \times\cos\theta/(192\ EhR^2) \end{aligned} \tag{11.91}$$

The characteristic ordinates of the last function are:

$$\begin{aligned} w(0,0) &= -g\left[L^2[5L^2 + 24(4+\nu)R^2] + 192R^4\right]/(192EhR^2) \\ w(L/2,0) &= -gR^2/(Eh) \end{aligned} \tag{11.92}$$

Notice that $|w(0,0)| > |w(L/2,0)|$.

Summarizing, the results of the analysis of the cylindrical shell of semicircular cross section with $\alpha = 90°$ are described by nine Equations (11.84), (11.85) and (11.91). The following features of the solution functions should be emphasized: (i) function $\cos\theta$ results in symmetric distribution with respect to the vertical plane $\theta = 0$ for the following fields: $n_x(x,\theta)$, $n_\theta(x,\theta)$, $\varepsilon_x(x,\theta)$, $\varepsilon_\theta(x,\theta)$, $u(x,\theta)$, $w(x,\theta)$ and (ii) function $\sin\theta$ implies antisymmetric distribution in circumferential direction of $n_{x\theta}(x,\theta)$, $\gamma_{x\theta}(x,\theta)$, $v(x,\theta)$.

In addition, the distribution of three membrane forces and three displacements in the direction of the x-axis along lines $\theta = 0°$ and $\theta = 90°$ should be commented on.

The description of membrane forces:

- along line $\theta = 0°$: compressive force n_x has parabolic distribution with minimum value $n_x(0, 0°) = -32.77$ kN/m, compressive force n_θ has constant value $n_\theta(x, 0°) = -32.77$ kN/m and force $n_{x\theta}(0, 0°) = 0.0$
- along line $\theta = 90°$: two forces n_x and n_θ are zero, force $n_{x\theta}$ depends linearly on x and $n_{x\theta}(\pm L/2, 90°) = \pm 65.53$ kN/m is its extreme value while $n_{x\theta}(0, 90°) = 0.0$

The description of displacement fields:

- along line $\theta = 0°$: displacement $u(x, 0°)$ is described by the function x^3, so $u(0, 0°) = 0$, displacement $v(x, 0°) = 0$, normal displacement $w(x, 0°)$ is described by the function x^4 and $|w(0, 0°)| > |w(L/2, 0°)|$
- along line $\theta = 90°$: $u(x, 90°) = 0$, $w(x, 90°) = 0$, $v(x, 90°)$ is the function of x^2 and x^4 and vanishes for $x = \pm L/2$, that is $v(\pm L/2, 90°) = 0.0$

Zero displacements $u(x, \theta)$ and $w(x, \theta)$ along lines $\theta = \pm 90°$ resulting from $\cos(\pm 90°) = 0$ can be interpreted as kinematic constraints resulting from the assumed support. Then zero value of the displacement w normal to the middle surface can be treated as a source of bending effects that requires a further analysis. On the connection line $x = \pm L/2$ between the roof and diaphragms the zero displacement $v(\pm L/2, \theta) = 0$ has just been taken into account as Equation (11.90). In fact, the additional constraint $w(\pm L/2, \theta) = 0$ should be postulated and would become another disturbance of the membrane state.

In Figure 11.20 the contour maps of three membrane forces and three displacements are presented on the cylinder surface for $\theta \in [90°, -90°]$ and $x \in [-7.62, 7.62]$. Characteristic values of membrane forces and displacements are included in Box 11.10.

11.7 Cylindrical Scordelis-Lo Roof in the Membrane-Bending State – Analytical and Numerical Solution

The shell considered in this section is a known benchmark called the Scordelis-Lo roof (see MacNeal and Harder 1985; Szabó and Sahrmann 1988). It is a segment of a cylinder with a limiting angle $\alpha = 40°$. The structure is loaded by self weight. Two curved edges rest on diaphragms and two straight boundaries remain unsupported. In the previous section the analytical solution for a cylindrical shell with a semicircular section ($\alpha = 90°$) was discussed. The consideration of the similar problem but for $\alpha \neq 90°$ turns out to be much more complicated. Obviously, in the case of numerical (FEM) analysis the value of α does not affect the complexity of the computations.

In Figure 11.21 the analysed configuration is shown. Note that the global Cartesian coordinate system (X, Y, Z) is introduced apart from the local (x, θ, z) one. In the global coordinate system the translations in the directions of axes X, Y, Z are denoted by UX, UY, UZ, respectively. In the local coordinate system the three respective displacements are u, v, w. Moreover, generalized shell forces are denoted using two indices x and θ.

In the cylindrical shell under self weight with kinematic (at two curved edges) and static (at two free unloaded straight edges) boundary conditions, a complex

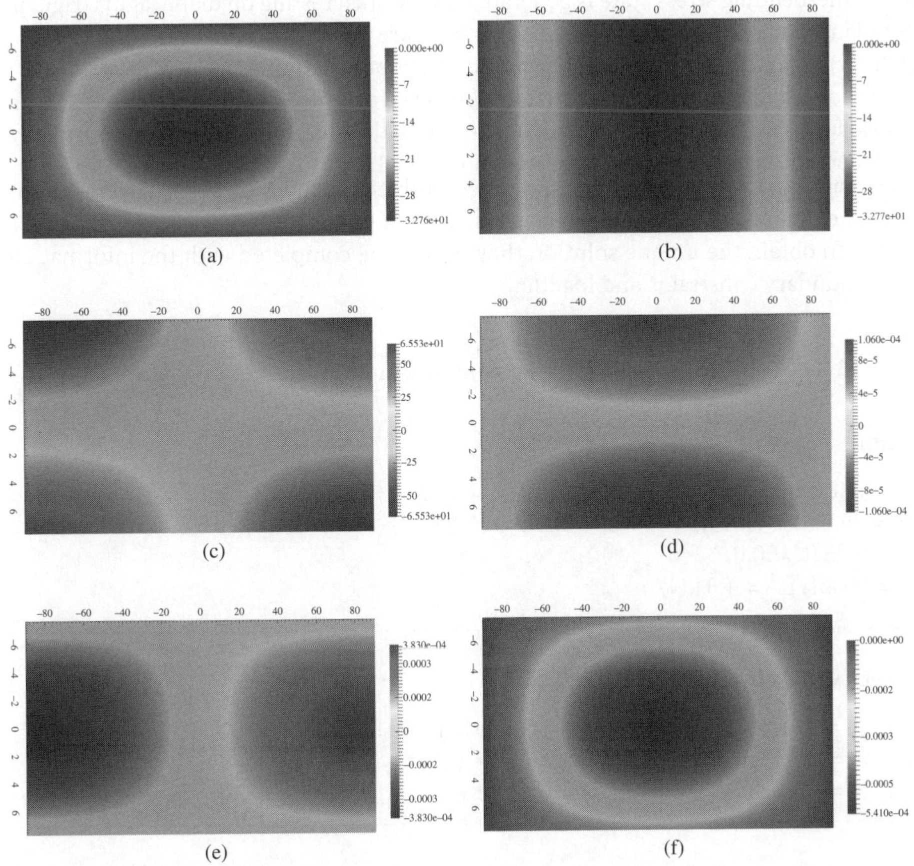

Figure 11.20 Cylindrical shell with semicircular cross section under self weight – contour maps of: (a)–(c) three membrane forces $n_x(x,\theta)$, $n_\theta(x,\theta)$, $n_{x\theta}(x,\theta)$ and (d)–(f) three translations $u(x,\theta)$, $v(x,\theta)$, $w(x,\theta)$; drawn on the basis of analytical solution

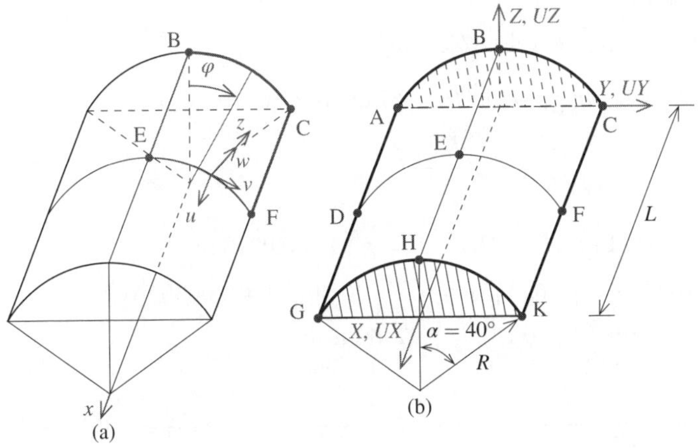

Figure 11.21 Cylindrical shell with marked characteristic points

membrane-bending stress state is generated. In the shell resting on diaphragms (rigid in their planes) displacements UY and UZ (or v, w) are prevented. However, displacement UX (or u) in the direction of cylinder meridian is not constrained.

The Scordelis-Lo roof is characterized by material parameters E, ν, dimensions R, L, h, α and load g. The adopted values with units used in the literature (Szabó and Sahrmann, 1988), and SI units in addition, are listed in Box 11.11.

The mathematical model of the considered problem includes three displacement differential equations (11.73) with derivatives of three unknown functions $u(x,\theta)$, $v(x,\theta)$, $w(x,\theta)$. To obtain the unique solution they have to be completed with the information about boundary constrains and loading.

Box 11.11 Cylindrical Scordelis-Lo roof in a membrane-bending state

Data

$E = 3.0 \times 10^7$ psi $= 2.07 \times 10^7$ kN/m^2, $\nu = 0$
$R = 25.0$ ft. $= 7.62$ m, $L = 50$ ft. $= 15.24$ m, $\alpha = 40°$,
$h = 0.25$ ft. $= 0.0762$ m
$g = 90$ lbf/ft^2 $= 4.3$ kN/m^2

Check values from analytical solution

$|UZ_F| = 3.70331$ in. $= 0.0938$ m – using shallow shell theory
$|UZ_F| = 3.533$ in. $= 0.0897$ m – using shell theory
$|UZ_E| = 0.5422$ in. $= 0.01377$ m

Check values from FEM solution

ANSYS, SHELL 93 FEs
$|UZ_E^{(A)}| = |UZ_E^{(R)}| = w_E = 0.014$ m, $|UX_H^{(A)})| = |u_H^{(A)}| = 0.000908$ m
$|UX_K^{(A)}| = |u_K^{(A)}| = 0.0038$ m
$n_{x,D}^{(A)} = n_{x,F}^{(A)} = 851$ kN/m, $n_{\theta,E}^{(A)} = -49.87$ kN/m
$|m_{x,E}^{(A)}| = 0.4089$ kNm/m, $|m_{x,D}^{(A)}| = |m_{x,F}^{(A)}| = 2.742$ kNm/m
$|m_{\theta,E}^{(A)}| = 9.16$ kNm/m
ROBOT
$|UZ_E^{(R)}| = |UZ_E^{(A)}| = w_E = 0.014$ m, $|UX_H^{(R)}| = |u_H^{(R)}| = 0.0010$ m
$|UZ_D^{(R)}| = |UZ_F^{(R)}| = 0.1045$ m, $|UX_K^{(R)}| = |UX_G^{(R)}| = |u_K^{(R)}| = 0.0035$ m
$|m_{x,E}^{(R)}| = 0.44$ kNm/m, $|m_{x,D}^{(R)}| = |m_{x,F}^{(R)}| = 2.97$ kNm/m
$|m_{\theta,E}^{(R)}| = 9.38$ kNm/m

Shell configuration shown in Figure 11.21

Due to two symmetry planes ($x = L/2$ and $\theta = 0°$) a quarter of the shell defined by points B–E–F–C–B is considered in the analytical analysis (see Figure 11.21). The following mixed boundary conditions are imposed at the four edges:

- edge B–C: $x = 0$, $\theta \in [0°, 40°]$ – resting on diaphragm:

$$w(0,\theta) = 0, \quad v(0,\theta) = 0, \quad n_x(0,\theta) = 0, \quad m_x(0,\theta) = 0 \tag{11.93}$$

- edge E–F: $x = L/2$, $\theta \in [0°, 40°]$ – curve edge in symmetry plane:

$$u(L/2,\theta) = 0, \quad \vartheta_x(L/2,\theta) = 0, \quad \tilde{n}_{x\theta}(L/2,\theta) = 0, \quad \tilde{t}_x(L/2,\theta) = 0 \tag{11.94}$$

- edge B–E: $\theta = 0°$, $x \in [0, L/2]$ – straight edge in symmetry plane:

$$v(x,0°) = 0, \quad \vartheta_\theta(x,0°) = 0, \quad \tilde{n}_{\theta x}(x,0°) = 0, \quad \tilde{t}_\theta(x,0°) = 0 \tag{11.95}$$

- edge C–F: $\theta = 40°$, $x \in [0, L/2]$ – straight free unloaded edge:

$$n_\theta(x,40°) = 0, \quad \tilde{n}_{\theta x}(x,40°) = 0, \quad m_\theta(x,40°) = 0, \quad \tilde{t}_\theta(x,40°) = 0 \tag{11.96}$$

The static boundary conditions have to be expressed using displacements and their derivatives. Together with the integration of displacement differential equations this is the difficulty in the analytical solution of the problem.

In Box 11.11 three characteristic displacements of the central point E and of the middle point F on edge $\theta = 40°$, computed in the global coordinate system, are quoted from the work of Szabó and Sahrmann (1988). The displacements are derived using either shallow or general shell theory and are going to be used to check the results of numerical analysis.

The numerical computations have been performed for the whole Scordelis-Lo roof domain using ANSYS and ROBOT computer codes. The obtained results are marked by the upper indices (A) and (R), respectively. The behaviour of the shell is described by the displacement field shown in Figure 11.22a as a vector representation. Contour maps of membrane forces n_x, n_θ, $n_{x\theta}$, bending moments m_x, m_θ, $m_{x\theta}$ and transverse shear forces t_x, t_θ are plotted in Figures 11.23a–d, 11.24a–d. In Figure 11.22b,c the distributions of horizontal displacement UX along lines A–D–G and G–H–K and vertical displacement UZ along the cross lines A–D–G and D–E–F are shown. The magnitudes and directions of principal stresses at the top and bottom surfaces are presented in Figure 11.25a,b. Analysing the generalized force distributions, the subdomains of pure membrane or membrane-bending state can be distinguished.

The characteristic values of numerical computations are listed in Box 11.11.

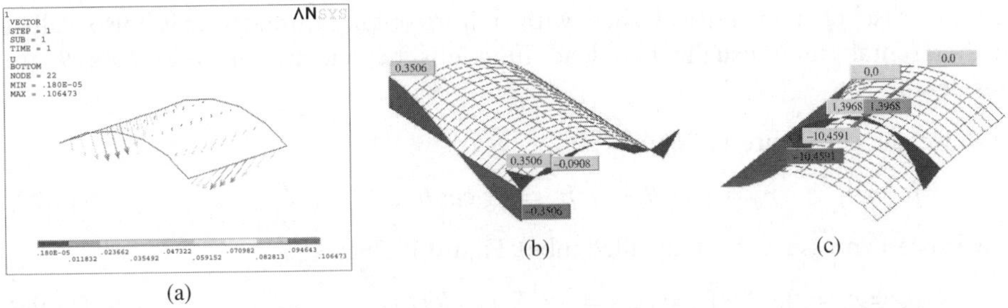

Figure 11.22 Scordelis-Lo roof: (a) vector presentation of displacement field [cm] – FEM results from ANSYS; graphs of displacements on four cross lines [m]: (b) UX and (c) UZ; FEM results from ROBOT.

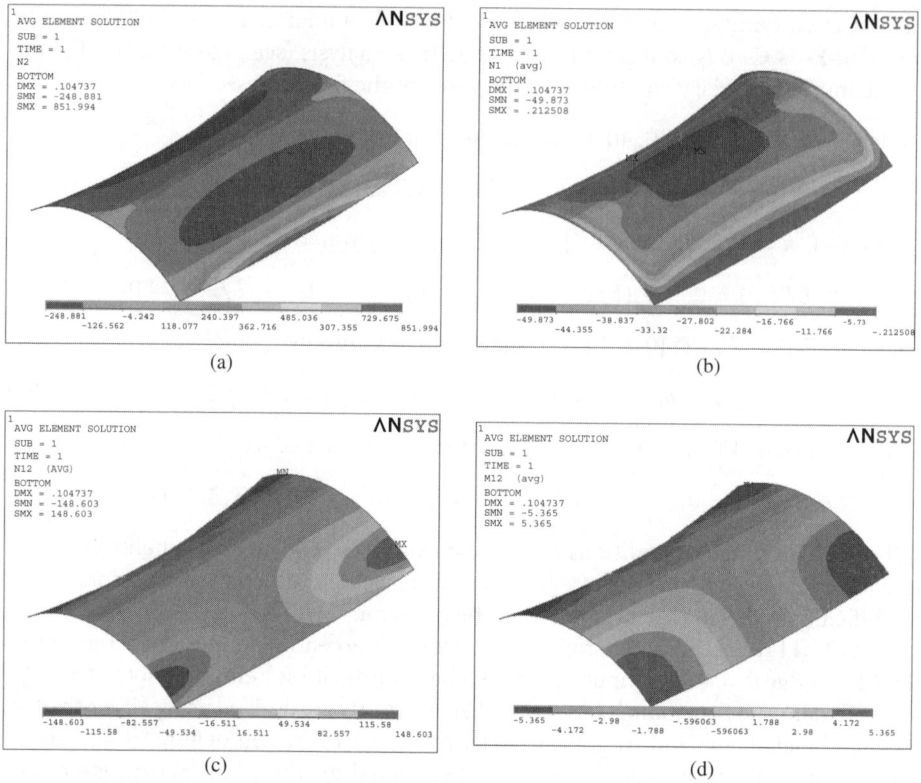

Figure 11.23 Scordelis-Lo roof – contour maps for: membrane forces [kN/m]: (a) n_x, (b) n_θ, (c) $n_{x\theta}$ and (d) twisting moment [kNm/m] $m_{x\theta}$; FEM results from ANSYS. *See Plate section for color representation of this figure.*

11.8 Single-Span Clamped Horizontal Cylindrical Shell under Self Weight

The considered engineering problem is referred to in many textbooks among others in Girkmann (1956) and Nowacki (1980). The example shows the complexity of the analytical solution. The obtained results are used to assess the accuracy of the numerical computations.

In the case of a cylindrical shell with a horizontal symmetry axis (also called the horizontal tube) usually two load functions dependent on coordinate θ are considered:

- self weight, see Figure 11.26b:

$$\hat{p}_x = 0, \qquad \hat{p}_\theta = g \sin\theta, \qquad \hat{p}_n = -g\cos\theta \tag{11.97}$$

- hydrostatic pressure (for fully filled tube), Figure 11.26c:

$$\hat{p}_x = \hat{p}_\theta = 0, \qquad \hat{p}_n = \hat{p}_{n,\mathrm{w}} = -\gamma_\mathrm{w} R + \gamma_\mathrm{w} R\cos\theta \tag{11.98}$$

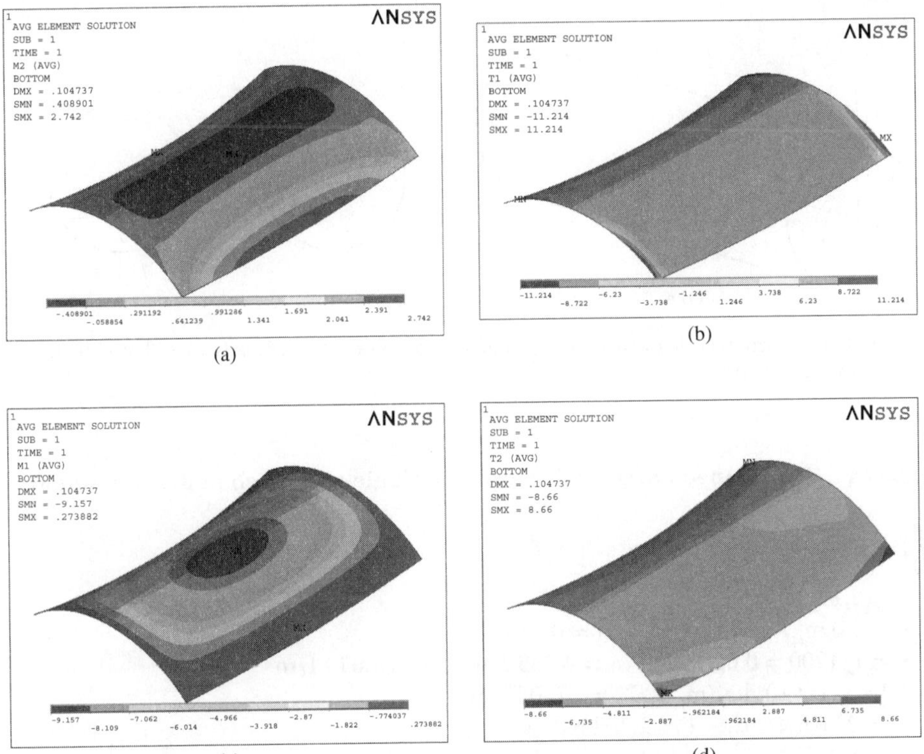

Figure 11.24 Scordelis-Lo roof – contour maps for bending moments [kNm/m]: (a) m_x, (c) m_θ, transverse shear forces [kN/m]: (b) t_x, (d) t_θ; FEM results from ANSYS. *See Plate section for color representation of this figure.*

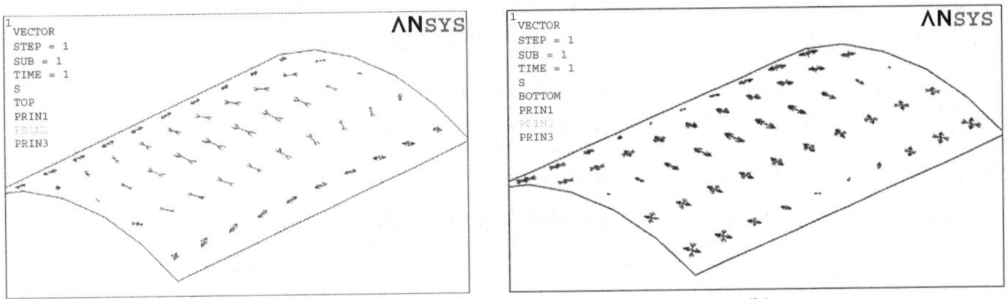

Figure 11.25 Scordelis-Lo roof – visualization of the directions and magnitudes of the principal stresses on surfaces: (a) bottom and (b) top; FEM results from ANSYS.

Here, the interest is limited to the first load case. The shell is a single-span horizontal tube. It is clamped at two circular ends $x = 0$, $x = L$. The analysed configuration is depicted in Figure 11.26a. The adopted values of material constants E, ν, γ_c and geometrical parameters R, L, h are included in Box 11.12.

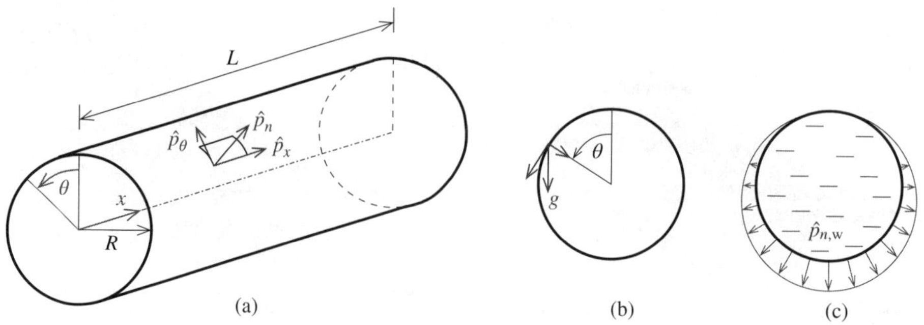

Figure 11.26 Horizontal cylindrical tube: (a) configuration, description of loads: (b) self weight and (c) hydrostatic pressure

Box 11.12 Clamped horizontal cylindrical single-span shell under self weight

Data

$E = 2.1 \times 10^7$ kN/m², $\nu = 0$
$R = 1.0$ m, $L = 10.0$ m, $h = 0.1$ m
$k = 1/1200 = 0.000833$, $\alpha = 4.283$ 1/m, $\beta = 4.043$ 1/m
$D^n = 2.1 \times 10^6$ kN/m, $D^m = 1750$ kNm
$\gamma_c = 24.0$ kN/m³, $g = 2.40$ kN/m²
$\hat{p}_x = 0$, $\hat{p}_\theta = 2.4 \sin\theta$ kN/m², $\hat{p}_n = -2.4 \cos\theta$ kN/m²

Check values from analytical solution

$\overline{n}_x(0, 0°) = 40.0$ kN/m, $\overline{n}_x(0, 180°) = -40.0$ kN/m
$\overline{n}_x(L/2, 0°) = -20.0$ kN/m, $\overline{n}_x(L/2, 180°) = 20.0$ kN/m
$n_x^{(\mathrm{II})}(0, 0°) = -0.389$ kN/m, $n_x^{(\mathrm{II})}(0, 180°) = 0.389$ kN/m
$n_x^{\mathrm{anal}}(0, 0°) = 39.611$ kN/m, $n_x^{\mathrm{anal}}(0, 180°) = -39.611$ kN/m
$n_x^{\mathrm{anal}}(L/2, 0°) = -20.0$ kN/m, $n_x^{\mathrm{anal}}(L/2, 180°) = 20.0$ kN/m
$m_x^{\mathrm{anal}}(0, 0°) = 0.389$ kNm/m, $m_x^{\mathrm{anal}}(0, 180°) = -0.389$ kNm/m

Check values from FEM solution

ANSYS, SHELL 93 FEs
$m_x^{\mathrm{FEM}(m)}(0, 0°) = 0.139 \text{ kNm/m} = 0.357\, m_x^{\mathrm{anal}}(0, 0°)$
$m_x^{\mathrm{FEM}(m)}(0, 180°) = -0.139 \text{ kNm/m} = 0.357\, m_x^{\mathrm{anal}}(0, 180°)$
$m_x^{\mathrm{FEM}(g)}(0, 0°) = 0.331 \text{ kNm/m} = 0.851\, m_x^{\mathrm{anal}}(0, 0°)$
$m_x^{\mathrm{FEM}(g)}(0, 180°) = -0.331 \text{ kNm/m} = 0.851\, m_x^{\mathrm{anal}}(0, 180°)$

Shell configuration shown in Figure 11.26

The considered shell could be also treated as a part of a multi-span pipeline. Depending on the way the pipe is connected with intermediate supports, two variants of boundary conditions are possible:

- fully clamped – kinematic: $u = 0, \quad v = 0, \quad w = 0, \ \vartheta_x = 0$
- clamped with translation in axial direction – mixed: $n_x = 0, \ v = 0, \ w = 0, \ \vartheta_x = 0$

11.8.1 Analytical Solution

The discussion of the analytical solution follows Girkmann (1956). The load function can be expressed by a trigonometric series expansion. The self weight load corresponds to the second component with $j = 1$. The expansion of the surface load function results in the dependence of forces, moments and displacements on coordinate θ. In particular, the fields: $n_x, n_\theta, m_x, m_\theta, t_x, u, w$ are expressed by $\cos\theta$ and the following four: $n_{x\theta}, m_{x\theta}, t_\theta, v$ depend on $\sin\theta$.

The supports and loading become the sources of local bending states in the vicinity of the two clamped edges. Thus, the solution is carried out in two stages. In stage I the distributions of three membrane forces $\overline{n}_x, \overline{n}_\theta, \overline{n}_{x\theta}$ and three displacements $\overline{u}, \overline{v}, \overline{w}$ are derived to describe the membrane state. In stage II the bending effects are taken into account. Certain simplifications are introduced to solve the boundary value problem of the cylindrical shell. The functions of transverse shear forces and bending moments are found and some functions obtained in the first stage are modified. To solve the problem, the set of equations resulting from the mathematical model is reduced to one eight-order differential equation with unknown function $w(x, \theta)$. Then, all generalized forces and other two displacements are expressed as functions of $w(x, \theta)$ and its derivatives with respect to x. The integration constants $\overline{C}_1, \dots, \overline{C}_4$ and C_1, C_2 are computed from the boundary conditions defined for quantities from both the membrane and bending state, respectively.

11.8.1.1 Stage I – analysis of the membrane state

After the integration of equilibrium equations of the membrane state (rewritten for the completeness of the algorithm):

$$\frac{\partial \overline{n}_x}{\partial x} + \frac{1}{R}\frac{\partial \overline{n}_{x\theta}}{\partial \theta} + \hat{p}_x = 0, \qquad \frac{\partial \overline{n}_{x\theta}}{\partial x} + \frac{1}{R}\frac{\partial \overline{n}_\theta}{\partial \theta} + \hat{p}_\theta = 0, \qquad \frac{\overline{n}_\theta}{R} - \hat{p}_n = 0 \tag{11.99}$$

polynomial-trigonometric functions for three membrane forces are obtained:

$$\begin{aligned} \overline{n}_x(x,\theta) &= \left(gx^2/R - \overline{C}_1 x/R + \overline{C}_2\right)\cos\theta \\ \overline{n}_\theta(x,\theta) &= -gR\cos\theta \\ \overline{n}_{x\theta}(x,\theta) &= \left(-2gx + \overline{C}_1\right)\sin\theta \end{aligned} \tag{11.100}$$

in which two integration constants $\overline{C}_1$ and $\overline{C}_2$ occur. A combination of kinematic and constitutive equations of the membrane state and consideration of these membrane forces lead to the set of three displacement differential equations. After integration, the following equations (with another two constants $\overline{C}_3$ and $\overline{C}_4$) are obtained for the displacements:

$$\overline{u}(x,\theta) = \frac{1}{D^n}\left(\frac{gx^3}{3R} - \overline{C}_1\frac{x^2}{2R} + \overline{C}_2 x + \overline{C}_3\right)\cos\theta \tag{11.101}$$

$$\overline{v}(x,\theta) = \frac{1}{D^n}\left[2g\left(\frac{x^4}{24R^2} - x^2\right) - \overline{C}_1\left(\frac{x^3}{6R^2} - 2x\right) + \overline{C}_2\frac{x^2}{2R} + \overline{C}_3\frac{x}{R} + \frac{\overline{C}_4}{R}\right]\sin\theta$$

$$\overline{w}(x,\theta) = -\frac{1}{D^n}\left[gR^2 + 2g\left(\frac{x^4}{24R^2} - x^2\right) - \overline{C}_1\left(\frac{x^3}{6R^2} - 2x\right)\right.$$
$$\left. + \overline{C}_2\frac{x^2}{2R} + \overline{C}_3\frac{x}{R} + \frac{\overline{C}_4}{R}\right]\cos\theta$$

and the derivative of function $\overline{w}$ reads:

$$\frac{\partial \overline{w}}{\partial x} = -\frac{1}{D^n}\left[2g\left(\frac{x^3}{6R^2} - 2x\right) - \overline{C}_1\left(\frac{x^2}{2R^2} - 2\right) + \overline{C}_2\frac{x}{R} + \frac{\overline{C}_3}{R}\right]\cos\theta \tag{11.102}$$

It is worth mentioning that constants $\overline{C}_1, \dots, \overline{C}_4$ from the membrane state are going to be coupled with constants C_1 and C_2 obtained in stage II and related to bending effects.

11.8.1.2 Stage II – analysis of the membrane-bending state

The source of bending effects in the vicinity of boundary $x = 0$ is taken into account. The analysed shell is a long one ($3\lambda < L$) and due to the symmetry it is not necessary to consider the second source of bending for $x = L$. The normal displacement $w(x,\theta)$ is described by the following exponential-trigonometric function

$$w(x,\theta) = W(x)\cos\theta = e^{-\alpha x}(C_1\cos\beta x + C_2\sin\beta x)\cos\theta \tag{11.103}$$

In further analysis, the amplitude function $W(x)$ and its four derivatives with respect to x are going to be used:

$$W = C_1, \qquad W_{,x} = -C_1\alpha + C_2\beta, \qquad W_{,xx} = C_1(\alpha^2 - \beta^2) - 2C_2\alpha\beta$$
$$W_{,xxx} = -C_1\alpha(\alpha^2 - 3\beta^2) - C_2\beta(\beta^2 - 3\alpha^2), \qquad W_{,xxxx} = C_1(\alpha^4 - 6\alpha^2\beta^2 + \beta^4) - 4 \tag{11.104}$$

The following two parameters quoted from Girkmann (1956):

$$\alpha = \frac{1}{R}\sqrt{\frac{1}{2}\sqrt{3 + \frac{1}{k}} + 1}, \qquad \beta = \frac{1}{R}\sqrt{\frac{1}{2}\sqrt{3 + \frac{1}{k}} - 1} \tag{11.105}$$

and simplified relations between membrane and bending stiffnesses (for Poisson's ratio $\nu = 0$):

$$D^n = \frac{Eh}{1-\nu} = Eh, \qquad D^m = \frac{Eh^3}{12(1-\nu^2)} = \frac{Eh^3}{12}$$
$$k = \frac{D^m}{D^n R^2} = \frac{h^2}{12R^2}, \qquad D^m = D^n k R^2, \qquad \frac{D^m}{D^n} = \frac{h^2}{12} \tag{11.106}$$

are used in further formulae.

Hereafter, the following relations (valid for $j = 1$ and $\nu = 0$) are needed:

$$u(x,\theta) = 2kRW_{,x}\cos\theta = 2\frac{D^m}{D^n R}W_{,x}\cos\theta = \frac{h^2}{6R}W_{,x}\cos\theta$$
$$v(x,\theta) = k\left[-2R^2 W_{,xx} + 3W\right]\sin\theta = \frac{h^2}{12R^2}\left[-2R^2 W_{,xx} + 3W\right]\sin\theta \tag{11.107}$$

$$
\begin{aligned}
n_x(x,\theta) &= D^n kRW_{,xx}\cos\theta = \frac{D^m}{R}W_{,xx}\cos\theta \\
n_\theta(x,\theta) &= -D^n k\left[R^3W_{,xxxx} - 2RW_{,xx}\right]\cos\theta = \frac{D^m}{R}\left[R^2W_{,xxxx} - 2W_{,xx}\right]\cos\theta \\
n_{x\theta}(x,\theta) &= D^n k\left[W_{,x} - R^2W_{,xxx}\right]\sin\theta = \frac{D^m}{R^2}\left[W_{,x} - R^2W_{,xxx}\right]\sin\theta
\end{aligned}
\tag{11.108}
$$

$$
\begin{aligned}
m_x(x,\theta) &= -D^n kR^2W_{,xx}\cos\theta = -D^m W_{,xx}\cos\theta, \qquad m_\theta = 0 \\
m_{x\theta}(x,\theta) &= D^n kRW_{,x}\sin\theta = \frac{D^m}{R}W_{,x}\sin\theta \\
t_x(x,\theta) &= -\frac{D^m}{R}(W_{,xxx} - W_{,x})\cos\theta, \qquad t_\theta(x,\theta) = \frac{D^m}{R}W_{,xx}\sin\theta
\end{aligned}
\tag{11.109}
$$

Taking advantage of the superposition of effects describing the membrane (stage I) and bending (stage II) states the relations resulting from static and kinematic boundary conditions are written to derive six integration constants $\overline{C}_1, \dots, \overline{C}_4$, C_1 and C_2:

- for the line in symmetry plane with $x = L/2$, $\theta \in [0°, 360°]$, that is in the subdomain of pure membrane state, two mixed boundary conditions are postulated and written in a general and detailed form:

$$
\begin{aligned}
&\overline{n}_{x\theta} = 0, && \overline{u} = 0 \\
&-2gL/2 + \overline{C}_1 = 0, && \frac{1}{D^n}\left(\frac{gL^3}{24R} - \overline{C}_1\frac{L^2}{8R} + \overline{C}_2\frac{L}{2} + \overline{C}_3\right) = 0
\end{aligned}
\tag{11.110}
$$

  Next, $\overline{C}_1$ is calculated and the relation between $\overline{C}_2$ and $\overline{C}_3$ is derived:

$$
\overline{C}_1 = gL, \qquad \overline{C}_3 = \frac{gL^3}{12R} - \frac{L}{2}\overline{C}_2
\tag{11.111}
$$

  and finally the formula for the value of displacement $\overline{u}$ at the tube end $x = 0$ is obtained:

$$
\overline{u}|_{x=0} = \frac{\overline{C}_3}{D^n} = \frac{1}{D^n}\left(\frac{gL^3}{12R} - \frac{L}{2}\overline{C}_2\right)
\tag{11.112}
$$

- for the clamped edge with $x = 0$, $\theta \in [0°, 360°]$, in the subdomain with a disturbance of the membrane state four kinematic boundary conditions are, in general, written as:

$$
\overline{u} + u = 0, \qquad \overline{v} + v = 0, \qquad \overline{w} + w = 0, \qquad \overline{w}_{,x} + w_{,x} = 0
\tag{11.113}
$$

  and their detailed form is:

$$
\begin{aligned}
&\frac{1}{D^n}\left(\frac{gL^3}{12R} - \frac{L}{2}\overline{C}_2\right) + 2kR(-C_1\alpha + C_2\beta) = 0 \\
&\frac{1}{D^nR}\overline{C}_4 + 2k\left(C_1\left[1.5 - R^2(\alpha^2 - \beta^2)\right] + 2C_2R^2\alpha\beta\right) = 0 \\
&-\frac{1}{D^nR}\left(gR^3 + \overline{C}_4\right) + C_1 = 0 \\
&-\frac{1}{2D^nR^2}\left[g\left(4R^2 + \frac{1}{6}L^2\right) - \overline{C}_2R\right] + (-C_1\alpha + C_2\beta) = 0
\end{aligned}
\tag{11.114}
$$

The final formulae for four integration constants are:

$$\begin{aligned}
C_1 &= \frac{gR^2}{D^n}\left(1-\frac{8\alpha kL}{2k+1}\right)\left(1+k\left[3+2R^2(\alpha^2+\beta^2)\right]\right)^{-1}\\
C_2 &= \frac{\alpha}{\beta}C_1+\frac{2gL}{D^n}\beta(2k+1)\\
\overline{C}_2 &= \frac{1}{6R}gL^2+\frac{8kgR}{2k+1}, \qquad \overline{C}_4 = D^nRC_1-gR^3
\end{aligned}\tag{11.115}$$

For the adopted input data the following values are calculated:

$$\begin{aligned}
&\alpha = 4.283\ 1/\text{m}, && \beta = 4.043\ 1/\text{m}\\
&\overline{C}_1 = 10g = 24.0\ \text{kN/m}, && \overline{C}_2 = 16.67g = 40.0\ \text{kN/m}\\
&\overline{C}_3 = 63.33g = 152\ \text{kN}, && \overline{C}_4 = -0.328g = -0.787\ \text{kNm}\\
&C_1 = 0.32109\times10^{-6}g = 0.77\times10^{-6}\ \text{m}\\
&C_2 = 2.692\times10^{-6}g = 6.461\times10^{-6}\ \text{m}
\end{aligned}\tag{11.116}$$

As a final solution of the problem, the functions of normal displacement $w(x,\theta)$, meridional moment $m_x(x,\theta)$ and additionally the meridional membrane force (found in stage I) $\overline{n}_x(x,\theta)$ are given next:

$$\begin{aligned}
w(x,\theta) &= W(x)\cos\theta = e^{-\alpha x}\left[C_1\cos\beta x + C_2\sin\beta x\right]\cos\theta\\
&= e^{(-4.283x)}\left[0.77\cos(4.043\,x)+6.461\sin(4.043\,x)\right]\times10^{-6}\ \cos\theta\\
m_x(x,\theta) &= -D^nkR^2W_{,xx}\cos\theta = -D^mW_{,xx}\cos\theta\\
&= e^{(-4.283x)}\left[-0.389\cos(4.043\,x)+0.06930\sin(4.043\,x)\right]\cos\theta\\
\overline{n}_x(x,\theta) &= \left[\overline{C}_2R-gx(L-x)\right]\cos\theta/R = \left[40.0-24.0x\,(1-0.1\,x)\right]\cos\theta
\end{aligned}\tag{11.117}$$

Their characteristic ordinates are listed in Box 11.12. Taking into account two equations:

$$m_x(x,\theta) = -D^mW_{,xx}\cos\theta, \qquad n_x^{(\mathrm{II})}(x,\theta) = D^mW_{,xx}\cos\theta/R \tag{11.118}$$

one can notice the simple relation between the meridional force in the bending state and the meridional bending moment:

$$n_x^{(\mathrm{II})}(x,\theta) = -m_x(x,\theta)/R \tag{11.119}$$

In Figure 11.27 three plots of: meridional membrane force $\overline{n}_x$, meridional bending moment $m_x(x)$ and final meridional force $n_x(x)$, being the sum of $\overline{n}_x$, and $n_x^{(\mathrm{II})}(x)$, are presented for two shell meridians $\theta = 0°$ and $\theta = 180°$.

11.8.2 FEM Solution

In Figures 11.28 and 11.29 the numerical results obtained using the ANSYS computer code are presented.

The bending state occurring in a vicinity of the curved clamped edges is characterized first by the meridional bending moment. In these regions the moment value increases rapidly while approaching the edges, in accordance with exponential-trigonometric function.

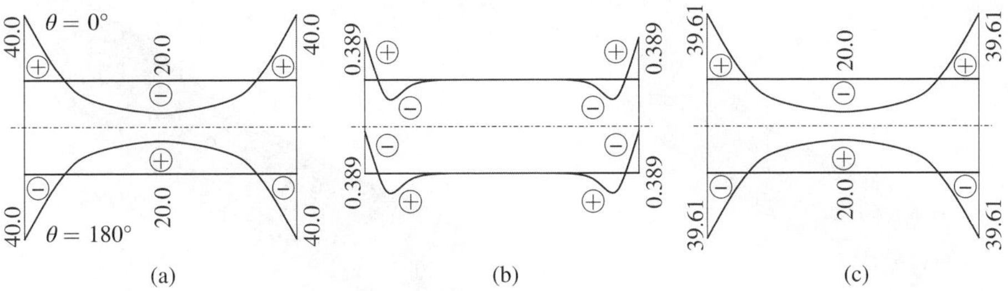

Figure 11.27 Horizontal cylindrical tube under self weight – graphs of three functions for two lines with coordinates $\theta = 0°, 180°$: (a) meridional membrane force $\overline{n}_x$ [kN/m], (b) meridional bending moment m_x [kNm/m] and (c) final meridional force n_x [kN/m]

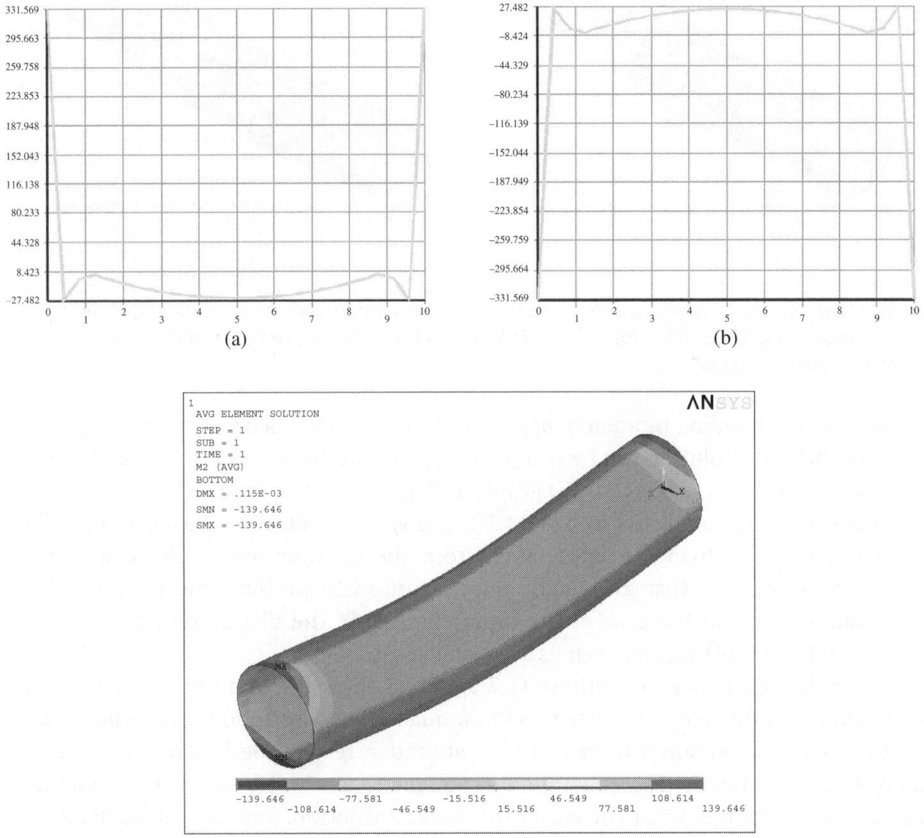

Figure 11.28 Horizontal tube under self weight – distribution of meridional bending moment m_x – graphs on lines: (a) $\theta = 0°$, (b) $\theta = 180°$ (characteristic value $m_x^{(g)}(0, 0°) = 331$ Nm/m $= 0.331$ kNm/m) and (c) contour map (characteristic value $m_x^{(m)}(0, 0°) = 139$ Nm/m $= 0.139$ kNm/m); FEM results from ANSYS.

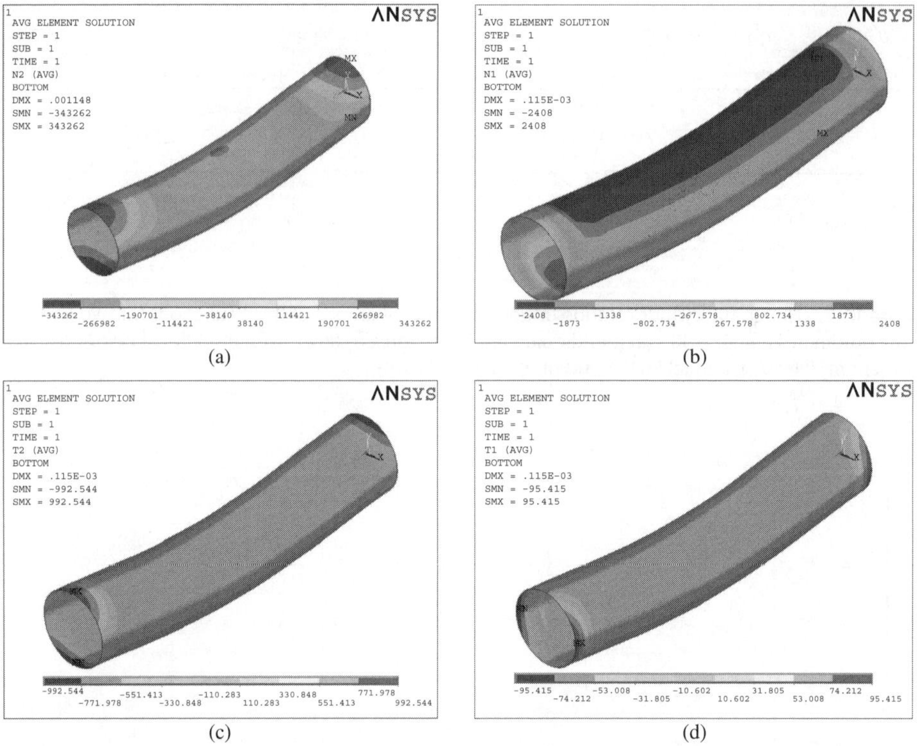

Figure 11.29 Horizontal tube under self weight – contour maps for membrane forces: (a) n_x, (b) n_θ and for transverse shear forces: (c) t_x, (d) t_θ [N,m]; FEM results from ANSYS. *See Plate section for color representation of this figure.*

The values of bending moments $m_x(0, 0°)$ and $m_x(0, 180°)$ necessary for comparison with the analytical solution can be read from: (i) contour map in Figure 11.28c (marked by upper index m in brackets) or (ii) graphs in Figure 11.28a,b (marked by upper index g in brackets). They are listed in Box 11.12. The values read from the plots are higher (in absolute value) than the ones taken from the contour maps. The discrepancy is caused by the fact that averaged values are used to produce the contour plots. As a consequence, in the case of the rapid changes in the distribution of quantities (meridional moment) extreme values are usually lost.

The two last examples in Sections 11.7 and 11.8 show the importance of the post-processing (contour plots in particular) in a qualitative description and evaluation of the shell state. To obtain a trustworthy quantitative results the FEs mesh should be improved. For a better discretization the obtained values of bending moments and transverse shear forces would be closer to the exact solution. The mesh densification is particulary recommended in the subdomains of a rapid change in function values.

References

ANKA 1993 ANKA – computer code for nonlinear analysis of structures: User's manual. Technical report, Cracow University of Technology, Cracow (in Polish).

ANSYS 2013 ANSYS Inc. PDF Documentation for Release 15.0. SAS IP, Inc.
Girkmann K 1956 *Flächentragwerke*. Springer-Verlag, Wien.
Kobiak J and Stachurski W 1991 *Reinforced Concrete Structures* vol. 4. Arkady, Warsaw (in Polish).
Kolkunov NV 1972 *Foundations of the Analysis of Elastic Shells*. Izdatel'stvo Vishaya Shkola, Moskow (in Russian).
MacNeal RH and Harder RL 1985 A proposed standard set of problems to test finite element accuracy. *Finite Elements in Analysis and Design* **1**(1), 3–20
Nowacki W 1980 *Plates and Shells*. PWN, Warsaw (in Polish).
ROBOT 2006 ROBOT Millennium: User's Guide. Technical report, RoboBAT, Cracow (in Polish).
Szabó BA and Sahrmann GJ 1988 Hierarchic plate and shell models based on p-extension. *International Journal for Numerical Methods in Engineering* **26**(8), 1855–1881
Timoshenko S and Woinowsky-Krieger S 1959 *Theory of Plates and Shells*. McGraw-Hill, New York-Auckland.

12

Shallow Shells

12.1 Equations for Shallow Shells

The shell is considered shallow when ratio f/L is small ($f/L < 1/5$) – see classification in Subsection 1.5.1. As shown in Figure 12.1 f denotes the rise of the shell (maximum distance between the shell and its projection onto the horizontal plane).

Introduction of relevant assumptions leads to a simplification of the mathematical model describing the case of shallow shells in comparison to the general shell theory – see Timoshenko and Woinowsky-Krieger (1959) and Csonka (1966).

The geometry of a shallow shell with rectangular projection can be described in the Cartesian coordinate system as follows:

$$\xi_1 = x, \quad \xi_2 = y, \quad z = z(x,y), \quad A_1 = \sqrt{1 + (z_{,x})^2} \approx 1, \quad A_2 = \sqrt{1 + (z_{,y})^2} \approx 1 \tag{12.1}$$

The area of surface segment $\mathrm{d}\Omega$ is related to the area of its projection $\mathrm{d}A = \mathrm{d}x\mathrm{d}y$ by:

$$\mathrm{d}\Omega = \sqrt{1 + (z_{,x})^2 + (z_{,y})^2}\,\mathrm{d}A \tag{12.2}$$

The curvature radii R_x, R_y used in further equations depend on the shape of shell $z = z(x,y)$.

In Figure 12.2 membrane forces, normal n_ξ, n_η and tangent $n_{\xi\eta}$, as well as their projections, n_x, n_y, n_{xy} on the horizontal plane parallel to (x,y) plane, are presented.

For simplification, a curved surface segment of shallow shell (Figure 12.2a) is replaced by ruled one (Figure 12.2b). The relations between membrane forces and their projections can be written as:

$$n_\xi = n_x \frac{\cos\beta}{\cos\alpha}, \qquad n_\eta = n_y \frac{\cos\alpha}{\cos\beta}, \qquad n_{\xi\eta} = n_{\eta\xi} = n_{xy} = n_{yx} \tag{12.3}$$

where trigonometric functions resulting from surface equation $z(x,y)$ are used:

$$\tan\alpha = z_{,x}, \qquad \tan\beta = z_{,y}, \qquad \cos\alpha = \frac{1}{\sqrt{1 + (z_{,x})^2}}, \qquad \cos\beta = \frac{1}{\sqrt{1 + (z_{,y})^2}} \tag{12.4}$$

Plate and Shell Structures: Selected Analytical and Finite Element Solutions, First Edition.
Maria Radwańska, Anna Stankiewicz, Adam Wosatko and Jerzy Pamin.
© 2017 John Wiley & Sons Ltd. Published 2017 by John Wiley & Sons Ltd.

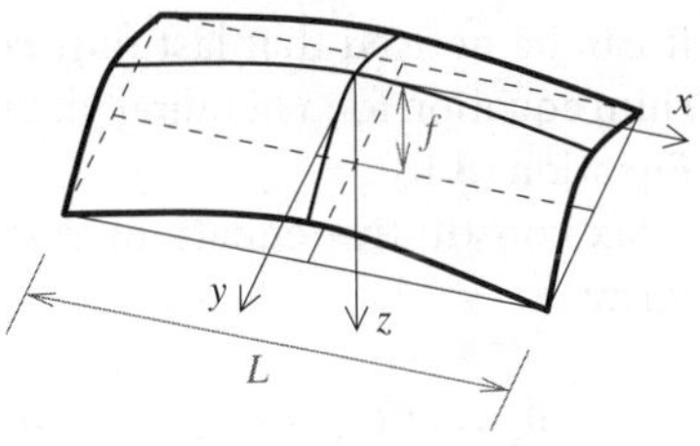

Figure 12.1 Model of shallow shell with rectangular projection

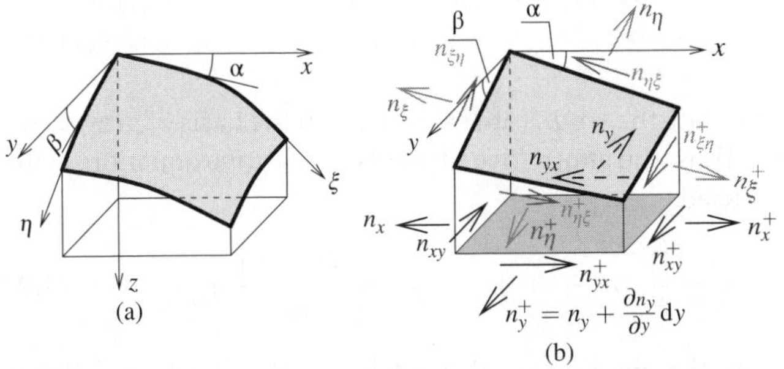

Figure 12.2 (a) Surface segment of doubly curved shallow shell and (b) ruled segment with membrane forces and their projections on horizontal plane

The assumption that deflection w is significantly larger than translations u_x and u_y leads to the reduced kinematic equations:

$$\varepsilon_x = \frac{\partial u_x}{\partial x} + \underline{\frac{w}{R_x}}, \qquad \varepsilon_y = \frac{\partial u_y}{\partial y} + \underline{\frac{w}{R_y}}, \qquad \gamma_{xy} = \frac{\partial u_x}{\partial y} + \frac{\partial u_y}{\partial x}$$
$$\kappa_x = -\frac{\partial^2 w}{\partial x^2}, \qquad \kappa_y = -\frac{\partial^2 w}{\partial y^2}, \qquad \chi_{xy} = -2\,\frac{\partial^2 w}{\partial x\,\partial y} \tag{12.5}$$

The simplification of equilibrium equations for the case of a shallow shell results from the fact that $R_\alpha \to \infty$ and thus $1/R_\alpha \to 0$. Preserving the components containing quotients $1/R_\alpha$ only in the third equilibrium equation below $(12.6)_3$, the following set of five equilibrium equations for a shallow shell is obtained:

$$\frac{\partial n_x}{\partial x} + \frac{\partial n_{yx}}{\partial y} + \hat{p}_x = 0, \qquad \frac{\partial n_{xy}}{\partial x} + \frac{\partial n_y}{\partial y} + \hat{p}_y = 0$$
$$\frac{\partial t_x}{\partial x} + \frac{\partial t_y}{\partial y} \underline{- \left(\frac{n_x}{R_x} + \frac{n_y}{R_y}\right)} + \hat{p}_z = 0 \tag{12.6}$$
$$\frac{\partial m_x}{\partial x} + \frac{\partial m_{yx}}{\partial y} - t_x = 0, \qquad \frac{\partial m_{xy}}{\partial x} + \frac{\partial m_y}{\partial y} - t_y = 0$$

The underlined components in the above kinematic and equilibrium equations couple the membrane and bending states. Equations $(12.6)_{3,4,5}$ can be reduced to one

$$\underline{-\left(\frac{n_x}{R_x} + \frac{n_y}{R_y}\right)} + \frac{\partial^2 m_x}{\partial x^2} + 2\,\frac{\partial^2 m_{xy}}{\partial x\,\partial y} + \frac{\partial^2 m_y}{\partial y^2} + \hat{p}_z = 0 \tag{12.7}$$

It can be noticed that last four components are exactly the same as in case of equilibrium equation for a bending plate, expressed in terms of moments, see Subsection 8.2.2, Equation (8.18).

Six constitutive equations referring to the membrane and bending states have the form:

$$\begin{aligned} n_x &= D^n(\varepsilon_x + \nu\,\varepsilon_y), \qquad n_y = D^n(\varepsilon_y + \nu\,\varepsilon_x), \qquad n_{xy} = D^n(1-\nu)\,\gamma_{xy}/2 \\ m_x &= D^m(\kappa_x + \nu\,\kappa_y), \qquad m_y = D^m(\kappa_y + \nu\,\kappa_x), \qquad m_{xy} = D^m(1-\nu)\,\chi_{xy}/2 \end{aligned} \tag{12.8}$$

In the analysis of the membrane state the strains are calculated on the basis of previously determined membrane forces. Thus, the alternative form of constitutive equations with strains on the left-hand side is used:

$$\varepsilon_x = \frac{1}{Eh}(n_x - \nu n_y), \qquad \varepsilon_y = \frac{1}{Eh}(n_y - \nu n_x), \qquad \gamma_{xy} = \frac{2(1+\nu)}{Eh}\,n_{xy} \tag{12.9}$$

As in the previous chapters the mathematical models for the local and global formulations of shallow shells are presented in Box 12.1. The local formulation includes 17 differential and algebraic equations: six kinematic, five equilibrium and six constitutive, completed by the information on boundary loads and kinematic constraints.

The lack of surface loads $\hat{p}_x$ and $\hat{p}_y$ is assumed. Additionally the strain compatibility equation is included, see Timoshenko and Woinowsky-Krieger (1959)

$$\frac{\partial^2 \varepsilon_x}{\partial y^2} + \frac{\partial^2 \varepsilon_y}{\partial x^2} - \frac{\partial^2 \gamma_{xy}}{\partial x\,\partial y} = \frac{1}{R_y}\frac{\partial^2 w}{\partial x^2} + \frac{1}{R_x}\frac{\partial^2 w}{\partial y^2} \tag{12.10}$$

Taking into account the relationships between stress function F and projections of membrane forces (see Chapter 6):

$$n_x = \frac{\partial^2 F}{\partial y^2}, \qquad n_y = \frac{\partial^2 F}{\partial x^2}, \qquad n_{xy} = -\frac{\partial^2 F}{\partial x\,\partial y} \tag{12.11}$$

a new form of the compatibility equation is derived and written next as Equation $(12.12)_1$. Now, equilibrium Equation (12.7) is reformulated introducing the stress function and combining kinematic Equations (12.5) and constitutive Equations (12.8) related only to the bending state. As a consequence the behaviour of shallow shell is described by a set of two displacement-force equations expressed by stress function $F(x,y)$ and deflection function $w(x,y)$:

$$\begin{aligned} &\nabla^2\nabla^2 F(x,y) - E\,h\left(\frac{1}{R_y}\frac{\partial^2 w}{\partial x^2} + \frac{1}{R_x}\frac{\partial^2 w}{\partial y^2}\right) = 0 \\ &D^m\,\nabla^2\nabla^2 w(x,y) + \left(\frac{1}{R_y}\frac{\partial^2 F}{\partial x^2} + \frac{1}{R_x}\frac{\partial^2 F}{\partial y^2}\right) - \hat{p}_z = 0 \end{aligned} \tag{12.12}$$

Box 12.1 Equations for shallow shells in a membrane-bending state

Local formulation – set of 17 equations and boundary conditions

(I) kinematic equations (6):

$$\varepsilon_x = \frac{\partial u_x}{\partial x} + \frac{w}{R_x}, \qquad \varepsilon_y = \frac{\partial u_y}{\partial y} + \frac{w}{R_y}, \qquad \gamma_{xy} = \frac{\partial u_x}{\partial y} + \frac{\partial u_y}{\partial x}$$

$$\kappa_x = -\frac{\partial^2 w}{\partial x^2}, \qquad \kappa_y = -\frac{\partial^2 w}{\partial y^2}, \qquad \chi_{xy} = -2\frac{\partial^2 w}{\partial x \partial y}$$

(II) equilibrium equations (5):

$$\frac{\partial n_x}{\partial x} + \frac{\partial n_{yx}}{\partial y} + \hat{p}_x = 0, \qquad \frac{\partial n_{xy}}{\partial x} + \frac{\partial n_y}{\partial y} + \hat{p}_y = 0$$

$$\frac{\partial t_x}{\partial x} + \frac{\partial t_y}{\partial y} - \left(\frac{n_x}{R_x} + \frac{n_y}{R_y}\right) + \hat{p}_z = 0$$

$$\frac{\partial m_x}{\partial x} + \frac{\partial m_{yx}}{\partial y} - t_x = 0, \qquad \frac{\partial m_{xy}}{\partial x} + \frac{\partial m_y}{\partial y} - t_y = 0$$

(III) constitutive equations (6):

$$n_x = D^n(\varepsilon_x + \nu\varepsilon_y), \qquad n_y = D^n(\varepsilon_y + \nu\varepsilon_x), \qquad n_{xy} = D^n(1-\nu)\,\gamma_{xy}/2$$

$$m_x = D^m(\kappa_x + \nu\kappa_y), \qquad m_y = D^m(\kappa_y + \nu\kappa_x), \qquad m_{xy} = D^m(1-\nu)\,\chi_{xy}/2$$

where $D^n = E\,h/(1-\nu^2)$ and $D^m = E\,h^3/[12(1-\nu^2)]$

(IV) boundary conditions:

a) kinematic: $u_\nu = \hat{u}_\nu,\ u_s = \hat{u}_s,\ w = \hat{w},\ \vartheta_\nu = -w_{,\nu} = \hat{\vartheta}_\nu$

b) static: $n_\nu = \hat{n}_\nu,\ n_{\nu s} = \hat{n}_{\nu s},\ \tilde{t}_\nu = t_\nu + \dfrac{\partial m_{\nu s}}{\partial s} = \hat{t}_\nu,\ m_\nu = \hat{m}_\nu$

c) mixed

Global formulation

Total potential energy $\quad \Pi = U^n + U^m - W$

Internal strain energy

$$U^n = \frac{D^n}{2}\int_A (n_x\varepsilon_x + n_{xy}\gamma_{xy} + n_y\varepsilon_y)\,\mathrm{d}x\,\mathrm{d}y$$

$$U^m = \frac{D^m}{2}\int_A (m_x\kappa_x + m_{xy}\chi_{xy} + m_y\kappa_y)\,\mathrm{d}x\,\mathrm{d}y$$

External load work

$$W = \int_A (\hat{p}_x u_x + \hat{p}_y u_y + \hat{p}_z w)\,\mathrm{d}x\,\mathrm{d}y + \int_{\partial A_\sigma} (\hat{n}_\nu u_\nu + \hat{n}_{\nu s} u_s + \hat{t}_\nu w + \hat{m}_\nu \vartheta_\nu)\,\mathrm{d}s$$

Introducing a new operator

$$\nabla_R^2(\,) = \frac{1}{R_y}\frac{\partial^2(\,)}{\partial x^2} + \frac{1}{R_x}\frac{\partial^2(\,)}{\partial y^2} \tag{12.13}$$

Equations (12.12) of a shallow shell are rewritten as:

$$\nabla^2\nabla^2 F(x,y) - E\,h\,\nabla_R^2\, w(x,y) = 0, \qquad D^m\,\nabla^2\nabla^2 w(x,y) + \nabla_R^2\, F(x,y) - \hat{p}_z = 0 \tag{12.14}$$

These two differential equations must be completed by boundary conditions and they represent the second variant of the mathematical model for shallow shells.

If $R_\alpha = \infty$ is taken into account in Equations (12.14) then two uncoupled equations for membrane (see Subsection 6.2.4) and bending plate (see Subsection 8.2.3) are obtained:

$$\nabla^2\nabla^2 F(x,y) = 0, \qquad D^m\nabla^2\nabla^2 w(x,y) - \hat{p}_z = 0 \tag{12.15}$$

The behaviour of shallow shells can also be described by a set of three differential equations expressed in terms of displacement functions only:

$$\begin{gathered}
\left(\frac{\partial^2}{\partial x^2} + \frac{1-\nu}{2}\frac{\partial^2}{\partial y^2}\right)u_x + \frac{1+\nu}{2}\frac{\partial^2 u_y}{\partial x\,\partial y} + \frac{\partial}{\partial x}\left(\underline{\frac{w}{R_x}} + \nu\,\underline{\frac{w}{R_y}}\right) = -\frac{\hat{p}_x}{D^n}\\
\frac{1+\nu}{2}\frac{\partial^2 u_x}{\partial x\,\partial y} + \left(\frac{\partial^2}{\partial y^2} + \frac{1-\nu}{2}\frac{\partial^2}{\partial x^2}\right)u_y + \frac{\partial}{\partial y}\left(\underline{\frac{w}{R_y}} + \nu\,\underline{\frac{w}{R_x}}\right) = -\frac{\hat{p}_y}{D^n}\\
\nabla^2\nabla^2 w - \frac{12}{h^2}\left[\left(\underline{\frac{1}{R_x^2}} + \underline{\frac{1}{R_x^2}} + \underline{\frac{1+\nu}{R_xR_y}}\right)w - \left(\underline{\frac{1}{R_x}} + \underline{\frac{\nu}{R_y}}\right)\frac{\partial u_x}{\partial x} - \left(\underline{\frac{1}{R_y}} + \underline{\frac{\nu}{R_x}}\right)\frac{\partial u_y}{\partial y}\right] = \frac{\hat{p}_z}{D^m}
\end{gathered} \tag{12.16}$$

For completeness of the formulation the boundary conditions have to be taken into account. It is necessary to express them by the displacement functions.

It should be emphasized that all these equations are related to the principal curvature lines and thus $1/R_{xy} = 0$.

12.2 Pucher's Equations for Shallow Shells in the Membrane State

The equations describing the membrane state of a shallow shell are presented in Box 12.2.

For a start the equilibrium equations in plane (x, y) are recalled:

$$\frac{\partial n_x}{\partial x} + \frac{\partial n_{yx}}{\partial y} + \hat{p}_x = 0, \qquad \frac{\partial n_{xy}}{\partial x} + \frac{\partial n_y}{\partial y} + \hat{p}_y = 0 \tag{12.17}$$

The shallow shell is usually loaded only by vertical surface load $\hat{p}_z$ with loads $\hat{p}_x = \hat{p}_y = 0$.

Note that in these equations the projections of membrane forces appear. Taking into account the directions of membrane forces n_ξ, n_η, $n_{\xi\eta}$ and the orientation of sides of

Box 12.2 Equations for shallow shells in a membrane state

Local formulation – set of nine equations and boundary conditions

(I) kinematic equations (3):

$$\varepsilon_x = \frac{\partial u_x}{\partial x} + \frac{w}{R_x}, \qquad \varepsilon_y = \frac{\partial u_y}{\partial y} + \frac{w}{R_y}, \qquad \gamma_{xy} = \frac{\partial u_x}{\partial y} + \frac{\partial u_y}{\partial x}$$

(II) equilibrium equations (3):

$$\frac{\partial n_x}{\partial x} + \frac{\partial n_{yx}}{\partial y} + \hat{p}_x = 0, \qquad \frac{\partial n_{xy}}{\partial x} + \frac{\partial n_y}{\partial y} + \hat{p}_y = 0$$

$$n_x \frac{\partial^2 z}{\partial x^2} + 2\, n_{xy} \frac{\partial^2 z}{\partial x\, \partial y} + n_y \frac{\partial^2 z}{\partial y^2} + \hat{p}_z = 0$$

(III) constitutive equations (3):

$$n_x = D^n(\varepsilon_x + \nu\varepsilon_y), \qquad n_y = D^n(\varepsilon_y + \nu\varepsilon_x), \qquad n_{xy} = D^n(1 - \nu)\, \gamma_{xy}/2$$

where $D^n = E\,h/(1 - \nu^2)$

(IV) boundary conditions:

a) kinematic: $u_\nu = \hat{u}_\nu$, $u_s = \hat{u}_s$

b) static: $n_\nu = \hat{n}_\nu$, $n_{\nu s} = \hat{n}_{\nu s}$

c) mixed

Global formulation

Total potential energy $\Pi = U^n - W$

Internal strain energy

$$U^n = \frac{D^n}{2} \int_A (n_x\, \varepsilon_x + n_{xy}\, \gamma_{xy} + n_y\, \varepsilon_y)\, dx\, dy$$

External load work

$$W = \int_A (\hat{p}_x\, u_x + \hat{p}_y\, u_y + \hat{p}_z\, w)\, dx\, dy + \int_{\partial A_\sigma} (\hat{n}_\nu\, u_\nu + \hat{n}_{\nu s}\, u_s)\, ds$$

the elementary ruled segment (Figure 12.2b) the third equilibrium equation $\Sigma P_z = 0$ is written as

$$n_x \frac{\partial^2 z}{\partial x^2} + 2\, n_{xy} \frac{\partial^2 z}{\partial x\, \partial y} + n_y \frac{\partial^2 z}{\partial y^2} + \hat{p}_z = 0 \tag{12.18}$$

This equation includes a sum of projections of all membrane forces on the z direction with factors depending on the second derivatives of the shallow shell surface function $z(x, y)$ assuming zero transverse shear forces $t_x = t_y = 0$.

The form of Equation (12.18) is analogous to three components of the equation

$$D^m \nabla^2 \nabla^2 w(x, y) - \underline{n_x \frac{\partial^2 w}{\partial x^2}} - \underline{2 n_{xy} \frac{\partial^2 w}{\partial x\, \partial y}} - \underline{n_y \frac{\partial^2 w}{\partial y^2}} = 0 \tag{12.19}$$

which describes the bending of rectangular plates and takes into account the influence of considerable membrane forces, see Section 14.2 and Equation (14.2). The underlined components represent a sum of projections of all membrane forces on the z direction including the factors dependent here on the second derivatives of the deformed plate middle surface function $w(x, y)$.

The projections of membrane forces, expressed by second derivatives of stress function $F(x, y)$, are introduced into Equation (12.18) and so-called Pucher's equation is obtained, see also Csonka (1966)

$$\frac{\partial^2 F}{\partial y^2}\frac{\partial^2 z}{\partial x^2} - 2\frac{\partial^2 F}{\partial x\,\partial y}\frac{\partial^2 z}{\partial x\,\partial y} + \frac{\partial^2 F}{\partial x^2}\frac{\partial^2 z}{\partial y^2} + \hat{p}_z = 0 \tag{12.20}$$

The boundary conditions related to lines $x = $ const. and/or $y = $ const. are required to complete the formulation and solution of Equation (12.20).

Knowing the load function $\hat{p}_z(x, y)$ and the shape of the shell described by function $z(x, y)$, the stress function $F(x, y)$ is derived from this equation. Then, the projections of membrane forces and the forces themselves are calculated.

In the book of Csonka (1966) examples of the analysis of many kinds of shallow shells in membrane state are discussed. Next, the analytical solution for one selected, typical shell is presented.

12.3 Hyperbolic Paraboloid with Rectangular Projection

12.3.1 Description of Geometry

The middle surface of the considered shell is a ruled hyperbolic paraboloid (see Figure 12.3), and is described in Subsection 1.3.4 and Subsection 3.2.3. The shell is analysed in the Cartesian coordinate system (with axis z pointing down) and defined by the equation

$$z(x, y) = kxy = \frac{f}{ab}\,xy, \qquad k = \frac{f}{ab}, \qquad z(a, b) = f \tag{12.21}$$

Adopting the dimensions f and $a = b = L$ (included in Box 12.3), the characteristic geometrical parameter k is equal

$$k = \frac{f}{a\,b} = \frac{f}{L^2} = 0.035\ 1/\text{m} \tag{12.22}$$

Now, the equation of the surface of the analysed shell is written as

$$z(x, y) = kxy = 0.035xy \tag{12.23}$$

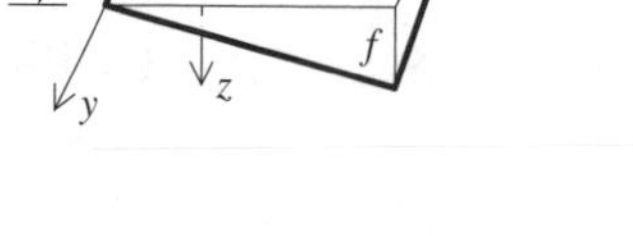

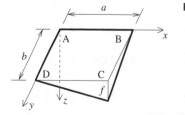

Figure 12.3 Hyperbolic paraboloid – configuration

The first and the second derivatives of function z are calculated as follows:

$$z_{,x} = ky = 0.035y, \qquad z_{,y} = kx = 0.035x$$
$$z_{,xx} = z_{,yy} = 0, \qquad z_{,xy} = k = 0.035\ 1/\mathrm{m}, \qquad k > 0 \tag{12.24}$$

and the area of elementary surface segment is

$$\mathrm{d}\Omega = \sqrt{1 + k^2x^2 + k^2y^2}\ \mathrm{d}x\mathrm{d}y \tag{12.25}$$

12.3.2 Analytical Solution of the Membrane State

The analytical solution starts with a recapitulation of the notation of membrane forces $n_\xi, n_\eta, n_{\xi\eta}$ and their projections n_x, n_y, n_{xy} on plane (x, y).

With zero values of second derivatives $z_{,xx} = z_{,yy} = 0$ Pucher's equation (12.20) simplifies considerably and reads

$$-\,2\,k\,\frac{\partial^2 F}{\partial x\,\partial y} + \hat{p}_z(x, y) = 0 \tag{12.26}$$

In the analysis two types of loading can be considered:

Case A – self weight per unit area in shell middle surface $g_0[\mathrm{kN/m^2}]$
Case B – self weight per unit area of horizontal projection $q_0[\mathrm{kN/m^2}]$

First, case A is considered. The shell is loaded by self weight $g_0 = \hat{p}_z = 2.82\ \mathrm{kN/m^2}$ per unit area in the shell middle surface. Using Equation (12.26) the second mixed derivative of the stress function is calculated as

$$\frac{\partial^2 F}{\partial x\,\partial y} = \frac{g_0}{2\,k} = 40.29\ \mathrm{kN/m} \tag{12.27}$$

and then the projection of tangent force n_{xy} is equal

$$n_{xy} = -\frac{g_0}{2k}\,\sqrt{1 + k^2\,x^2 + k^2\,y^2} \tag{12.28}$$

Box 12.3 Hyperbolic paraboloid in a membrane state

Data

$a = b = L = 10.0$ m, $\quad f = 3.5$ m
$k = 0.035\ 1/\mathrm{m}$, $\quad g_0 = 2.82\ \mathrm{kN/m^2}$ – per unit area in shell middle surface

Check values from analytical solution

$n_x|_{\mathrm{ext}} = -4.585$ kN/m, $\quad n_\xi|_{\mathrm{ext}} = -4.858$ kN/m
$n_y|_{\mathrm{ext}} = -4.585$ kN/m, $\quad n_\eta|_{\mathrm{ext}} = -4.858$ kN/m
$n_{\xi\eta\ \mathrm{B,D}} = -42.682$ kN/m
$n_{\xi\eta\ \mathrm{A}} = -40.286$ kN/m, $\quad n_{\xi\eta\ \mathrm{C}} = -44.95\ \mathrm{kN/m} \approx -45.00$ kN/m
$n_{\mathrm{I\,A}} = -n_{\mathrm{II\,A}} = |n_{\xi\eta}| = 40.286$ kN/m, $\quad n_{\mathrm{I\,C}} = -n_{\mathrm{II\,C}} = |n_{\xi\eta}| = 45.00$ kN/m

Shell configuration shown in Figure 12.3

Now, from the two equilibrium equations (12.17) the final formulae for projections n_x and n_y are derived:

$$n_x = 0.5\, g_0\, y\, \log\left(\frac{k\,x + \sqrt{1 + k^2\,x^2 + k^2\,y^2}}{k\,a + \sqrt{1 + k^2\,a^2 + k^2 y^2}}\right)$$
$$n_y = 0.5\, g_0\, x\, \log\left(\frac{k\,y + \sqrt{1 + k^2\,x^2 + k^2\,y^2}}{k\,b + \sqrt{1 + k^2\,x^2 + k^2 b^2}}\right) \qquad (12.29)$$

The membrane forces n_ξ, n_η, $n_{\xi\eta}$ are calculated from Equations (12.3), taking into account parameter k resulting from the geometry of the shell:

$$n_\xi = n_x\,\frac{\cos\beta}{\cos\alpha},\quad n_\eta = n_y\,\frac{\cos\alpha}{\cos\beta},\quad \text{where}\quad \tan\alpha = z_{,x} = k\,y,\quad \tan\beta = z_{,y} = k\,x$$

$$\cos\alpha = \frac{1}{\sqrt{1 + (z_{,x})^2}} = \frac{1}{\sqrt{1 + k^2\,y^2}},\qquad \cos\beta = \frac{1}{\sqrt{1 + (z_{,y})^2}} = \frac{1}{\sqrt{1 + k^2\,x^2}} \qquad (12.30)$$

$$n_\xi = n_x\,\frac{\sqrt{1 + k^2\,y^2}}{\sqrt{1 + k^2\,x^2}},\qquad n_\eta = n_y\,\frac{\sqrt{1 + k^2\,x^2}}{\sqrt{1 + k^2\,y^2}},\qquad n_{\xi\eta} = n_{xy}$$

Next, the set of detailed formulae for the three forces and their projections for the adopted values of a, b, f, k (see Box 12.3) is presented.

- Normal force $n_\xi(x, y)$ and its projection $n_x(x, y)$:

$$n_\xi(x, y) = n_x\,\sqrt{\frac{1 + 0.035^2\,y^2}{1 + 0.035^2\,x^2}}$$
$$n_x(x, y) = 1.41\,y\,\log\left(\frac{0.035\,x + \sqrt{1 + 0.035^2\,x^2 + 0.035^2\,y^2}}{0.35 + \sqrt{1 + 0.035^2 \times 100 + 0.035^2\,y^2}}\right) \qquad (12.31)$$

  Extreme forces $n_x|_{\text{ext}} = -4.585$ kN/m and $n_\xi|_{\text{ext}} = -4.858$ kN/m occur at point D with coordinates $x = 0$, $y = L$. Forces n_x and n_ξ are equal to zero along the lines $y = 0$ and $x = L$.

- Normal force $n_\eta(x, y)$ and its projection $n_y(x, y)$:

$$n_\eta(x, y) = n_y\,\sqrt{\frac{1 + 0.035^2 x^2}{1 + 0.035^2 y^2}}$$
$$n_y(x, y) = 1.41\,x\,\log\left(\frac{0.035\,y + \sqrt{1 + 0.035^2\,x^2 + 0.035^2 y^2}}{0.35 + \sqrt{1 + 0.035^2 \times 100 + 0.035^2 x^2}}\right) \qquad (12.32)$$

  Extreme forces $n_y|_{\text{ext}} = -4.585$ kN/m and $n_\eta|_{\text{ext}} = -4.858$ kN/m occur at point B with coordinates $x = L, y = 0$. These forces are equal to zero along the lines $x = 0$ and $y = L$.

- Tangent force $n_{\xi\eta}(x, y)$ and its projection $n_{xy}(x, y)$

$$n_{\xi\eta}(x, y) = n_{xy}(x, y) = -40.29\,\sqrt{1 + 0.035^2\,x^2 + 0.035^2 y^2} \qquad (12.33)$$

Extreme force $n_{\xi\eta}|_{\text{ext}} = -44.95$ kN/m ≈ -45.00 kN/m occurs at corner C with coordinates $x = y = L$. At point C also extreme principal forces in diagonal directions occur $n_{\text{I C}} = -n_{\text{II C}} = |n_{\xi\eta}| = 45.00$ kN/m. At three other corners A, B, D – tangent forces have the following values: $n_{\xi\eta\ \text{A}} = -40.286$ kN/m, $n_{\xi\eta\ \text{B,D}} = -42.682$ kN/m.

All characteristic check values can also be found in Box 12.3.

In Figure 12.4 the contour plots of distributions of projections of membrane forces are presented. Note that the distribution of tangent force corresponds with shape of shell with characteristic symmetry along the diagonal $x = y$ in shell with square projection.

When loading case B is considered (self weight load $q = q_0 =$ const. defined per unit area of horizontal projection) for hyperbolic paraboloidal shell, the state of pure uniform shear occurs with one constant nonzero projection of tangent force field and with zero projections $n_x = n_y = 0$

$$n_{xy}(x,y) = -\frac{q_0}{2\,k} = -\frac{a\,b}{2\,f}\,q_0 = \text{const.} \tag{12.34}$$

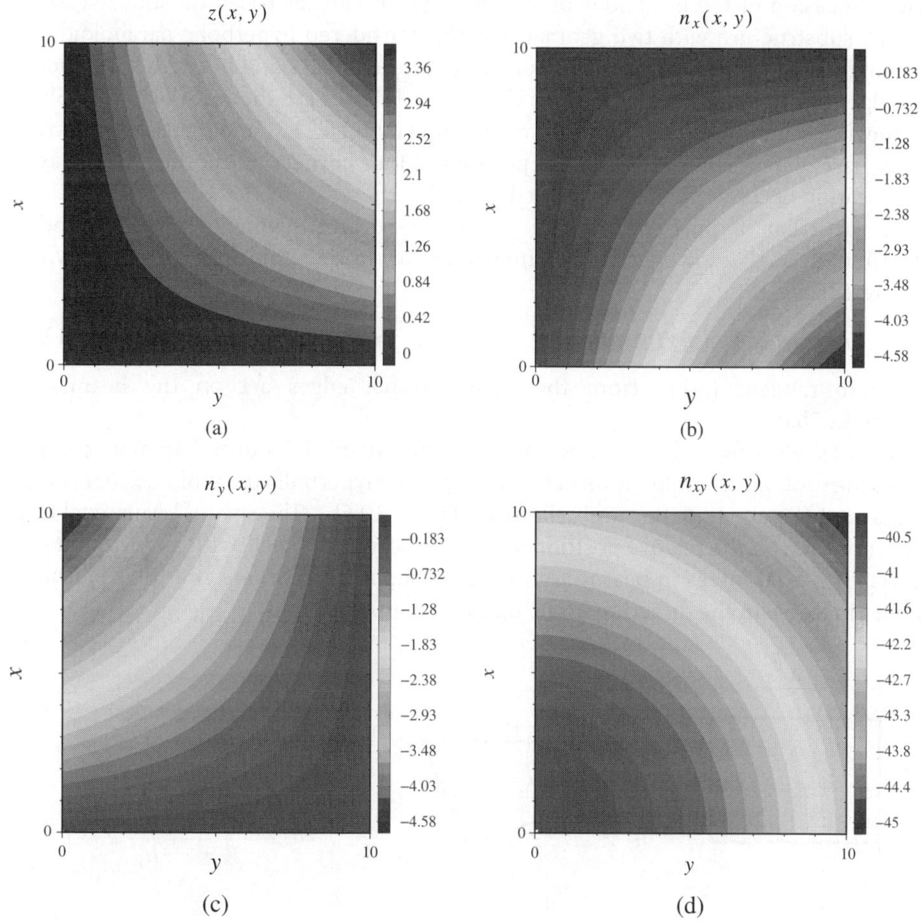

Figure 12.4 Hyperbolic paraboloid – contour plots of: (a) middle surface $z(x, y)$ and analytically determined distributions of projections of membrane forces: (b) $n_x(x, y)$, (c) $n_y(x, y)$ and (d) $n_{xy}(x, y)$

The tangential forces are related as follows

$$n_{\xi\eta} = n_{\eta\xi} = n_{xy} = n_{yx} \tag{12.35}$$

The principal membrane forces can also be calculated and have the following values

$$n_{\mathrm{I}} = -n_{\mathrm{II}} = |n_{xy}| = \frac{q_0}{2\,k} = \frac{a\,b}{2\,f}\,q_0 = \text{const.} \tag{12.36}$$

For the shallow shell over a square, the angles defining the directions of principal forces are equal to 45° and 135° for n_{I} and n_{II}, respectively.

Formulae (12.29) to (12.33) can also be found in the book by Kobiak and Stachurski (1991). The analysed hyperbolic paraboloidal shell is a part of a cross-roof, shown in Figure 12.6a.

12.4 Remarks on Engineering Applications

The discussion of the behaviour of shallow shells is further based on similar cases of a shell substructure with two geometries. The considered hyperbolic paraboloid with rectangular projection can be constructed in different ways and in engineering design its edges can be supported on beams. Two possible cases are shown in Figure 12.5b,c.

The distributions of membrane forces, denoted by n_ξ, n_η, $n_{\xi\eta}$, and their projections n_x, n_y, n_{xy} are dependent on the relative position of four corners of the shell as well as on the location of the supported or free edges and corners.

In particular, a pure shear state occurs in surface segments under uniformly distributed load $q_0 = \text{const.}$ (per unit area of horizontal projection) shown in Figure 12.5b,c:

$$n_{xy}(x,y) = n_{\xi\eta}(x,y) = \text{const.}, \qquad n_{\mathrm{I}} = -n_{\mathrm{II}} = |n_{xy}(x,y)| \tag{12.37}$$

Nonzero tangent forces from the shell external edges act on the beams, see Figure 12.5b,c.

In roof shell design edge and ridge beams are frequently introduced. In more complex cases the roofs are a combination of several separate hyperbolic paraboloids, for example cross-roof (Figure 12.6a) or umbrella-roof (Figure 12.6b). The sign of beam axial force depends then on the relative position of four corners of the shell, with and without supports. The axial forces in beams are zero at unsupported corners and their (absolute) values increase in the direction of the opposite supported end.

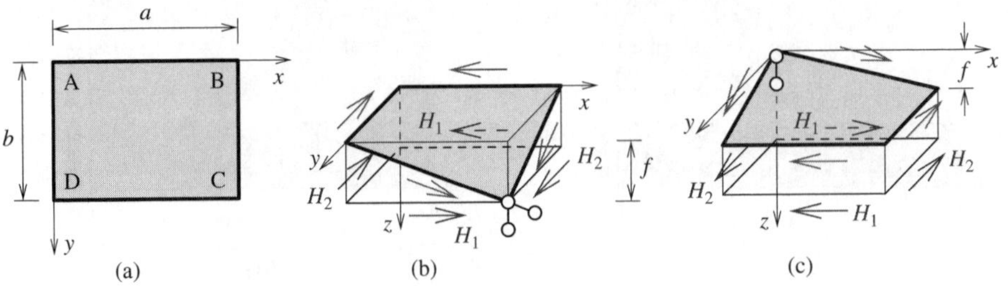

Figure 12.5 Two cases of ruled shells with point supports as well as visualization of actions along edge beams, and their projections H_1, H_2 on horizontal plane

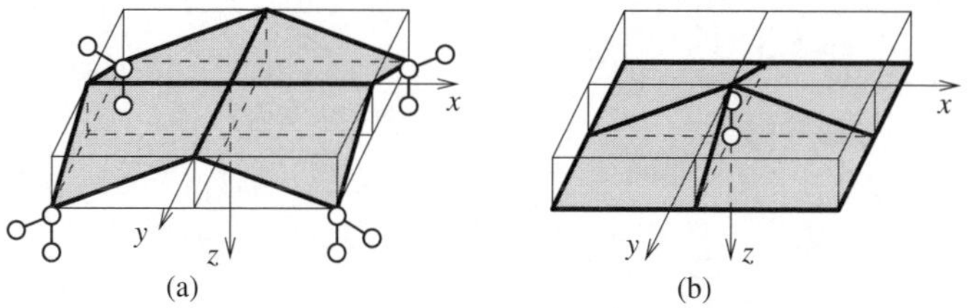

Figure 12.6 Complex roofs: (a) cross-roof and (b) umbrella-roof

For complex roof structures, the sharp bends that are generated at the interface of different surfaces and the introduction of joints between the covering and ridge or edge beams become an inevitable source of bending effects.

In the past, when the design process was not computer-aided, it was necessary to perform static calculations in stages. Separated substructures were analysed and then their mutual connections were considered. External actions were transferred to the zones of supports through interactions of separated parts (subdomains of shell, ridge and edge beams).

Nowadays, a general availability of computer packages makes it possible to generate a comprehensive model of a structure (set of shell segments, beams and supports) and perform computations taking simultaneously into account all interacting parts.

References

Csonka P 1966 *Membranschalen*. W. Ernst & Son, Berlin, München (in German).

Kobiak J and Stachurski W 1991 *Reinforced Concrete Structures* vol. 4. Arkady, Warsaw (in Polish).

Timoshenko S and Woinowsky-Krieger S 1959 *Theory of Plates and Shells*. McGraw-Hill, New York-Auckland.

Waszczyszyn Z and Radwańska M 1985 *Plates and Shells*. Cracow University of Technology, Cracow (in Polish).

13

Thermal Loading of Selected Membranes, Plates and Shells

13.1 Introduction

Two simplest types of thermal loads are distinguished, that is two kinds of temperature change distribution along the thickness:

- uniform, see Figure 13.1a
- linear, see Figure 13.1b

It is assumed that the temperature change does not depend on the surface coordinates ξ_1, ξ_2.

The uniform distribution of temperature change (independent of coordinate z) for all surfaces equidistant from the middle surface is expressed as

$$\Delta T_0 = T_0 - T_{\mathrm{r}} = \text{const.} \tag{13.1}$$

where the temperature change ΔT_0 is defined in relation to the reference temperature T_{r}. In this case thermal membrane strains occur

$$\varepsilon_{11,\Delta T_0} = \varepsilon_{22,\Delta T_0} = \alpha_T \, \Delta T_0 = \text{const.} \tag{13.2}$$

where α_T is the thermal expansion coefficient. They are included in the constitutive equations (underlined components) together with the mechanical strains:

$$\begin{aligned} n_{11} &= D^n[(\varepsilon_{11} + \nu\varepsilon_{22}) - \underline{(1+\nu)\, \alpha_T \, \Delta T_0]} \\ n_{22} &= D^n[(\varepsilon_{22} + \nu\varepsilon_{11}) - \underline{(1+\nu)\, \alpha_T \, \Delta T_0]} \end{aligned} \tag{13.3}$$

The linear distribution of the temperature change is a function of coordinate z

$$\Delta T(z) = k\, z, \quad \text{where} \quad k = \Delta T_{\mathrm{h}}/h, \quad \Delta T_{\mathrm{h}} = \Delta T_{\mathrm{ext}} - \Delta T_{\mathrm{int}} \tag{13.4}$$

For the middle surface the temperature change is $\Delta T(0) = 0$ and for the limiting surfaces with $z = \pm h/2$ the absolute values of temperature changes are equal $|\Delta T_{\mathrm{int}}| = |\Delta T_{\mathrm{ext}}|$.

Plate and Shell Structures: Selected Analytical and Finite Element Solutions, First Edition.
Maria Radwańska, Anna Stankiewicz, Adam Wosatko and Jerzy Pamin.
© 2017 John Wiley & Sons Ltd. Published 2017 by John Wiley & Sons Ltd.

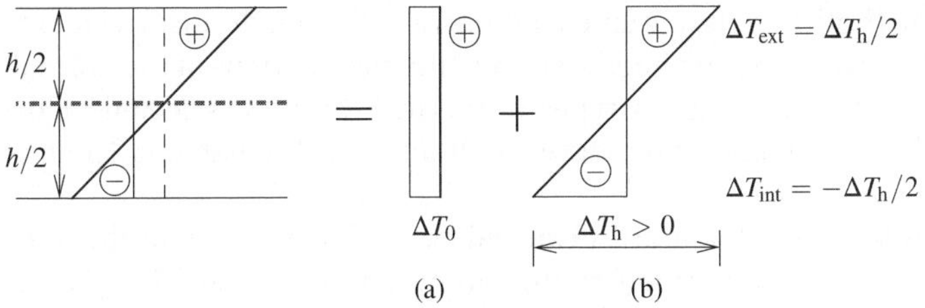

Figure 13.1 Distribution of temperature changes along the thickness: (a) uniform and (b) linear

In the considered case curvatures occur in plates (plane structures) or changes of curvatures in curved shells

$$\kappa_{11,\Delta T_h} = \kappa_{22,\Delta T_h} = \alpha_T \, k = \alpha_T \, \Delta T_h / h = \text{const.} \tag{13.5}$$

They are taken into account in the general constitutive equations as:

$$\begin{aligned} m_{11} &= D^m[(\kappa_{11} + \nu\kappa_{22}) - \underline{(1+\nu)\,\alpha_T\,\Delta T_h/h}] \\ m_{22} &= D^m[(\kappa_{22} + \nu\kappa_{11}) - \underline{(1+\nu)\,\alpha_T\,\Delta T_h/h}] \end{aligned} \tag{13.6}$$

In our considerations the subscripts ΔT_0 and ΔT_h are used to indicate the type of thermal load.

The selected formulae describing thermal effects in shell structures are included in Box 13.1. Note that in-plane shear strain and warping are zero.

It should be emphasized that the response induced by the thermal load depends on the geometry and supports of a structure. The boundary conditions can either permit a

Box 13.1 Selected formulae for shells in a membrane-bending state with thermal loads

Uniform distribution of temperature change along thickness: $\Delta T_0 = T_0 - T_r$

- Normal strains

$$\varepsilon_{11,\Delta T_0} = \varepsilon_{22,\Delta T_0} = \alpha_T \, \Delta T_0$$

- Constitutive equations

$$\begin{aligned} n_{11} &= D^n[(\varepsilon_{11} + \nu\varepsilon_{22}) - \underline{(1+\nu)\,\alpha_T\,\Delta T_0}] \\ n_{22} &= D^n[(\varepsilon_{22} + \nu\varepsilon_{11}) - \underline{(1+\nu)\,\alpha_T\,\Delta T_0}] \end{aligned}$$

Linear distribution of temperature change along thickness: $\Delta T_h = \Delta T_{ext} - \Delta T_{int}$

- Changes of curvature

$$\kappa_{11,\Delta T_h} = \kappa_{22,\Delta T_h} = \alpha_T \, k = \alpha_T \, \Delta T_h / h$$

- Constitutive equations

$$\begin{aligned} m_{11} &= D^m[(\kappa_{11} + \nu\kappa_{22}) - \underline{(1+\nu)\,\alpha_T\,\Delta T_h/h}] \\ m_{22} &= D^m[(\kappa_{22} + \nu\kappa_{11}) - \underline{(1+\nu)\,\alpha_T\,\Delta T_h/h}] \end{aligned}$$

free thermal deformation (both membrane and bending) or constrain the displacements in the direction tangent or normal to the middle surface and the rotations of the normal.

The essential effects of the two simplest types of thermal loads for selected membranes, plates and shells with different boundary conditions are discussed in Sections 13.2 and 13.3.

In Krzyś and Życzkowski (1962) the more advanced case of distribution of the temperature $T(r,z)$ with linear dependence on coordinates z and r in a circular plate is considered

$$T(r,z) = T_{\mathrm{r}} + \Delta T_{\mathrm{h}}\, z\, (1 + c\, r)/h \tag{13.7}$$

More results of the analytical and numerical analysis can be found, among others, in Timoshenko and Woinowsky-Krieger (1959) and Reddy (1999).

13.2 Uniform Temperature Change along the Thickness

When the temperature change along the thickness $\Delta T_0 = T_0 - T_{\mathrm{r}} = \text{const.}$, that is its distribution is uniform, membrane thermal strains $\varepsilon_{11,\Delta T_0}$ and $\varepsilon_{22,\Delta T_0}$ and respective displacements occur in the case of unrestricted deformation, and they are accompanied by zero membrane forces. However, when displacements are constrained then nonzero membrane strains induce nonzero forces.

In example calculations three kinds of shell structures are taken into account: circular membrane, cylindrical and hemispherical shells (see Figure 13.2). The subscripts 11, 22 are replaced by appropriate indices resulting from the introduced coordinate system.

13.2.1 Circular Membrane with Free Radial Displacements on the External Contour

The temperature increment $\Delta T_0 > 0$ (Figure 13.2a) causes:

- radial and circumferential extensions

$$\varepsilon_{r,\Delta T_0} = \varepsilon_{\theta,\Delta T_0} = \alpha_T\, \Delta T_0 = \text{const.} \tag{13.8}$$

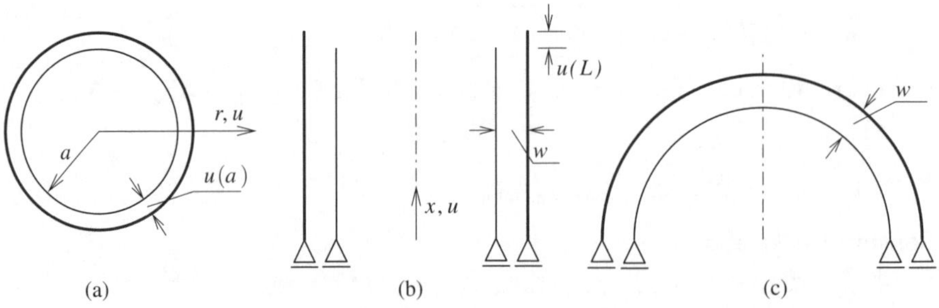

Figure 13.2 Configurations of three structures and their thermal deformations under uniform distribution of temperature change along the thickness $\Delta T_0 > 0$: (a) circular membrane, (b) cylindrical shell and (c) hemispherical shell

- radial displacement field (positive towards the boundary of the membrane), derived by integration of kinematic equation:

$$u' = \varepsilon_{r,\Delta T_0} \quad \text{with} \quad u(0) = 0, \qquad \text{where} \quad (\ )' = \mathrm{d}(\ )/\mathrm{d}r$$
$$u_{\Delta T_0}(r) = \varepsilon_{r,\Delta T_0}\, r = \alpha_T\, \Delta T_0\, r \tag{13.9}$$

with maximum value on external contour

$$u_{\Delta T_0}(a) = \alpha_T\, \Delta T_0\, a \tag{13.10}$$

- zero membrane forces (both radial and circumferential)

$$n_{r,\Delta T_0} = n_{\theta,\Delta T_0} = 0 \tag{13.11}$$

Conclusion: in a circular membrane (see Figure 13.2a) with a uniform distribution of temperature change nonzero thermal axisymmetric radial translations and membrane strains are associated with zero membrane forces.

13.2.2 Simply Supported Cylindrical Shell with Free Horizontal Movement of the Bottom Edge

For each point of the middle surface of a unidirectionally curved cylindrical shell (Figure 13.2b) uniformly heated ($\Delta T_0 > 0$) the following effects occur in the axisymmetric membrane state:

- extensions in both meridional and circumferential directions

$$\varepsilon_{x,\Delta T_0} = \varepsilon_{\theta,\Delta T_0} = \alpha_T\, \Delta T_0 = \text{const.} \tag{13.12}$$

- translations tangent and normal to the meridian:

$$u' = \varepsilon_{x,\Delta T_0} \quad \text{with} \quad u(0) = 0, \qquad \text{where} \quad (\ )' = \mathrm{d}(\ )/\mathrm{d}x$$
$$u_{\Delta T_0}(x) = \alpha_T\, \Delta T_0\, x, \quad w_{\Delta T_0} = R\, \varepsilon_{\theta,\Delta T_0} = R\, \alpha_T\, \Delta T_0 = \text{const.} \tag{13.13}$$

with maximum value u on top contour $x = L$

$$u_{\Delta T_0}(L) = \alpha_T\, \Delta T_0 L \tag{13.14}$$

- two zero membrane forces

$$n_{x,\Delta T_0} = n_{\theta,\Delta T_0} = 0 \tag{13.15}$$

Conclusion: in a cylindrical shell with a uniform distribution of temperature change along the thickness and boundary conditions shown in Figure 13.2b, nonzero kinematic fields (membrane strains and translations) are accompanied by zero membrane forces.

13.2.3 Simply Supported Hemispherical Shell with Free Horizontal Movement of the Bottom Edge

For each point of the middle surface of the hemispherical shell with a uniform distribution of temperature change $\Delta T_0 > 0$ (Figure 13.2c), the following effects appear:

Box 13.2 Simply supported hemispherical shell with a uniform temperature change along the thickness

Data

$R = 50.0$ m, $\alpha = 90°$, $\Delta T_0 = 10$ deg, $\alpha_T = 10^{-5}$ 1/deg

---

Check values from analytical solution

$\varepsilon_{\varphi,\Delta T_0} = \varepsilon_{\theta,\Delta T_0} = 0.0001 = \text{const.}$, $u_{\Delta T_0} \equiv 0$, $w = 0.005$ m $= \text{const.}$

$n_{\varphi,\Delta T_0} = n_{\theta,\Delta T_0} = 0$

Shell configuration shown in Figure 13.2c

- constant extensions in meridional and circumferential directions

$$\varepsilon_{\varphi,\Delta T_0} = \varepsilon_{\theta,\Delta T_0} = \alpha_T \, \Delta T_0 = \text{const.} \tag{13.16}$$

- constant displacements:

$$w_{\Delta T_0} = R \, \varepsilon_{\theta,\Delta T_0} = R \, \alpha_T \, \Delta T_0 = \text{const.}, \qquad u_{\Delta T_0} \equiv 0 \tag{13.17}$$

- zero membrane forces

$$n_{\varphi,\Delta T_0} = n_{\theta,\Delta T_0} = 0 \tag{13.18}$$

In Box 13.2 the input data (R, α, ΔT_0, α_T) and the results describing the fields of strains, displacements and membrane forces are listed.

Conclusion: in the considered hemispherical shell with a uniform distribution of temperature change, constant nonzero membrane strains and displacements normal to the middle surface are observed, but the displacements tangent to a meridian and the membrane forces are zero. Similar results are obtained for a spherical shell with a uniform temperature change ΔT_0.

13.2.4 Hemispherical Shell Clamped on the Bottom Edge – Analytical and Numerical Solutions

In this subsection a more complex case of deformation is considered. This time, the thermal deformation is constrained due to the clamped edge $\varphi = \alpha$ of the considered hemispherical shell. The analytical solution and the numerical one obtained using package ANSYS (2013) are discussed. In Box 13.3 the adopted values of E, ν, R, h, α, ΔT_0 and α_T are included. The following geometric-material parameter ς is used in the calculations (see Chapter 11)

$$\varsigma = \sqrt{\frac{R}{h}} \sqrt[4]{3(1 - \nu^2)} \tag{13.19}$$

Box 13.3 Hemispherical shell clamped at the bottom edge – uniform temperature change along the thickness

Data

$E = 2.1 \times 10^7$ kN/m^2, $\nu = 0.1667$, $R = 50.0$ m, $h = 0.10$ m, $\alpha = 90°$

$\varsigma = 29.222$, $\Delta T_0 = 100.0$ deg, $\alpha_T = 10^{-5}$ 1/deg

Check values from analytical solution

$m^{\text{anal}}_{\varphi,\Delta T_0}(\alpha) = -61.48$ kNm/m $= M$, $t^{\text{anal}}_{\varphi,\Delta T_0}(\alpha) = -71.86$ kN/m$= H$

Check values from FEM solution using ANSYS (FEs SHELL93)

$n^{\text{FEM}}_{\varphi,\Delta T_0}(\alpha) = 4.006$ kN/m, $n^{\text{FEM}}_{\theta,\Delta T_0}(\alpha) = -2160$ kN/m

$m^{\text{FEM}}_{\varphi,\Delta T_0}(\alpha) = 58.27$ kNm/m $= -0.948\, m^{\text{anal}}_{\varphi,\Delta T_0}(\alpha)$

$m^{\text{FEM}}_{\varphi,\Delta T_0,\min} = -12.99$ kNm/m

$t^{\text{FEM}}_{\varphi,\Delta T_0}(\alpha) = 70.87$ kN/m $= -0.986\, t^{\text{anal}}_{\varphi,\Delta T_0}(\alpha)$

Shell configuration shown in Figure 13.3

For a start, a brief discussion of the analytical solution is provided. As shown in Figure 13.3 axisymmetric thermal effects are coupled with occurrence of uniformly distributed boundary loads M [kNm/m] and H [kN/m] on the bottom contour. The senses of these loads are assumed as shown in Figure 13.3b,c. In Subsection 11.3.2 Equations (11.57) and (11.60) describing the effects of loads M and H at the bottom boundary $\varphi = \alpha = 90°$ are derived.

The solution of the example provides the values of boundary moment M and force H and functions describing the distribution of all generalized forces in the axisymmetric membrane-bending state along the meridional section $\theta =$ const.

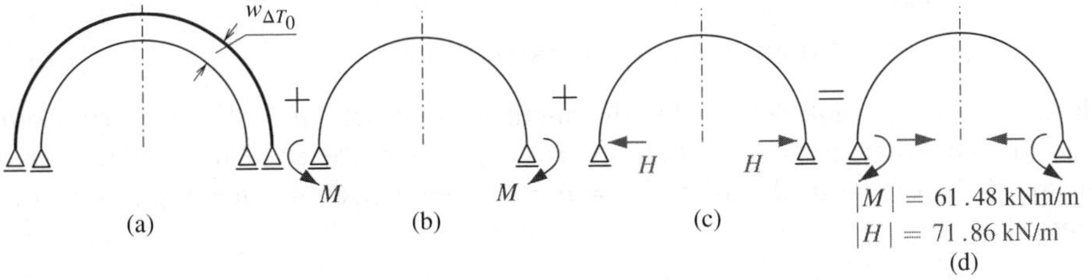

Figure 13.3 Hemispherical shell – idea of superposition of states induced sequentially by: (a) thermal load, distributed boundary loads (b) M and (c) H; (d) configuration of the shell with boundary loads corresponding to clamped edge

For a clamped boundary of the hemispherical shell the values of displacement $w(\alpha)$ and rotation $\vartheta(\alpha) = \vartheta_r(\alpha)$ are zero. Initially, the shell is assumed to be simply supported with unrestricted horizontal movement (see Figure 13.3).

Using the idea of superposition of states, two effects $w(\alpha)$ and $\vartheta(\alpha)$ of three different sources are determined:

- thermal load ΔT_0 in the whole domain results in:

$$\begin{aligned} w_{\Delta T_0}(\alpha) &= R\ \alpha_T\ \Delta T_0 = 0.05\ \text{m} \\ \vartheta_{\Delta T_0}(\alpha) &= 0 \end{aligned} \tag{13.20}$$

- uniformly distributed boundary moment $M = 1$ kNm/m causes:

$$\begin{aligned} w_M(\alpha) &= -2\ \varsigma^2 M/(Eh) = -0.813 \times 10^{-3}\ \text{m} \\ \vartheta_M(\alpha) &= 4\ \varsigma^3 M/(ERh) = 0.951 \times 10^{-3} \end{aligned} \tag{13.21}$$

- horizontal boundary force $H = 1$ kN/m induces:

$$\begin{aligned} w_H(\alpha) &= 2\ R\ \varsigma H/(Eh) = 1.392 \times 10^{-3}\ \text{m} \\ \vartheta_H(\alpha) &= -2\ \varsigma^2 H/(Eh) = -0.813 \times 10^{-3} \end{aligned} \tag{13.22}$$

Both the total translation $w^{\text{total}}(\alpha)$ and rotation $\vartheta^{\text{total}}(\alpha)$ (i.e. the sum of effects of H, M, ΔT_0) at the bottom edge of the clamped shell have to be zero. This leads to two linear algebraic equations in the following form:

$$\begin{aligned} w^{\text{total}} &= 2\ R\ \varsigma/(Eh)\,H - 2\ \varsigma^2/(Eh)\,M + R\ \alpha_T\ \Delta T_0 = 0 \\ \vartheta^{\text{total}} &= -2\ \varsigma^2/(Eh)\,H + 4\ \varsigma^3/(ERh)\,M + 0 = 0 \end{aligned} \tag{13.23}$$

For the numerical data included in Box 13.3 the six numbers computed here are the coefficients of two algebraic equations with unknowns H and M

$$10^{-3}\begin{bmatrix} 1.392 & -0.813 \\ -0.813 & 0.951 \end{bmatrix} \begin{bmatrix} H \\ M \end{bmatrix} = \begin{bmatrix} -0.05 \\ 0.0 \end{bmatrix} \tag{13.24}$$

The values of distributed boundary moment M and horizontal force H are determined as:

$$M = -61.48\ \text{kNm/m}, \qquad H = -71.86\ \text{kN/m} \tag{13.25}$$

and shown in Figure 13.3d.

The function of meridional bending moment is defined by the formula known from the description of the local bending state in spherical caps (see Section 11.3) in the vicinity of the bottom edge

$$m_\varphi(\varphi_1) = \mathrm{e}^{-\varsigma\ \varphi_1}[61.48\ \sin(\varsigma\varphi_1) - 61.48\ \cos(\varsigma\varphi_1)] \tag{13.26}$$

with the boundary value derived for the angular coordinate $\varphi_1 = 0°$ (φ_1 starts from the bottom edge). Taking into account $\varphi_1 = \alpha - \varphi$ (where the angular coordinate φ is measured from the vertical axis) and $\varphi = \alpha = 90°$ the meridional bending moment is expressed as

$$m_\varphi(\varphi_1)|_{\varphi_1=0°} = M = -61.48\ \text{kNm/m} = m_\varphi(\varphi)|_{\varphi=\alpha} \tag{13.27}$$

The meridional transverse shear force on the shell edge has the value

$$t_\varphi(\varphi_1)|_{\varphi_1=0°} = -H = 71.86\ \text{kN/m} = -t_\varphi(\varphi)|_{\varphi=\alpha} \tag{13.28}$$

The numerical analysis of the considered example has been performed using package ANSYS. The diagrams of membrane forces $n_\varphi(\varphi)$, $n_\theta(\varphi)$, bending moments $m_\varphi(\varphi)$, $m_\theta(\varphi)$ and transverse shear force $t_\varphi(\varphi)$ are shown in Figure 13.4 for $\varphi \in [0°, 90°]$. The selected values obtained from the numerical computations are listed in Box 13.3 and used to compare results of analytical and numerical analysis: (i) extreme meridional bending moments extending the internal fibres at the bottom contour and compressing the external fibres above the support and (ii) meridional transverse shear force at the boundary.

Note that the meridional bending moments m_φ^{anal} and m_φ^{FEM} as well as transverse shear forces t_φ^{anal} and t_φ^{FEM} have opposite signs due to different sign convention in our theoretical consideration and ANSYS computations, see Box 13.3. Analysing the presented diagrams one can notice that the bending effects occur only in a close neighbourhood of the clamped edge, while in the remaining part of the shell zero membrane and transverse shear forces as well as zero bending moments occur.

13.3 Linear Temperature Change along the Thickness – Analytical Solutions

The second type of thermal load is defined as temperature change function $\Delta T(z)$, linear in the thickness direction and described by the given value of temperature increment ΔT_{h}:

$$\Delta T(z) = k\, z = \Delta T_{\text{h}}\, z/h, \qquad k = \Delta T_{\text{h}}/h \tag{13.29}$$

The following notation is used:

- for bending plates with axis z directed downwards (see Figure 13.5a):

$$\begin{aligned} &\Delta T_{\text{b}}(h/2) = \Delta T_{\text{h}}/2, \qquad \Delta T_{\text{t}}(-h/2) = -\Delta T_{\text{h}}/2 \\ &\Delta T_{\text{h}} = \Delta T_{\text{b}}(h/2) - \Delta T_{\text{t}}(-h/2) > 0 \end{aligned} \tag{13.30}$$

- for curved shells with axis z directed outwards (see Figure 13.5b):

$$\begin{aligned} &\Delta T_{\text{ext}}(h/2) = \Delta T_{\text{h}}/2, \qquad \Delta T_{\text{int}}(-h/2) = -\Delta T_{\text{h}}/2 \\ &\Delta T_{\text{h}} = \Delta T_{\text{ext}}(h/2) - \Delta T_{\text{int}}(-h/2) \end{aligned} \tag{13.31}$$

The linear change of temperature $\Delta T(z)$ induces thermal curvatures in plates and changes of curvatures in shells

$$\kappa_{11,\Delta T_{\text{h}}} = \kappa_{22,\Delta T_{\text{h}}} = \alpha_T\, k = \alpha_T\, \Delta T_{\text{h}}/h = \text{const.} \tag{13.32}$$

On any surface equidistant from the middle surface, that is for $z \in [-h/2, +h/2]$ the following strains occur:

$$\varepsilon_{11}(z) = \varepsilon_{22}(z) = z\, \kappa_{11,\Delta T_{\text{h}}} = z\, \kappa_{22,\Delta T_{\text{h}}} = z\, \alpha_T\, k = \alpha_T\, \Delta T_{\text{h}}\, z/h \tag{13.33}$$

In the next subsections the effects induced by the linear temperature change along the thickness in circular and rectangular plates, cylindrical and spherical shells are described. The subscripts 11, 22 are replaced by indices appropriate for the coordinate system used.

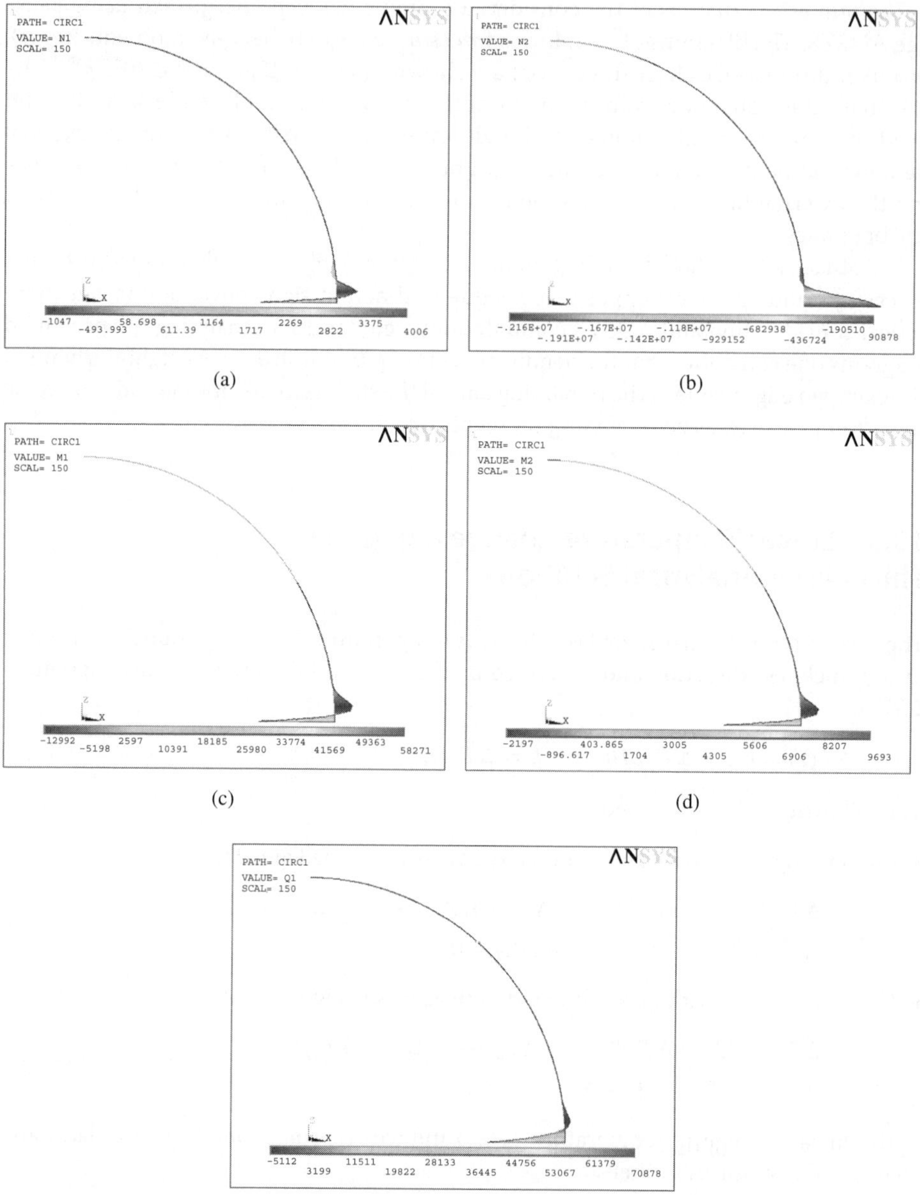

Figure 13.4 Hemisphere clamped on bottom edge and uniformly heated – diagrams of: (a) meridional membrane force $n_\varphi(\varphi)$, extreme value $n_\varphi^{\rm FEM}(\alpha) = 4006$ N/m, (b) circumferential membrane force $n_\theta(\varphi)$, extreme value $n_\theta^{\rm FEM}(\alpha) = -2\,160\,000$ N/m, (c) meridional bending moment $m_\varphi(\varphi)$, extreme value $m_\varphi^{\rm FEM}(\alpha) = 58\,271$ Nm/m, (d) circumferential bending moment $m_\theta(\varphi)$, extreme value $m_\theta^{\rm FEM}(\alpha) = 9693$ Nm/m and (e) meridional transverse shear force $t_\varphi(\varphi)$, extreme value $t_\varphi^{\rm FEM}(\alpha) = 70\,878$ N/m; FEM results from ANSYS

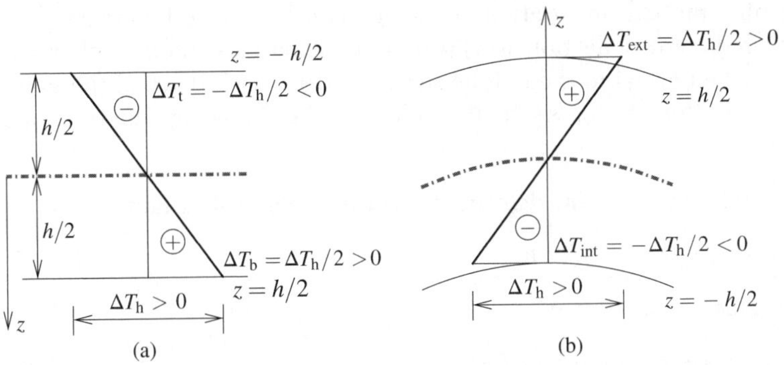

Figure 13.5 Notation for linear distribution of temperature change along thickness for: (a) flat plate and (b) curved shell

13.3.1 Simply Supported and Clamped Circular Plates

The following formulae describe the distribution of temperature:

$$\begin{aligned} &\Delta T(z) = k\,z = \Delta T_h\, z/h = (\Delta T_b - \Delta T_t)\, z/h \\ &\Delta T_b(h/2) = \Delta T_h/2 > 0, \qquad \Delta T_t(-h/2) = -\Delta T_h/2 < 0 \end{aligned} \tag{13.34}$$

The increment $\Delta T(0) = 0$ valid for the middle surface with $z = 0$ corresponds to the lack of thermal membrane effects.

Two types of circular plates with different constraints on external contours are considered, see also Timoshenko and Woinowsky-Krieger (1959) and Krzyś and Życzkowski (1962):

- case A – simply supported edge, see Figure 13.6a
- case B – clamped edge, see Figure 13.6b

13.3.1.1 Case A – circular plate with simply supported external contour

In an axisymmetric circular plate that is simply supported at the external contour, unrestricted axisymmetric thermal bending deformation occurs with the following thermal curvature in diametrical section $\theta =$ const.

$$\kappa_{r,\Delta T_h}(r) = -w'' = \alpha_T\, k = \alpha\, \Delta T_h/h = \text{const.}, \quad \text{where} \quad (\,)' = \mathrm{d}(\,)/\mathrm{d}r \tag{13.35}$$

Formula (13.35) can be treated as a second-order differential equation and its integral describes a shape of thermally deformed middle surface of the plate

$$w(r) = -\alpha_T\, k\, r^2/2 + C_1\, r + C_2 \tag{13.36}$$

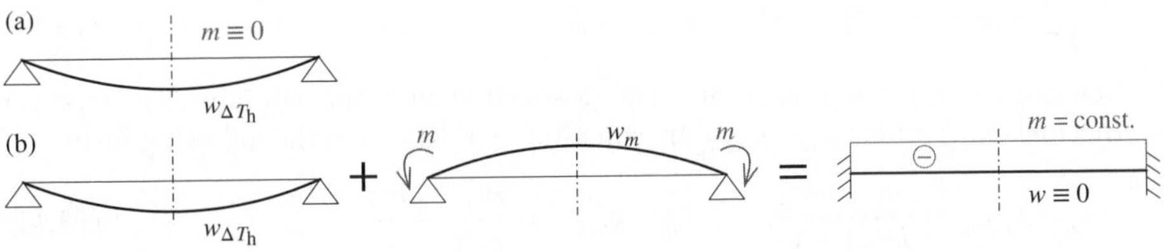

Figure 13.6 Influence of thermal loads for two analysed circular plates with: (a) simply supported edge and (b) clamped edge

In order to determine the two integration constants, two kinematic boundary conditions are formulated: zero deflection $w(a) = 0$ at the external contour with $r = a$ and zero rotation of the normal to the middle surface $\vartheta_r(0) = -w'(0) = 0$ at the point on the symmetry axis (for $r = 0$). As to the solution, the following functions are determined:

- deflection function for the plate (its deformed shape is a spherical surface)

$$w_{\Delta T_h}(r) = \alpha_T \, \Delta T_h (a^2 - r^2)/(2\,h) \tag{13.37}$$

with maximum value

$$w_{\Delta T_h,\max} = w(0) = \alpha_T \, \Delta T_h \, a^2/(2\,h) \tag{13.38}$$

- function of normal rotation

$$\vartheta_{r,\Delta T_h}(r) = -w'(r) = -\alpha_T \, \Delta T_h \, r/h \tag{13.39}$$

and its value at external contour

$$\vartheta_{r,\Delta T_h}(a) = -w'(a) = -\alpha_T \, \Delta T_h \, a/h \tag{13.40}$$

- constant curvatures:

$$\begin{aligned} \kappa_{r,\Delta T_h}(r) &= -w'' = \alpha_T \, \Delta T_h/h = \text{const.} \\ \kappa_{\theta,\Delta T_h}(r) &= -w'/r = \alpha_T \, \Delta T_h/h = \text{const.} \end{aligned} \tag{13.41}$$

- two bending moments identically equal to zero

$$m = m_{r,\Delta T_h}(r) = m_{\theta,\Delta T_h}(r) \equiv 0 \tag{13.42}$$

Conclusion: the thermal plate deformation determined for case A involves nonzero deflections and curvatures, despite the fact that the bending moments and stresses equal zero.

13.3.1.2 Case B – circular plate with clamped external contour

The analysis of the clamped plate is more complex. In order to obtain zero rotation of normal at the external clamped contour $\vartheta_r(a) = 0$ in the final deformation state, nonzero bending moments (radial and circumferential) have to occur in the plate. They reduce the rotation $\vartheta_{r,\Delta T_h}(a)$, caused by thermal load, see Figure 13.6b.

In the axisymmetric state of the circular plate bending moments $m_r(r)$ and $m_\theta(r)$ have to satisfy the differential equilibrium equation (homogeneous in the case of zero transverse shear forces)

$$\frac{\mathrm{d}}{\mathrm{d}r}[m_r(r) + m_\theta(r)] = -(1 - \nu)\, t_r(r) = 0 \tag{13.43}$$

The bending moments $m_r(r)$ and $m_\theta(r)$ are associated with final curvatures $\kappa_r(r)$, $\kappa_\theta(r)$ and thermal curvatures $\kappa_{r,\Delta T_h}$, $\kappa_{\theta,\Delta T_h}$ by constitutive relations in the following form:

$$\kappa_r - \kappa_{r,\Delta T_h} = \frac{m_r - \nu m_\theta}{D^m(1-\nu)}, \qquad \kappa_\theta - \kappa_{\theta,\Delta T_h} = \frac{m_\theta - \nu m_r}{D^m(1-\nu)} \tag{13.44}$$

$$m_r(r) + m_\theta(r) = (\kappa_r - \kappa_{r,\Delta T_h} + \kappa_\theta - \kappa_{\theta,\Delta T_h})\, D^m(1+\nu) \tag{13.45}$$

Substituting the sum of these moments into the differential equilibrium Equation (13.43) the following relationship is obtained

$$\frac{\mathrm{d}}{\mathrm{d}r}(\kappa_r + \kappa_\theta) = \frac{\mathrm{d}}{\mathrm{d}r}(\kappa_{r,\Delta T_\mathrm{h}} + \kappa_{\theta,\Delta T_\mathrm{h}}) \tag{13.46}$$

On the right-hand side of Equation (13.46) constant curvatures $\kappa_{r,\Delta T_\mathrm{h}} = \kappa_{\theta,\Delta T_\mathrm{h}} = \alpha_T\, k$ appear, hence their derivatives are equal to zero. Curvatures $\kappa_r(r) = -w''$ and $\kappa_\theta(r) = -w'/r$ can be related to the final deflection function $w(r)$, leading to a new form of Equation (13.46)

$$\frac{\mathrm{d}}{\mathrm{d}r}\left(-\frac{\mathrm{d}^2 w}{\mathrm{d}r^2} - \frac{1}{r}\frac{\mathrm{d}w}{\mathrm{d}r}\right) = 0 \tag{13.47}$$

Triple integration of Equation (13.47) results in the following form of the deflection function

$$w(r) = C_1\, r^2/4 + C_2\, \log\, r + C_3 \tag{13.48}$$

For the circular plate clamped at its external contour, two kinematic boundary conditions $w(a) = 0$ and $w'(a) = 0$ have to be satisfied and an additional condition on the finite value of deflection $w(0)$ at the point on the symmetry axis must be ensured. As a result, three zero constants $C_1 = C_2 = C_3 = 0$ are determined and thus the deflection is equal to zero in the whole plate domain

$$w(r) \equiv 0 \tag{13.49}$$

To make the concept clear, next we provide the reasoning based on the superposition principle, see Figure 13.6b. Two thermal curvatures $\kappa_{r,\Delta T_\mathrm{h}}$ and $\kappa_{\theta,\Delta T_\mathrm{h}}$ correspond to the (positive) deflection function

$$w_{\Delta T_\mathrm{h}}(r) = \alpha_T\, \Delta T_\mathrm{h}\, (a^2 - r^2)/(2\, h) \tag{13.50}$$

and two zero bending moments $m_{r,\Delta T_\mathrm{h}}$ and $m_{\theta,\Delta T_\mathrm{h}}$. In the plate additional bending moments occur

$$m = m_r = m_\theta = -D^m(1+\nu)\, \Delta T_\mathrm{h}\, \alpha_T/h = -\frac{E\, \alpha_T\, \Delta T_\mathrm{h}\, h^2}{12(1-\nu)} = \text{const.} \tag{13.51}$$

and induce a (negative) deflection field

$$w_m(r) = -\alpha_T\, \Delta T_\mathrm{h}\, (a^2 - r^2)/(2\, h) \tag{13.52}$$

Upon superposition the analysed plate has zero deflection

$$w(r) = w_{\Delta T_\mathrm{h}}(r) + w_m(r) \equiv 0 \tag{13.53}$$

The negative sign of both the radial and circumferential moments as in Equation (13.51), with the adopted changes of temperature $\Delta T_\mathrm{t} < 0$ and $\Delta T_\mathrm{b} > 0$, indicates tensile stresses

Box 13.4 Simply supported (case A) and clamped (case B) circular plates with a linear distribution of temperature change along the thickness

Data

$E = 1.5 \times 10^7$ kN/m^2, $\nu = 0.1667$, $r_{\text{ext}} = a = 3.0$ m, $h = 0.12$ m

$\Delta T_{\text{h}} = 10$ deg, $\Delta T_{\text{b}} = 5$ deg, $\Delta T_{\text{t}} = -5$ deg, $\alpha_T = 10^{-5}$ 1/deg

---

Check values from analytical solution

A $\quad w_{\Delta T_{\text{h}}}(0) = 0.00375$ m, $\quad \kappa_{\Delta T_{\text{h}}} = 0.000835$ 1/m

$m_{r,\Delta T_{\text{h}}}(r) = m_{\theta,\Delta T_{\text{h}}}(r) \equiv 0$

B $\quad w(r) \equiv 0$

$m_r(r) = m_\theta(r) = -2.161$ kNm/m = const.

$\sigma_r^{\max} = \sigma_\theta^{\max} = 900.0$ kN/m^2

Circular plate configurations shown in Figure 13.6

at the top surface $z = -h/2$

$$\sigma_{r,\Delta T_{\text{h}}}^{\max} = \sigma_{\theta,\Delta T_{\text{h}}}^{\max} = \sigma(-h/2) = -\frac{6}{h^2}\, m = \left(-\frac{6}{h^2}\right)\left(-\frac{E\,\alpha_T\,\Delta T_{\text{h}}\,h^2}{12(1-\nu)}\right) = \frac{E\,\alpha_T\,\Delta T_{\text{h}}}{2(1-\nu)} \tag{13.54}$$

and compression at the bottom surface $z = h/2$.

The values of E, ν, a, h, ΔT_{h}, α_T used in the calculations for both the cases and their selected numerical results are included in Box 13.4.

Summary:

- The behaviour of simply supported circular plate is characterized by nonzero deflection $w(r)$ and zero bending moments $m_{r,\Delta T_{\text{h}}}(r)$, $m_{\theta,\Delta T_{\text{h}}}(r)$, see Box 13.4.
- In the clamped circular plate zero deflection w in the whole plate is obtained and two nonzero constant moments $m_r(r)$, $m_\theta(r)$ (resulting in extreme stresses at the limiting surfaces $z = \pm h/2$) take place, see Box 13.4.

13.3.2 Simply Supported Rectangular Plate

In the case of thermal load described by $\Delta T(z)$ the analytical method based on expansions in double and single trigonometric series (see Section 8.7) can also be used. The analysis starts from the fourth-order differential equation of bending plate

$$\nabla^2\nabla^2 w(x,y) = -(1-\nu)\,\alpha_T\,\nabla^2[\Delta T_{\text{h}}(x,y)/h] \tag{13.55}$$

with an appropriate right-hand side. A lot of example solutions can be found in the work of Reddy (1999), among others, for an isotropic simply supported plate subjected to a temperature field constant with respect to coordinates x, y and linearly dependent on coordinate z, as defined in Equation (13.29).

Omitting detailed derivations the final deflection function according to Reddy (1999) reads

$$w(x,y) = \frac{16\alpha_T\ \Delta T_{\rm h}(1+\nu)\ b^2}{\pi^4 h} \sum_{i=1}^{I}\sum_{j=1}^{J} \frac{1}{(m^2 s^2 + n^2)\ m\ n} \sin\frac{m\pi x}{a} \sin\frac{n\pi y}{b} \tag{13.56}$$

where $s = a/b$. Taking the following data:

$$\begin{aligned} &E = 1.5\times 10^7\ \text{kN/m}^2, \quad \nu = 0.25, \quad a = b = 3.0\ \text{m}, \quad h = 0.12\ \text{m} \\ &\Delta T_{\rm h} = 10\,\text{deg}, \quad \alpha_T = 10^{-5}\ 1/\text{deg} \end{aligned} \tag{13.57}$$

two characteristic values are calculated using the formulae and table from Reddy (1999):

- maximum deflection of the plate at the central point equals

$$w_C = 0.00921\ \alpha_T\ \Delta T_{\rm h}\ a^2/h = 0.000069\ \text{m} \tag{13.58}$$

- the values of stresses at limiting surfaces with $z = \pm h/2$ are

$$|\sigma_{xx}| = |\sigma_{yy}| = 0.0024\,E\ \alpha_T\ \Delta T_{\rm h} = 36.0\ \text{kN/m}^2 \tag{13.59}$$

Regarding the effectiveness of the method, according to the work of Reddy (1999) the results of calculations almost do not change for the number of series components larger than 49.

13.3.3 Simply Supported Cylindrical Shell with Horizontal Movement of the Bottom Edge

The thermal load is described in Figure 13.7a by the following increments of temperature:

$$\Delta T_{\rm h} = \Delta T_{\rm ext} - \Delta T_{\rm int}, \qquad \Delta T_{\rm ext} = \Delta T_{\rm h}/2 > 0, \qquad \Delta T_{\rm int} = -\Delta T_{\rm h}/2 < 0 \tag{13.60}$$

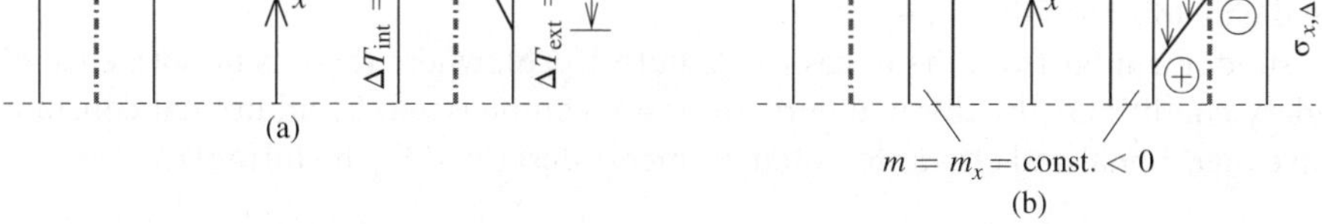

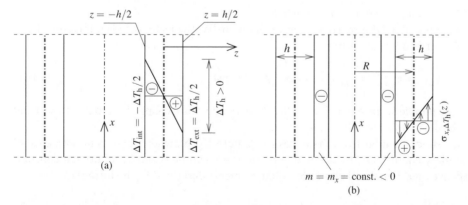

Figure 13.7 Cylindrical shell in stage I – subdomain far from edges: (a) description of linear change of temperature along thickness and (b) diagrams of constant meridional moment m_x along meridian and linear distribution of stress $\sigma_{x,\Delta T_{\rm h}}(z)$ along thickness

Box 13.5 Simply supported cylindrical shell with horizontal movement at the bottom edge admitted – linear distribution of temperature change along the thickness

Data

$E = 1.5 \times 10^7$ kN/m², $\nu = 0.1667$, $R = 10$ m, $L = 20.0$ m, $h = 0.12$ m

$\beta = 1.193$ 1/m, $\lambda = 2.633$ m

$\Delta T_{\mathrm{h}} = 10$ deg, $\Delta T_{\mathrm{ext}} = 5$ deg, $\Delta T_{\mathrm{int}} = -5$ deg, $\alpha_T = 10^{-5}$ 1/deg

---

Check values from analytical solution

(I) $\kappa_{\Delta T_{\mathrm{h}}} = 0.00083$ 1/m

$m_{x,\Delta T_{\mathrm{h}}} = m_{\theta,\Delta T_{\mathrm{h}}} = -2.161$ kNm/m = const.

$\sigma_{x,\Delta T_{\mathrm{h}},\mathrm{int}} = \sigma_{\theta,\Delta T_{\mathrm{h}},\mathrm{int}} = 900.0$ kN/m²

$\sigma_{x,\Delta T_{\mathrm{h}},\mathrm{ext}} = \sigma_{\theta,\Delta T_{\mathrm{h}},\mathrm{ext}} = -900.0$ kN/m²

(II) $w(0) = -0.000342$ m $n_\theta(0) = -61.5$ kN/m

$M = +2.161$ kNm/m $= -m_{x,\Delta T_{\mathrm{h}}}$, $m_x(0) = 0$, $t_x(0) = 0$

$\sigma_{\theta,\mathrm{ext}}(0) = -1262$ kN/m²

Shell configuration shown in Figure 13.7

This issue is discussed in the work of Timoshenko and Woinowsky-Krieger (1959). The dimensions and parameters describing the material and thermal load are included in Box 13.5.

The solution of the problem is obtained in two stages (I and II).

13.3.3.1 Stage I – effect of thermal load

First, unrestricted thermal deformation of the shell with unsupported ends is considered and described by curvatures

$$\kappa_{\Delta T_{\mathrm{h}}} = \kappa_{x,\Delta T_{\mathrm{h}}} = \kappa_{\theta,\Delta T_{\mathrm{h}}} = \alpha_T \, \Delta T_{\mathrm{h}}/h = \text{const.} \tag{13.61}$$

On the middle surface ($z = 0, x \in [0, L]$) the temperature change is zero $\Delta T(0) = 0$ and hence the membrane effects do not occur.

In a subdomain far from the edges (see Figure 13.7b), where there is no influence of boundary conditions, the thermal curvatures are compensated by additional constant negative meridional and circumferential moments described by the formulae

$$\begin{aligned} m = m_{x,\Delta T_{\mathrm{h}}} = m_{\theta,\Delta T_{\mathrm{h}}} &= -D^m(1+\nu)\,\kappa_{\Delta T_{\mathrm{h}}} \\ &= -D^m(1+\nu)\,\alpha_T\,\Delta T_{\mathrm{h}}/h = -E\,\alpha_T\,\Delta T_{\mathrm{h}}\,h^2/[12(1-\nu)] \end{aligned} \tag{13.62}$$

The bending moments are the source of:

- tension on the internal surface ($z = -h/2$) both in meridional and circumferential direction

$$\sigma_{x,\Delta T_{\rm h},{\rm int}} = \sigma_{\theta,\Delta T_{\rm h},{\rm int}} = \frac{E\,\alpha_T\,\Delta T_{\rm h}}{2(1-\nu)} \tag{13.63}$$

- compressive stresses on the external surface ($z = h/2$)

$$\sigma_{x,\Delta T_{\rm h},{\rm ext}} = \sigma_{\theta,\Delta T_{\rm h},{\rm ext}} = -\frac{E\,\alpha_T\,\Delta T_{\rm h}}{2(1-\nu)} \tag{13.64}$$

The calculated characteristic values of thermal curvatures, bending moments and stresses on the limiting surfaces for stage I are listed in Box 13.5.

13.3.3.2 Stage II – effect of thermal load and boundary conditions

Now, the boundary support of the shell is taken into account. If the bottom edge of the shell ($x = 0$) is simply supported with $u(0) = 0$ then the normal displacement $w(0)$ and rotation $w'(0)$ are nonzero and two homogeneous static boundary conditions have to be satisfied, where $(\ )' = {\rm d}(\ )/{\rm d}x$. It is required that the final values of the meridional moment and meridional transverse shear force are equal to zero:

$$m_x(0) = 0, \qquad t_x(0) = 0 \tag{13.65}$$

In the previous stage the constant value of meridional moment induced by temperature change in the whole domain of the shell and thus along edge $x = 0$ has been determined

$$m_{x,\Delta T_{\rm h}} = -\frac{E\,\alpha_T\,\Delta T_{\rm h}\,h^2}{12(1-\nu)} = {\rm const.} \tag{13.66}$$

Now, it has to be balanced by uniformly distributed boundary moment M at contour $x = 0$

$$M = \frac{E\,\alpha_T\,\Delta T_{\rm h}\,h^2}{12(1-\nu)} \tag{13.67}$$

which has opposite sign in comparison to $m_{x,\Delta T_{\rm h}}$ to obtain $m_x(0) = m_{x,\Delta T_{\rm h}} + M = 0$.

The boundary moment M becomes a source of local bending state, coupled with normal displacement field described by function (see Chapter 11)

$$w_M(x) = -e^{-\beta x}[B_1\cos(\beta x) + B_2\sin(\beta x)], \qquad \text{where} \qquad \beta = \sqrt{\frac{1}{R\,h}}\,\sqrt[4]{3(1-\nu^2)} \tag{13.68}$$

The formulae given next are consistent with the sign of boundary moment M shown in Figure 13.8b. Only two integration constants B_1 and B_2 are considered if $3\lambda < L$ for a long shell, where $\lambda = \pi/\beta$. They are determined from two static homogeneous boundary conditions:

$$m_x(0) = 0, \qquad t_x(0) = [m_x(x)]'|_{x=0} = 0 \tag{13.69}$$

The action of boundary moment $M = 2.161$ kNm/m results in the following effects in the neighbourhood of the bottom edge $x = 0$, described by five formulae written for the

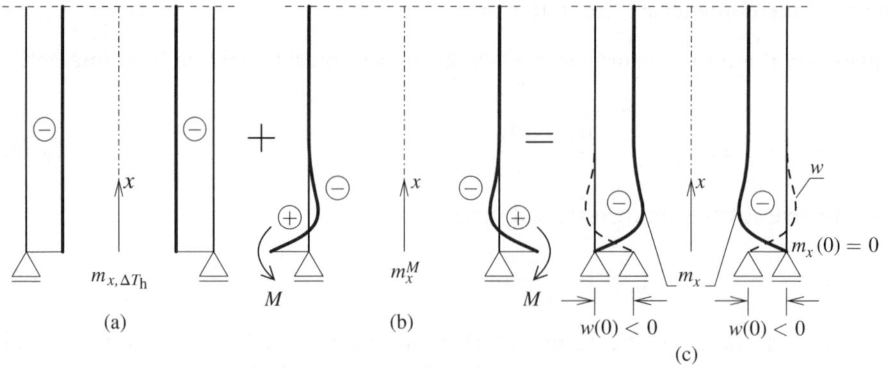

Figure 13.8 Cylindrical shell (subdomain near bottom edge) – superposition of effects resulting from: (a) thermal load $\Delta T_h > 0$, (b) boundary moment M and (c) two functions of meridional moment $m_x(x)$ and normal displacement $w(x)$ in the neighbourhood of simply supported moveable contour

data included in Box 13.5:

$$\begin{aligned}
w_M(x) &= -\frac{1}{2\,D^m\,\beta^2}\,e^{-\beta x}[\cos(\beta x) - \sin(\beta x)]\,M \\
&= -0.000342\,e^{(-1.193\,x)}\,[\cos(1.193\,x) - \sin(1.193\,x)] \\
m_x^M(x) &= e^{-\beta x}[\cos(\beta x) + \sin(\beta x)]\,M \\
&= 2.161\,e^{(-1.193\,x)}\,[\cos(1.193\,x) + \sin(1.193\,x)] \\
m_\theta^M(x) &= \nu\,m_x^M(x) \\
t_x^M(x) &= -2\,\beta\,e^{-\beta x}\,\sin(\beta x)\,M = -5.156\,e^{(-1.193\,x)}\,\sin(1.193\,x) \\
n_\theta^M(x) &= \frac{E\,h}{R}\,w_M(x) = -\frac{E\,h}{2\,R\,D^m\,\beta^2}\,e^{-\beta x}[\cos(\beta x) - \sin(\beta x)]\,M \\
&= -61.5\,e^{(-1.193\,x)}\,[\cos(1.193\,x) - \sin(1.193\,x)]
\end{aligned} \tag{13.70}$$

The following characteristic final boundary magnitudes are calculated for $x = 0$:

- negative horizontal translation (inwards the shell)

$$w(0) = -\frac{R\,\alpha_T\,\Delta T_h\,\sqrt{1-\nu^2}}{2\,(1-\nu)\sqrt{3}} \tag{13.71}$$

- zero value of meridional bending moment and transverse shear force:

$$m_x(0) = m_{x,\Delta T_h} + M = 0, \qquad t_x(0) = 0 \tag{13.72}$$

- value of circumferential force

$$n_\theta(0) = \frac{E\,h}{R}\,w(0) = -\frac{E\,h\,\alpha_T\,\Delta T_h\,\sqrt{1-\nu^2}}{2\,(1-\nu)\sqrt{3}} \tag{13.73}$$

It is worthwhile to compute the final value of circumferential stress σ_θ at the bottom edge ($x = 0$) and at the external surface ($z = h/2$)

$$\sigma_{\theta,\text{ext}}(0) = \sigma_\theta^{(1)}(0) + \sigma_\theta^{(2)}(0) + \sigma_\theta^{(3)}(0) = -\frac{E\,\alpha_T\,\Delta T_h}{2(1-\nu)}\left[1 - \nu + \frac{\sqrt{1-\nu^2}}{\sqrt{3}}\right] \tag{13.74}$$

The three components result from:

- thermal stress from stage I, see Equation (13.64)

$$\sigma_\theta^{(1)}(0) = -\frac{E\,\alpha_T\,\Delta T_\mathrm{h}}{2(1-\nu)} = -900\ \mathrm{kN/m^2} \tag{13.75}$$

- action of circumferential moment $m_\theta(0) = \nu\, m_x(0) = \nu\, M$ caused by boundary moment M

$$\sigma_\theta^{(2)}(0) = \frac{6\, m_\theta(0)}{h^2} = \frac{\nu\, E\,\alpha_T\,\Delta T_\mathrm{h}}{2(1-\nu)} = 150\ \mathrm{kN/m^2} \tag{13.76}$$

- circumferential force $n_\theta(0)$ coupled with boundary moment M

$$\sigma_\theta^{(3)}(0) = \frac{n_\theta(0)}{h} = -\frac{E\,\alpha_T\,\Delta T_\mathrm{h}\,\sqrt{1-\nu^2}}{2(1-\nu)\sqrt{3}} = -512\ \mathrm{kN/m^2} \tag{13.77}$$

To finish the section we briefly consider an alternative case of boundary conditions. If the clamped shell is considered two stages must also be analysed. However, the change of supports affects the solution in stage II. In stage I the meridional and circumferential curvatures induced by the thermal load are the same as in the case of simply supported shell, see Equation (13.61). In stage II, the bending state resulting from kinematic constraints $w(0) = 0$ and $w'(0) = 0$ imposed at the clamped edge have to be taken into account. Thus, the analytical solution for the thermal load is analogous to that shown in Subsection 11.1.6, obtained for a static load.

13.3.4 Stresses in a Spherical Shell

Considering a shell with cooled external surface ($\Delta T_\mathrm{ext} < 0$) and heated internal surface ($\Delta T_\mathrm{int} > 0$), with zero temperature change on the middle surface, the following effects are observed:

- no change in the geometry of a spherical shell when its centre does not move
- thermal curvatures caused by $\Delta T_\mathrm{h} = \Delta T_\mathrm{ext} - \Delta T_\mathrm{int} < 0$ have to be eliminated by the curvatures generated by additional moments expressed as

$$m_{\varphi,\Delta T_\mathrm{h}} = m_{\theta,\Delta T_\mathrm{h}} = -D^m(1+\nu)\,\alpha_T\,\Delta T_\mathrm{h}/h = -\frac{E\,\alpha_T\,\Delta T_\mathrm{h}\,h^2}{12(1-\nu)} = \mathrm{const.} > 0 \tag{13.78}$$

The behaviour of a spherical shell with numerical data listed in Box 13.6 and for any value of radius R is described by constant meridional and circumferential moments, see Equation (13.78), and by the following tension stress at the external surface and compressive stress at the internal surface:

$$\begin{aligned} \sigma^{\max}_{\varphi,\Delta T_\mathrm{h},\mathrm{ext}} &= \frac{6\, m_{\varphi,\Delta T_\mathrm{h}}}{h^2} = \sigma^{\max}_{\theta,\Delta T_\mathrm{h},\mathrm{ext}} = \frac{6\, m_{\theta,\Delta T_\mathrm{h}}}{h^2} = -\frac{E\,\alpha_T\,\Delta T_\mathrm{h}}{2(1-\nu)} > 0 \\ \sigma^{\min}_{\varphi,\Delta T_\mathrm{h},\mathrm{int}} &= \sigma^{\min}_{\theta,\Delta T_\mathrm{h},\mathrm{int}} = \frac{E\,\alpha_T\,\Delta T_\mathrm{h}}{2(1-\nu)} < 0 \end{aligned} \tag{13.79}$$

At the end of this chapter we emphasize that using available software one has to be very careful about the notation and sign convention adopted in a particular computer package. In fact, the analytical solutions presented in this chapter can be useful for readers preparing the input data for shell structures under thermal loading and interpreting numerical results.

Box 13.6 Spherical shell with a linear temperature change along the thickness

Data

$E = 2.1 \times 10^7\ \text{kN/m}^2,\ \ \nu = 0.1667,\quad R - \text{optional},\quad h = 0.1\ \text{m},\ \ D^m = 1800\ \text{kNm}$

$\Delta T_\text{h} = -10\,\text{deg},\quad \Delta T_\text{ext} = -5\,\text{deg},\quad \Delta T_\text{int} = 5\,\text{deg},\quad \alpha_T = 10^{-5}\ 1/\text{deg}$

Check values from analytical solution

$m_{\varphi,\Delta T_\text{h}} = m_{\theta,\Delta T_\text{h}} = 2.1\ \text{kNm/m} = \text{const.}$

$\sigma_{\varphi,\Delta T_\text{h},\text{ext}} = \sigma_{\theta,\Delta T_\text{h},\text{ext}} = 1260\ \text{kN/m}^2,\quad \sigma_{\varphi,\Delta T_\text{h},\text{int}} = \sigma_{\theta,\Delta T_\text{h},\text{int}} = -1260\ \text{kN/m}^2$

References

ANSYS 2013 ANSYS Inc. PDF Documentation for Release 15.0. SAS IP, Inc.
Krzyś A and Życzkowski M 1962 *Elasticity and Plasticity*. PWN, Warsaw (in Polish).
Reddy JN 1999 *Theory and Analysis of Elastic Plates*. Taylor & Francis.
Timoshenko S and Woinowsky-Krieger S 1959 *Theory of Plates and Shells*. McGraw-Hill, New York-Auckland.

Part 4

Stability and Free Vibrations

14

Stability of Plates and Shells

14.1 Overview of Plate and Shell Stability Problems

The branch of mechanics related to the analysis of structural stability (or rather instability) is important and extensive. It contains many difficult terms and concepts that are necessary to describe complex phenomena connected with the loss of stability. In this book, we present selected key aspects of stability analysis of shell structures, because thin plates and shells are particularly exposed to the risk of instability. Stability considerations are crucial for engineers, supporting rational design of new structures and verification of the safety of existing ones.

The analysis of complex problems of structural stability is now possible due to progress of both the theory and computational methods (first of all the Finite Element Method). Theoretical considerations as well as analytical or numerical calculations require the formulation of complex models and the development of efficient calculation algorithms. The goal is achieved owing to advanced theoretical works and new capacities of numerical methods and computational tools.

In principle, two approaches are known in the structural stability analysis:

- classical linear buckling analysis
- nonlinear analysis of stability

The most general approach based on the analysis of motion stability (important in mechanical engineering) is not considered here. We focus on the stability of equilibrium state for conservative systems, which is usually valid for civil engineering structures.

In the mechanical description of structural stability problems several specific notions are introduced, see Waszczyszyn et al. (1994):

- structures are ideal (perfect) or with imperfections (imperfect)
- loads are spatially fixed (dead load) or follow the changing configuration of a structure (follower load)
- imperfections can have the character of geometrical or material inaccuracy or be related to a disturbing load
- the equilibrium state can be stable, unstable or neutral
- the loss of stability can occur by buckling (bifurcation) or by reaching a limit state with an associated snap-through phenomenon
- equilibrium paths in the load-displacement space are either primary (fundamental, prebuckling) or secondary (postbuckling, postcritical)

Plate and Shell Structures: Selected Analytical and Finite Element Solutions, First Edition.
Maria Radwańska, Anna Stankiewicz, Adam Wosatko and Jerzy Pamin.
© 2017 John Wiley & Sons Ltd. Published 2017 by John Wiley & Sons Ltd.

- different points on the equilibrium path can be distinguished: bifurcation or limit (upper, lower) points
- loads are called critical (buckling) or limit (maximum, minimum)
- linear or geometrically nonlinear analysis can be performed
- calculation methodology involves usually either a linear algebraic eigenvalue problem or an incremental-iterative (step-by-step) algorithm for tracing a nonlinear equilibrium path in the load-displacement space
- the simulation of deformation process is performed under either load, displacement or arc-length control

The most important task in the stability analysis is to determine bifurcation points or limit points on equilibrium paths (see Figure 14.1) and to calculate the respective critical and limit loads. These points are found as the first ones encountered on fundamental equilibrium paths, starting from the origin of the system (Λ, w_r) for an ideal structure. Moreover, a nonlinear fundamental path is shown, which has been traced using geometrically nonlinear analysis of a shell.

The bibliography of the structural stability analysis is very extensive: books and chapters in books, papers in scientific journals and conference proceedings. From the historical point of view, we need to cite one of the pioneering works – the doctoral thesis by Koiter (1945). The thesis was defended in Delft (The Netherlands) in 1945, but it became more commonly known after the English translation was published in 1967 as NASA Report TT F-10, 833, and in 1970 as AFFDL Report TR 70-25. It deals with the modern theory of structural stability, with theoretical investigations of postbuckling structural behaviour in the vicinity of critical states, as well as with the analysis of sensitivity of structures to initial imperfections.

A detailed discussion of the fundamental theoretical and computational issues associated with structural instability can be found in Timoshenko and Gere (1961), Brush and Almroth (1975), Waszczyszyn (1987), Waszczyszyn et al. (1994), where several models and methods are described both for linear and nonlinear analysis. The monograph on *Optimal Structural Design under Stability Constraints* by Gajewski and Życzkowski (1988) is dedicated to scientists and engineering designers.

The behaviour of a loaded shell structure depends on many factors. Therefore, the commonly used load-displacement space is sometimes expanded to a so-called event space. There, besides basic variables, some new variables and effects are included. There may be geometrical or material parameters, measures of imperfections, variants of load or support constraints and variants of control parameters. Many different links and relations between the variables and structural behaviour are analysed in either

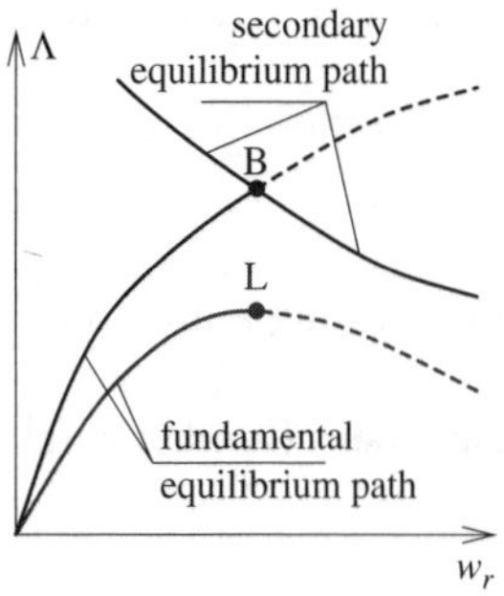

Figure 14.1 Example equilibrium paths with either bifurcation point B or limit point L, obtained in geometrically nonlinear analysis; Λ denotes a load factor and w_r is a selected representative displacement

multi-dimensional space, or on some intersections, or projections into a subspace. Various phenomena characterizing the structural behaviour are investigated via a parametric analysis, in which a chosen variable is changed in order to examine its influence. For example, various types of instability points on equilibrium paths are found this way, see Radwańska (1990), ANKA (1993).

Next, selected geometrical parameters are listed, which are crucial for the stability of shell structures. The following parameters are defined for plates: slenderness h/L, ratio of basic dimensions for rectangular plates L_x/L_y or for annular plates r_{int}/r_{ext}. For shells one uses the thickness to length ratio h/L, for shallow shells, the dimensionless rise f/L, and for cylindrical shells: thickness to radius ratio h/R, ratio R/L, as well as a so-called Batdorf's parameter $Z = L^2\sqrt{1-\nu}/(R\ h)$. For rotationally symmetric shells, the Gaussian curvature $K = k_1\ k_2 = 1/(R_1\ R_2)$ is also employed.

Different loading processes can be taken into account. In particular, one can consider:

- one-parameter load, parametrized by a single load factor Λ; this is either a load $\Lambda\mathbf{p}^*$ proportional to so-called reference load $\mathbf{p}^*$ or a sum of loads $\mathbf{p}_1 + \Lambda\mathbf{p}_2^*$ (with $\mathbf{p}_1$ fixed)
- multi-parameter load, which is a combination of some loads $\Lambda_i\mathbf{p}_i^*$, with independently growing factors Λ_i, see Waszczyszyn et al. (1994)

In the analysis of imperfect plates and shells the dimensionless amplitude of initial displacements w_0/h is used as a measure of geometrical imperfections. Alternatively various types of disturbing loads are applied.

In this chapter two model examples are chosen to present the essence of two different approaches to stability analysis (see Sections 14.2 and 14.3). Section 14.2 describes the buckling-type stability loss, relying on a linear theory. The detailed description of the numerical simulation of the loading process associated with an instability of a shallow cylindrical shell and a membrane/plate under unidirectional compression is presented in Section 14.3. It comprises the consideration and solution of both theoretical and computational problems. Some interesting examples of buckling analysis, selected because of the limited volume of this book, are presented in Section 14.4. At the end of this chapter, in Section 14.5, the results of advanced FEM nonlinear analysis of a shallow cylindrical shell loaded with a normal concentrated force are given.

14.2 Basis of Linear Buckling Theory, Assumptions and Computational Models

The considerations in this section are limited to rectangular membranes/plates and the Cartesian coordinate system. An example of a membrane under unidirectional compression is presented in Figure 14.2a. For this case (which will further be referred to as case C from Compression) we present the analytical solution in Subsection 14.2.1 and an approximate solution using the Finite Difference Method in Subsection 14.2.2.

Further cases of membranes under uniform shear (case S from Shear) or under in-plane bending (case B from Bending) are presented in Figures 14.2c,d and in Subsection 14.4.2.

To describe the buckling phenomenon one applies the static criterion of stability, analysing the types of equilibrium states of the structure. Buckling is a result of the stability loss which involves a bifurcation of equilibrium states, that is two qualitatively different equilibrium states (membrane and membrane-bending) become possible. At

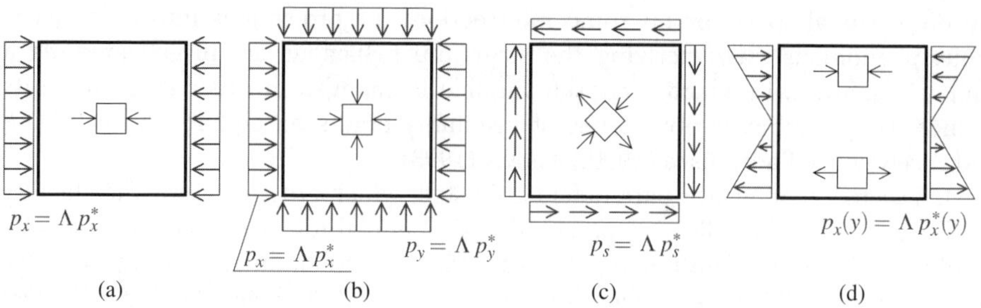

Figure 14.2 Examples of membranes susceptible to buckling under various loads

the bifurcation point the primary (fundamental) equilibrium path splits into secondary (postbuckling) paths, see Figure 14.1. In fact, the postbuckling path $\Lambda(w_r)$, where w_r is a representative displacement, can be determined only using the equations of geometrically nonlinear analysis.

In engineering practice the first bifurcation point encountered while tracing the equilibrium path under a growing load (with increasing load parameter Λ) is the most interesting. The minimum load for which the bifurcation takes place is called critical.

In theory, see for instance Brush and Almroth (1975), the nonlinear interaction between membrane forces and rotations characterizing the bending state is considered. The equations representing the balance of forces and moments must be derived for a membrane-plate element in a slightly deformed configuration (the assumption of small deformations is dropped). The angles of rotation are still small, and sines and cosines of the angles may be replaced by the angles themselves and by unity, respectively.

In Chapter 8 the equilibrium equation $\sum P_z = 0$ for a rectangular plate in bending has been derived, which is the linear displacement differential equation (8.19). Now, the summation of forces in the z direction contains the contribution of membrane forces represented by substitute transverse load $\overline{p}_z$

$$\overline{p}_z = n_x \frac{\partial^2 w}{\partial x^2} + 2n_{xy} \frac{\partial^2 w}{\partial x\, \partial y} + n_y \frac{\partial^2 w}{\partial y^2} \tag{14.1}$$

Equation (8.19) is thus modified by adding $\overline{p}_z$ and setting the transverse surface load $\hat{p}_z$ to zero. This leads to the following equation of a plate with in-plane loading

$$D^m\, \nabla^2\nabla^2 w(x,y) - \left(n_x \frac{\partial^2 w}{\partial x^2} + 2n_{xy} \frac{\partial^2 w}{\partial x\, \partial y} + n_y \frac{\partial^2 w}{\partial y^2} \right) = 0 \tag{14.2}$$

It is a homogeneous equation with variable coefficients $n_x(x,y), n_{xy}(x,y), n_y(x,y)$. The equation is nonlinear if the coefficients depend on deflection $w(x,y)$. In our consideration the membrane forces do not depend on w and are determined from the linear equations of the membrane state.

The linear analysis of buckling is based on the following three assumptions:

- Load is conservative (spatially fixed) and grows with one-parameter, that is a monotonic increase of one parameter Λ causes a proportional change of the whole applied load; hence Λ is a multiplier of a so-called reference load $\mathbf{p}^*$; for example, in bi-directional compression (Figure 14.2b) the loads p_x, p_y are proportional to one parameter Λ:

$$p_x = \Lambda p_x^*, \quad p_y = \Lambda p_y^*, \quad \mathbf{p}^* = [p_x^*, p_y^*] \tag{14.3}$$

- Before buckling the membrane is perfect, that is it has an ideal middle plane and the load acts exactly in this plane.
- When a bifurcation of equilibrium states occurs membrane forces, membrane strains and in-plane displacements do not change, however, bending effects appear additionally: deflections, curvatures and bending moments. The membrane forces are independent of deflection function $w(x, y)$ and are calculated as products of load parameter Λ and the reference membrane forces induced by reference load $\mathbf{p}^*$:

$$n_x = \Lambda\, n_x^*, \quad n_y = \Lambda\, n_y^*, \quad n_{xy} = \Lambda\, n_{xy}^* \tag{14.4}$$

In the general case of prebuckling membrane state, the equation of buckling for a rectangular plate takes the form

$$D^m\, \nabla^2\nabla^2 w(x,y) - \Lambda \left(n_x^* \frac{\partial^2 w}{\partial x^2} + 2 n_{xy}^* \frac{\partial^2 w}{\partial x\, \partial y} + n_y^* \frac{\partial^2 w}{\partial y^2} \right) = 0 \tag{14.5}$$

The load parameter Λ and the buckling mode – deflection function $w(x, y)$ are the unknowns in this equation. The minimum value of Λ is usually called the critical load parameter Λ_{cr}

$$\Lambda_{cr} = \Lambda_{min} \tag{14.6}$$

In summary, the buckling problem is solved in two stages:

- Stage I is the analysis of the membrane prebuckling state, which serves the purpose of determination of the reference membrane forces n_x^*, n_y^*, n_{xy}^* in the membrane state caused by reference load $\mathbf{p}^*$.
- Stage II is the buckling analysis in which the linear homogeneous fourth-order differential equation (14.5) is solved, taking into account the boundary constraints for the bending state.

For this linear differential equation with the right-hand side of zero value, the formulation of the existence condition for nonzero solution $w(x, y)$ is required. This way one arrives at the so-called stability equation. Most frequently, only the critical (minimum) value of the load parameter Λ_{cr} and the buckling mode are computed from this equation.

The linearized Equation (14.5) provides the exact values of load parameters for consecutive bifurcation points, but the deflection modes are determined with an arbitrary multiplier. Considering the plane (Λ, w_b), see Figure 14.3, the postcritical path for the linear analysis is the horizontal line $\Lambda = \Lambda_{cr}$. It is an envelope of nonlinear equilibrium paths representing membrane-bending states for imperfect plates (initially deflected due to geometrical imperfection or transverse load).

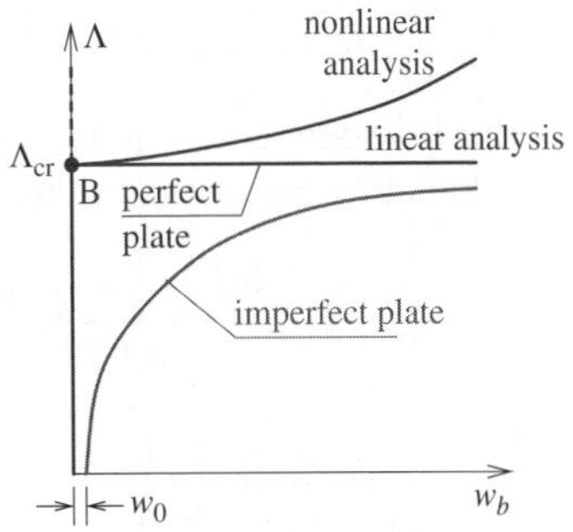

Figure 14.3 Example equilibrium paths for a plate: pre- and post-buckling paths for a perfect plate and membrane-bending path for an imperfect plate with initial deflection w_0; w_b denotes the active postbuckling deflection

14.2.1 Analytical Solution to the Buckling Problem for a Square Membrane under Unidirectional Compression

The solution of the buckling problem for the analysed membrane is divided into two stages described next.

Stage I (analysis of membrane prebuckling state): in the case of rectangular membrane under unidirectional compression with reference load p_x^*, the only nonzero membrane force is $n_x^* = -p_x^*$.

Stage II (buckling analysis): for the considered case the buckling Equation (14.5) is significantly simplified after the introduction of this membrane force, and takes the form

$$D^m \, \nabla^2\nabla^2 w(x,y) - \Lambda \left(n_x^* \frac{\partial^2 w}{\partial x^2} \right) = 0 \tag{14.7}$$

The solution is obtained in terms of a double trigonometric series that satisfies the boundary conditions of simple support along the entire contour of the plate in the post-buckling bending state

$$w(x,y) = \sum_{i=1}^{I} \sum_{j=1}^{J} W_{ij} \sin(\alpha_i x) \sin(\beta_j y), \quad \text{where} \quad \alpha_i = i\pi/a, \;\; \beta_j = j\pi/b \tag{14.8}$$

After this function is substituted into the differential equation, the homogeneous algebraic equation is obtained

$$\sum_{i=1}^{I} \sum_{j=1}^{J} \left[D^m \left(\alpha_i^2 + \beta_j^2 \right)^2 - \Lambda \, \alpha_i^2 \, p_x^* \right] W_{ij} \sin(\alpha_i \, x) \sin(\beta_j \, y) = 0 \tag{14.9}$$

with the following unknowns: values of load parameter Λ and scalar factors W_{ij} being amplitudes of the deflection functions, associated with a combination of the numbers of half-waves i and j (in directions x and y, respectively) of the curved middle surface of the plate.

For each pair (i, j) two solutions (strictly connected with the defined two stages) are possible:

(I) $W_{ij} = 0$, which leads to zero deflections corresponding to the membrane prebuckling state

(II) $W_{ij} \neq 0$, which involves nonzero postbuckling deflections in the bending state, but then the condition of zero value of the expression in square brackets in Equation (14.9) must be imposed

$$D^m(\alpha_i^2 + \beta_j^2)^2 - \Lambda \, \alpha_i^2 \, p_x^* = 0 \tag{14.10}$$

This homogeneous equation is the so-called stability equation and serves the purpose of calculating the critical value of the load factor

$$\begin{aligned} \Lambda_{\text{cr}} &= D^m \frac{(\alpha_i^2 + \beta_j^2)^2}{\alpha_i^2 \, p_x^*} = \frac{D^m \, \pi^2 \, a^2}{i^2 \, p_x^*} \left(\frac{i^2}{a^2} + \frac{j^2}{b^2} \right)^2 \\ &= \frac{D^m \, \pi^2}{p_x^* \, b^2} \left(i \frac{b}{a} + \frac{a}{i \, b} \right)^2 = \frac{D^m \, \pi^2}{p_x^* \, b^2} \left(\frac{i}{\mu} + \frac{\mu}{i} \right)^2 = D \, k(\mu; i) \end{aligned} \tag{14.11}$$

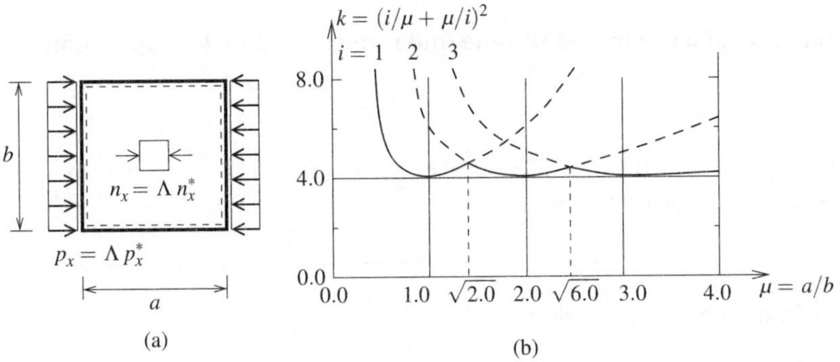

Figure 14.4 Rectangular membrane under unidirectional compression: (a) configuration and (b) diagram of function $k(\mu; i)$

where:

$$\mu = \frac{a}{b}, \quad j = 1, \quad D = \frac{D^m \pi^2}{p_x^* \, b^2}, \quad k(\mu; i) = \left(\frac{i}{\mu} + \frac{\mu}{i}\right)^2 \tag{14.12}$$

The substitution of $j = 1$ (one half-wave in direction y) leads to function $k(\mu; i)$ presented in Figure 14.4b as diagrams $k(\mu)$ for a sequence of values $i = 1, 2, 3, \ldots$ As can be seen in the parabolic diagrams $k_{\min} = 4.0$ independently of the number i of half-waves in direction x and this value of k for a membrane with dimension proportion $\mu = a/b = i$ leads to the critical load parameter Λ_{cr} and respective critical load p_{cr}. For the entire set of parabolic curves ($i = 1, 2, 3, \ldots$) their lowest parts corresponding to the minimum value of k for a given range of $\mu = a/b$ are plotted in Figure 14.4b with a solid line.

The critical load is calculated as the product

$$p_{\mathrm{cr}} = \Lambda_{\mathrm{cr}}\, p_x^*, \quad \text{where} \quad p_x^* = 1.0\,\mathrm{kN/m} \tag{14.13}$$

A particular solution requires the information about material parameters E, ν, dimensions a, b, h, reference load p_x^*, bending stiffness D^m; example values are presented in Box 14.1. The critical load for a square membrane ($\mu = a/b = 1.0$) and for $i = j = 1$ (see the upper indices) calculated on the basis of $k(\mu; i) = k_{\min}(1.0; 1) = 4$ is

$$\Lambda_{\mathrm{cr}}^{(1,1)} = D\, k(1.0; 1) = 4.0\, D \tag{14.14}$$

$$p_{\mathrm{cr}}^{(1,1)\ \mathrm{anal}} = \Lambda_{\mathrm{cr}}^{(1,1)}\, p_x^* \tag{14.15}$$

Higher loads correspond to other buckling modes ($i = 2, 3, j = 1$) and are calculated on the basis of $k(1.0;2)$ and $k(1.0;3)$. Box 14.1 contains respective values of parameters $k(\mu; i)$ and loads for three pairs $(i, j) = (1, 1), (2, 1), (3, 1)$ describing the buckling modes.

14.2.2 Approximate Solution Obtained using FDM

If an analysed structure is discretized by the finite difference method FDM (or FEM), then the calculation model transforms into a set of homogeneous linear algebraic equations, representing the algebraic eigenvalue problem. Its solution is a set of pairs

Box 14.1 Buckling of a square membrane under unidirectional compression

Data

$E = 2.0 \times 10^8\,\text{kN/m}^2$, $\nu = 0.3$, $a = b = 1.0\,\text{m}$, $\mu = a/b = 1.0$, $h = 0.01\,\text{m}$
$D^m = 18.315\,\text{kNm}$, $\hat{p}_x^* = 1.0\,\text{kN/m}$

Check values from analytical solution

$\Lambda = 180.762\, k(1.0, i)$
$i = 1, j = 1,\; k(1.0; 1) = 4.00,\quad \Lambda_{cr} = 723.047,\quad p_{cr}^{(1,1)\,\text{anal}} = \Lambda_{cr}\, p_x^* = 723.05\,\text{kN/m}$
$i = 2, j = 1,\; k(1.0; 2) = 6.25,\quad \Lambda = 1129.76,\quad p^{(2,1)\,\text{anal}} = \Lambda\, p_x^* = 1129.76\,\text{kN/m}$
$i = 3, j = 1,\; k(1.0; 3) = 11.11,\quad \Lambda = 2008.09,\quad p^{(3,1)\,\text{anal}} = \Lambda\, p_x^* = 2008.09\,\text{kN/m}$

Check values from FDM solution

$\lambda = a/4,\quad p_{cr}^{(1,1)\,\text{FDM}} = 686.6\,\text{kN/m} = 0.950\, p_{cr}^{(1,1)\,\text{anal}}$
$\lambda = a/8,\quad p_{cr}^{(1,1)\,\text{FDM}} = 713.6\,\text{kN/m} = 0.987\, p_{cr}^{(1,1)\,\text{anal}}$

Plate configuration shown in Figure 14.4a

of eigenvalues and eigenvectors. The lowest eigenvalue corresponds to the critical load parameter, and the eigenvector coupled with it represents the first postbuckling deformation mode.

The description of the approximate solution of the differential equation

$$D^m\, \nabla^2\nabla^2 w(x, y) - \Lambda \left(n_x^* \frac{\partial^2 w}{\partial x^2} \right) = 0 \tag{14.16}$$

with only one nonzero membrane force $n_x^* = -p_x^* = -1.0\,\text{kN/m}$ is presented next. Figure 14.5 illustrates a set of grid points, distributed at the distance $\lambda^{\text{FDM}} = a/4$. The figure indicates the numbering of internal, boundary and external (fictitious) points, taking into account the boundary conditions for simply supported edges and double symmetry of the expected first buckling mode. For four basic grid points (1)–(4) the finite difference equations are written. For example, for the central point (1) the following algebraic equation is obtained

$$D^m \left[\frac{20\, w_1 - 8\,(2 w_2 + 2 w_4) + 2(4\, w_3) + 4\, w_0}{\lambda^4} \right] - \Lambda\,(-1.0) \left[\frac{w_4 - 2 w_1 + w_4}{\lambda^2} \right] = 0 \tag{14.17}$$

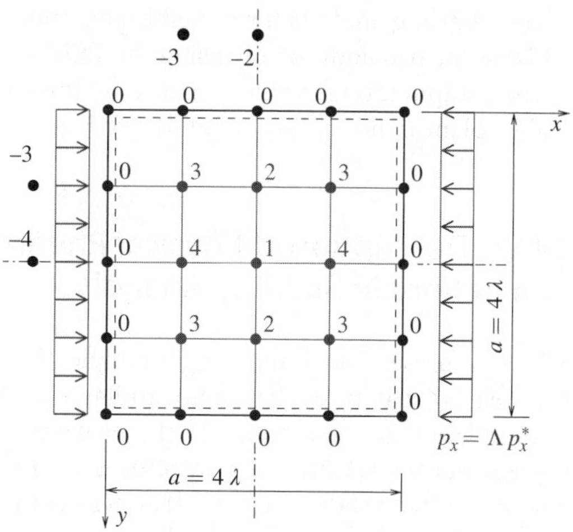

Figure 14.5 Square membrane under unidirectional compression – FDM discretization

The complete set of four homogeneous algebraic equations, after some rearrangements, can be written in the matrix form

$$(\mathbf{A}_{(4\times4)} - \tilde{\Lambda}\,\mathbf{B}_{(4\times4)})\,\mathbf{W}_{(4\times1)} = \mathbf{0}_{(4\times1)}, \quad \text{where} \quad \tilde{\Lambda} = \Lambda\,\lambda^2/D^m \tag{14.18}$$

$$\left(\begin{bmatrix} 10 & -8 & 4 & -8 \\ -8 & 20 & -16 & 4 \\ 4 & -16 & 40 & -16 \\ -8 & 4 & -16 & 20 \end{bmatrix} - \tilde{\Lambda} \begin{bmatrix} 1 & 0 & 0 & -1 \\ 0 & 2 & -2 & 0 \\ 0 & -2 & 4 & 0 \\ -1 & 0 & 0 & 2 \end{bmatrix}\right) \begin{bmatrix} w_1 \\ w_2 \\ w_3 \\ w_4 \end{bmatrix} = \begin{bmatrix} 0 \\ 0 \\ 0 \\ 0 \end{bmatrix} \tag{14.19}$$

Equation (14.19) is the nonstandard algebraic eigenvalue problem. Its solution for $N = 4$ consists of four pairs, each containing eigenvalue $\tilde{\Lambda}_n$ and eigenvector $\mathbf{W}_n$, for $n = 1, 2, 3, 4$. The lowest eigenvalue, that is the critical load factor $\tilde{\Lambda}_{cr} = \tilde{\Lambda}_1$, and the corresponding buckling mode are of most practical importance

$$(\tilde{\Lambda}_1, \mathbf{W}_1) = (2.3431, [0.6667, 0.4714, 0.3333, 0.4714]^{\mathrm{T}}) \tag{14.20}$$

Note that the eigenvector describes the shape of the buckling mode, that is it can be scaled by any nonzero constant. The components of the eigenvector associated with the lowest eigenvalue indicate equality $w_2 = w_4$, which means that the buckling mode has two additional diagonal planes of symmetry and this is in agreement with the number of half-waves in two directions $i = j = 1$.

Taking into account the following substitution and data (also presented in Box 14.1)

$$p_{cr} = \tilde{\Lambda}_{cr} \frac{D^m}{\lambda^2}\, p_x^* \tag{14.21}$$

one obtains the approximation

$$\begin{aligned} p_{cr}^{(1,1)\ \mathrm{FDM}} &= 2.3431\,\frac{D^m}{\lambda^2}\,1.0 = 2.3431\,D^m \left(\frac{4}{a}\right)^2 1.0 = 3.798\frac{D^m\,\pi^2}{a^2} \\ &= 687.46\ \mathrm{kN/m} = 0.9508\ p_{cr}^{\mathrm{anal}}, \quad \text{where} \quad p_{cr}^{\mathrm{anal}} = 723.05\,\mathrm{kN/m} \end{aligned} \tag{14.22}$$

The comparison of FDM versus analytical results is presented in Box 14.1.

The buckling analysis for a membrane under unidirectional compression (load type C) is performed additionally using the FEM – it is presented in Subsection 14.4.2 next to the computational results concerning the membrane under pure shear (load type S) and in-plane bending (load type B), see Figures 14.2c,d.

14.3 Description of Physical Phenomena and Nonlinear Simulations in Stability Analysis

We now present an interesting example that illustrates the loss of stability of two thin-walled structures that seem similar – a shallow cylindrical shell and a flat membrane/plate that corresponds to the projection of the shell onto the horizontal plane, see Radwańska (1990). Both structures are in unidirectional compression and simply supported (with sliding admitted) along their contours (Figure 14.6a,b).

The values of parameters used in the calculations, that is material constants E, ν, dimensions of the plate (L_x, L_y, h) and shell (L_x, L_y, f, R, h), reference load p_y^*, as well as the optional load imperfection P_C are given in Box 14.2.

The example starts from the description of the buckling phenomenon of a perfect structure. The change of load p_y is performed using load parameter Λ, which scales the fixed reference load p_y^*. During the loading process, for $\Lambda < \Lambda_{cr}$ both the shell and the membrane (perfect, i.e. without lateral load $P_C = 0$) are in the prebuckling membrane state until the load parameter reaches value Λ_{cr}. Then a bifurcation of the equilibrium path occurs and, beside the fundamental membrane state, a new, qualitatively different, bending state becomes possible. In the linear buckling analysis an eigenvalue problem is solved. The determined lowest eigenvalue gives the critical load as $\Lambda_{cr} p_y^*$, whereas the eigenvector associated with it represents the postbuckling deformation mode $w(x, y)$. The values of critical loads are computed using ANKA for the shallow cylindrical shell ($p_{cr}^{shell} = 1481\,\text{kN/m}$) and for the plate ($p_{cr}^{plate} = 1136\,\text{kN/m}$). The postbuckling deformed

Box 14.2 Stability analysis of a shallow cylindrical shell and membrane under unidirectional compression

Data

$E = 3.4 \times 10^6\,\text{kN/m}^2$, $\nu = 0.2$, $h = 0.1\,\text{m}$
for shell: $f = 0.15\,\text{m}$, $R = 83.3\,\text{m}$, for shell and plate: $L_x = L_y = 10.0\,\text{m}$
$D^m = 295.139\,\text{kNm}$, $p_y^* = 1.0\,\text{kN/m}$, $P_C = \pm 0.001\,\text{kN}$

Check values from FEM solution using ANKA

shell: $p_{cr}^{FEM} = 1481\,\text{kN/m}$
plate: $p_{cr}^{FEM} = 1136\,\text{kN/m}$ $\quad p_U = 1280\,\text{kN/m}$ for $P_C = -0.001\,\text{kN}$

Shell and membrane configurations shown in Figure 14.6

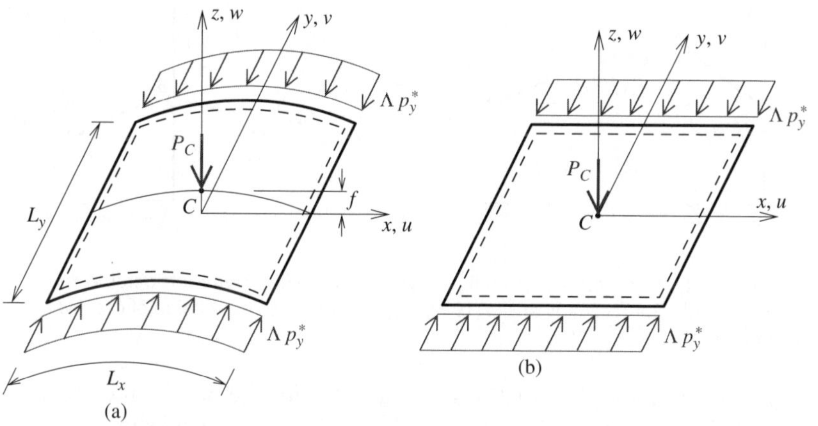

Figure 14.6 (a) Shallow cylindrical shell and (b) its projection onto the horizontal plane – geometry, load $\Lambda\, p_y^*$ and load imperfection P_C

surface exhibits one half-wave in both directions and has two planes of symmetry. These results allow one to mark two bifurcation points: $B_{SS}\,(0, p_{cr}^{plate})$ and $B_A\,(0, p_{cr}^{shell})$ on the vertical axis of the load-displacement plane (p_y, w_C), where $w_C = w_r = w_b$, see Figure 14.7.

It is important to know the type of a bifurcation point. In general there are three options: B_A – asymmetric, B_{SS} – symmetric stable and B_{SU} – symmetric unstable, see ANKA (1993), Radwańska (1990). Nonlinear FEM computations are necessary to examine the type of a bifurcation point. In the ANKA computer code an appropriate algorithm is incorporated. In this algorithm the positions of two points L and R in the vicinity of the bifurcation point B are determined. They indicate the shape of the postbuckling path emerging from a bifurcation point and simultaneously its type, see Figure 14.8, in which w_b denotes the prominent buckling displacement.

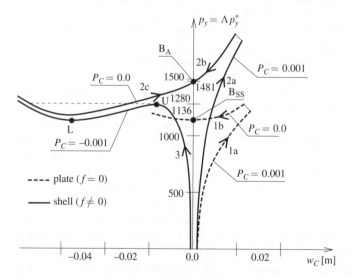

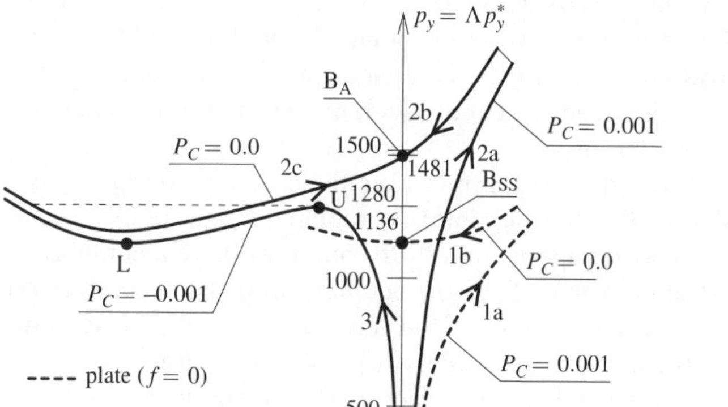

Figure 14.7 Shallow cylindrical shell ($f \neq 0$) and flat membrane ($f = 0$) – relations $p_y(w_C)$ for load imperfection cases $P_C = 0, +0.001, -0.001$

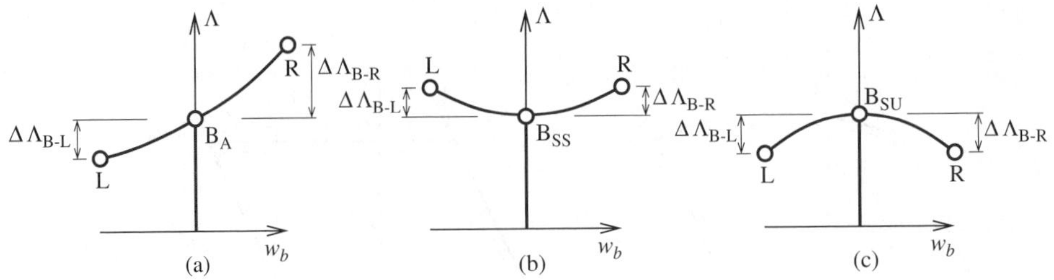

Figure 14.8 Three types of bifurcation points: (a) B_A – asymmetric, (b) B_{SS} – symmetric stable and (c) B_{SU} – symmetric unstable. Source: Radwańska (1990). Reproduced with permission of Cracow University of Technology.

In our example the nonlinear calculations indicate that for the perfect membrane a symmetric stable bifurcation point B_{SS} appears, whereas for the perfect cylindrical shell an asymmetric point B_A occurs (see Figure 14.7). Two rising postcritical paths, symmetric with respect to the vertical axis p_y, start from the B_{SS} point. Two paths emerge from the asymmetric point B_A, but one is rising (for $w_C > 0$) and the other is declining (for $w_C < 0$).

The knowledge of the type of the bifurcation point is helpful in predicting (or confirming) the sensitivity of a structure to the types and magnitudes of imperfections associated with either geometrical, material inaccuracy or initial displacements, disturbing loads or stresses. Obviously, this problem is crucial for structural safety, and hence it is addressed in design recommendations and codes of engineering practice.

Subsequent nonlinear calculations are performed for imperfect structures with load imperfection in the form of a vertical force P_C applied at the centre and having different senses. In the case of simultaneous action of both the variable boundary traction Λp_y^* and the fixed lateral force P_C, the loading process $P_C + \Lambda p_y^*$ is implemented.

In order to determine the relevant parts of a nonlinear equilibrium path, representing membrane-bending states, the incremental-iterative Newton–Raphson method is used (Crisfield 1981, Waszczyszyn 1987, Waszczyszyn et al. 1994). This way, more and more advanced membrane-bending equilibrium states are found. The nonlinear path is traced step-by-step using either load control with the load factor as the control parameter called fictitious time $\tau = \Lambda$, or displacement control with increasing central deflection playing the role of fictitious time $\tau = w_C$.

Very important results are obtained in the parametric nonlinear analysis (applying two senses and various values of force P_C). It is possible to combine the signs of increments of the control parameter $\Delta\tau$ and load imperfection P_C to compute different equilibrium paths. While reducing the absolute value of P_C, the equilibrium paths become closer and closer to the primary path, which is a part of the vertical axis p_y, as well as to the bifurcation point and then to the postcritical path of the perfect structure.

As shown in Figure 14.7, for negative load imperfection $P_C < 0$ and $w_C < 0$ – upper U and lower L points are found. They correspond to the upper limit load $p_U = \Lambda_U\, p_y^* = 1280\,\text{kN/m}$ and the lower limit load $p_L = \Lambda_L\, p_y^*$, respectively. Both limit values p_U and p_L are far below the bifurcation load $p_B = p_{cr} = \Lambda_{cr}\, p_y^*$. The essence of instability of the imperfect shallow shell consists in the following phenomenon: at the upper limit point U a sudden increase of the deflection absolute value can appear without a change of the load value. This is referred to as a snap-through phenomenon, which is another form of instability – this phenomenon is very dangerous for real shell structures since $p_U < p_B$.

An important conclusion is drawn from the description of the behaviour of the imperfect shallow cylindrical shell. An imperfection reducing the rise of the shell ($P_C < 0$) is very unfavourable. It causes the appearance of the limit point U lying below the bifurcation point B, which means $p_U < p_B$. We observe the particular sensitivity of the shallow cylindrical shell to the sense of the force imperfection, in particular an imperfection diminishing the rise of the shell is dangerous.

The data and computation results are given in Box 14.2.

14.4 Analytical and Numerical Buckling Analysis for Selected Plates and Shells

14.4.1 Remarks on Bifurcation of Equilibrium States

The buckling instability occurs for the so-called critical load. Then, at the bifurcation point two qualitatively different equilibrium states are observed, that is the prebuckling one and the postbuckling (postcritical) state close to it. In the case of plates and shells various changes can take place at the bifurcation point, that is during the transition from the prebuckling to the postbuckling state:

- next to the membrane state a membrane-bending state can occur
- axially symmetric membrane state is possible beside a deformation mode with waves in meridional and/or circumferential direction
- axially symmetric membrane-bending state can be accompanied by a nonsymmetric deformation state

A few examples are presented in the next subsections.

14.4.2 Buckling of Rectangular Plates for Three Load Cases

In this subsection buckling of rectangular plates undergoing three types of loading is considered, see Figure 14.9. The plates (initially membranes) are either unidirectionally compressed (case *C*) or subjected to a load causing uniform shear (case *S*) or in-plane bending (case *B*).

For an example analysis material parameters E, ν, geometrical data $L_x = a$, $L_y = b$, h, reference load p^* and bending stiffness D^m listed in Box 14.3 are adopted. In the prebuckling state the reference load p^*, corresponding to $\Lambda = 1$ (Figure 14.9a), results in the following membrane forces:

(C) – unidirectional compression $\quad p_x^* = 1.0\,\text{kN/m}$

$$n_x^* = -1.0\,\text{kN/m}, \qquad n_y^* = 0, \qquad n_{xy}^* = 0 \tag{14.23}$$

(S) – uniform shear $\quad p_s^* = 1.0\,\text{kN/m}$

$$n_x^* = 0, \qquad n_y^* = 0, \qquad n_{xy}^* = -1.0\,\text{kN/m} \tag{14.24}$$

(B) – in-plane bending $\quad p_{xK}^* = 1.0\,\text{kN/m}$

$$n_x^* = -\frac{2y}{b}\,p_{xK}^*, \qquad n_y^* = 0, \qquad n_{xy}^* = 0 \tag{14.25}$$

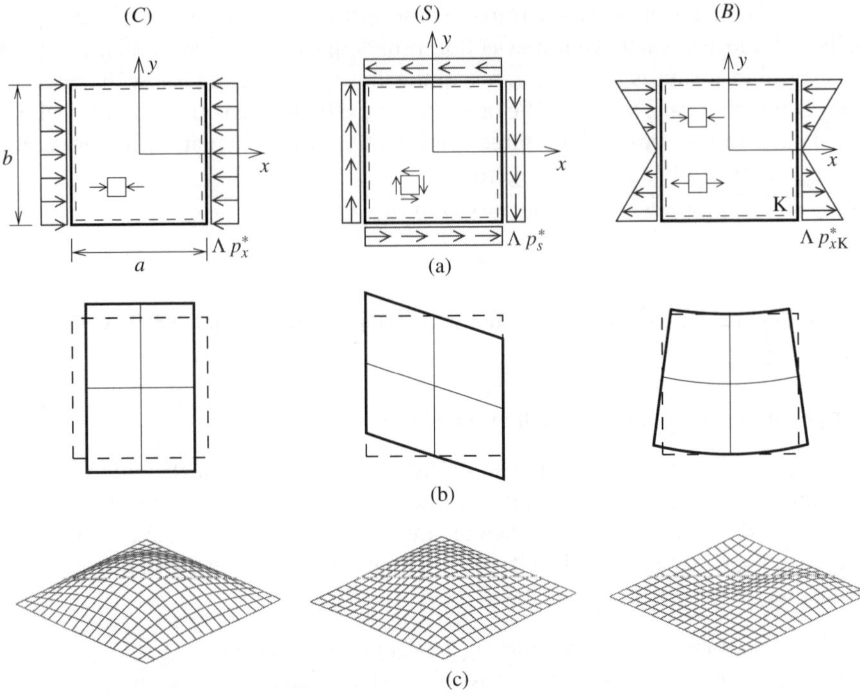

Figure 14.9 (a) Membranes subjected to three types of load (*C*, *S*, *B*), (b) prebuckling deformations in middle plane and (c) primary buckling modes. Source: Waszczyszyn et al. (1994). Reproduced with permission from Elsevier.

and prebuckling deformations shown in Figure 14.9b. The critical (lowest) loads are determined analytically (the solution for load case C is presented in Section 14.2) and the following three coefficients k_{min} are calculated in Timoshenko and Gere (1961) for $\mu = a/b = 1.0$ and for each of the load cases C, S, B, respectively:

$$k_{\text{min}}^{(C)} = 4.00, \quad k_{\text{min}}^{(S)} = 9.34, \quad k_{\text{min}}^{(B)} = 25.6 \tag{14.26}$$

Using the general formula

$$p_{\text{cr}}^{\text{anal}} = D\,k\,p^*, \quad \text{where} \quad D = \frac{D^m\,\pi^2}{p^*\,b^2}, \quad b = L_y \tag{14.27}$$

the critical (minimum) load values are calculated and presented in Box 14.3. They are used for the estimation of the accuracy of FEM results.

Next, a numerical analysis is performed using the algorithm described in Chapter 4 (Section 4.3). In Stage I of calculations the prebuckling membrane state is considered, the vector of nodal generalized displacements of a structure is computed $\mathbf{Q}^* = \mathbf{K}_0^{-1}\mathbf{P}^*$ for the reference load $\mathbf{P}^*$ and vectors of stresses σ^* are computed for each FE. Next, in Stage II the algebraic eigenproblem is solved, often calculating only the critical load parameter (minimum eigenvalue) and the buckling mode (eigenvector coupled with the minimum eigenvalue). However, we are also interested in higher load parameters, for which buckling with further deflection modes takes place.

Box 14.3 Critical loads of a square membrane for load cases *C*, *S*, *B*

Data

$E = 2.0 \times 10^8$ kN/m^2, $\nu = 0.3$, $a = b = 1.0$ m, $\mu = a/b = 1.0$, $h = 0.01$ m
$D^m = 18.315$ kNm, $D = 180.76$ kN/m

(*C*) unidirectional compression $p_x^* = 1.0$ kN/m

(*S*) uniform shear $p_s^* = 1.0$ kN/m

(*B*) in-plane bending $p_{x\,\mathrm{K}}^* = 1.0$ kN/m

Check values from analytical solution

(*C*) unidirectional compression $k_{\min}^{(C)} = 4.00$, $p_{\mathrm{cr}}^{(C)\ \mathrm{anal}} = 723.047$ kN/m

(*S*) uniform shear $k_{\min}^{(S)} = 9.34$, $p_{\mathrm{cr}}^{(S)\ \mathrm{anal}} = 1688.32$ kN/m

(*B*) in-plane bending $k_{\min}^{(B)} = 25.6$, $p_{\mathrm{cr}}^{(B)\ \mathrm{anal}} = 4627.50$ kN/m

---

Check values from FEM solution using ANKA

(*C*) unidirectional compression

$$p_{\mathrm{cr}}^{(C)\ \mathrm{FEM}} = \Lambda_{\mathrm{cr}}^{(C)\ \mathrm{FEM}}\, p_x^* = 723\ \mathrm{kN/m} \approx p_{\mathrm{cr}}^{(C)\ \mathrm{anal}}$$

(*S*) uniform shear

$$p_{\mathrm{cr}}^{(S)\ \mathrm{FEM}} = \Lambda_{\mathrm{cr}}^{(S)\ \mathrm{FEM}}\, p_s^* = 1685\ \mathrm{kN/m} = 0.998\, p_{\mathrm{cr}}^{(S)\ \mathrm{anal}}$$

(*B*) in-plane bending

$$p_{\mathrm{cr}}^{(B)\ \mathrm{FEM}} = \Lambda_{\mathrm{cr}}^{(B)\ \mathrm{FEM}}\, p_{x\,\mathrm{K}}^* = 4601\ \mathrm{kN/m} = 0.994\, p_{\mathrm{cr}}^{(B)\ \mathrm{anal}}$$

Membrane configurations shown in Figure 14.9a

In the FEM approach, the buckling phenomenon is analysed as an initial or linearized buckling problem, with the use of two alternative matrix equations (Waszczyszyn et al. 1994)

$$[\mathbf{K}_0 + \Lambda\ \mathbf{K}_\sigma^*(\sigma^*)]\mathbf{V} = \mathbf{0} \qquad \text{or} \qquad [\mathbf{K}_0 + \Lambda\ (\mathbf{K}_\sigma^*(\sigma^*) + \mathbf{K}_u^*(\mathbf{u}^*))]\mathbf{V} = \mathbf{0} \tag{14.28}$$

Beside the linear stiffness matrix $\mathbf{K}_0$, stress stiffness matrix $\mathbf{K}_\sigma^*(\sigma^*)$ and displacement stiffness matrix $\mathbf{K}_u^*(\mathbf{u}^*)$ are used. The last matrix takes into account the initial displacement gradients or rotations in the prebuckling state.

Code ANKA is employed with a four-node membrane-plate (conforming) finite elements PMK3. A sufficient accuracy of the solution, measured by the ratio $p_{\rm cr}^{\rm FEM}/p_{\rm cr}^{\rm anal}$ (given in Box 14.3), is obtained with the number of elements $NSE = 16 \times 16$ for the three analysed load cases. Both the prebuckling and the postbuckling modes are presented in the three Figures 14.10–14.12.

While running the algorithm of the numerical buckling analysis, code ANKA solves the algebraic eigenvalue problem using the subspace iteration method. Calculating a selected number of eigenvalues (and eigenvectors) from an indicated range of eigenvalue spectrum, the eigenvalues from the initial part of the spectrum interval are obtained with the best accuracy. Each subsequent value is determined with a bigger error. For

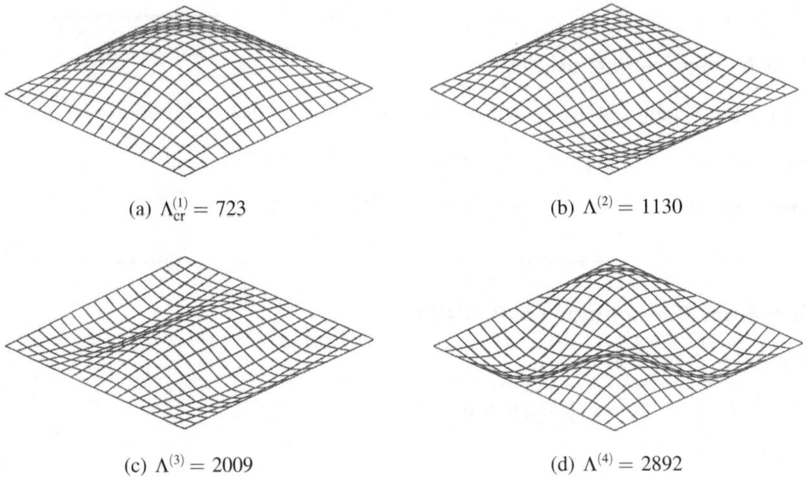

Figure 14.10 Buckling modes for unidirectionally compressed square plate; FEM results from ANKA. Source: Waszczyszyn et al. (1994). Reproduced with permission from Elsevier.

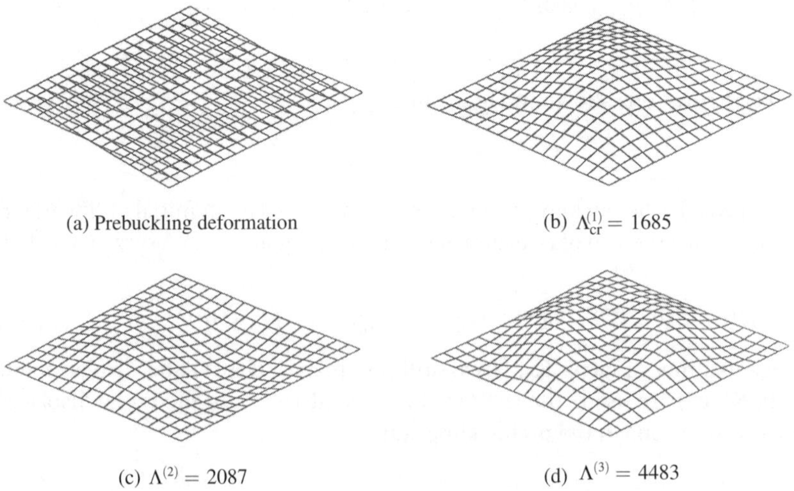

Figure 14.11 Prebuckling deformation and buckling modes for square plate under uniform shear; FEM results from ANKA. Source: Waszczyszyn et al. (1994). Reproduced with permission from Elsevier.

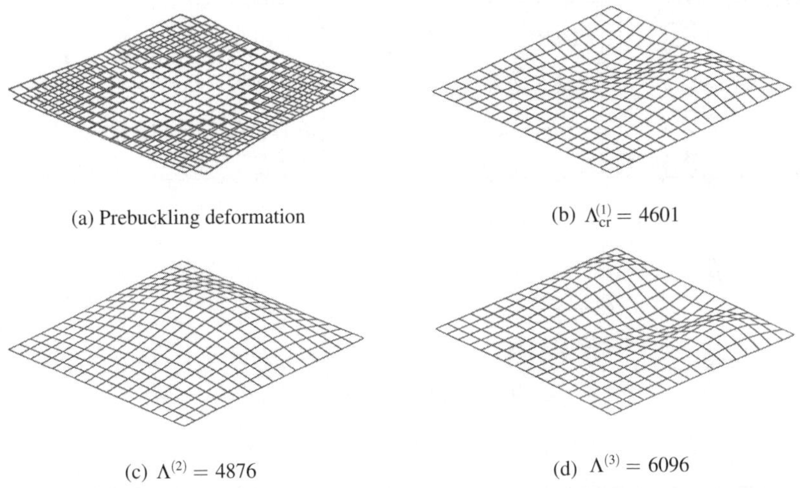

Figure 14.12 Prebuckling deformation and buckling modes for square plate under in-plane bending; FEM results from ANKA. Source: Waszczyszyn et al. (1994). Reproduced with permission from Elsevier.

instance, in order to determine two eigenvalues exactly four or five eigenvalues should be calculated using the subspace iteration method.

A wide parametric study was performed when the computer code for stability analysis ANKA (1993) was programmed, and when the monograph by Radwańska (1990) and the book by Waszczyszyn et al. (1994) were written. Several types of FEs (L4, S8, H8/9 Heterosis, L9, SRK) were used. Various aspects of numerical static and buckling analysis are discussed in Chapters 17 and 18. Numerical integration options, so-called locking phenomenon and parasitic deformation modes are discussed there in detail.

In the considered example, full, reduced and selective numerical integration quadratures (abbreviations FI, RI and SI are, respectively, used), as well as the value of ratio a/h, describing the plate slenderness, were changed and the influence of the number of FEs was tested.

We now focus on three different incorrect solutions that were observed in stage I (membrane state) of the pilot calculations of the square membrane under unidirectional compression, see Figure 14.13a–c. They are referred to as spurious deformation modes. When only one element S8/RI (2×2 Gauss points) was used for the entire domain, an 'hourglass' mode was found (a), when a quarter of the membrane was discretized by means of one (b) or four (c) elements L9/RI, parasitic so-called 'Escher modes' appeared.

Numerical experiments indicate that the finite elements S8 with RI and finite elements H8/9 with SI are recommended to be used, especially for thin membranes/plates. Additional information about some numerical problems in FEM buckling analysis is presented in Section 17. As can be seen in Table 17.3, an overestimation of the critical load and the singularity of the stiffness matrix occurs for two thin membrane/plate cases.

Next, we discuss one more numerical aspect of the analysis. For the load cases S and B (plate in uniform shear and in-plane bending, respectively) the eigenvalue problem is solved by means of an appropriate algorithm, including a shift of eigenvalue spectrum, which is incorporated in the ANKA code. The reason is that the pairs of eigenvalues with the same absolute value but opposite signs occur (i.e. the eigenvalue spectrum is symmetric with respect to zero). This is physically justified – the buckling similarly

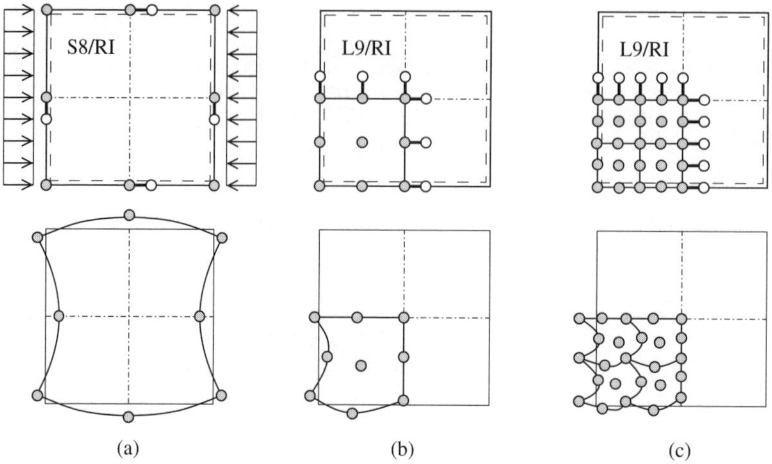

Figure 14.13 Membrane under unidirectional compression – spurious modes resulting from the use of (a) S8 and (b) and (c) L9 finite elements with reduced integration (RI). Source: Radwańska (1990). Reproduced with permission of Cracow University of Technology.

occurs for the load parameter Λ of positive or negative sign. This means that the change of the sign of Λ results in the interchange of the stress signs in the respective membrane subdomains. In that case the critical eigenvalue is understood as the absolute value closest to zero.

14.4.3 Buckling of a Circular Plate under Radial Compression

The boundary load directed to the axis of symmetry of a circular plate results in compressive membrane forces, both in the radial and circumferential direction (Figure 14.14). The kinematic constraints imposed on boundary $r = a$ allow for radial displacements in prebuckling state. In the postbuckling state two types of boundary conditions (clamped – case A and simply supported – case B) are considered.

In case A where the outer edge is clamped the vertical deflection $w(a)$ and angle of rotation $\vartheta_r(a)$ are restrained (the radial movement is not constrained).

In the theory of circular plates (see Section 9.5) the third-order differential equation for the axisymmetric bending state is used

$$D^m \frac{\mathrm{d}}{\mathrm{d}r}\left[\frac{1}{r}\frac{\mathrm{d}}{\mathrm{d}r}\left(r\,\frac{\mathrm{d}w(r)}{\mathrm{d}r}\right)\right] = -t_r(r) \tag{14.29}$$

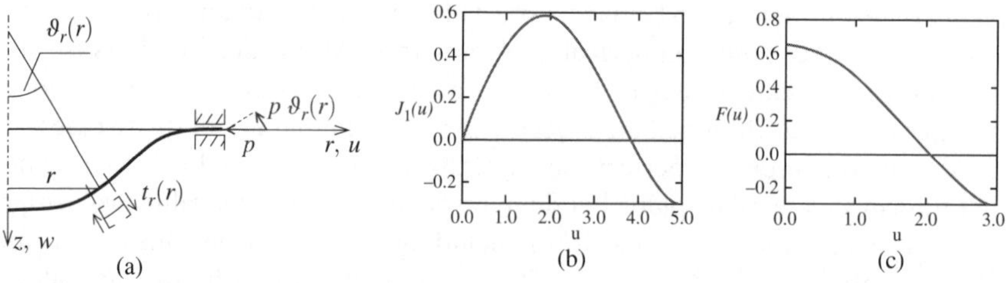

Figure 14.14 Circular plate under radial compression: (a) configuration and deformation for case A, (b) Bessel function $J_1(u)$, see Equation (14.35) and (c) function $F(u)$ for case B, see Equation (14.40)

In the postbuckling configuration the transverse force $t_r(r)$ appears as a component of the boundary load, which is normal to the deformed surface of the plate. It is expressed by means of the radial pressure p and rotation angle $\vartheta_r(r)$, see Figure 14.14a, in the following way

$$t_r(r) = -p\,\vartheta_r(r) \tag{14.30}$$

After three substitutions:

$$\vartheta_r(r) = -\frac{\mathrm{d}w(r)}{\mathrm{d}r}, \quad \alpha^2 = \frac{p}{D^m}, \quad u(r) = \alpha\, r \tag{14.31}$$

the governing Equation (14.29) takes the new form of second-order differential stability equation with rotation $\vartheta_r(r)$ as the unknown function beside load p hidden in function u

$$\frac{\mathrm{d}^2\vartheta_r}{\mathrm{d}u^2} + \frac{1}{u}\frac{\mathrm{d}\vartheta_r}{\mathrm{d}u} + \left(1 - \frac{1}{u^2}\right)\vartheta_r = 0 \tag{14.32}$$

The general integral of this equation is expressed as a combination of two Bessel functions: $J_1(u)$ – function of the first order and first kind, and $Y_1(u)$ – function of the first order and second kind

$$\vartheta_r(u) = C_1\, J_1(u) + C_2\, Y_1(u) \tag{14.33}$$

Since $\lim_{u\to 0} Y_1 = -\infty$, constant C_2 is assumed to be zero in order to exclude a singular solution. The homogeneous differential equation describing the axisymmetric postbuckling state must be completed with boundary conditions. It is important that the kinematic constraint is used to find the critical (lowest) load parameter. The condition of zero rotation angle on contour $r = a$ contains the load parameter hidden in the dimensionless argument of the Bessel function

$$\vartheta_r(u) = C_1\, J_1(u)|_{u=\alpha a} = 0\,, \quad \text{where} \quad u = \alpha\, a = \sqrt{\frac{p}{D^m}}\, a \tag{14.34}$$

This homogeneous equation of multiplicative form can be satisfied in two ways. This ambiguity reflects the physical description of the bifurcation phenomenon. First, $C_1 = 0$ can be assumed and this means the occurrence of the membrane state I (as the prebuckling one), with $\vartheta_r(r) = 0$. Alternatively, if $C_1 \neq 0$ and $\vartheta_r(r) \neq 0$, in the postbuckling bending state II the factor $J_1(u)$ needs to vanish for $u = \alpha\, a$

$$J_1(u) = 0 \tag{14.35}$$

Solving this equation is equivalent to searching for a root of function $J_1(u)$. The smallest value $u_{\min}$ of the root is sought and the critical load p_{cr} is determined. Using appropriate tables of Bessel functions or a mathematical package, first the minimum root of the equation, and next the parameter α are determined:

$$u_{\min} = 3.832, \quad \alpha = \frac{u_{\min}}{a} = \frac{3.832}{1.0} = 3.832\,1/\mathrm{m} \tag{14.36}$$

For the given parameters (E, ν, a, h, D^m – see Box 14.4) the critical value of the load is calculated with the formula

$$p_{\mathrm{cr}}^{\mathrm{anal}} = \alpha^2\, D^m = 33\,610\,\mathrm{kN/m} \tag{14.37}$$

and shown in Box 14.4.

Box 14.4 Buckling of a clamped circular plate under radial compression

Data

$E = 2.0 \times 10^8$ kN/m^2, $\nu = 0.3$, $a = 1.0$ m, $h = 0.05$ m
$D^m = 2889.40$ kNm

Check values from analytical solution

$p_{\text{cr}}^{(j=0)\ \text{anal}} = 33\,610$ kN/m

Check values from FEM solution

elements SRK, ANKA computer code
$p_{\text{cr}}^{\text{FEM}} = 33\,612$ kN/m $\approx p_{\text{cr}}^{\text{anal}}$
SHELL93 FEs, ANSYS computer code
$p_{\text{cr}}^{\text{FEM}} = 33\,264$ kN/m $= 0.9897\, p_{\text{cr}}^{\text{anal}}$

Plate configuration shown in Figure 14.14a

In case B, when the circular plate is simply supported, the boundary condition $m_r = 0$ can be written as the following differential relation between the radial moment m_r and the angle function $\vartheta_r(r)$

$$m_r = D^m(\kappa_r + \nu\kappa_\theta) = D^m\left(-\frac{\mathrm{d}\vartheta_r}{\mathrm{d}r} - \frac{\nu}{r}\,\vartheta_r\right) = 0 \tag{14.38}$$

This homogeneous static boundary condition can thus be written in the simplified form

$$\frac{\mathrm{d}\vartheta_r}{\mathrm{d}r} + \frac{\nu}{r}\,\vartheta_r = 0 \tag{14.39}$$

Having considered the general integral (14.33) and the rules of differentiation of Bessel functions, this boundary condition is written in terms of function $F(u)$

$$F(u) = \left[J_0(u) - \frac{1-\nu}{u}J_1(u)\right]_{u=\alpha a} = 0 \tag{14.40}$$

depending on two Bessel functions J_0 and J_1. The function $F(u)$ is shown in Figure 14.14c, and its first root $u_{\min} = 2.049$ leads to the following value of the critical load

$$p_{\text{cr}}^{\text{anal}} = \alpha^2\, D^m = 9610\,\text{kN/m} \tag{14.41}$$

The FEM analysis is performed for the clamped plate (case A) using two types of FEs: one-dimensional element SRK in code ANKA and two-dimensional element SHELL93 in code ANSYS (2013). A set of buckling loads $p^{(j)}$ calculated using ANKA for the number of circumferential waves $j = 0, 1, 2, 3$ is presented in the following vector

$$\mathbf{p}^{\text{ANKA}} = [\underline{33\,612},\ 60\,380,\ 93\,191,\ 131\,830]^{\mathrm{T}}\,\text{kN/m} \tag{14.42}$$

with the critical (minimum) load corresponding to the axisymmetric buckling mode ($j = 0$). Using ANSYS, four eigenvalues are found and grouped in the vector

$$\mathbf{p}^{\text{ANSYS}} = [\underline{33\,264},\ 59\,186,\ 59\,186,\ 90\,326]^{\text{T}}\ \text{kN/m} \tag{14.43}$$

The critical (minimum) value is underlined. The values of the critical buckling loads, coming from the calculations with the use of ANKA and ANSYS codes, are additionally given in Box 14.4.

Noteworthy is the appearance of double eigenvalues, which are related to similar buckling modes, but rotated by 180° with respect to each other. This is often encountered in the numerical analysis of both buckling and free vibrations (see Chapter 15) of axisymmetric structures. Again, the attention must be drawn to the proper way of domain discretization in the vicinity of the centre of a circular membrane/plate when two-dimensional FEs are used (as in the static analysis – see Chapter 7 and 9, and also Figure 9.4).

14.4.4 Buckling of Cylindrical Shells under Axial Compression or External Pressure – Theoretical and Numerical Analysis

In Chapter 5 of (Brush and Almroth 1975) we can find detailed information about the stability of a circular cylindrical shell. The shell is an important example in which the prebuckling deformation of the ideal structure involves rotations of elementary segments of the shell middle surface. When the rotations are taken into account, the prebuckling deformation analysis is nonlinear and hence the fundamental equilibrium path is nonlinear, see Figure 14.1.

On the other hand, when the terms containing the prebuckling rotations are neglected and the membrane forces do not depend on function $w(x, \theta)$ in the description of the membrane state, we obtain the following three linear equilibrium equations:

$$R\frac{\partial n_x}{\partial x} + \frac{\partial n_{x\theta}}{\partial \theta} = 0, \quad R\frac{\partial n_{x\theta}}{\partial x} + \frac{\partial n_\theta}{\partial \theta} = 0$$

$$D^m\,\nabla^4 w + \frac{1}{R}\,n_\theta - \underline{\left(n_x\frac{\partial^2 w}{\partial x^2} + \frac{2}{R}\,n_{x\theta}\frac{\partial^2 w}{\partial x\,\partial \theta} + \frac{1}{R^2}\,n_\theta\frac{\partial^2 w}{\partial \theta^2}\right)} = 0 \tag{14.44}$$

These equations must be completed by three constitutive and three kinematic equations related to the membrane state. The underlined terms represent the effective transverse load $\overline{p}_n(x)$ and indicate the fact that the last equation is written for the deformed elementary shell segment taking into account the rotations that appear after buckling. This linear stability equation is analogical to Equation (14.2) for buckling of a rectangular plate.

We consider two cases of loading of a cylindrical shell (marked further A and B), presented in Figure 14.15.

Case A – shell subjected to axial compressive load $p_x = \Lambda p_x^*$

In the prebuckling state (stage I) a uniformly distributed boundary load Λp_x^* results in a constant meridional compressive force:

$$n_x(x) = -p_x = -\Lambda\,p_x^*, \quad n_\theta(x) = n_{x\theta}(x) = 0 \tag{14.45}$$

In case A the effective transverse load normal to the deformed surface of a shell has only one component, see Equation (14.44)

$$\overline{p}_n(x) = n_x\frac{\partial^2 w}{\partial x^2} = -\Lambda\,p_x^*\frac{\partial^2 w}{\partial x^2} \tag{14.46}$$

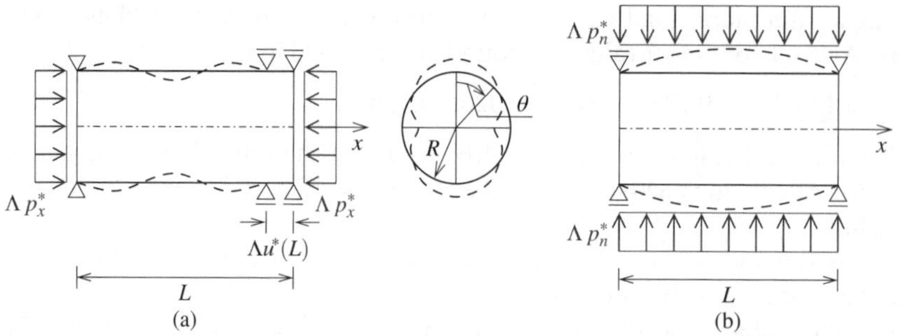

Figure 14.15 Cylindrical shells: (a) under axial compression and (b) under external pressure

In stage II the bifurcation is considered, in which displacements $w(x, \theta)$ normal to the middle surface appear.

Omitting detailed derivations, but quoting Brush and Almroth (1975), one comes to the buckling equation of axially compressed cylindrical shell in the form of eighth-order differential equation with constant coefficients

$$D^m \nabla^8 w + \frac{D^n(1-\nu^2)}{R^2}\frac{\partial^4 w}{\partial x^4} + \Lambda p_x^* \nabla^4 \left(\frac{\partial^2 w}{\partial x^2}\right) = 0 \tag{14.47}$$

Taking into account simple support conditions for the edges (with free axial displacement at one of the ends, see Figure 14.15a) and the possible formation of i half-waves along the meridian and of j complete cosine waves in the circumferential direction, the following form of normal displacement function $w(x, \theta)$ is assumed

$$w(x,\theta) = A \sin\frac{i\,\pi\,x}{L}\cos(j\,\theta), \quad \text{where} \quad \theta = \frac{y}{R} \tag{14.48}$$

The relations between p_{cr}, i, j, R, L, h that guarantee the existence of a nonzero solution of the differential equation (nonzero displacement $w(x, \theta)$) as well as indicate the critical load and the respective mode of postbuckling deformation, were derived in Brush and Almroth (1975)

$$p_{cr} = \frac{E\,h^2}{R\sqrt{3(1-\nu^2)}}, \quad \left(i^2 + \frac{L^2}{\pi^2 R^2} j^2\right)^2 = \frac{\sqrt{12\,(1-\nu^2)}}{\pi^2}\frac{L^2}{Rh} \tag{14.49}$$

Assuming $\nu = 0.3$, the formula very often quoted in the literature is obtained

$$p_{cr} = 0.605\frac{E\,h^2}{R} \tag{14.50}$$

The data used for calculation, as in Brush and Almroth (1975), are presented in Box 14.5. They result in the following values of the shell parameters:

$$\frac{L}{R} = 1.0, \quad \frac{L}{h} = 100, \quad Z = \frac{L}{R}\frac{L}{h}\sqrt{1-\nu^2} = 100 \times 0.954 \approx 100 \tag{14.51}$$

With the use of this formula and data, the first approximation of the critical load, denoted by upper index (1), is

$$p_{cr}^{(1)} = \frac{E\,h^2}{R\sqrt{3(1-\nu^2)}} = 12\,105 \text{ psi} \times \text{in.} = 2122\,\text{kN/m} \tag{14.52}$$

Box 14.5 Buckling of a cylindrical shell under axial compression

Data

$E = 10^7$ psi $= 6.90 \times 10^7$ kPa, $\nu = 0.3$
$R = L = 20$ in. $= 0.5080$ m, $h = 0.2$ in. $= 0.005080$ m, $Z \approx 100$
$D^m = 0.7326 \times 10^4$ psi $\times$ in.^{3} $= 0.8284$ kNm

Check values from analytical solution

$p_{cr}^{(1)} = 12\,105$ psi $\times$ in. $= 2122$ kN/m
$p_{cr}^{(2)} = 12\,653$ psi $\times$ in. $= 2218$ kN/m

Check values from FEM solution

$NSE = 30$, SRK elements, ANKA computer code
$p_{cr}^{(1,7)\ \mathrm{FEM}} = 2077$ kN/m

Shell configuration shown in Figure 14.15a

Another formula, helpful in the estimation of the critical pressure, employs a function $k = k(Z)$ of Batdorf's parameter Z, the graph of which is given in Figure 14.16

$$p_{cr} = \sigma_{cr}\, h = \frac{\pi^2 D^m}{L^2}\, k(Z) \tag{14.53}$$

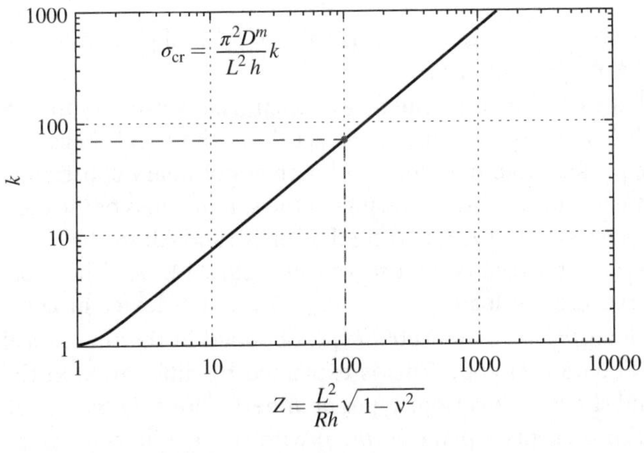

Figure 14.16 Relation $k(Z)$ between dimensionless pressure and the Batdorf's parameter for cylindrical shell under axial compression. Source: Batdorf (1947), NASA.

For $Z \approx 100$ the value $k \approx 70$ is read from the graph and hence the second approximation of the critical boundary load is calculated

$$p_{\mathrm{cr}}^{(2)} = 12\,653\ \mathrm{psi} \times \mathrm{in.} = 2218\,\mathrm{kN/m} \tag{14.54}$$

Next, the numerical analysis of the load case A is discussed. The literature and the authors' own experience indicate computational difficulties connected with the numerical analysis of buckling of axially compressed cylindrical shells. A poor convergence is observed when the algebraic eigenvalue problem is solved using the subspace iteration method. This is associated with the existence of very similar eigenvalues that are coupled with significantly different eigenvectors, corresponding to different numbers of circumferential waves. The ANKA computer code with the SRK finite elements is used for calculations. The algebraic eigenproblem is solved for consecutive values of $j = 0, \ldots, 8$ (number of circumferential waves) given in the input data. The obtained eigenvalues are the load multipliers Λ for the reference load $p_x^* = 1.0\,\mathrm{kN/m}$. The associated eigenvectors (in particular normal displacements of meridional nodes) reproduce i meridional waves, whereas number j determines the shape of deformed shell in the circumferential direction. Thirty one-dimensional elements SRK are applied to discretize the shell meridian ($NSE = 30$). It turns out that the subspace iteration method does not converge for $j = 1, 4, 9$. Namely, the number of iterations exceeds the (quite large) limit value. The values of load $p^{(i,j)}$, obtained after converged iteration process, are given with the buckling mode identified by numbers i and j (as upper indices):

$$\begin{aligned} &p^{(6,0)} = 2131\,\mathrm{kN/m}, \quad p^{(6,2)} = 2135\,\mathrm{kN/m}, \quad p^{(6,3)} = 2141\,\mathrm{kN/m} \\ &p^{(5,5)} = 2133\,\mathrm{kN/m}, \quad p^{(5,6)} = 2121\,\mathrm{kN/m} \\ &\underline{p^{(1,7)} = 2077\,\mathrm{kN/m},} \quad p^{(4,8)} = 2117\,\mathrm{kN/m} \end{aligned} \tag{14.55}$$

Very close values of loads, coupled with different buckling modes, are observed. The lowest value, underlined previously in Equation (14.55), is accepted as the critical boundary load for $p_x^* = 1\,\mathrm{kN/m}$

$$p_{\mathrm{cr}}^{(1,7)\ \mathrm{FEM}} = 2077\,\mathrm{kN/m} \tag{14.56}$$

and the buckling form is described by $i = 1, j = 7$. This value $p_{\mathrm{cr}}^{\mathrm{FEM}}$ is slightly different than the two estimates $p_{\mathrm{cr}}^{(1)}$ and $p_{\mathrm{cr}}^{(2)}$ presented earlier.

The axially compressed cylindrical shell is a structure of particular sensitivity to various types of imperfections. It is not easy to achieve perfect thin cylindrical shells in a manufacturing process. Also, perfect boundary load and perfect boundary constraints require great precision. Many experimental study results of the stability loss of the considered shell are described in Harris et al. (1957) and cited by Brush and Almroth (1975).

A large scatter of the experimental results is observed in Figure 14.17a. The most important is that the measured critical loads $p_{\mathrm{cr}}^{\mathrm{exp}} = \sigma_{\mathrm{exp}} h$ are much lower in comparison with the bifurcation load $p_{\mathrm{cr}}^{\mathrm{anal}} = p_B = 0.605Eh^2/R$ obtained in the theoretical considerations according to Equation (14.50). This is explained by differences in the behaviour of a perfect shell and shells with various types of imperfections (geometrical, material or load). Figure 14.17b presents a plane $(p, u(L))$ with two equilibrium paths for a perfect shell. There is a primary (prebuckling) path and secondary (postbuckling) one. Both paths intersect at the bifurcation point B at the level of buckling load p_{B}. For

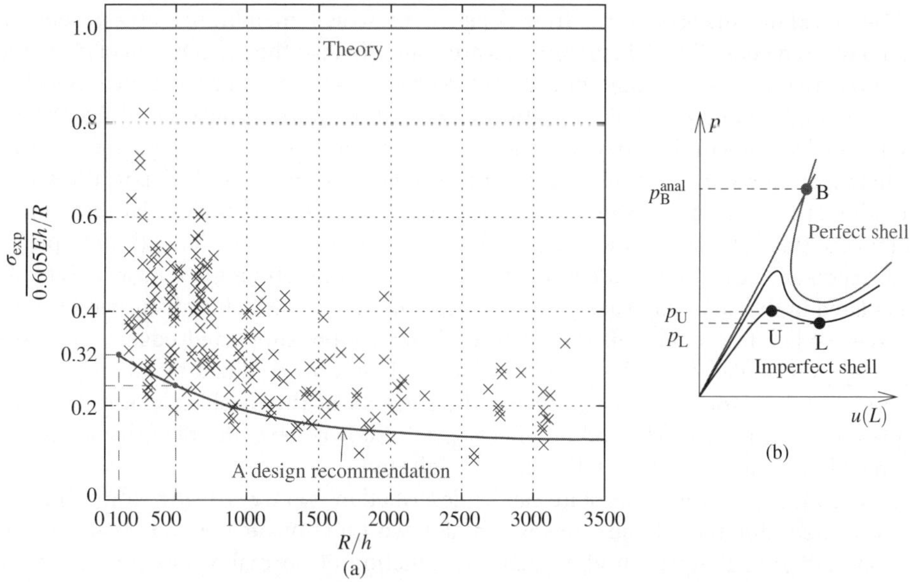

Figure 14.17 (a) Distribution of experimental buckling test data for cylindrical shells subjected to axial compression; (b) characteristic equilibrium paths of perfect and imperfect axially compressed cylindrical shells. Source: Harris et al. (1957). Subfigure (a) reproduced with permission of AIAA.

an imperfect shell, a nonlinear path of membrane-bending equilibrium starts from the beginning of the loading process and reaches the upper limit point, which corresponds to the load $p_U < p_B$, much lower than the bifurcation load p_B.

Further examination of the test data set in Figure 14.17a allows for drawing an appropriate lower-bound curve through the points. The ordinate of each point on this curve is interpreted as a reduction factor for the bifurcation critical load within a large range of ratio $R/h = a/h$. In a sense, the reduction factor reflects the sensitivity of axially compressed cylindrical shells to initial imperfections. It is taken into account in design recommendations.

A simple calculation may be enlightening. First, we use Equation (14.50). Next, for example for a shell with $R/h = 100$, Figure 14.17a shows a point on the lower-bound curve, which determines the reduction factor $n \approx 0.32$. This means that for safety reasons, the value $p_{cr} = n\, p_{cr}^{anal} \approx 0.194Eh^2/R$ ought to be considered in engineering design instead of p_{cr}^{anal}. That is why industrial organizations, which possess the knowledge on the experimental results for real structures, exert influence on design criteria related to shell instability. The analysed thin cylindrical shell is a structure of particular imperfection sensitivity. Another important conclusion from the theoretical consideration is that the behaviour of an imperfect shell strictly depends on the shape of the secondary equilibrium path related to an analogous perfect shell (see Figure 14.17b).

Case B – shell loaded with external lateral pressure $p_n = \Lambda p_n^*$

Under such loading the prebuckling membrane-bending deformation of the shell is axisymmetric. The critical pressure p_{cr} is defined as the lowest pressure, for which another quantitatively different equilibrium state occurs, exhibiting waves in two directions.

The buckling analysis starts from stage I, in which membrane forces, bending moments and generalized displacements are computed for the reference load p_n^*. Some essential assumptions are then introduced in order to create a computational model in the form of differential stability equation. Namely, the rotation angles φ_x of the normal to the shell surface in the axisymmetric membrane-bending state are neglected. In this state tangent force $n_{x\theta} = 0$ and, due to unrestricted motion in the shell axial direction, meridional force also vanishes $n_x = 0$. The only nonzero circumferential force $n_\theta(x)$ enters the stability equation with a multiplier equal to the second derivative of normal displacement with respect to coordinate θ. The circumferential force $n_\theta(x)$ is initially calculated in the analysis of axisymmetric membrane state, which is disturbed by local sources of bending effects in the vicinity of two simply supported edges. The force $\overline{n}_\theta = -p_n R$, calculated for the membrane state, should be modified by adding a component $n_\theta(x) = Ehw(x)/R$, coming from the solution of the displacement differential equation describing local bending effects. As is shown next, the disturbance of the prebuckling membrane state in the shell is neglected.

In stage II the postbuckling state can be described in two ways, using either (i) three displacement differential equations with unknown functions $u(x,\theta), v(x,\theta), w(x,\theta)$, or (ii) one differential equation with unknown function of normal displacement $w(x,\theta)$. In both cases the equations are homogeneous and contain a component depending on constant circumferential membrane force $n_\theta(x) = \overline{n}_\theta = -p_n R$ found at stage I. The condition of existence of a nonzero solution is employed to obtain the critical value of load parameter.

Regarding case (i), the following three displacement differential equations describe the buckling problem:

$$\begin{aligned}
&\frac{\partial^2 u}{\partial x^2} + \frac{1-\nu}{2R^2}\frac{\partial^2 u}{\partial \theta^2} + \frac{1+\nu}{2R}\frac{\partial^2 v}{\partial x\,\partial\theta} + \frac{\nu}{R}\frac{\partial w}{\partial x} = 0\\
&\frac{1+\nu}{2R}\frac{\partial^2 u}{\partial x\,\partial\theta} + \frac{1-\nu}{2}\frac{\partial^2 v}{\partial x^2} + \frac{1}{R^2}\frac{\partial^2 v}{\partial \theta^2} + \frac{1}{R^2}\frac{\partial w}{\partial\theta} = 0\\
&\left(\frac{\nu}{R}\frac{\partial u}{\partial x} + \frac{1}{R^2}\frac{\partial v}{\partial\theta} + \frac{w}{R^2}\right) + \frac{h^2}{12}\nabla^4 w - \frac{1-\nu^2}{EhR^2}\,n_\theta\,\frac{\partial^2 w}{\partial\theta^2} = 0
\end{aligned} \tag{14.57}$$

The third equation includes a component dependent on $n_\theta(x)\ (\partial^2 w/\partial\theta^2)$. For the case of simple support of both shell ends, the following trigonometric functions of coordinates x and θ with amplitudes U, V, W are chosen:

$$\begin{aligned}
u(x,\theta) &= U\ \sin\frac{\pi\,x}{L}\ \cos(j\theta)\\
v(x,\theta) &= V\ \cos\frac{\pi\,x}{L}\ \sin(j\theta)\\
w(x,\theta) &= W\ \sin\frac{\pi\,x}{L}\ \cos(j\theta)
\end{aligned} \tag{14.58}$$

One half-wave in the meridional direction and j complete sine/cosine waves in the circumferential direction are assumed. The stability equation, ensuring the existence of nonzero amplitudes U, V, W, results in a very complex relation between the critical load p_{cr}, numbers i, j and parameters k and λ rewritten from Brush and Almroth (1975):

$$p_{cr}\frac{(1-\nu^2)\,R}{E\,h} = \frac{(1-\nu^2)}{(j^2-1)\,(1+\lambda^2\,j^2)^2} + k\left(j^2 - 1 + \frac{2j^2-1-\nu}{1+\lambda^2\,j^2}\right)$$

$$\text{where} \quad k = \frac{h^2}{12\,R^2}, \quad \lambda = \frac{L}{\pi\,R} \tag{14.59}$$

Regarding case (ii), the following eighth-order differential equation with variable coefficient $n_\theta(x)$ is presented in Brush and Almroth (1975)

$$D^m\,\nabla^8 w + \frac{D^n(1-\nu^2)}{R^2}\frac{\partial^4 w}{\partial x^4} - \frac{1}{R^2}\,\nabla^4\left[n_\theta(x)\,\frac{\partial^2 w}{\partial \theta^2}\right] = 0 \tag{14.60}$$

If the circumferential force is determined based solely on the linear prebuckling membrane state with constant force $n_\theta(x) = \overline{n}_\theta = -p_n\,R$, the stability equation has the constant coefficients

$$D^m\,\nabla^8 w + \frac{D^n(1-\nu^2)}{R^2}\frac{\partial^4 w}{\partial x^4} + \frac{p_n}{R}\,\nabla^4\left[\frac{\partial^2 w}{\partial \theta^2}\right] = 0 \tag{14.61}$$

The set of normal displacement functions $w(x,\theta;j)$, each satisfying the boundary conditions of simply supported edges:

$$w|_{x=0,L} = 0, \qquad \left.\frac{\partial^2 w}{\partial x^2}\right|_{x=0,L} = 0 \tag{14.62}$$

is proposed for a solution of the differential equation (14.61)

$$w(x,\theta) = W\,\sin\frac{\pi x}{L}\,\sin(j\,\theta), \quad j = 1,2,3,\ldots \tag{14.63}$$

The analytical treatment enables one to formulate the relations between dimensionless load $\overline{p}$, the Batdorf's parameter Z and number j:

$$\overline{p} = \frac{(1+\overline{j}^{2})^2}{\overline{j}^{2}} + \frac{1}{\overline{j}^{2}(1+\overline{j}^{2})^2}\,\frac{12}{\pi^4}\,Z^2, \quad \text{where} \quad Z = \frac{L^2}{R\,h}\sqrt{1-\nu^2}, \quad \overline{j} = \frac{j\,L}{\pi\,R} \tag{14.64}$$

The minimization of the load function with respect to j results in the dependence $p_{cr}(Z)$, presented in Figure 14.18. An approximate calculation of the critical pressure is performed for the data presented in Box 14.6.

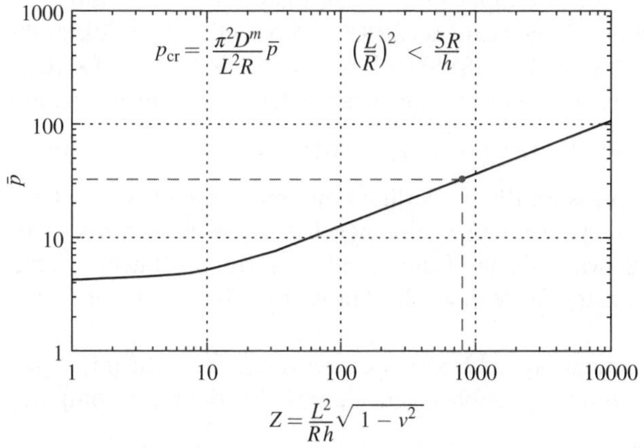

Figure 14.18 Diagram of relation $\overline{p}(Z)$ between dimensionless critical pressure and Batdorf's parameter for cylindrical shells with external lateral pressure. Source: Batdorf (1947), NASA.

Box 14.6 Buckling of a cylindrical shell under external pressure

Data

$E = 3.0 \times 10^7$ kPa, $\nu = 0.1667$, $R = 20$ m, $L = 60$ m, $h = 0.22$ m
$Z \approx 806.73$, $D^m = 27\,381$ kNm

Check values from analytical solution

$p_{cr}^{anal} = 120\ \text{kN/m}^2$

Check values from FEM solution

ANKA, elements SRK, $NSE = 20$, $NEN = 4$, $NNDOF = 2$, $NEDOF = 8$
$p_{cr}^{FEM} = 118.64\ \text{kN/m}^2$, $j = 5$
ANSYS, elements SHELL93, $NSE = 10 \times 16$
$p_{cr}^{FEM} = 153.82\ \text{kN/m}^2$

Shell configuration shown in Figure 14.15b

The value of the Batdorf's parameter $Z = 806.73$ (computed for the data in Box 14.6) is used to read out the dimensionless pressure $\overline{p} \approx 32$ from the graph of relation $\overline{p}(Z)$ in Figure 14.18, and the critical pressure is determined as

$$p_{cr}^{anal} = \frac{\pi^2 D^m}{L^2 R}\,\overline{p} = 120\text{kN/m}^2 \qquad (14.65)$$

The approximate value of the critical load p_{cr}^{anal} is confronted with the results of the numerical buckling analysis, using two computer codes ANKA and ANSYS (see output data in Box 14.6). As in the previous example, in the FEM analysis the calculations are performed using either one-dimensional elements SRK (ANKA) with $NSE = 20, 30, 42$, or two-dimensional elements SHELL93 (ANSYS) with $NSE = NS_x \times NS_\theta = 10 \times 16$. In the first case four eigenvalues $p^{(j)}$ and eigenvectors are used in the description of buckling

$$[p^{(4)}, p^{(5)}, p^{(6)}, p^{(7)}]^T = [159.31, \underline{118.64}, 139.32, 178.61]^T\text{kN/m}^2 \qquad (14.66)$$

In the set of loads the critical pressure (the underlined minimum value) is found and is related to $j = 5$. The normal displacements of the meridian discretization nodes are used to determine the one half-wave shape of the meridian in the postbuckling state (Figure 14.19). Calculations repeated for $NSE = 30, 42$ prove that the results for $NSE = 20$ are of sufficient accuracy.

Using two-dimensional discretization in ANSYS, six eigenvalues from the initial part of the spectrum are detected. Omitting double eigenvalues we list the three computed values of the pressure

$$[p^{(5)}, p^{(6)}, p^{(7)}]^T = [\underline{153.82}, 161.93, 207.23]^T\text{kN/m}^2 \qquad (14.67)$$

and the first (underlined) value is the critical one, associated with $j = 5$.

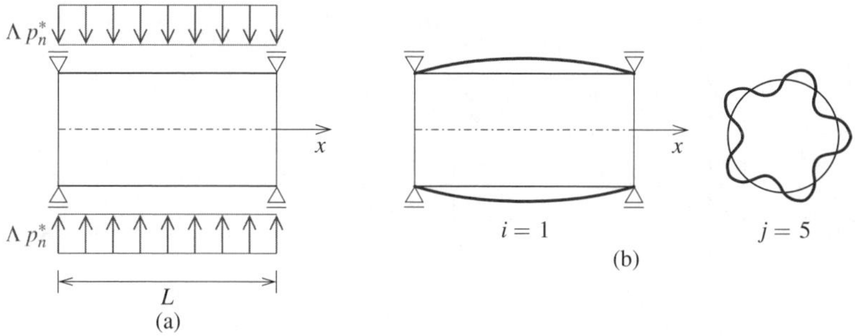

Figure 14.19 Cylindrical shell under external pressure: (a) configuration and (b) postbuckling mode for $i = 1$ and $j = 5$

The graphical postprocessor allows one to present the postbuckling modes determined by eigenvectors for the entire shell. The way of discretization of two-dimensional structures is particularly important in the buckling analysis, since it affects the accuracy of determination of the coupled eigenvalues and eigenvectors. It is emphasized that the number of FEs adopted in the circumferential direction is particularly important for the description of deformation modes with several waves.

Finally, we summarize important comments of Brush and Almroth (1975) on the behaviour of cylindrical shells under external pressure, related to Figure 5.7 on p. 166 in that book.

Three geometrical parameters of cylindrical shells R, L, h, in particular their ratios L/R and R/h, influence the value of the critical pressure and the associated number of circumferential waves j. The relationship between the dimensionless critical pressure $pR/(Eh) \times 10^{-4}$ and parameter $L/(\pi R) \in [0.1, 100]$, for three selected values $R/h =$ 100, 200, 500 are presented in the mentioned figure. For instance, we set the value of R. We consider longer and longer shells with $L/(\pi R) = 1, 10{,}100$. Furthermore, we take into account three thinner and thinner shells with $R/h = 100, 200, 500$.

We can read from the diagram in Figure 5.7 of Brush and Almroth (1975) the dimensionless critical pressure (which corresponds to $j = 5$) for the short shell with $L/(\pi R) = 1$ and $R/h = 100$. We note that reducing the thickness ($R/h = 200, 500$) causes the decrease of the critical pressure, while the number j grows to $j = 6, 7$. Next, for $R/h = 100$, increasing the length of the shell in the range $L/(\pi R) \in [10{,}100]$ (moderately long and long shells), the same value of critical pressure and $j = 2$ is obtained. Now, reducing the thickness ($R/h = 200, 500$) the critical pressure obviously decreases but the buckling mode is always described by $j = 2$. We can observe that number j decreases as we reduce the length of the shell.

Finally, Brush and Almroth (1975) notice that the analysis of long cylindrical shells with ratio $L/(\pi R) > 10$, based on the Donnell–Mushtari–Vlasow quasi-shallow shell equations, yields 25% higher values of the critical pressure (with $j = 2$) in comparison with the results of the Sanders–Koiter equations for nonshallow cylindrical shells with any value of slenderness h/R.

14.4.5 Buckling of Shells of Revolution with Various Signs of Gaussian Curvature – FEM Results

The following axisymmetrical shells of various Gaussian curvatures $K = 1/R_1\,R_2$ are considered: barrel-shaped (case B) with $K > 0$, cylindrical (case C) with $K = 0$ and

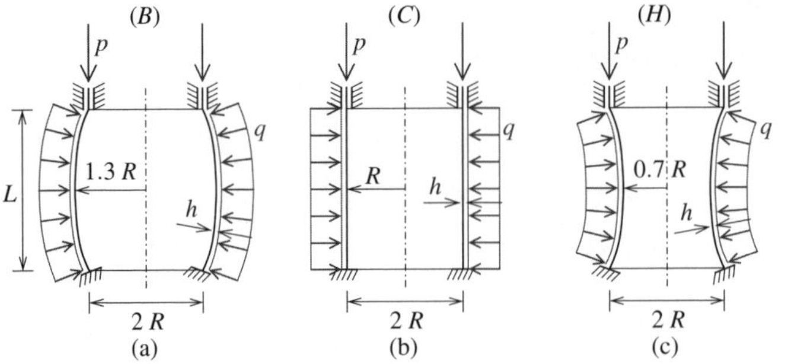

Figure 14.20 Axisymmetric shells with various shapes of meridian (*B*, *C*, *H*) and two types of loads *q* and *p*

Box 14.7 Buckling of three shells of revolution under external pressure or boundary load

Data

$E = 3.65 \times 10^6$ kPa, $\nu = 0.35$, $R = 0.225$ m, $L = 0.60$ m, $h = 0.0016$ m
$D^m = 27381$ kNm

Check values from FEM

$NSE = 40$, elements SRK, ANKA computer code
(*B*) – barrel shell:
$q_{cr}^{(j=15)} = 30.53\ \text{kN/m}^2$, $p_{cr}^{(j=0)} = 25.21\ \text{kN/m}$
(*C*) – cylindrical shell:
$q_{cr}^{(j=6)} = 7.24\ \text{kN/m}^2$, $p_{cr}^{(j=10)} = 26.36\ \text{kN/m}$
(*H*) – hyperboloidal shell:
$q_{cr}^{(j=5)} = 18.71\ \text{kN/m}^2$, $q_{cr}^{(j=6)} = 18.13\ \text{kN/m}^2$, $p_{cr}^{(j=5)} = 5.37\ \text{kN/m}$

Shell configuration shown in Figure 14.20

hyperboloidal (case *H*) with $K < 0$, see Figure 14.20. Please note that the case names *B* and *C* formerly used in Section 14.2 have been redefined and now mark the shell shape. They are compressed by external pressure q or uniformly distributed boundary load p. A parametric numerical analysis is carried out using the geometrically one-dimensional elements SRK and the ANKA code. The shells are described by data presented in Box 14.7.

The eigenvalue problem, being the computational model for the linear buckling analysis, is solved for successive numbers of circumferential waves j taken from an interval predefined in the input data. The critical load values from among the calculated sets of

$q^{(j)}$ or $p^{(j)}$ are respectively calculated as:

$$q_{\mathrm{cr}} = \inf_j q^{(j)}, \quad p_{\mathrm{cr}} = \inf_j p^{(j)} \tag{14.68}$$

When the SRK elements are used, the buckling mode is determined by the displacements of meridional nodes (constituting the eigenvector) and the number j of waves in the circumferential direction.

The results of FEM calculations are presented in Box 14.7. They allow one to draw some conclusions. The highest critical pressure $q_{\mathrm{cr},B}^{(j=15)} = 30.53\,\mathrm{kN/m^2}$ (corresponding to $j = 15$) in the case of external pressure load, is observed for the barrel-shaped shell. The cylindrical shell can carry the pressure nearly four times lower, $q_{\mathrm{cr},C}^{(j=6)} = 7.24\,\mathrm{kN/m^2}$. The lowest value of the critical load in axial compression $p_{\mathrm{cr},H}^{(j=5)} = 5.37\,\mathrm{kN/m}$ is exhibited by the hyperboloidal shell. The critical boundary load $p_{\mathrm{cr},H}$ for this shell is about five times lower than the critical loads $p_{\mathrm{cr},C}$ for cylindrical and $p_{\mathrm{cr},B}$ for barrel-shape shells.

14.5 Snap-Through and Snap-Back Phenomena Observed for Elastic Shallow Cylindrical Shells in Geometrically Nonlinear Analysis

In this section a shallow cylindrical shell is analysed under a central transverse force. The shell has two straight edges simply supported without sliding and two curved edges free, see Figure 14.21.

This example is presented by many authors including Sabir and Lock (1972), Crisfield (1981), Radwańska (1990) and Waszczyszyn et al. (1994). It is treated as a benchmark test for the examination of the robustness of the nonlinear calculation methodology and FEs. The geometry, load and material data are accepted as:

$$E = 3.4 \times 10^6\,\mathrm{kN/mm^2}, \quad \nu = 0.3$$

$$L_x = L_y = 504\,\mathrm{mm}, \qquad R = 2540\,\mathrm{mm}, \qquad f = 12.5\,\mathrm{mm}$$

$$P = \Lambda\,|P^*|, \quad P^* = -1.0\,\mathrm{kN} \tag{14.69}$$

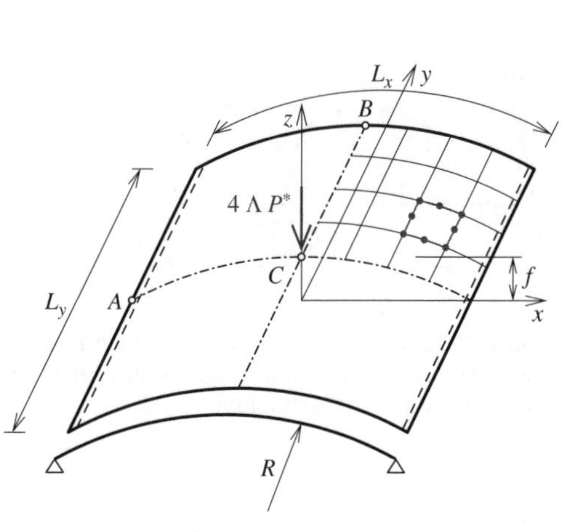

Figure 14.21 Cylindrical shell – configuration with load and boundary conditions; FEM mesh for shell quarter

A quarter of the shell is discretized by means of $NSE = 4 \times 4$ eight-noded elements SQR1 (ANKA).

On the load-displacement plane (P, w_C) the deformation process is represented with sequences of points on nonlinear equilibrium paths. Tracing the equilibrium paths requires a suitable monotonically increasing control parameter τ. The results of nonlinear analysis, obtained with incremental-iterative algorithm, are presented in Figure 14.22 in the form of relationship $P(w_C; h^{(i)})$ between the load P and the central deflection w_C. It turns out that the behaviour of the shell under the load and the equilibrium path significantly depend on the thickness.

Two cases from among the five analysed in Radwańska (1990) are described here. They are related to values $h^{(3)} = 12.70$ mm and $h^{(1)} = 6.35$ mm (see Figure 14.22), for which different diagrams $P(w_C; h^{(i)})$ are obtained.

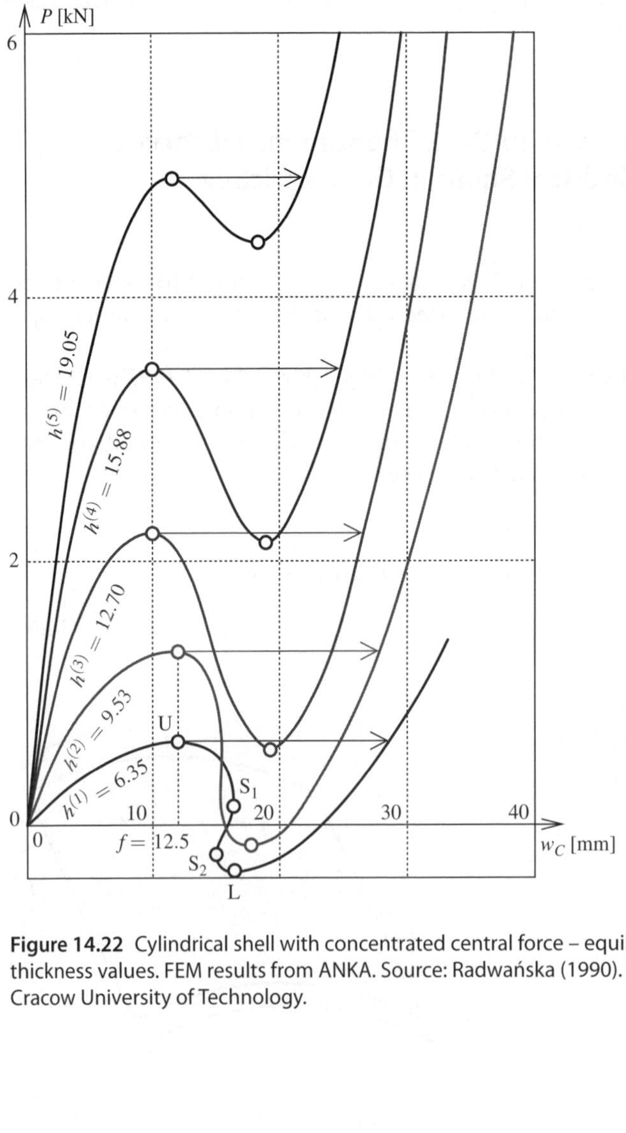

Figure 14.22 Cylindrical shell with concentrated central force – equilibrium paths $P(w_C, h^{(i)})$ for five thickness values. FEM results from ANKA. Source: Radwańska (1990). Reproduced with permission of Cracow University of Technology.

In the case of shell with thickness $h^{(3)} = 12.70\,\text{mm}$, the so-called upper limit point U (maximum) at the load level P_U as well as the lower limit point L (minimum) with the load P_L occur. In the continuation process of the numerical analysis the entire path $P(w_C; h^{(3)})$ is traced with the use of displacement control, with $\tau = w_C$ as the control parameter. For a monotonically increasing load (e.g. in experiments or computer simulations with load control $\tau = P$) the snap-through phenomenon marked by horizontal arrow occurs at the level of load P_U, with abrupt transition to another state of advanced displacements, but without qualitative changes of the deformation.

A more complex path $P(w_C; h^{(1)})$ is observed for $h^{(1)} = 6.35\,\text{mm}$. Beside points U and L, there are two so-called snap-back points S_1 and S_2. Here, the arc length control parameter ($\tau = s$) is applied in calculations, due to ambiguity with respect to both the vertical and horizontal axes.

At the end, an important issue in the incremental-iterative algorithm is addressed. A criterion of the termination of iterations is required for each incremental step. For this purpose the value of limit error measure with respect to both the norm of the displacement increment vector and of the residual force vector is given as input data. The details, related to algorithms used in nonlinear analysis, can be found in many works dedicated to computation methodology in nonlinear numerical analysis.

References

ANKA 1993 ANKA – computer code for nonlinear analysis of structures: User's manual. Technical report, Cracow University of Technology, Cracow (in Polish).

ANSYS 2013 ANSYS Inc. PDF Documentation for Release 15.0. SAS IP, Inc.

Batdorf SB 1947 A simplified method of elastic-stability analysis for thin cylindrical shells. Technical Report No. 874, NASA.

Brush DO and Almroth BO 1975 *Buckling of Bars, Plates and Shells.* McGraw-Hill.

Crisfield MA 1981 A fast incremental/iterative solution procedure that handles 'snap-through'. *Computers & Structures* **13**(1–3), 55–62.

Gajewski A and Życzkowski M 1988 *Optimal Structural Design under Stability Constraints.* Kluwer Academic Publishers, Dordrecht/Boston.

Harris LA, Suer HS, Skene WT and Benjamin RJ 1957 The stability of thin-walled unstiffened circular cylinders under axial compression, including the effects of internal pressure. *Journal of the Aeronautical Sciences* **24**(8), 587–596.

Koiter WT 1945 *On the Stability of Elastic Equlibrium* PhD dissertation Delft University of Technology Delft (in Dutch).

Radwańska M 1990 Analysis of stability and large displacements of shell structures using FEM. Technical Report Monograph 105, Cracow University of Technology, Cracow (in Polish).

Sabir AB and Lock AC 1972 The application of finite elements to the large deflection geometrically non-linear behaviour of cylindrical shells. In Brebbia CA and Tottenham H (eds), *Variational Methods in Engineering*, vol. 7. Southampton University Press, Southampton. pp. 66–75.

Timoshenko S and Gere JM 1961 *Theory of Elastic Stability*, 2nd edn. McGraw-Hill, New York.

Waszczyszyn Z (ed.) 1987 *Selected Problems of Structural Stability*. Ossolineum, Wroclaw.

Waszczyszyn Z and Radwańska M 1985 *Plates and Shells*. Cracow University of Technology, Cracow (in Polish).

Waszczyszyn Z, Cichoń C and Radwańska M 1994 *Stability of Structures by Finite Element Methods*. Elsevier, Amsterdam.

15

Free Vibrations of Plates and Shells

15.1 Introduction

Similarly to 1D and 3D structures one can consider plates and shells subjected to dynamic loads. Dynamic analyses of shell structures are described for instance in the books by Reddy (1999, 2007) and Geradin and Rixen (2015). The load can change in time, that is its value, direction or point of application can vary, causing dynamic effects. Loads can be applied to a structure and then suddenly removed (e.g. of impact type), which causes free vibrations.

In the case of loads changing in time, we are interested in the determination of the response of the structure. Either the state of a real structure is monitored on site or a numerical simulation is performed for a suitable model of the structure. The change of the load character from static to dynamic can lead to excessive loading of a structure. Therefore, in structural design the assessment of dynamic influences is crucial and must be performed; for instance for ship hulls in naval engineering, for car bodies in mechanical engineering, for aerplane wings and fuselages in aerospace engineering and for roof shells or cooling towers in civil engineering.

To perform a dynamic analysis we augment the equilibrium equations with the inertia forces and the vibration damping forces generated by acceleration and velocity fields, respectively. They also depend on geometrical and material characteristics of the structure.

In a general analysis of continuous structural models the continuous distribution of mass is assumed. Using d'Alambert principle, the so-called initial-boundary value problem (IBVP) is formulated, that is a set of partial differential equations completed with limit (boundary and initial) conditions. In the displacement differential equations we have partial derivatives of displacements with respect to two surface coordinates ξ_α, $\alpha = 1, 2$ and with respect to time t. Obviously, the inertia forces are functions corresponding to the second derivatives of displacements with respect to time, and the damping forces are functions of the first derivatives.

To derive equilibrium equations for a two-dimensional structure it is necessary to reduce three-dimensional description to the middle surface, which results the emergence of inertia forces and moments (similar to surface body forces and moments in statics). The inertia forces are expressed by second-order derivatives of displacements with respect to time using the surface mass density called inertia and denoted by I_0,

Plate and Shell Structures: Selected Analytical and Finite Element Solutions, First Edition.
Maria Radwańska, Anna Stankiewicz, Adam Wosatko and Jerzy Pamin.
© 2017 John Wiley & Sons Ltd. Published 2017 by John Wiley & Sons Ltd.

and second order time derivatives of rotation angles using a so-called rotational (rotatory) inertia I_2. The surface mass density μ [kg/m^2] per unit area of middle surface of a two-dimensional structure with thickness h is used to define the dynamic properties of the plate/shell cross section:

$$I_0 = \mu = \rho\, h = \gamma\, h/g, \qquad I_2 = \mu\, h^2/12 = \rho h^3/12 \tag{15.1}$$

In these equations ρ [kg/m^3] is the material density (specific mass), γ [N/m^3] is the unit weight and g [m/s^2] is the gravitation constant. In calculations material density ρ must be adopted, for example for steel $\rho_s = 7800$ kg/m^3 = 7800 Ns^2/m^4 and for normal-weight concrete $\rho_c = 2400$ kg/m^3 = 2400 Ns^2/m^4.

With regards to the type of two-dimensional thin-walled structures the following vibration phenomena are distinguished:

- vibrations of a membrane in its middle plane with translations $u(\xi_\alpha, t)$, $v(\xi_\alpha, t)$
- transverse vibrations of plates with deflection $w(\xi_\alpha, t)$
- complex vibrations of shells with displacements $u(\xi_\alpha, t)$, $v(\xi_\alpha, t)$, $w(\xi_\alpha, t)$

For each type of shell structure, appropriate inertia terms appear either in the equilibrium equations or in the displacement differential equations, associated with one-, two-, three- or five-parameter theories. The terms related to I_2 can be omitted in the case of thin plates/shells with $L/h > 10$ (see Reddy 1999).

Taking into account only undamped, synchronous and harmonic vibrations, functions appearing in dynamic relations can be written in a multiplicative form, with one factor dependent on two spatial coordinates and the other one on the time. In this way we introduce functions of amplitudes multiplied by an appropriate function of time.

In the three-parameter K–L theory the vector of shell displacements has three components, described by their amplitude functions U, V, W and the assumed function of time $T(t) = \sin \omega t$ (note that here T is not temperature):

$$\begin{aligned} &u(\xi_1, \xi_2, t) = U(\xi_1, \xi_2)\, T(t), \qquad v(\xi_1, \xi_2, t) = V(\xi_1, \xi_2)\, T(t) \\ &w(\xi_1, \xi_2, t) = W(\xi_1, \xi_2)\, T(t) \end{aligned} \tag{15.2}$$

where angular frequency ω of the vibration appears. In what follows, differential equations with spatial derivatives of amplitude functions U, V, W will be applied, which describe the boundary value problem (BVP), supplemented as usual by boundary conditions. Furthermore, three components of inertia forces can be defined as:

$$\begin{aligned} F_1 &= I_0\, \ddot{u} = \mu\, U(\xi_1, \xi_2)\, \ddot{T}(t) = -\omega^2\, \mu\, U(\xi_1, \xi_2)\, T(t) \\ F_2 &= I_0\, \ddot{v} = \mu\, V(\xi_1, \xi_2)\, \ddot{T}(t) = -\omega^2\, \mu\, V(\xi_1, \xi_2)\, T(t) \\ F_n &= I_0\, \ddot{w} = \mu\, W(\xi_1, \xi_2)\, \ddot{T}(t) = -\omega^2\, \mu\, W(\xi_1, \xi_2)\, T(t) \end{aligned} \tag{15.3}$$

where the superscript dot $(\dot{\ })$ denotes the derivative with respect to time.

As in static analysis, in dynamic considerations discrete models of the structures are applied, taking into account a finite number of dynamic degrees of freedom (*NDDOF*). For years, discrete models have been created by the introduction of concentrated masses, which made it possible to construct a physical model of a structure with a limited number of *NDDOF*. The model with concentrated (lumped) masses is particularly simple in the case of bar structures (beams, frames) and the calculation algorithms related to this approach are broadly applied in *structural mechanics* courses.

In the dynamic modelling of 1D, 2D and 3D structures using FEM we obtain the discrete systems according to the usual algorithm and matrix definitions, whereby so-called consistent mass matrix is computed on the basis of finite element shape functions. In a FEM model *NDDOF* is equal to the standard *NSDOF*. A large number of *NDDOF* makes it possible to simulate dynamic processes for a real structure with a sufficient accuracy. Now, taking into account the potential of hardware and software, there are practically no limitations to *NDDOF*.

We emphasize that what is first analysed in the scope of dynamics are the natural (free) vibrations of a structure, on which no external forces act. Damping is also usually neglected in this fundamental problem, on which we focus in this chapter. It requires the solution of an algebraic eigenproblem leading to the calculation of an (angular) frequency spectrum and associated vibration modes (eigenvalues and eigenvectors). For discrete models of structures we usually calculate just a few couples of natural frequency and vibration mode. The smallest frequency is often called the fundamental. The eigenvalues and eigenvectors are treated as a dynamic characteristic of the vibrating structure. It is very important that any complex vibration can be decomposed in the space of eigenvectors, using the so-called modal decomposition method.

For vibrations generated by changing loads (whether harmonic or impulsive) the simulation of the dynamic process requires the integration of the differential equations of motion, which can be performed analytically or numerically using suitable algorithms of integration in the time domain, see Geradin and Rixen (2015). We mention that the issue of proper damping representation is out of the scope of this book.

In Section 15.2 an example analytical solution of the free vibrations problem for rectangular plates using Navier's method is presented. The results of numerical analysis are shown for typical plates (Subsection 15.2.2) and shells (Subsection 15.4.2). The influence of the rotational inertia on the results for thin and moderately thick plates is discussed in Section 15.3.

15.2 Natural Transverse Vibrations of a Thin Rectangular Plate

15.2.1 Analytical Solution

Transverse natural (free) vibrations occur with no transverse load. In the case of a thin plate with uniform thickness and material properties, described in the Cartesian coordinate system, which exhibits deflection

$$w(x, y, t) = W(x, y)\, T(t) \tag{15.4}$$

the following field of inertia forces is defined

$$F_n(x, y, t) = \mu \frac{\partial^2 w(x, y, t)}{\partial t^2} \tag{15.5}$$

This field is introduced into the displacement differential equation for a plate (including fourth-order derivatives with respect to coordinates x, y and second time derivatives)

$$\nabla^2\nabla^2 w(x, y, t) + \frac{\mu}{D^m} \frac{\partial^2 w(x, y, t)}{\partial t^2} = 0 \tag{15.6}$$

It is mentioned that in this commonly employed equation the term $-I_2 \nabla^2 \ddot{w}$ has been omitted on the left-hand side. This equation must be completed by initial and boundary

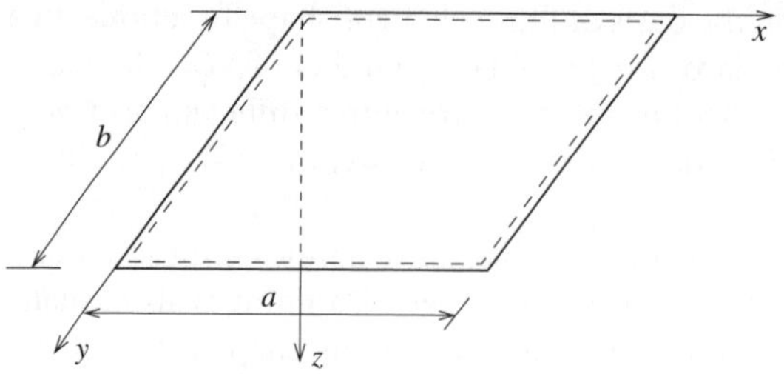

Figure 15.1 Configuration of a square plate with all simply supported edges

conditions. Having adopted the amplitude function $W(x, y)$ and the time function $T(t) = \sin \omega t$, the initial conditions for $t = 0$ are:

$$T(0) = 0, \qquad \dot{T}(0) = \omega \tag{15.7}$$

After the multiplicative decomposition with respect to spatial and time variables, a homogeneous fourth-order differential equation with only partial spatial derivatives is obtained

$$\nabla^2\nabla^2 W(x, y) - \frac{\mu}{D^m}\,\omega^2\, W(x, y) = 0 \tag{15.8}$$

with an unknown deflection amplitude function $W(x, y)$ and unknown angular frequency of vibrations ω.

The analytical solution is based on Navier's method of double trigonometric series (similar to the static case described in Section 8.7). Taking into account the boundary conditions of the simply supported plate contour, see Figure 15.1, the function of deflection amplitude is written as

$$W(x, y) = \sum_{i=1}^{\infty}\sum_{j=1}^{\infty} A_{ij}\, \sin\frac{i\pi x}{a}\, \sin\frac{j\pi y}{b} \tag{15.9}$$

When the boundary conditions on the entire edge:

$$W|_{x=0,a} = W|_{y=0,b} = 0, \qquad \left.\frac{\partial^2 W}{\partial x^2}\right|_{x=0,a} = 0, \qquad \left.\frac{\partial^2 W}{\partial y^2}\right|_{y=0,b} = 0 \tag{15.10}$$

are satisfied by the assumed function $W(x, y)$, the differential equation (15.8) is reworked into

$$\sum_{i=1}^{\infty}\sum_{j=1}^{\infty} A_{ij}\left[\pi^4\left(\frac{i^2}{a^2}+\frac{j^2}{b^2}\right)^2 - \frac{\mu}{D^m}\,\omega^2\right]\sin\frac{i\pi x}{a}\,\sin\frac{j\pi y}{b} = 0 \tag{15.11}$$

From the condition of a nontrivial solution of the homogeneous equation

$$\pi^4\left(\frac{i^2}{a^2}+\frac{j^2}{b^2}\right)^2 - \frac{\mu}{D^m}\,\omega^2 = 0 \tag{15.12}$$

angular frequencies $\omega_{(i,j)}$ are derived for each combination of numbers i and j defines the numbers of half-waves in a vibration mode in the x and y directions, respectively

$$\omega_{(i,j)} = \pi^2\left(\frac{i^2}{a^2}+\frac{j^2}{b^2}\right)\sqrt{\frac{D^m}{\mu}} = \frac{\pi^2}{a^2}\,(i^2+\beta^2 j^2)\sqrt{\frac{D^m}{\mu}} = \frac{\pi^2}{a^2}\,\alpha_{(i,j)}\sqrt{\frac{D^m}{\mu}} \tag{15.13}$$

where

$$\alpha_{(i,j)} = i^2 + \beta^2 j^2, \qquad \beta = a/b \tag{15.14}$$

For a square plate with $\beta = a/b = 1$ a new form of Equation (15.13) is obtained

$$\omega_{(i,j)} = \frac{\pi^2}{a^2}(i^2 + j^2)\sqrt{\frac{D^m}{\mu}} \tag{15.15}$$

Very often the fundamental angular frequency for $i = j = 1$ and $\beta = a/b = 1$ is calculated using the formula

$$\omega_{(1,1)} = \frac{2\,\pi^2}{a^2}\sqrt{\frac{D^m}{\mu}} \tag{15.16}$$

We emphasize an analogy between the natural vibration Equation (15.8) for a plate (and the method of its solution) and the buckling Equation (14.7), written for a rectangular simply supported plate under unidirectional compression (see Subsection 14.2.1).

For a square plate described by material constants (E, ν, ρ) and geometrical parameters (a, h), whose values are shown in Box 15.1, we calculate three angular frequencies for couples $(i,j) = (1,1), (1,2), (2,2)$ on the basis of coefficients $\alpha_{(i,j)}$ equal to, respectively: $\alpha_{(1,1)} = 2$, $\alpha_{(1,2)} = 5$, $\alpha_{(2,2)} = 8$. The values of angular frequency $\omega_{(i,j)}$ [rad/s] as well as frequency $f_{(i,j)} = \omega_{(i,j)}/(2\pi)$ [Hz] are also presented in Box 15.1.

It is worth noting that, in addition to the exact solutions, the Rayleigh–Ritz method can be applied to obtain the approximate solution of the free vibration problem for continuous systems (in Section 8.12 the Ritz method is used to approximate the

Box 15.1 Free vibrations of a simply supported square plate

Data

$E = 1.47 \times 10^8$ kN/m^2, $\nu = 0.3$, $\rho = 10^4$ kg/m^3
$a = b = 1.2$ m, $h = 0.0151$ m, $D^m = 46.35$ kNm

Check values from analytical solution

$i = 1,\ j = 1,\quad \alpha_{(1,1)} = 2,\quad \omega_{(1,1)} = 240.16$ rad/s, $\quad f^{\text{anal}}_{(1,1)} = 38.22$ Hz
$i = 1,\ j = 2,\quad \alpha_{(1,2)} = 5,\quad \omega_{(1,2)} = 600.40$ rad/s, $\quad f^{\text{anal}}_{(1,2)} = 95.56$ Hz
$i = 2,\ j = 2,\quad \alpha_{(2,2)} = 8,\quad \omega_{(2,2)} = 960.64$ rad/s, $\quad f^{\text{anal}}_{(2,2)} = 152.89$ Hz

---

Check values from FEM solution

$NSE = 20 \times 20$, ANSYS computer code
$f^{\text{FEM}}_{(1,1)} = 37.99$ Hz $= 0.994\, f^{\text{anal}}_{(1,1)}$, $\quad f^{\text{FEM}}_{(1,2)} = 95.11$ Hz $= 0.995\, f^{\text{anal}}_{(1,2)}$
$f^{\text{FEM}}_{(2,2)} = 151.77$ Hz $= 0.993\, f^{\text{anal}}_{(2,2)}$

Plate configuration shown in Figure 15.1

static response of a plate under bending). In the analysis of rectangular plates specific kinematically admissible basis functions are applied. We can consider, for example, a cantilever plate, and the representation of the two-dimensional mode shapes can be constructed using mode shapes of a cantilever beam in one direction and the translation and rotation rigid-body modes of an unconstrained beam (with free ends) in the second direction. This method of function selection takes advantage of the orthogonality properties of mode shapes for a beam and thus leads to simplifications in the evaluation of the stiffness and mass matrices.

15.2.2 Results of FEM Analysis

To describe the natural (free) vibrations using FEM it is necessary to solve the following homogeneous algebraic matrix equation

$$(\mathbf{K} - \omega^2\, \mathbf{M})\mathbf{V} = \mathbf{0} \tag{15.17}$$

where **K** – stiffness matrix, **M** – inertia (mass) matrix and **V** – vector of amplitudes of nodal displacements describing the vibration mode. This equation represents an algebraic eigenproblem, whose solution consists of the calculation of eigencouples (eigenvalue and eigenvector) that have a physical interpretation of the squared angular frequency and vibration mode.

From the numerical analysis using $NSE = 20 \times 20$ finite elements and computer code ANSYS (2013) we present three frequency values in Box 15.1, comparing them with analytical results. In the case of structures with symmetry planes the occurrence of double eigenvalues is possible. The graphical postprocessor makes it possible to vizualize the vibration modes using different techniques (Figure 15.2).

15.3 Parametric Analysis of Free Vibrations of Rectangular Plates

For the Kirchhoff plate model there are many analytical solutions, for several domain shapes (e.g. rectangular, circular) and for many cases of boundary conditions. On the contrary, in many FE codes, the formulation based on M–R plate theory is often applied. For example, Al Janabi et al. (1989) present a parametric study of the natural vibrations for square plates using Mindlin plate FEs with nine-node Lagrange shape functions and enhanced assumed shear strain interpolation (see Subsection 18.3.9). The starting point of theoretical considerations is the work of Mindlin (1951) on the application of moderately thick plate theory, which includes the effects of rotatory inertia and transverse shear deformation. Two mass representation options are considered: (i) consistent mass matrix is computed using 3×3 Gauss NI quadrature or (ii) diagonal matrix of lumped masses is obtained with the 3×3 Lobatto integration rule. Many numerical solutions presented in Al Janabi et al. (1989) can be treated as benchmark tests for free vibration analysis. The results for various cases (e.g. boundary conditions) and values of essential parameters (e.g. thickness-to-span ratios) are considered and compared with the results obtained using the 3D formulation and thin plate theory.

On the basis of Reddy (2007) we discuss the dependence of natural frequencies on: (i) representation of transverse shear deformation, including two values of shear

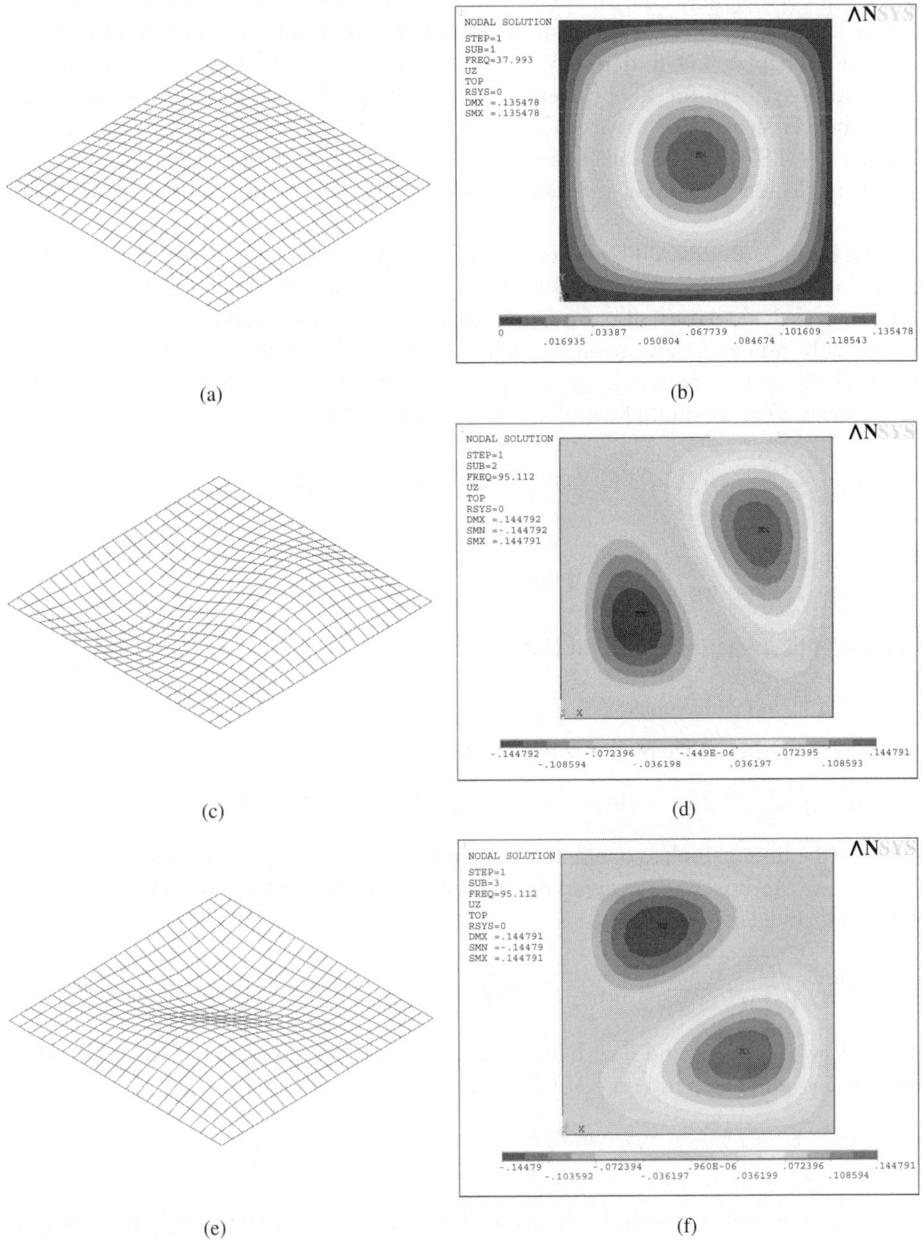

Figure 15.2 Simply supported square plate – three first modes of transverse natural vibrations: (a), (b) for frequency $f^{(1)} = f_{(1,1)} = 37.99$ Hz, (c)–(f) for equal frequencies $f^{(2)} = f^{(3)} = f_{(1,2)} = 95.11$ Hz; two techniques of mode presentation, ROBOT (2006) and ANSYS. *Subfigures (b) and (d) shown in Plate section for color representation of this figure.*

correction coefficient k and (ii) rotational inertia (abbreviated to ROTI). Please, note that in Reddy (2007) the abbreviation RI is used for rotational inertia, but in this book (see Chapter 17) it denotes the so-called reduced numerical integration, so we substitute it here for ROTI. Depending on the plate thickness and the ratio L/h either the classical plate theory (CPT model) or the first-order shear deformation plate theory (FSDT model) are commonly applied. The influence of rotational inertia can be neglected (which is marked as 'w/o ROTI') or included (marked as 'with ROTI').

In the book by Reddy (2007) a host of different plates are considered. We focus our discussion on a square isotropic simply supported plate ($L = a = b$). The analytical and numerical calculation of natural angular frequencies $\omega_{(i,j)}$ and associated vibration modes with indices i, j are given, for instance, in books by Reddy (2007), Batoz and Dhatt (1990), and in the article by Batoz et al. (1980). Different notation is proposed in these works. Next we quote formulae starting with Reddy (2007):

$$\omega_{(1,1)} = \tilde{\omega}^{\mathrm{R}}_{(1,1)}\, A, \quad \text{where} \quad \tilde{\omega}^{\mathrm{R}}_{(1,1)} = \overline{\omega}^{\mathrm{R}}_{(1,1)}\, \pi^2, \quad A = \frac{1}{L^2}\sqrt{\frac{D^m}{\rho h}} \tag{15.18}$$

$$\omega_{(1,1)} = \tilde{\tilde{\omega}}^{\mathrm{R}}_{(1,1)}\, B, \quad \text{where} \quad B = \frac{h}{L}\sqrt{\frac{E}{\rho}} \tag{15.19}$$

from the book by Batoz and Dhatt (1990)

$$\omega_{(i,j)} = \lambda^{\text{B-D}}_{(i,j)}\, B \tag{15.20}$$

as well as from the article by Batoz et al. (1980)

$$\omega^2_{(i,j)} = \lambda^{\text{B-B-H}}_{(i,j)}\, A^2\,, \quad \text{where} \quad \lambda^{\text{B-B-H}}_{(i,j)} = (i^2 + j^2)^2\, \pi^4 \quad \text{for} \quad \beta = a/b = 1 \tag{15.21}$$

We write the dimensionless coefficients using upper indices R, B-D and B-B-H to mark the sources. Moreover, we provide relations between the coefficients employed in the previous formulae:

$$\tilde{\omega}^{\mathrm{R}}_{(1,1)}\sqrt{\frac{1}{12(1-\nu^2)}} = \tilde{\tilde{\omega}}^{\mathrm{R}}_{(1,1)}, \qquad \tilde{\tilde{\omega}}^{\mathrm{R}}_{(i,j)} = \lambda^{\text{B-D}}_{(i,j)}, \qquad \tilde{\omega}^{\mathrm{R}}_{(i,j)} = \sqrt{\lambda^{\text{B-B-H}}_{(i,j)}} \tag{15.22}$$

We recall two factors A and B with unit [rad/s], defined in Equations (15.18) and (15.19), respectively, which depend on plate dimensions and material data:

$$A = \frac{1}{L^2}\sqrt{\frac{D^m}{\rho\, h}}, \qquad B = \frac{h}{L}\sqrt{\frac{E}{\rho}} \tag{15.23}$$

Considering the first example, we read the dimensionless coefficients $\overline{\omega}_{(1,1)}$ for a moderately thick plate characterized by $L/h = 10$ from Table 9.2.1 in Reddy (2007), optionally taking into account the rotational inertia. Using Equation $(15.18)_2$ we obtain:

- w/o ROTI: $\overline{\omega}^{\mathrm{R}}_{(1,1)} = 2.00$, $\tilde{\omega}^{\mathrm{R}}_{(1,1)} = 19.73$
- with ROTI: $\overline{\omega}^{\mathrm{R}}_{(1,1)} = 1.984$, $\tilde{\omega}^{\mathrm{R}}_{(1,1)} = 19.58$

Box 15.2 contains the material and geometrical data for the square simply supported isotropic plate, as well as the values of the lowest angular frequency $\omega_{(1,1)}$ for the considered cases, computed from Equation $(15.18)_1$.

Box 15.2 Fundamental natural frequencies for a moderately thick simply supported square plate

Data

$E = 1.47 \times 10^{11}$ N/m^2, $\nu = 0.3$, $\rho = 10^4$ kg/m^3 = 10^4 N s^2/m^4
$L = 1.0$ m, $h = 0.1$ m, $L/h = 10$, $\mu = \rho\, h = 10^3$ N s^2/m^3
$D^m = 1.346 \times 10^7$ Nm
$A = L^{-2}\,[D^m/(\rho\, h)]^{1/2} = 116$ rad/s
$B = (h/L)\,(E/\rho)^{1/2} = 383$ rad/s

Analytical values calculated on the basis of Equation (15.18)$_1$ $\omega_{(1,1)} = \tilde{\omega}^{\mathrm{R}}_{(1,1)}\,A$

w/o RI: $\tilde{\omega}_{(1,1)} = 19.73$, $\omega_{(1,1)} = 2288.68$ rad/s
with RI: $\tilde{\omega}_{(1,1)} = 19.58$, $\underline{\omega_{(1,1)} = 2271.28 \text{ rad/s}}$

Analytical values using Equation (15.19)$_1$ $\omega_{(1,1)} = \tilde{\tilde{\omega}}^{\mathrm{R}}_{(1,1)}\,B$

(1) CPT w/o ROTI: $\tilde{\tilde{\omega}}^{\mathrm{R/K-L}}_{(1,1)} = 5.973$, $\omega_{(1,1)} = 2288$ rad/s
(2) CPT with ROTI: $\tilde{\tilde{\omega}}^{\mathrm{R/K-L}}_{(1,1)} = 5.925$, $\underline{\omega_{(1,1)} = 2269 \text{ rad/s}}$
(3) FSDT $k = 5/6$ w/o ROTI: $\tilde{\tilde{\omega}}^{\mathrm{R/M-R}}_{(1,1)} = 5.812$, $\omega_{(1,1)} = 2226$ rad/s
(4) $k = 5/6$ with ROTI: $\tilde{\tilde{\omega}}^{\mathrm{R/M-R}}_{(1,1)} = 5.769$, $\omega_{(1,1)} = 2210$ rad/s
(5) FSDT $k = 2/3$ w/o ROTI: $\tilde{\tilde{\omega}}^{\mathrm{R/M-R}}_{(1,1)} = 5.773$, $\omega_{(1,1)} = 2211$ rad/s
(6) $k = 2/3$ with ROTI: $\tilde{\tilde{\omega}}^{\mathrm{R/M-R}}_{(1,1)} = 5.732$, $\omega_{(1,1)} = 2197$ rad/s

We notice that for the limit value of ratio $L/h = 10$ some decrease in the lowest angular frequency occurs when the rotational inertia is taken into account.

In Box 15.2 we compare the results obtained using Equation (15.19)$_1$ and parameter B. We consider a moderately thick plate. Cases (1),(2) are first analysed according to CPT and cases (3)–(6) are modelled by FSDT. The dimensionless coefficients $\tilde{\tilde{\omega}}_{(1,1)}$ are read from Table 10.2.3 in Reddy (2007) and the actual fundamental frequencies $\omega_{(1,1)}$ are calculated.

Obviously, when one assumes $\nu = 0.3$ and incorporates the rotational inertia, Equations (15.18) and (15.19) provide the same results (within truncation error):

$$\begin{aligned} \omega_{(1,1)} &= \tilde{\omega}^{\mathrm{R}}_{(1,1)}\,A = 19.58 \times 116 \text{ rad/s} \approx 2270 \text{ rad/s} \\ \omega_{(1,1)} &= \tilde{\tilde{\omega}}^{\mathrm{R}}_{(1,1)}\,B = 5.925 \times 383 \text{ rad/s} \approx 2270 \text{ rad/s} \end{aligned} \tag{15.24}$$

The results for this case are underlined in Box 15.2.

For the plate with the ratio $L/h \leq 10$ the enhancement of the plate theory from CPT to FSDT, as well as the incorporation of ROTI, leads to a decrease in the fundamental angular frequency.

Box 15.3 Natural frequencies of a thin and moderately thick simply supported square plate

Data

$E = 1.47 \times 10^{11}\ \text{N/m}^2, \quad \nu = 0.3, \quad \rho = 10^4\ \text{kg/m}^3 = 10^4\ \text{N s}^2/\text{m}^4$
$L = 1.0\ \text{m}, \quad B^{(i)} = (h^{(i)}/L)\,(E/\rho)^{1/2} = h^{(i)} \times 3830\ \text{rad/s}$

Analytical values using Equation (15.20) $\omega_{(i,j)} = \lambda_{(i,j)}^{\text{B-D}}\, B^{(i)}$

$h^{(1)} = 0.001\ \text{m}, \quad L/h^{(1)} = 1000, \quad B^{(1)} = 3.830\ \text{rad/s}$

K–L $\quad \lambda_{(1,1)}^{\text{B-D/K-L}} = 5.97, \quad \omega_{(1,1)} = 22.86\ \text{rad/s}$

$\lambda_{(3,3)}^{\text{B-D/K-L}} = 53.76, \quad \omega_{(3,3)} = 205.9\ \text{rad/s}$

$h^{(2)} = 0.1\ \text{m}, \quad L/h^{(2)} = 10, \quad B^{(2)} = 383.0\ \text{rad/s}$

M–R $\quad \lambda_{(1,1)}^{\text{B-D/M-R}} = 5.77, \quad \omega_{(1,1)} = 2209.9\ \text{rad/s}$

$\lambda_{(3,3)}^{\text{B-D/M-R}} = 42.38, \quad \omega_{(3,3)} = 16\,231.5\ \text{rad/s}$

Next, we read the dimensionless coefficients denoted by $\lambda_{(i,j)}^{\text{B-D}}$ from Table 4.1.9 in Batoz and Dhatt (1990), see Equation (15.20). For a very thin plate with $L/h = 1000$ the upper index K–L is additionally used, as opposed to the case of a moderately thick plate with $L/h = 10$, which is marked by index M–R.

Using Equation (15.20), which contains the coefficient B, dependent among otherthings on thickness h, we calculate two selected angular frequencies $\omega_{(1,1)}$ and $\omega_{(3,3)}$ associated with vibration modes with two symmetry planes, that is with the following numbers of half-waves $i = j = 1$ and $i = j = 3$. The results are listed in Box 15.3. The value of $\omega_{(i,j)}$ is proportional to h, so we notice that a plate that is 100 times thicker vibrates with about 100 times larger fundamental frequency. We also mention that, for the thin plate, the frequency associated with $i = j = 3$ is about nine times larger than for $i = j = 1$.

Next we consider a second example. In Batoz et al. (1980) the results of numerical analysis are compared with analytical ones. One quarter of a square plate with the size $L = 2\,a$ is considered. The influence of the triangular FE model, mesh orientation and mesh density on the simulation results is presented in Tables I and III in Batoz et al. (1980). Two FE formulations are employed: Discrete Kirchhoff Triangle (DKT) – see Subsection 18.3.3 and Hybrid Stress Triangle (HST) – see Subsection 18.2.6. Two different orientations of regular meshes are applied: elements in mesh A have hypotenuses parallel to the plate diagonal direction and elements in mesh B have hypotenuses normal to the diagonal. The mesh density is described by numbers $N = 1, 2, 4, 8$ defining the quarter domain edge division.

In Batoz et al. (1980) Equation (15.21) is used in which $\tilde{\omega}_{(i,j)}^2 = \lambda_{(i,j)}^{\text{B-B-H}}$. This relation is a result of the form of matrix equation of free vibrations (15.17), where ω^2 is denoted by λ.

Table 15.1 Values of relative errors for natural frequencies $\tilde{\omega}^{\text{FEM}}_{(i,j)}/\tilde{\omega}^{\text{anal}}_{(i,j)}$ for eight cases

(i,j)	FEs	Mesh	$N=1$	$N=2$	$N=4$	$N=8$	$\tilde{\omega}^{\text{anal}}_{(i,j)}$
(1, 1)	DKT	A	0.927	0.981	0.994		19.74
		B	0.988	0.980	0.993		
	HST	A	1.129	1.028	1.007		
		B	1.087	1.021	1.005		
(3, 3)	DKT	A		0.678	0.925	0.982	177.65
		B		0.725	0.947	0.987	
	HST	A		1.018	1.005	1.000	
		B		1.124	1.023	1.006	

Using Equation (15.21) we calculate for $i = j = 1$ and $i = j = 3$:

- $\tilde{\omega}^2_{(1,1)} = \lambda^{\text{B-B-H}}_{(1,1)} = 389.6 \quad \rightarrow \quad \tilde{\omega}^{\text{anal}}_{(1,1)} = 19.74$
- $\tilde{\omega}^2_{(3,3)} = \lambda^{\text{B-B-H}}_{(3,3)} = 31\,560.55 \quad \rightarrow \quad \tilde{\omega}^{\text{anal}}_{(3,3)} = 177.65$

These values are used to compute the relative error of FEM results in terms of dimensionless quantities $\tilde{\omega}^{\text{FEM}}_{(i,j)}/\tilde{\omega}^{\text{anal}}_{(i,j)}$.

On the basis of Tables I and III from Batoz et al. (1980) we have computed the relative error of numerical simulations, listed in Table 15.1. The error is monitored for the first frequency $i = j = 1$ and the frequency with $i = j = 3$. For element DKT convergence from below is observed, while for element HST it is from above. The convergence is faster for the lowest frequency $i = j = 1$ and element HST seems more robust.

The analysis of free vibrations of thin and moderately thick plates should be based on the classical Kirchhoff–Love theory and the Mindlin–Reissner theory, respectively. The analytical results of vibration analysis for thin and moderately thick plates are confronted with numerical computations. It is observed that the application of FEs based on K–L theory for moderately thick plates gives overestimated values of natural frequencies. The effect is analogical to the numerical analysis of buckling of moderately thick plates where too large critical loads are predicted if the K–L theory is employed.

15.4 Natural Vibrations of Cylindrical Shells

15.4.1 Analytical Solution of the Displacement Differential Equation for Free Vibrations

We now consider the free vibrations of thin cylindrical shells described by the equations of three-parameter K–L theory. Inertia forces F_x, F_θ, F_n, expressed by second time derivatives of three displacements $u_x = u$, $u_\theta = v$, $u_n = w$ and by inertia $I_0 = \mu$:

$$\begin{aligned} F_x(x,\theta,t) &= I_0\,\ddot{u}(x,\theta,t) \\ F_\theta(x,\theta,t) &= I_0\,\ddot{v}(x,\theta,t) \\ F_n(x,\theta,t) &= I_0\,\ddot{w}(x,\theta,t) \end{aligned} \tag{15.25}$$

are introduced into the set of three equations of equilibrium written in terms of displacements:

$$\begin{aligned}
&\frac{\partial^2 u}{\partial x^2} + \frac{1-\nu}{2R^2}\frac{\partial^2 u}{\partial \theta^2} + \frac{1+\nu}{2R}\frac{\partial^2 v}{\partial x\,\partial \theta} + \frac{\nu}{R}\frac{\partial w}{\partial x} + \frac{1-\nu^2}{Eh}F_x = 0\\
&\frac{1+\nu}{2R}\frac{\partial^2 u}{\partial x\,\partial \theta} + \frac{1-\nu}{2}\frac{\partial^2 v}{\partial x^2} + \frac{1}{R^2}\frac{\partial^2 v}{\partial \theta^2} + \frac{1}{R^2}\frac{\partial w}{\partial \theta} + \frac{1-\nu^2}{Eh}F_\theta = 0\\
&\left(\frac{\nu}{R}\frac{\partial u}{\partial x} + \frac{1}{R^2}\frac{\partial v}{\partial \theta} + \frac{w}{R^2}\right) + \frac{h^2}{12}\left(\frac{\partial^4 w}{\partial x^4} + \frac{2}{R^2}\frac{\partial^4 w}{\partial x^2\,\partial \theta^2} + \frac{1}{R^4}\frac{\partial^4 w}{\partial \theta^4}\right) + \frac{1-\nu^2}{Eh}F_n = 0
\end{aligned} \tag{15.26}$$

After the separation of two spatial coordinates x, θ and time t the functions of displacement amplitudes $U(x,\theta)$, $V(x,\theta)$, $W(x,\theta)$ appear, which are multiplied by the function $T(t)$:

$$u(x,\theta,t) = U(x,\theta)\,T(t), \quad v(x,\theta,t) = V(x,\theta)\,T(t), \quad w(x,\theta,t) = W(x,\theta)\,T(t) \tag{15.27}$$

In the next step for consideration, we use constant coefficients $\tilde{U}$, $\tilde{V}$, $\tilde{W}$ (maximum displacement values) and appropriate trigonometric functions dependent on coordinate x and angular coordinate θ or arc variable $y = R\theta$:

$$\begin{aligned}
&U(x,y) = \tilde{U}\cos\frac{i\pi x}{L}\cos\frac{jy}{R}, \quad V(x,y) = \tilde{V}\sin\frac{i\pi x}{L}\sin\frac{jy}{R}\\
&W(x,y) = \tilde{W}\sin\frac{i\pi x}{L}\cos\frac{jy}{R}
\end{aligned} \tag{15.28}$$

The functions of displacement amplitudes proposed in this way can describe the vibrations of a cylindrical shell with the following boundary conditions: (i) a shell connected at two ends with diaphragms stiff in their planes or (ii) a shell simply supported on two curved edges (sliding parallel to cylinder axis is possible). For these two cases of boundary conditions, both displacements normal and tangents to the circular edges in their planes are restricted. These conditions are satisfied for $x = 0, L$ owing to the adoption of function $\sin(i\pi x/L)$. After the introduction of this displacement amplitudes $\tilde{U}$, $\tilde{V}$, $\tilde{W}$ into the three displacement differential Equations (15.26) and the elimination of trigonometric functions a set of three homogeneous linear algebraic equations is obtained with the unknowns angular frequency ω and coefficients $\tilde{U}$, $\tilde{V}$, $\tilde{W}$. The condition of the occurrence of vibrations described by nonzero displacement amplitudes requires that the determinant of the matrix for the set of equations vanishes. We obtain a third degree equation in dimensionless frequency, which has three real and positive roots. Each root corresponds to an angular vibration frequency and a respective ratio of amplitudes $\tilde{U} : \tilde{V} : \tilde{W}$. Three cases can be considered where one parameter dominates. When $\tilde{U}$ is much larger compared with $\tilde{V}$, $\tilde{W}$ then longitudinal vibrations in direction x occur, when $\tilde{V}$ dominates then torsional vibrations appear, when $\tilde{W}$ is the largest then transverse oscillations take place.

In the displacement differential Equation $(15.26)_3$ the inertia terms $-I_2\,(\ddot{\varphi}_x)_{,x}$ and $-I_2\,(\ddot{\varphi}_\theta)_{,\theta}$ have been neglected since we are dealing with thin shells. For moderately thick shells rotational inertia terms should be considered with rotation angles treated as independent fields according to five-parameter shell theory.

15.4.2 Natural Vibrations of Cylindrical Shells Clamped on the Bottom Edge – FEM Solution

Two cylindrical shells differing only in length (so-called long or short shells) are characterized by material and geometrical data shown in Box 15.4. The shell is classified as thin since $h/R = 0.12/10 = 0.012 < 0.05$, see Section 1.5. Computer code ANSYS and two types of two-dimensional elements SHELL43 and SHELL93 are used to carry out the numerical analysis.

In the computation of free vibrations for the axially symmetric shells (similarly to the buckling analysis) double eigenvalues occur. They are associated with qualitatively similar displacement modes, rotated with respect to each other. In Box 15.4 the set of calculated frequencies is presented for long and short shells (double eigenvalues are considered only once) and for the two elements SHELL43 and SHELL93. Three initial qualitatively different free vibration modes of the long and short shells are shown in Figure 15.3. In the calculations using two types of FEs (and with different numbers of elements in the model), close values of frequencies are obtained. In the case of the long shell the fundamental vibration mode associated with the lowest frequency exhibits four circumferential waves. On the other hand, the first eigenmode for the short shell exhibits seven circumferential waves. We notice that, for the short shell, three frequency

Box 15.4 Free vibrations of long and short cylindrical shells

Data

$E = 1.47 \times 10^7\ \text{kN/m}^2,\quad \nu = 0.167,\quad \gamma = 24.0\ \text{kN/m}^3$
$R = 10.0\ \text{m},\quad h = 0.12\ \text{m},\quad L_\text{L} = 20.0\ \text{m},\quad L_\text{S} = 5.0\ \text{m}$

Check values from FEM solution

ANSYS computer code

- long shell discretized by:
  - SHELL43 FEs with $NSE = 3672$
    $f^{(j=4)} = 90.824$ Hz, $f^{(j=3)} = 106.70$ Hz, $f^{(j=5)} = 114.21$ Hz
  - SHELL93 FEs with $NSE = 11016$
    $f^{(j=4)} = 90.681$ Hz, $f^{(j=3)} = 106.65$ Hz, $f^{(j=5)} = 113.82$ Hz
- short shell discretized by:
  - SHELL43 FEs with $NSE = 972$
    $f^{(j=7)} = 364.15$ Hz, $f^{(j=6)} = 374.78$ Hz,
    $f^{(j=8)} = 387.34$ Hz, $f^{(j=5)} = 424.17$ Hz
  - SHELL93 FEs with $NSE = 2916$
    $f^{(j=7)} = 363.27$ Hz, $f^{(j=6)} = 374.04$ Hz,
    $f^{(j=8)} = 385.95$ Hz, $f^{(j=5)} = 423.41$ Hz

Vibration modes for short and long shells shown in Figure 15.3

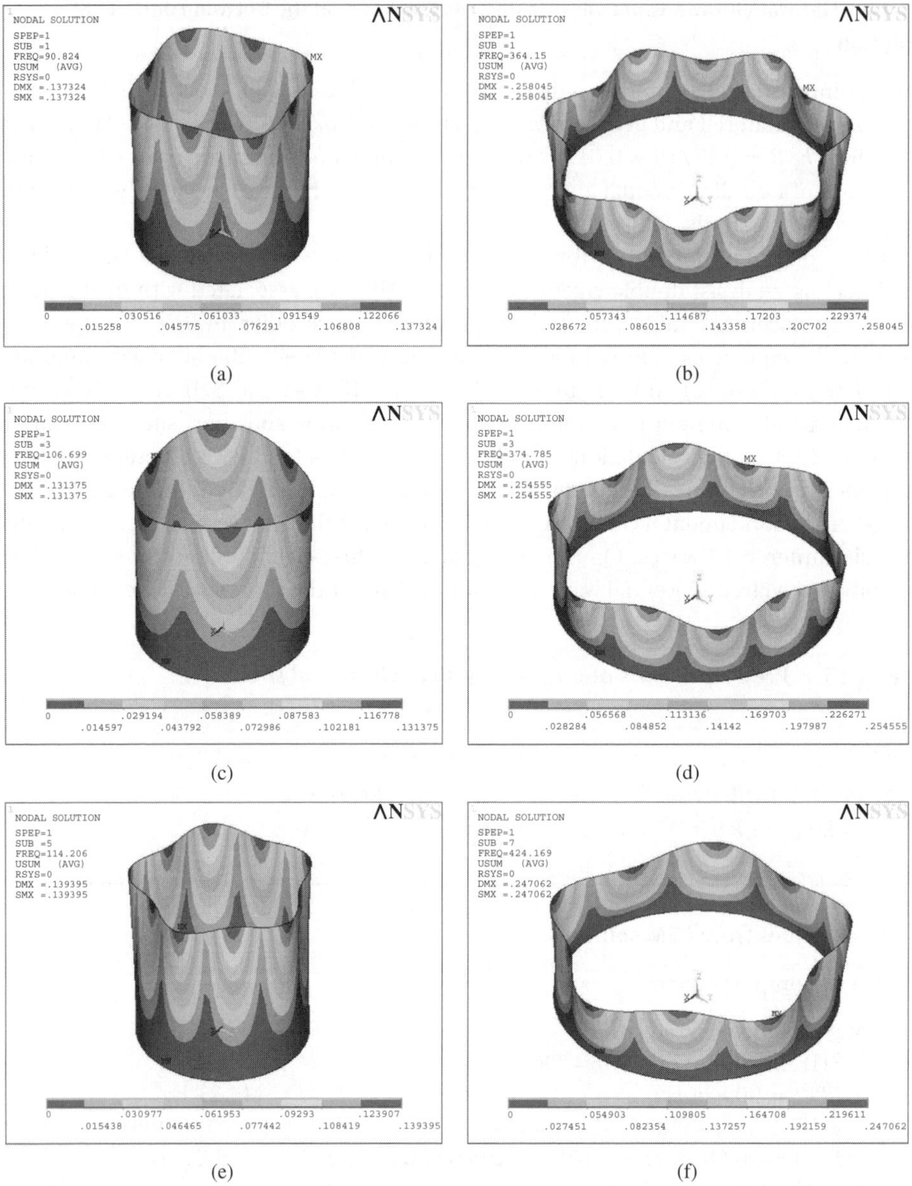

Figure 15.3 Three natural vibration modes related to three initial frequencies for a long shell with $j = 4, 3, 5$ (first column) and for a short shell with $j = 7, 6, 5$ (second column); FEM results from ANSYS. *See Plate section for color representation of this figure.*

values for $j = 7, 6, 8$ are very close to one another, while the frequency for $j = 5$ is much larger.

Next, we provide some comments on the analogy between the computational models and ensuing eigenproblems of free vibrations and buckling. In the book by Reddy (2007), in the same Section 11.5, one can find information about both vibration and buckling analysis of cylindrical shells. It turns out that in both these problems one observes similar qualitative and quantitative effects, related to the solution of eigenvalue problems.

In the previous Chapter 14 in Subsection 14.4.4 the influence of geometrical parameters L/R and R/h on the critical load and the associated buckling mode is discussed. Here we emphasize that fundamental frequency and vibration mode depend on the same parameters.

Considering shells with $R/h = 60$ and $L/R = 1$ and using the formulation based on FSDT we obtain the fundamental frequency that is related to $j = 6$ and is lower in comparison with frequency $f^{(j=0)}$ connected with an axisymmetric mode. If the shell is extremely short, the frequency $f^{(j=0)}$, is always much larger than the fundamental frequency related to a higher number j of circumferential waves, see Reddy (2007).

In parametric analysis different cases of boundary conditions are taken into account and their influence on the frequencies as well as on respective mode shapes (especially on the number of circumferential waves) is similar to the buckling analysis. If, beside the constraint $w = \varphi_\alpha = 0$ the condition of zero translations $u_\alpha = 0$ at two clamped ends of a cylindrical shell is additionally imposed, then the fundamental frequency is reduced.

15.5 Remarks

In this section we have described some basic problems of dynamic analysis, focusing on free vibrations of plates and shells without damping. Here we provide some final remarks.

First of all we recommend the FEM calculations to be performed for the whole structure (not a quarter or a segment). This is because, considering for instance a specific segment of a shell of revolution we must assume appropriate boundary conditions on the lines generated by vertical section planes, limiting in this way the set of possible vibration modes.

Different methods are implemented in computer codes to solve the algebraic eigenproblem of free vibrations. In the analysis of both the natural vibrations and buckling eigenvalues close to one another or identical occur in the solutions of eigenvalue problems, while the eigenvectors associated with them describe completely different deformation modes. In such cases proper procedures to solve the eigenproblem are required and a sufficiently dense discretization must be used to reproduce complex deformation modes involving many waves (for shells of revolution along the circumference and/or the meridian).

Although the current computational capabilities of the codes hardly limit the number of *NDDOF*, usually only some eigenvalues lying in an interval, defined by the user in input data, are computed. One often decides to calculate the initial part of the eigenvalue spectrum (the lowest frequencies) because structures most often vibrate with low frequencies, but not necessarily with the simplest modes. On the other hand it is advisable to pay attention to the frequency of forced vibrations corresponding to the dynamic load assumed to be harmonic. In engineering structures resonance phenomena must be avoided, so the frequency of the applied force should not be close to any natural vibration frequency. In the case of detection that the load frequency is similar to a natural frequency, the structure should be tuned by changing its mass or stiffness, which determine its dynamic characteristics.

References

Al Janabi BS, Hinton E and Vuksanovic D 1989 Free vibrations of Mindlin plates using the finite element method: Part 1. Square plates with various edge conditions. *Eng. Comput.* **6**, 90–96

ANSYS 2013 ANSYS Inc. PDF Documentation for Release 15.0. SAS IP, Inc.

Batoz JL and Dhatt G 1990 *Modélisation des Structures par Élément Finis.* Hermes, Paris.

Batoz JL, Bathe KL and Ho LW 1980 A study of three-node triangular plate bending elements. *International Journal for Numerical Methods in Engineering* **15**(12), 1771–1812

Geradin M and Rixen D 2015 *Mechanical Vibrations: Theory and Application to Structural Dynamics* 3rd edn. John Wiley & Sons, Ltd, Chichester, UK.

Mindlin RD 1951 Influence of rotatory intertia and shear on flexural motions of isotropic, elastic plates. *ASME Journal of Applied Mechanics* **73**, 31–38

Reddy JN 1999 *Theory and Analysis of Elastic Plates.* Taylor & Francis, London.

Reddy JN 2007 *Theory and Analysis of Elastic Plates and Shells* 2nd edn. CRC Press/Taylor & Francis, Boca Raton-London-New York.

ROBOT 2006 ROBOT Millennium: User's Guide. Technical report, RoboBAT, Cracow (in Polish).

Part 5

Aspects of FE Analysis

16

Modelling Process

16.1 Advantages of Numerical Simulations

Computational mechanics of structures makes it possible to obtain solutions of complex problems, in particular for plates and shells. In fact, such problems stimulated the fast progress of FEM. Analytical solutions known from the literature, including this book, cover simple examples (standard structural elements, loads and boundary conditions). In advanced engineering projects designers often encounter complex structures, various external actions and intricate forms of deformation. Fortunately, most of these aspects can be represented in FE modelling. While performing such advanced analyses one should pay special attention to the following issues: convex/concave domain with corners, concentrated loads, line and/or point supports, connections between substructures (possibly having different dimensions – 1D, 2D and 3D). All these issues can be the source of increased discretization errors.

Modern computational technologies make it possible to reconstruct certain structural states and provide a computer simulation of complex mechanical phenomena or processes. However, without mechanical knowledge we would not be able to propose a reasonable model for computations. We often begin the analysis from running 'pilot' computations for a simple model, and then we upgrade it. Basing on physical arguments we are able to determine crucial factors, the correction of which leads to a more accurate description of the problem at hand, for example we indicate places in the structure that should be examined with particular attention.

In the computational analysis of plates and shells one often performs a parametric study to investigate the structural response to various external actions and the influence of selected model parameters. For instance the ratio L/h (length to thickness) is varied to examine thin and moderately thick plates or the ratio L/R (length to radius) is changed to analyse long and short cylindrical shells.

In engineering education the finite element toolbox (CALFEM 2004) for MATLAB environment is frequently used at numerous universities. The toolbox makes it possible to implement a selected finite element, new solution algorithm or postprocessing routine, using suitable numerical methods, as well as to perform computations for varying input data (i.e. geometrical or material data, loading or boundary conditions).

Professional finite element codes offer FE libraries, leaving it up to the user to select a proper element for the problem at hand. A further decision is related to the selection of the order of approximation and of the FE mesh density. In this process the user

Plate and Shell Structures: Selected Analytical and Finite Element Solutions, First Edition.
Maria Radwańska, Anna Stankiewicz, Adam Wosatko and Jerzy Pamin.
© 2017 John Wiley & Sons Ltd. Published 2017 by John Wiley & Sons Ltd.

must exhibit consciousness and creativity based on the knowledge of mechanical and numerical aspects of the problem to be solved. Fortunately, modern FE packages provide advanced and reliable preprocessors, solvers and postprocessors. Nowadays, user-friendly preprocessors make the change of input parameters easy. The efficiency of processors is very high, enabling numerous repetitions of the simulations. Advanced postprocessors support a fast qualitative and quantitative analysis of results, but the user must select suitable variables to be checked. Moreover, contemporary software often makes use of adaptive mesh refinement techniques, coupled with appropriate discretization error estimation (see, for instance, Zienkiewicz and Zhu 1987).

16.2 Complexity of Shell Structures Affecting FEM

One has to know the mechanics of plates and shells (assumptions and equations of the theories of thin and moderately thick surface structures) to be able to apply the FE technology in a conscious manner (Cook et al. 1989, Cook 1995). It is trivial to say that the user of FE software must read the manual, together with its theoretical part. Note that the formulations must be derived from either the continuum mechanics (3D) or from the theory of plates and shells. If the reader has studied Parts 1–4 of this book and is embarking on the task of performing simulations using a selected FE package, then this Part 5 is useful in providing information on both the process of FE modelling and numerous recent developments concerning FE formulations for the considered field of application.

In (geometrically and physically) linear analysis of thin-walled structures, three basic mechanical states can be separated out: membrane (n), bending (flexure) (m) and transverse shear (t). Complex stress states are understood as a superposition of the three components having different range and intensity. Using the cross-section model for thin and moderately thick shells we employ generalized strains as well as resultant forces and moments (stresses integrated over the thickness). They are mutually related by respective stiffnesses:

$$D^n = \frac{Eh}{1-\nu^2}, \qquad D^m = \frac{Eh^3}{12(1-\nu^2)}, \qquad D^t = \frac{kEh}{2(1+\nu)}, \qquad k = \frac{5}{6} \tag{16.1}$$

In the case of thin shells the membrane and transverse shear stiffnesses (proportional to h) are much larger than the bending stiffness (proportional to h^3). We also know from the shell theory that the bending effects in structures can have a local character and involve abrupt variations in the distributions of characteristic fields, including sign changes (see Sections 11.1 and 11.2). The effects of transverse shear must be assessed in the analysis of thin/moderately thick plates/shells (see Subsection 8.5.1 and Box 8.2 for limits of L/h ratio).

It is necessary to be aware of the relations between displacements, strains and stresses in the adopted physical and mathematical model in order to be able to understand FE formulations, involving one, two or many approximated fields. The FE models are based on various approximation methods of growing complexity. Some information in this respect is given further in Chapter 18. For example we recall that in the one-field displacement-based model the employed polynomial must be complete and at least of degree p. The integer p is the highest order of a derivative present in the global formulation based on potential energy, which is a functional of the displacement vector function. It is simultaneously related to order $2p$ of the displacement differential equation occurring in the local formulation of the problem: for instance for a bending thin plate $p = 2$ (Zienkiewicz et al. 2005). Obviously, the formulations incorporate differential kinematic relations and this means that the approximation of strains (as secondary field) is determined by the approximation of displacements.

As noted already on p. 139 in the book by Cook 1995, 'some commercial programs are self-adaptive, which means that they are able to estimate the error of a FE solution, revise the mesh, reanalyse, and repeat this cycle automatically until a prescribed convergence tolerance is met.' It seems that, up to now, this possibility is not often available. The user usually is able to choose between h-refinement and p-refinement, in which the mesh density or the order of interpolation polynomial is upgraded, respectively. To be precise, in the h-refinement the element size decreases to provide a better approximation without changing the polynomial order (e.g. preserving linear interpolation). In the p-refinement the approximation order is increased without changing the element size h. In the latter case nodes are added to simple elements and/or existing nodes are augmented with additional degrees of freedom. Classical or hierarchical shape functions can be used (Zienkiewicz et al. 2005). Some very modern programs use finite element approximations based on a combination of the h-refinement and p-refinement, that is hp-refinement (see Subsection 18.3.10).

Before embarking on a computational analysis it is crucial to answer a couple of questions concerning the problem at hand. In the work by Bischoff et al. 2004 the following questions are posed. 'Which mechanical effects should be included and which can be neglected? Is it better to start from a shell theory or develop continuum-based elements along the lines of the degenerated solid approach? Which simplifications are useful – and admissible? Which consequences does the formulation have for the finite elements model? Which parametrization of degrees of freedom is sensible for the applications one has in mind? Should drilling degrees of freedom be included in a shell formulation or not?' As the authors stress, more questions can be rightfully asked and there are many publications providing answers relevant for plate and shell problems.

In this part of the book we will limit ourselves to a brief overview of existing advanced finite element formulations, thus summarizing of the progress of FEM for the considered field of applications and addressing these questions.

16.3 Particular Requirements for FEs in Plate and Shell Discretization

In Sections 1.3 and 1.4 the description of the geometry, as well as of displacement and strain fields for shell structures, have been provided. Several sets of equations in the local formulation have been given, including kinematic and constitutive relations, equilibrium equations and kinematic and/or static boundary conditions. These sets are the mathematical models of typical plates and shells. Using FEM one must represent the continuous model of physical reality by a discrete approximation. Therefore, finite elements should satisfy suitable requirements.

The discrete representation of the surface geometry can be added to the description of the other mechanical fields (displacements, strains and stresses). When the geometry is approximated using the same shape functions as the displacements the so-called isoparametric elements are formulated that can represent shells of arbitrary shape. Subparametric or superparametric combinations are also possible, in which the order of interpolation polynomials used for the representation of the geometry is lower or higher than the order of shape functions used for the approximation of the displacements, respectively. FEs applied to discretize two-dimensional surface structures should satisfy the condition of geometrical isotropy, which requires the geometrical variables ξ_1 and ξ_2 to be used in the same way in the polynomial approximation functions.

The deformation can be decomposed into fundamental deformation modes and approximation of each of them must satisfy certain conditions. In particular, FEs must be capable of reproducing the states of constant strain and representing strainless motion (translation and/or rotation of a finite element as a rigid body), see Section 17.5. This is related to the requirements that the differential operators in the kinematic relations impose on the displacement approximation. Testing an FE involves the assessment of the properties of its stiffness matrix, which depend, among others, on the type of numerical integration used (this issue is addressed in Section 17.3).

The deformation reproduced by an FE can be analysed in the space of eigenvectors of the stiffness matrix, determined in spectral analysis (see Section 17.2). As an example we now describe four-node element L4 with bilinear approximation of membrane displacements. Its stiffness matrix has the dimensions 8×8 and one Gauss point is used for integration. Figure 16.1 presents eight independent displacement modes of this element. The first three represent rigid-body motions (two translations and in-plane rotation). The next three modes represent the fundamental constant strain modes (two combinations of one-directional extension and/or compression and a shear mode). The last two modes no. 7 and 8 incorrectly reproduce in-plane bending (see Section 6.5). The reason is that in this element only bilinear shape functions are employed and it cannot represent curved edge lines related to in-plane bending (see also Subsection 17.3.1). This is an example of the influence of FE approximation on the reproduction of elementary deformation states.

It is obvious to say that the FE approximation must be consistent with the type of analysed structure and the mathematical model for the adopted theory representing the mechanical problem we would like to solve. In the context of plates and shells it should be emphasized that the five-parameter Mindlin–Reissner (M–R) theory is more general than the three-parameter Kirchhoff–Love (K–L) shell theory. Therefore, finite elements based on the former theory can be used to model both thin and moderately thick shells, taking into account the membrane (n), bending (m) and transverse shear (t) effects. In fact, universal finite elements should describe shells of arbitrary shape and with thickness from a wide range, providing proper results for a combination of

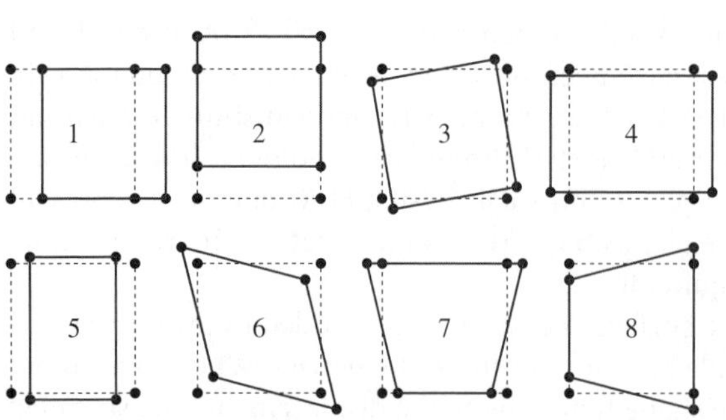

Figure 16.1 Independent displacement modes of bilinear membrane element

membrane (n), bending (m) and transverse shear (t) states. In Chapter 18 we will show that such elements exist.

Moreover, the fields approximated by FEs must satisfy suitable continuity requirements on interelement boundaries: for the K–L plate theory we must guarantee C^1-continuity of the deflection function and for the M–R theory C^0-continuity of the three independently approximated fields of the deflection and rotation angles. This means that in the former case the function approximating the plate deflection w must be continuous and possess continuous slopes in the direction normal and tangent to the element boundaries ($w,_\nu$, $w,_s$). For shells one must additionally guarantee the continuity of the two translations tangent to the middle surface.

The other essential requirement, which must by satisfied by an FE formulation, is that tractions along interelement boundaries must be in equilibrium. This requirement can be satisfied exactly or in a weak (variational) sense. This will be explained in Chapter 18.

Next, a FE should not be sensitive to shape distortion. Nevertheless, it is necessary to avoid the following shape defects of finite elements (see Cook 1995): large aspect ratio (see the quadrilateral and triangle in Figure 16.2a,b), near-triangle quadrilateral (see subfigure c), highly skewed shape (d), triangular quadrilateral (e), strongly curved sides (f) and off-centre node on an edge of 8/9-node FEs (g).

When FE software is developed several convergence tests are performed. In the tests the influence of FE mesh, that is node density and element shapes, on the results is examined. Advanced programs that make use of mesh adaption techniques involve iterative enhancement of results. In particular, during postprocessing the quality of the solution is examined (discretization error is estimated) and next during repeated preprocessing a better mesh is generated. This can be done automatically or the user can determine the point or line around which (or a subdomain in which) the mesh should be refined.

Finally, FE algorithms and their implementation should be optimized to provide results in a reasonable time of computations, especially when a sequence of calculations is needed. This becomes less and less important with the present progress of computer hardware, but in fact a part of these requirements has already been stated in the pioneering paper by Irons (1976).

We also stress that the knowledge and experience of mechanics and computer science specialists, who are users of FE packages, are the key to a successful modelling process and assessment of computational results.

Figure 16.2 Disadvantageous finite element shapes

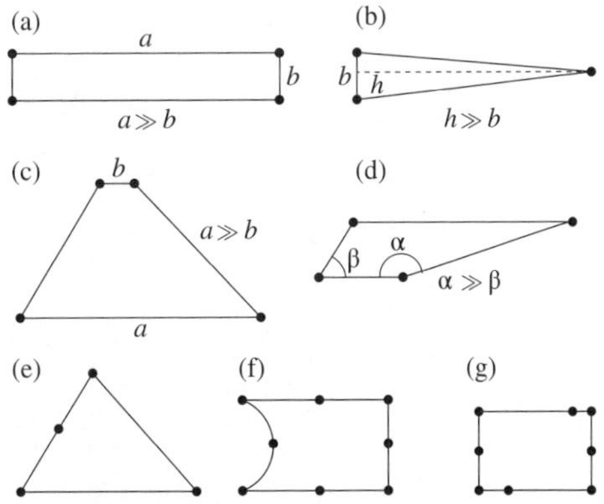

References

Bischoff M, Wall WA, Bletzinger KU and Ramm E 2004 Models and finite elements for thin-walled structures. In Stein E, de Borst R and Hughes TJR (eds), *Encyclopedia of Computational Mechanics: Solids and Structures*, vol. 2. John Wiley & Sons, Ltd, Chichester, UK, chapter 3, pp. 59–137.

CALFEM 2004 CALFEM – a finite element toolbox. Technical report, Structural Mechanics, Lund University, Lund.

Cook RD 1995 *Finite Element Modeling for Stress Analysis*. John Wiley & Sons Ltd, Chichester, UK.

Cook RD, Malkus DS and Plesha ME 1989 *Concepts and Applications of Finite Element Analysis* 3rd edn. John Wiley & Sons, Inc., New York.

Irons W 1976 The semiloof shell element. In Ashwell DG and Gallagher RH (eds), *Finite Elements for Thin Shells and Curved Members*. John Wiley & Sons, Ltd, Chichester, UK. pp. 197–222.

Zienkiewicz OC and Zhu JZ 1987 A simple error estimator and adaptative procedure for practical engineering analysis. *International Journal for Numerical Methods in Engineering* **24**(2), 337–357.

Zienkiewicz OC, Taylor RL and Zhu JZ 2005 *The Finite Element Method: Its Basis and Fundamentals*, 6th edn. Elsevier Butterworth-Heinemann.

17

Quality of FEs and Accuracy of Solutions in Linear Analysis

17.1 Order of Approximation Function versus Order of Numerical Integration Quadrature

In FEM, which is an approximate numerical method, one needs to guarantee the quality of FEs and the accuracy of the employed numerical routines. In what follows, some concepts of assessment of FEs and computation results are presented. Sections 17.1–17.3 concern mainly the displacement-based finite elements, since at the initial stage of FEM development this formulation was mostly used.

The order of polynomials in the shape functions which describe the displacements and the order of derivatives in the kinematic relations determine the order of polynomials describing the strain field. Numerical integration (NI) is usually necessary in the process of computation of element matrices and vectors expressed as surface integrals, while the order of interpolation polynomial implies which quadrature is required. When the number of Gauss points guarantees exact integration for a given order of the integrand then the quadrature is called full integration (FI). The use of a smaller number of integration points than in FI is reduced integration (RI). If this reduced order quadrature is applied to all mechanical actions the reduced integration is called uniform (URI). If the integrand is a sum of contributions and different quadrature orders are used for particular components then so-called selective integration (SI) is carried out, see Table 17.1.

In Figure 17.1 the positions of Gauss points used in FI and RI are shown for four- and eight-noded reference elements representing isoparametric quadrilateral ones denoted by Q4 and Q8, respectively,

17.2 Assessment of Element Quality via Spectral Analysis

First a physical interpretation is given for the spectral analysis of the stiffness matrix of a single finite element (the following also holds for a relevant set of elements, see Cook et al. 1989). We consider an FE without kinematic constraints and loads. It is known that the vector of nodal displacements is proportional to the vector of nodal forces. Using this relation we obtain the algebraic eigenproblem

$$\mathbf{k}\mathbf{q} = \mathbf{f} = \lambda\mathbf{q} \rightarrow (\mathbf{k} - \lambda\mathbf{I})\mathbf{q} = \mathbf{0} \tag{17.1}$$

Plate and Shell Structures: Selected Analytical and Finite Element Solutions, First Edition.
Maria Radwańska, Anna Stankiewicz, Adam Wosatko and Jerzy Pamin.
© 2017 John Wiley & Sons Ltd. Published 2017 by John Wiley & Sons Ltd.

Table 17.1 Numbers of zero eigenvalues for plate FEs based on M–R theory: L4, S8, L9, S12, L16

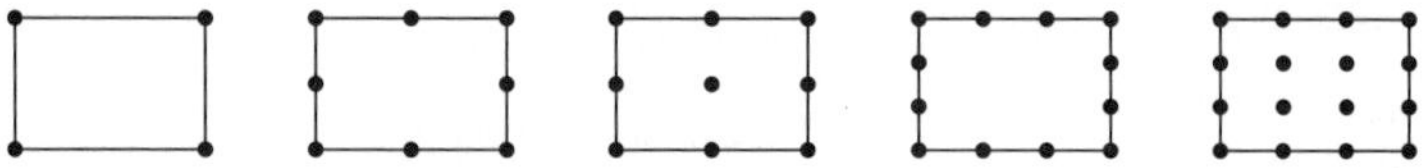

	L4	S8	L9	S12	L16
Shape functions	**1st**	**2nd**	**2nd**	**3th**	**3th**
NI			**Number of Gauss points**		
FI	2×2	3×3	3×3	4×4	4×4
RI	1×1	2×2	2×2	3×3	3×3
SI^t	1×1	2×2	2×2	3×3	3×3
SI^m	2×2	3×3	3×3	4×4	4×4
NI			**Number of zero eigenvalues**		
FI	3	3	3	3	3
SI	5	3	4	3	4
RI	7	4	7	3	7

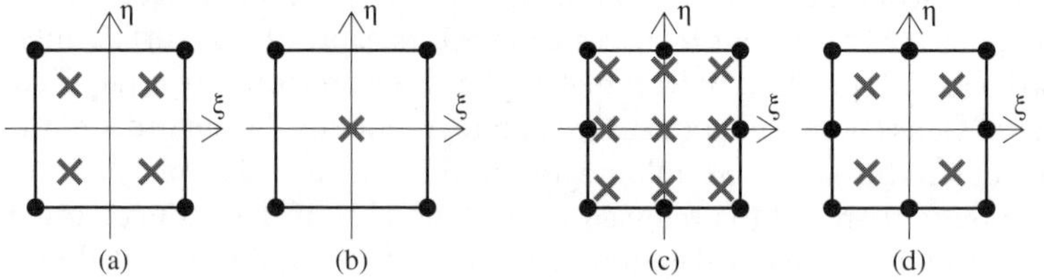

Figure 17.1 Positions of Gauss points in 2D FEs: (a) Q4 with FI (2×2), $\xi_i, \eta_i = \pm0.57735$; (b) Q4 with RI ($1\times1$), $\xi_1, \eta_1 = 0.0$; (c) Q8 with FI (3×3), $\xi_i, \eta_i = \pm0.77460, 0.0$ and (d) Q8 with RI (2×2)

whose solution is n pairs of eigenvalues and eigenvectors

$$(\lambda_i, \mathbf{q}_i), \quad i = 1, \dots, n \tag{17.2}$$

where n is the number of element dofs, that is the size of the stiffness matrix. For normalized eigenvectors $\mathbf{v}_i = \mathbf{q}_i/|\mathbf{q}_i|$ the elastic strain energy U_i of the FE in the state represented by eigenvector $\mathbf{v}_i$ is expressed by the following quadratic form and eigenvalue λ_i

$$U_i = \frac{1}{2}\mathbf{v}_i^{\mathrm{T}}\mathbf{k}^e\mathbf{v}_i = \frac{1}{2}\lambda_i \tag{17.3}$$

The stiffness matrix of a correctly formulated displacement-based FE is positive semidefinite. The zero eigenvalues are associated exclusively with eigenvectors describing rigid-body motions of the FE. The other eigenvalues are real and positive.

This spectral analysis of the stiffness matrix of a single element or of a suitable set of elements is used to determine the number of additional zero eigenvalues that are related to so-called spurious deformation modes. In other words, when the number of zero eigenvalues is larger than the number of rigid-body modes then the difference gives the number of so-called parasitic modes and the order of rank deficiency of the matrix. The spurious modes occur when a too low order of numerical integration quadrature is applied. They can propagate over the whole model unless boundary constraints prevent it.

Both the rigid-body modes and the parasitic ones involve zero strain energy, since they are related to zero eigenvalues, see Equation (17.3).

We now focus on the features of selected membrane and plate elements. In Figure 17.2, based on Bićanić and Hinton (1979), the initial eight (out of total 18) eigenmodes for the membrane nine-node Lagrange element L9 obtained for FI, SI and RI are shown. For the membrane state SI means that FI is used for the stiffness components related to normal strains and RI for the shear stiffness terms. As can be seen, for FI and SI the first three modes are associated with rigid-body motions. Six unexpected modes are obtained for RI and they result from the singularity of the stiffness matrix. In fact, the modes are combinations of three rigid-body motions and three parasitic deformations. They are represented by eigenvectors associated with zero eigenvalues. These modes, known in the literature as 'Escher modes', occurred already in our solution of the problem of membrane buckling under unidirectional compression, see Subsection 14.4.2.

The results of spectral analyses for plate FEs formulated according to the Mindlin–Reissner theory are presented in Table 17.1, based on the information provided in Chapter 3 of Hinton and Owen (1984). In particular the following FEs are included: L4 (4-node element with bilinear shape functions), S8/S12 (8- or 12-node FE with Serendipity shape functions) and L9/L16 (9- and 16-node FE with Lagrange shape functions). In the table, the numbers of zero eigenvalues are given for stiffness matrices without kinematic constraints imposed, for three orders of approximation polynomials and for different numbers of Gauss points employed in FI, RI or SI. Note that in the selective integration case RI is used for transverse shear (t) and FI for bending (m).

We next emphasize the fact that in the displacement-based FE model the use of FI often leads to an excessive stiffness of the element and element assembly, which is called locking, and as a result too small displacements are computed. This unacceptable effect is described in the next section. On the other hand RI reduces the stiffness but, as explained in this section, can imply the matrix rank deficiency, that is its singularity

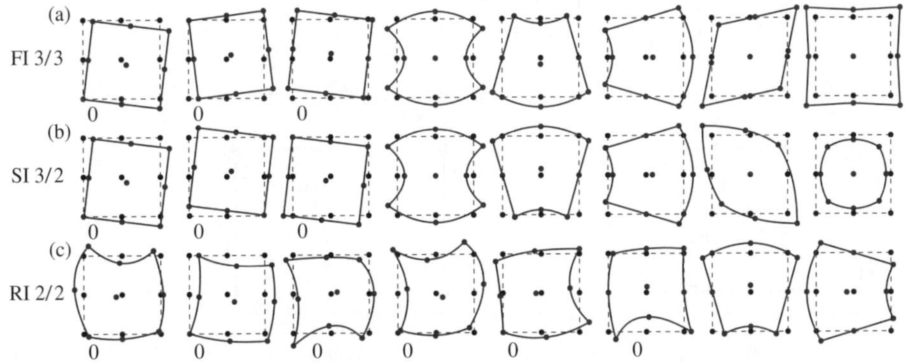

Figure 17.2 Initial eight deformation modes of single nine-node Lagrangian membrane FE for different NI quadratures: (a) FI – 3/3, (b) SI – 3/2 and (c) URI – 2/2; some eigenmodes are associated with zero eigenvalues

and associated spurious zero-energy deformation modes. Finally, in Subsection 17.3.4 the so-called Heterosis plate/shell finite element is presented, in which both the negative phenomena are reduced.

17.3 Numerical Effects of Shear Locking and Membrane Locking

Overstiffness (locking) is a phenomenon that occurs in the analysis of both 1D beams and 2D structures. Next, we describe the shear locking (SL) phenomenon that occurs in membrane finite elements, as well as in plate elements, and membrane locking (ML) encountered in shell finite elements.

17.3.1 Shear Locking in Pure In-Plane Bending

Four-node bilinear membrane element Q4 does not reproduce the pure in-plane bending state in a proper way, see Cook et al. (1989) and Hartmann and Katz (2007).

As shown in Figure 17.3b, all points in the four-node element undergo only nonzero horizontal displacements $u(x, y)$, while the proper displacement and strain state is shown in Figure 17.3a (see the detailed description in Section 6.5). On the basis of the displacement functions retrieved in FE:

$$u^{\text{FEM}}(x, y) = -\frac{2\hat{p}}{Eah}xy = u^{\text{anal}}(x, y) \qquad v^{\text{FEM}}(x, y) = 0 \tag{17.4}$$

the strain components calculated using the kinematic equations read:

$$\epsilon_x^{\text{FEM}}(x, y) = -\frac{2\hat{p}}{Eah}y, \qquad \epsilon_y^{\text{FEM}}(x, y) = 0, \qquad \gamma_{xy}^{\text{FEM}}(x, y) = -\frac{2\hat{p}}{Eah}x \neq 0 \tag{17.5}$$

The last equation represents the parasitic nonzero shear strain in pure in-plane bending. In Figure 17.3 two end moments are also shown, which rotate the two edges of the element by the angle $\theta = 2\hat{p}/(Eh)$ in a bilinear element with aspect ratio a/b. One can compare moments M^{anal} and M^{FEM} causing the same rotation angle θ. In the membrane described in Section 6.5 and in the four-node element the following relation holds (Hartmann and Katz 2007)

$$M^{\text{FEM}} = \frac{1}{1+\nu}\left[\frac{1}{1-\nu} + \frac{1}{2}\left(\frac{a}{b}\right)^2\right] M^{\text{anal}} \tag{17.6}$$

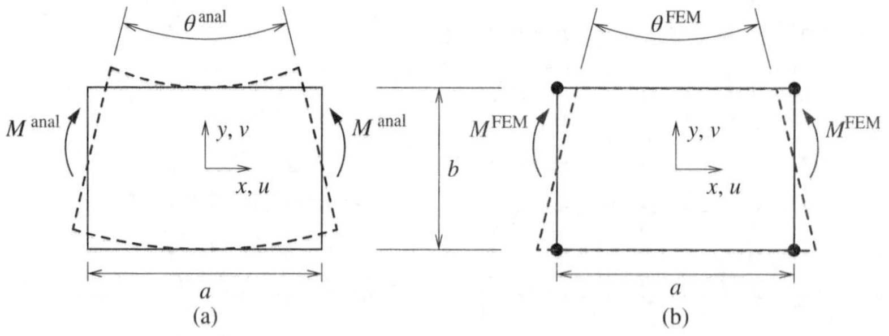

Figure 17.3 Rectangular membrane in pure in-plane bending: (a) proper deformation and (b) deformation mode reproduced by one bilinear finite element Q4

Since $M^{\text{FEM}} > M^{\text{anal}}$ we have a too stiff element (displacement v is zero) and we deal with the shear locking phenomenon occurring in the membrane state. Relation (17.6) additionally shows that M^{FEM} grows with the ratio a/b squared.

The moments applied to the edges $x = \pm a/2$ are statically equivalent to linearly varying traction $p_x(y) = -2\hat{p}y/b$ adopted in the analytical solution of the pure in-plane bending state (see Section 6.5).

The ensuing spurious shear strain γ_{xy}^{FEM} is zero on line $x = 0$, in particular at the element centre. The application of selective integration (SI) for the four-node membrane finite element consists in the separation of the stiffness contributions of normal strains ϵ_x and ϵ_y from the contribution of shear strain γ_{xy}. In the former case 2×2 Gauss points are used (FI) and in the latter just one point (RI). This integration scheme results in a proper reproduction of the energy balance (including the zero energy related to the shear strain) and provides the correct stiffness of the four-node FE. This for instance leads to satisfactory results for the cantilever beam, which is an important benchmark test (see Section 6.6 and Subsection 17.5.1).

17.3.2 Transverse Shear Locking in the Bending State

To describe plates and shells in bending two theories are usually used due to Kirchhoff–Love or Mindlin–Reissner. In principle, the M–R theory incorporates the K–L theory, since it is enough to impose the condition of zero transverse shear strains valid for thin plates/shells in the formulation of FE (M–R).

The search for a finite element that could serve the purpose of plate discretization in a wide range of plate thicknesses faced certain obstacles in the FEM progress. The main problem consists of the fact that an M–R plate does not behave like a K–L plate when its thickness is reduced ($h \to 0$). The results obtained using the FE (M–R) do not converge to the results expected for thin plates: the deflections are too small. The phenomenon is called transverse shear locking. On the other hand, an important advantage of FE (M–R) should be emphasized, namely that it enables a reduction of the continuity requirements. Instead of C^1-continuity of the deflection function we have C^0-continuity of the deflection and of the angles of rotation of the normal vector, which are approximated independently of the deflection.

In the literature (e.g. Cook et al. 1989) one can find the presentation of linear static analysis results for a uniformly loaded square plate, clamped on all edges. The numerical results obtained using different elements for the thin plate and increasing ratio L/h (h goes down) illustrate the shear locking phenomenon.

In Figure 15.3-2 of Cook et al. (1989), the diagrams of the dependence of the dimensionless central deflection (which can be treated as a locking measure) $\tilde{w} = w_{\text{max}}^{\text{FEM}} / w_{\text{max}}^{\text{anal}}$ on the varying span-to-thickness L/h ratios are presented. The computations are performed for a broad range of values $L/h = 10$–1000. The diagrams significantly decaying below 1.0 are shown for the mesh of 8×8 using various finite elements. Here, we quote the results for two FEs: bilinear (L4) with FI and Serendipity (S8) with three variants of numerical integration. For element L4 with FI and $L/h = 20$ the locking measure $\tilde{w} = 0.82$ indicates a strong underestimation of the deflection. Element S8 behaves better for moderate L/h ratios, but the deflection decreases rapidly for $L/h > 100$. When FI is used then for $L/h = 300$ we have $\tilde{w} = 0.86$. Both for RI and SI a similar locking measure is observed for $L/h = 500$.

In Subsection 17.3.4 the displacement-based element H8/9 Heterosis will be described, which makes use of both Serendipity and Lagrange shape functions and is superior to standard elements.

17.3.3 Membrane Locking

The phenomenon of membrane locking (ML) is related to the occurrence of membrane effects in pure bending of shell FEs. This means there are problems with reproducing inextensional bending.

Linear plane triangular elements are completely free from ML. The pure bending mode cannot be represented exactly by standard curved displacement-based elements. The parasitic membrane strains correspond to shape functions used to approximate displacement/rotation fields and curved geometry. The parasitic energy associated with nonzero spurious membrane strains leads to overestimation of the stiffness of the shell and underestimation of the displacements.

The phenomena of SL and ML intensify each other in curved thin shells. However, like transverse shear locking, membrane locking occurs when the shell thickness approaches zero. Fortunately, the standard finite elements are practically free from ML in the range of thicknesses encountered in practice.

The popular methods to completely avoid SL and ML are the reduced integration (Zienkiewicz et al. 1971, Stolarski and Belytschko 1982, 1983) and enhanced assumed strain method (Chang et al. 1989, Andelfinger and Ramm 1993) – see Chapter 18. The majority of new shell elements exhibit the ML phenomenon, which is confirmed by the satisfaction of, for example, an inextensional bending test for a 90° section of a cylindrical shell (see Subsection 17.4.1).

17.3.4 Heterosis Finite Element

In the analysis of plates and shells next to elements S8 and L9 the displacement-based element called Heterosis and denoted by H8/9 is used, see (Hughes and Cohen 1978). The symbol means that the Serendipity approximation with eight nodes is used for three translations (displacements normal and tangent to the shell surface) in contrast with the Lagrangian approximation with nine nodes that is used for two rotations. In Table 17.2 the characteristics of three shell elements, S8, H8/9 and L9, are presented. In the table, the number of spurious modes marked with an asterisk concerns those modes that do not transmit from one element to its neighbours when the mesh contains at least two elements. It can be seen that the number of parasitic mechanisms is reduced when SI is used instead of RI. Element H8/9 has a proper rank of matrices and, when used in the analysis of plate bending for $h \to 0$, it is free from transverse shear locking (SL). However, the element does not pass the patch test of plate bending when its shape is not a rectangle or parallelogram.

The behaviour of elements Q8, H9, Q8/9 Heterosis was also tested in the problem of critical load determination for plate buckling, see Radwańska (1990). In the numerical analysis of a unidirectionally compressed plate the accuracy of the solution was examined and the results were presented in terms of normalized critical load $\tilde{p} = p_{cr}^{FEM}/p_{cr}^{anal}$ for different options of numerical integration (FI, RI, SI) and two values of the inverse of plate slenderness ($L/h = 80, 800$), see Table 17.3. For element S8/FI the value $\tilde{p}_{cr} = 6.366$ demonstrates overestimation of the critical load due to too large stiffness of the plate model. On the other hand, for L9/RI, computations fail due to the singularity of the stiffness matrix.

Table 17.2 Number of spurious modes for various FEs: S8, H8/9, L9 and NI: RI, SI

	FE	**S8**	**H8/9**	**L9**
		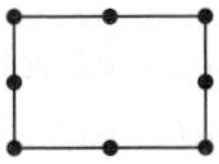	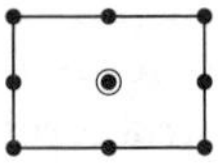	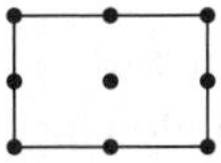
	Shape functions			
	for u, v, w	**Serendipity**	**Serendipity**	**Lagrangian**
	for φ_1, φ_2	**Serendipity**	**Lagrangian**	**Lagrangian**
			Number of spurious modes	
RI	in membrane state	1^*	1^*	$2+1^*$
	in bending state	1^*	$2+1^*$	$3+1^*$
SI	in membrane state		1^*	$2+1^*$
	in bending state			1

RI – 2×2 Gauss points
SI – 2×2 for extension/shear, 3×3 for bending
$(\,)^*$ – without propagation of spurious modes if $NSE > 2$

Table 17.3 Values of normalized critical loads for plate under unidirectional compression for two L/h ratios and three types of NI

	FE	NI	$\tilde{p}_{cr} = p_{cr}^{FEM}/p_{cr}^{anal}$ $h = 0.1$ m, $L/h = 80$	$h = 0.01$ m, $L/h = 800$
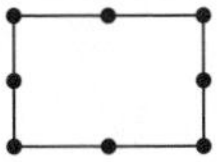	S8	FI	1.077	6.366
		RI	0.971	1.031
		SI	0.971	1.031
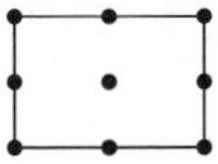	L9	FI	1.023	1.027
		RI	matrix rank deficiency (singularity)	
		SI	0.965	0.967
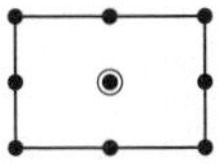	H8/9	FI	1.024	1.027
		RI	0.964	0.967
		SI	0.965	0.967

17.4 Examination of Element Quality – One-Element and Patch Tests

17.4.1 Single-Element Tests

The one-element test is used to verify the quality of representation of characteristic states for the following shell structures: membranes (unidirectional tension, compression, shear or in-plane bending), plates (bending or twisting, or coupled bending with transverse shear) and shells (constant inextensional bending of an infinitely long cylindrical shell), see (Cook et al. 1989). Moreover, a single element test can also verify whether the displacements representing rigid-body motion (pure translation or rotation) do not involve strains.

The test for in-plane bending in Figure 17.4a is performed to examine the sensitivity of a FE to the aspect ratio L_x/L_y.

The coupling of bending and transverse shear in a plate, see Figure 17.4b, is tested in order to detect the tendency of an element to shear locking when the plate gets thinner (L/h grows). Simultaneously, one can assess whether the transverse shear effects are correctly incorporated when the plate becomes thicker (L/h decreases). If software allows that, the user can perform a parametric study by selecting the type of integration quadrature next to changing the plate thickness.

The examination of twisting of a flat strip with unit width, loaded by two lateral forces acting in opposite directions (Figure 17.4c) or by two equivalent twisting moments (Figure 17.4d), aims at verification of the correctness of a pure warping state. For $E = 10^7$, $\nu = 0.25$ and $h = 0.05$ one can for instance check the deflection of the plate corner, which should be equal $w = (3L - 0.6) \times 10^{-3}$ (Batoz 1982).

The test of inextensional bending of a cylindrical shell (Figure 17.4e) consists in the check whether the zero membrane and transverse shear effects (strains and forces) and constant circumferential bending moment $m_\theta = \hat{m}$ are reproduced. The analytical value of the horizontal translation of the straight free edge is equal to $|u| = \hat{m}R^2/D^m$.

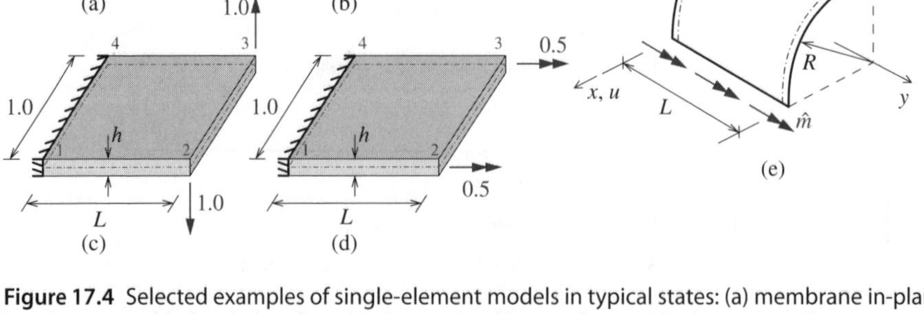

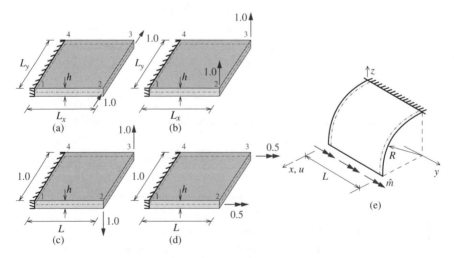

Figure 17.4 Selected examples of single-element models in typical states: (a) membrane in-plane bending state, (b) plate in bending, (c) plate twisted by two forces, (d) plate twisted by two moments and (e) pure bending of infinitely long cylindrical shell

In the parametric study the ratio R/h and the (regular or irregular) mesh density can be changed. For instance, in Andelfinger and Ramm 1993 the following data are adopted; $L = 10$, $R = 10$, $E = 1000$, $\nu = 0.3$, and calculations are performed for one element as well as regular and irregular meshes with different densities. Note that for zero Poisson's coefficient this problem transforms into the one-dimensional bending of a curved beam (Bischoff et al. 2004) and to examine the convergence to the exact solution it is enough to refine the mesh only in the circumferential direction.

17.4.2 Patch Tests

The patch test (PT) concept was introduced by Irons and Razzaque (1972). The satisfaction of PT is a sufficient condition for convergence of the displacement-based formulation (Taylor et al. 1986) and mixed formulations (Zienkiewicz et al. 1986). To perform a patch test (PT) a set of elements of arbitrary shape is selected to discretize a regular domain. The set must contain at least one internal node or at least one internal element (Figure 17.5).

The beginning of a patch test is the definition of displacement field (PT-D) or loading (PT-L). In PT-D we derive an appropriate function that represents the displacement field and, according to this function, we determine the displacement values at the nodes on the perimeter of the patch. Further, the displacements of the internal nodes are computed and they have to comply with the assumed displacement function. Moreover, it is verified if the strains in the elements are consistent with the displacement function. In PT-L an appropriate boundary loading is imposed, which implies a specific displacement field coupled to an associated strain and/or resultant stress fields.

In the literature, see for instance Batoz and Dhatt (1990) and Yang (1986), one can find detailed information on various patch tests for membranes as well as Kirchhoff–Love or Mindlin–Reissner plates and shells. We describe first some PT-D examples for a K–L or M–R plate. The geometry of the element patch is defined in Figure 17.5b.

For the K–L thin plate represented by the FE patch we adopt the following deflection function

$$w = C(x^2 + xy + y^2)/2 \tag{17.7}$$

which corresponds to constant curvatures and warping:

$$\kappa_x(x,y) = -w_{,xx} = -C, \quad \kappa_y(x,y) = -w_{,yy} = -C, \quad \chi_{xy}(x,y) = -2w_{,xy} = -C \tag{17.8}$$

$E = 10^6$ $\nu = 0.25$ $a = 0.24$ $b = 0.12$ $h = 0.001$

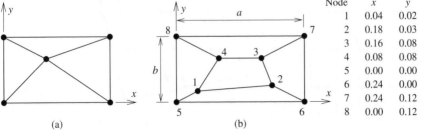

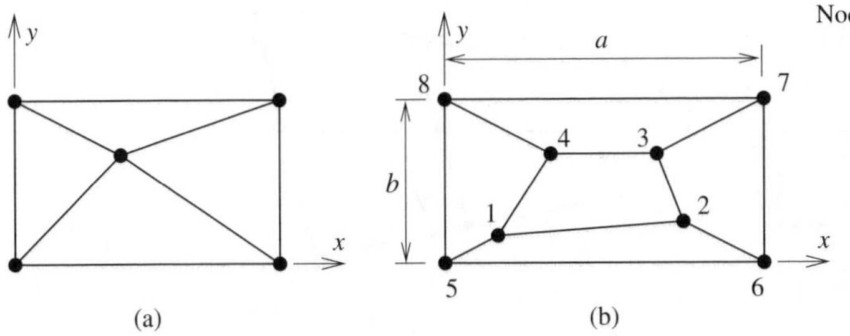

Node	x	y
1	0.04	0.02
2	0.18	0.03
3	0.16	0.08
4	0.08	0.08
5	0.00	0.00
6	0.24	0.00
7	0.24	0.12
8	0.00	0.12

Figure 17.5 (a) Patch with one internal node and (b) patch with one internal element; material data, dimensions and values of coordinates of all nodes are given

Next, analysing the FE patch for the M–R moderately thick plate, two sets with three functions for the deflection and two rotations are derived. They respectively represent:

- the state of constant bending and twisting:

$$\begin{aligned} & w = 10^{-3}(x^2 + xy + y^2)/2, \quad \vartheta_x = 10^{-3}(-x - y/2), \quad \vartheta_y = 10^{-3}(-x/2 - y) \quad \rightarrow \\ & \kappa_x = \vartheta_{x,x} = -10^{-3}, \quad \kappa_y = \vartheta_{y,y} = -10^{-3}, \quad \chi_{xy} = \vartheta_{x,y} + \vartheta_{y,x} = -10^{-3} \\ & m_x = m_y = -10^{-3} D^m (1 + \nu), \qquad m_{xy} = -10^{-3} D^m (1 - \nu)/2 \\ & \gamma_{xz} = \gamma_{yz} = 0, \qquad t_x = t_y = 0 \end{aligned} \tag{17.9}$$

- the state of constant transverse shear:

$$\begin{aligned} & w = 10^{-3} x, \qquad \vartheta_x = \vartheta_y = 0 \qquad \rightarrow \\ & \gamma_{xz} = w_{,x} + \vartheta_x = 10^{-3}, \qquad \gamma_{yz} = w_{,y} + \vartheta_y = 0 \\ & t_x = -10^{-3} D^t, \qquad t_y = 0 \\ & \kappa_x = \kappa_y = \chi_{xy} = 0, \qquad m_x = m_y = m_{xy} = 0 \end{aligned} \tag{17.10}$$

In PT-L tests (Yang 1986), one examines the plates shown in Figure 17.6. The uniformly distributed moment loads $\hat{m}$ on two parallel edges cause constant moment and curvature fields in the domain (Figure 17.6a):

$$\begin{aligned} & m_x = \hat{m}, \qquad m_y = m_{xy} = 0 \\ & \kappa_x = -w_{,xx} = \frac{12\hat{m}}{Eh^3}, \qquad \kappa_y = -w_{,yy} = -\frac{12\nu\hat{m}}{Eh^3}, \qquad \chi_{xy} = 0 \end{aligned} \tag{17.11}$$

Four appropriate point loads $\hat{P}$ acting at corners result in the state of constant twisting moments and warping (Figure 17.6b):

$$m_{xy} = -\hat{P}/2, \qquad \chi_{xy} = -2w_{,xy} = -\frac{12\hat{P}(1+\nu)}{Eh^3} \tag{17.12}$$

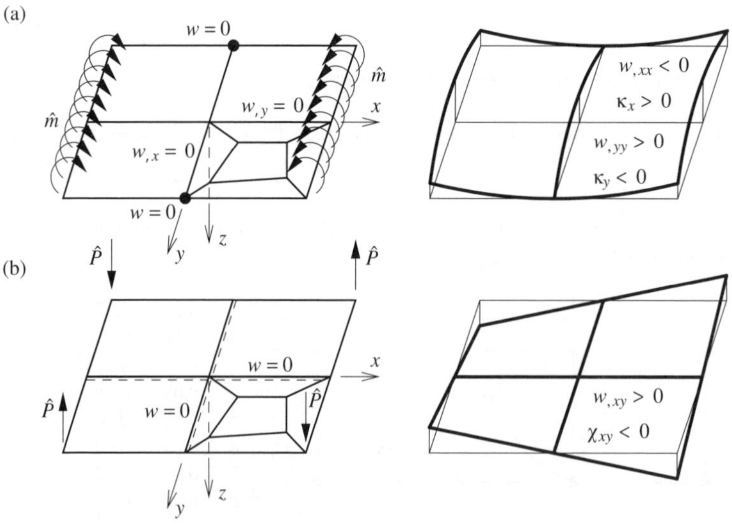

Figure 17.6 Patch tests with loading (PT-L) for examination of the following states: (a) constant curvatures and (b) constant twist (warping)

17.5 Benchmarks for Membranes and Plates

Standard benchmarks are gathered in the well-known article by MacNeal and Harder (1985) titled 'Proposed standard set of problems to test finite element accuracy'. In many papers it is possible to find more advanced test problems, relevant for selected plate and shell structures. For instance, Sze et al. (2004) contains a set of tests for geometrically nonlinear problems.

Each publication describing a new finite element contains the results of test computations, which confirm the correctness of a given FE formulation and the accuracy of benchmark solutions. The description of a new FE should always be completed with proofs of its accuracy and robustness. They are provided by comparing the results of benchmarks with the results obtained using available formulations.

In what follows, we give the aim and contents of selected verification tests for membranes and plates.

17.5.1 Cantilever Beam

In Section 6.6 we presented the analytical and numerical solutions for a cantilever beam with unit thickness, shown in Figure 17.7. We emphasize that this test is considered in the literature with various dimensions and material parameters. Moreover, the way the loading is applied (concentrated vertical load in the middle of the left edge or parabolic distribution of tangent traction) and the selection of kinematic constraints on the right supported edge influence the deformation and stress fields. The impact of element type, element size (including aspect ratio) and mesh layout on the analysis results is examined in the problem.

The formulae for the vertical displacement on the free edge are given in Section 6.6 for two cases: taking into account shear effects (Timoshenko theory) and neglecting the shear effects (Bernoulli–Euler beam). In Cook (1995), Waszczyszyn and Radwańska (1995) and Reddy (2007), numerous comparisons of the test results for different elements CST, LST, Q4(L4), QM6, Q8(S8), Q9(L9) can be found. The exact reproduction of the membrane state, in particular of in-plane bending with/without the shear strain, requires appropriate approximation improvements in the displacement-based FE models, for instance the introduction of additional drilling dof (in-plane rotations, see Subsection 18.3.1).

17.5.2 Swept Panel

The test concerns a short cantilever beam called swept panel, with uniformly distributed tangential load along the right edge, see Figure 17.8. The dimensions are given in the

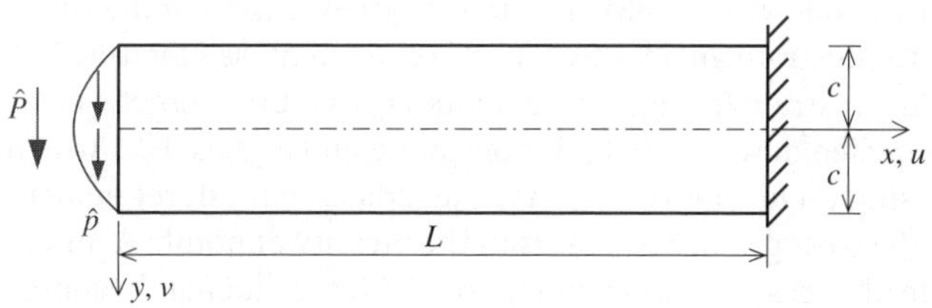

Figure 17.7 Cantilever beam

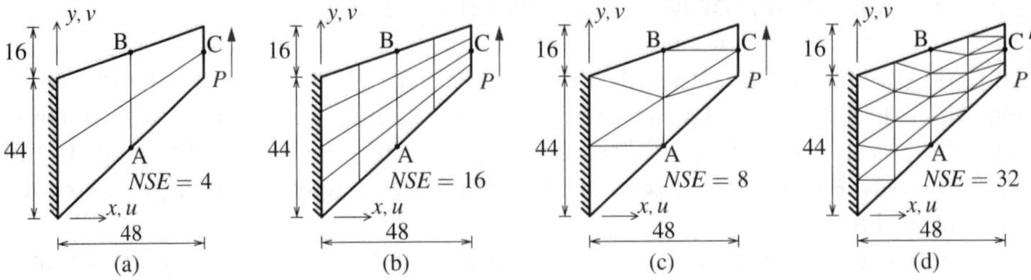

Figure 17.8 Swept panel – dimensions, load, boundary conditions and different discretization. Source: Cook et al. (1989). Reproduced with permission of John Wiley & Sons.

figure, the Poisson's ratio $\nu = 0.333$, unit values of: the thickness, load sum P and Young's modulus E are adopted. Since the configuration is distorted, quadrilateral elements have general shape and in the derivation of isoparametric element matrices complex integrands occur, which requires appropriate numerical integration. There is no analytical solution for the problem, but its better and better approximation can be obtained using a sequence of meshes with increasing density.

The accuracy of the solution can be monitored using the vertical displacement v_C at point C (we will refer next to nondimensional value $\tilde{v}_C = v_C/v_C^{\text{exact}}$, which should converge to 1) and the horizontal normal stresses at point A (maximum) and at point B (minimum). The results of calculations are normalized with respect to the converged FE solution, which is treated as exact $v_C^{\text{exact}} = 23.91$ (Jang and Pinsky 1987).

In Taylor et al. (1976), Cook et al. (1989) and Yuqiu and Yin (1994), the problem is solved using different quadrilateral and triangular finite elements. Standard quadrilateral Q4 is confronted with upgraded finite elements Q6 and QM6 in which displacement approximation is augmented by incompatible modes and modified selective numerical integration. For instance it is shown that for $NSE = 4$ (Figure 17.8a) $\tilde{v}_C = 0.498$ for Q4 and $\tilde{v}_C = 0.884$ for QM6, while for $NSE = 16$ (Figure 17.8b) $\tilde{v}_C = 0.769$ for Q4 and $\tilde{v}_C = 0.967$ for QM6. Moreover, classical triangular finite elements CST, LST are compared with membrane elements with drilling degrees of freedom (see Subsection 18.3.2). For example it is shown that for $NSE = 8$ (Figure 17.8c) $\tilde{v}_C = 0.502$ for CST and $\tilde{v}_C = 0.852$ for the element with drilling dofs, while for $NSE = 32$ (Figure 17.8d) $\tilde{v}_C = 0.765$ for CST (very bad result) and $\tilde{v}_C = 0.954$ for the enriched triangle.

The authors of Cook et al. (1989) concluded that the upgraded finite elements are by far superior to the classical ones. The results of this benchmark can also be found in Chang et al. (1989) and Felippa (2003).

17.5.3 Square Cantilever

In Cook (1995) and Zhu and Zienkiewicz (1988), a square cantilever membrane (deep cantilever beam) is analysed to illustrate an adaptive mesh refinement, see Figure 17.9. For a coarse mesh with eight linear strain triangular elements T6, a global energetic error measure η, defined in Zhu and Zienkiewicz (1988), is computed and equals 0.27, which is inadmissible. Therefore, basing on local error estimate, an adaptive mesh refinement is performed in regions of high stress gradients, that is in the vicinity of points A and B. Two steps of this procedure lead to results shown in Figure 17.9b,c, reducing the global error η to 0.04. We note that corners A and B are sources of stress concentration. It would be better to avoid such singularities in the domain definition.

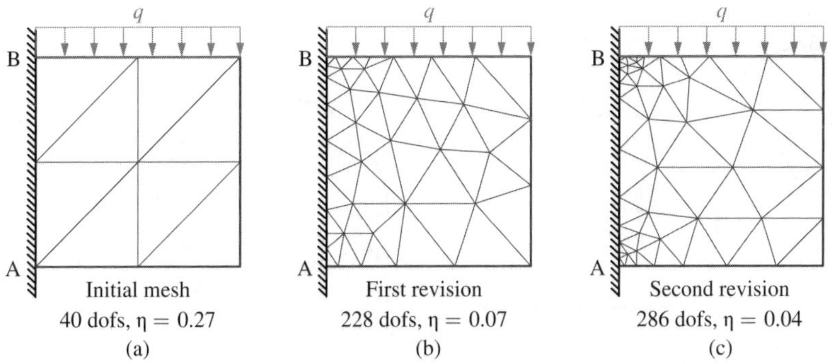

Figure 17.9 Plane square cantilever – results of an adaptive solution in a plane region using linear strain triangles. Source: Cook (1995). Reproduced with permission of John Wiley & Sons.

17.5.4 Square Plate in Bending

The basic benchmark for plate bending is a square plate with edge dimension L, uniformly supported on the whole perimeter (either simply supported or clamped). The deflection at the centre is treated as a check value. For two cases of loading (uniform load q or concentrated central force P) the maximum deflection is calculated from the following formulae (Timoshenko and Woinowsky-Krieger 1959):

$$w_C = \alpha q L^4 / D^m \qquad \text{or} \qquad w_C = \alpha P L^2 / D^m \tag{17.13}$$

The values of coefficient α derived for $\nu = 0.3$ and four combinations of loading and boundary conditions are: (i) for q and a simply supported plate $\alpha = 0.00406$, (ii) for P and a simply supported plate $\alpha = 0.0116$, (iii) for q and a clamped plate $\alpha = 0.00126$, and (iv) for P and a clamped plate $\alpha = 0.00560$.

In MacNeal and Harder (1985) the values of α coefficient are listed for the square plate and also for a rectangular one with a side ratio equal to 5. Moreover, a comparison of numerical results is provided for a host of two- and three-dimensional finite elements.

17.6 Benchmarks for Shells

Three well-known benchmarks are presented in MacNeal and Harder (1985), Cook et al. (1989) and in Figure 17.10: (a) a cylindrical shell roof, (b) a pinched cylinder and (c) a hemisphere. Figure 17.10 contains the configurations, the loading specification as well as input data and expected check values Δ_A.

The tests are performed to examine various aspects of numerical analysis: (i) approximation of the geometry of shells with different shapes using flat or curved FEs, (ii) representation of simple and complex deformation states and (iii) description of local bending effects near point loads or supports.

17.6.1 Cylindrical Shell Roof

This example was first presented in Scordelis and Lo (1969) and hence is called the Scordelis-Lo roof, see Figure 17.10a. The shell is loaded by self weight, supported by two diaphragms stiff in their planes and has two free straight edges. Different data and

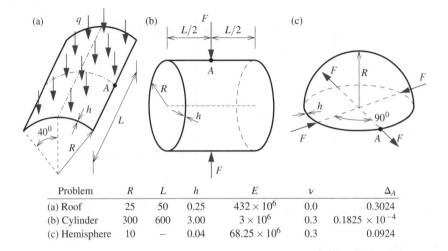

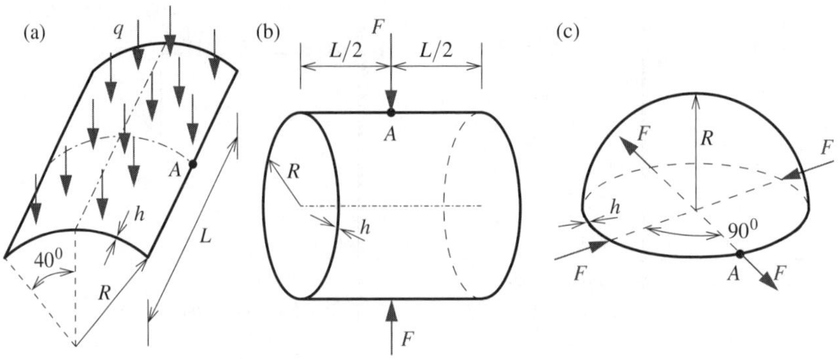

Problem	R	L	h	E	ν	Δ_A
(a) Roof	25	50	0.25	432×10^6	0.0	0.3024
(b) Cylinder	300	600	3.00	3×10^6	0.3	0.1825×10^{-4}
(c) Hemisphere	10	–	0.04	68.25×10^6	0.3	0.0924

Figure 17.10 Three shell configurations, input data, verification results: (a) cylindrical shell roof, (b) pinched cylinder and (c) hemisphere

units are encountered in the literature. Our computations presented in Section 11.7 are performed for the input data defined in Szabó and Sahrmann (1988).

In this typical roof structure we deal with a complex and nonuniform membrane-bending state. The general case of equilibrium involves three membrane forces, three moments and two shear forces. In the pioneering work of Ahmad et al. (1970) the diagrams of selected generalized resultant forces and translations are plotted for two central cross sections (along straight and curved lines) and for a cross section close to the diaphragm. Nowadays, the results can also be presented as contour maps of some field quantities, see Section 11.7 and Figures (11.22)–(11.25).

The value monitored in MacNeal and Harder (1985) and Cook et al. (1989) is the vertical displacement of the centre of the free straight edge $\Delta_A = 0.3024$. This test has been analysed by numerous authors, including MacNeal and Harder (1985), Bathe and Dvorkin (1986), Szabó and Sahrmann (1988), Chang et al. (1989), Andelfinger and Ramm (1993), using different codes and finite elements.

17.6.2 Pinched Cylinder

The pinched cylinder in Figure 17.10b is supported on two circular end diaphragms and loaded with two unit concentrated forces. This configuration involves a coupled bending and membrane response, in particular around the central section of the cylinder. The check value is the vertical displacement at the point of force application $\Delta_A = |w_A| = 0.1825 \times 10^{-4}$.

17.6.3 Hemisphere

The hemisphere is loaded by four uniformly spaced unit radial forces acting in the equatorial plane, see Figure 17.10c. The test verifies the ability of the element to represent the geometry of a doubly curved shell as well as its complex deformation. The check value is the normal displacement at the point of load application, equal to $\Delta_A = |w_A| = 0.0924$ (note that the sign of the displacement is consistent with the sense of the force).

17.7 Comparison of Analytical and Numerical Solutions, Application of Various FE Formulations

A comparative analysis of exact (analytical) and approximate (numerical) results is performed either when one generates new FE software, in particular a new finite element, or when available software is tested, although one expects that commercial packages should have passed all required tests, for instance according to NAFEMS recommendations (NAFEMS is National Agency for Finite Element Methods and Standards), see MacNeal and Harder (1985).

In the literature one can find several examples of examination of convergence of numerical analyses. After a series of computations it is proven that a satisfactory accuracy of the solution is obtained for a dense mesh of simple FEs or a coarse mesh of FEs based on advanced formulations.

The results of convergence tests are usually presented in a diagram of the relationship between a normalized check quantity (i.e. ratio of numerically obtained value and its analytical counterpart) and a parameter characterizing the mesh density. We expect an asymptotic convergence of the diagram to 1.0. Alternatively, one can compute an error measure that should converge to 0.0.

For example, in numerical analysis of a square thin plate one monitors the convergence to 1.0 of the ratio of either maximum deflection at the centre of the plate $w_C^{\text{FEM}}/w_C^{\text{anal}}$ or maximum bending moment $m_C^{\text{FEM}}/m_C^{\text{anal}}$. Alternatively, the relative error (in percent) of the maximum deflection $(w_C^{\text{FEM}} - w_C^{\text{anal}})/w_C^{\text{anal}}$ is checked.

A comparative study should be carried out if no analytical results are available for a given problem. One possibility is to perform computations several times using just one package and one finite element: first it is done for a simplified model and then for more and more advanced models, for example with increased mesh density or upgraded approximation, whereby asymptotic convergence of results is expected upon mesh refinement. Therefore, if we do not have an analytical solution, the check value obtained as a result of such a convergence study serves as a reference solution and is used instead of the analytical one to compute the error $(w_C^{\text{FEM}} - w_C^{\text{ref}})/w_C^{\text{ref}}$. Another method of verification is to compare the results obtained using different software packages.

In Figure 17.11 the results from the book by Yang (1986) for a square simply supported plate under central force are presented. The comparison for various FE formulations is performed. We have introduced the following symbols for the compared FEs: conforming displacement-based elements D-BFS (Bogner et al. 1966), D-FV (Fraeijs de Veubeke 1968), D-CF (Clough and Felippa 1968), non-conforming displacement-based element D-ACM (Adini and Clough 1961, Melosh 1963), hybrid displacement-based HD-P (Pian 1966), hybrid stress HS-C (Cook 1972) and mixed elements M1-BD, M2-BD (Bron and Dhatt 1972). Depending on the FE model, one can observe convergence of results to the zero error line from above or below, and for one case the sign of the error changes.

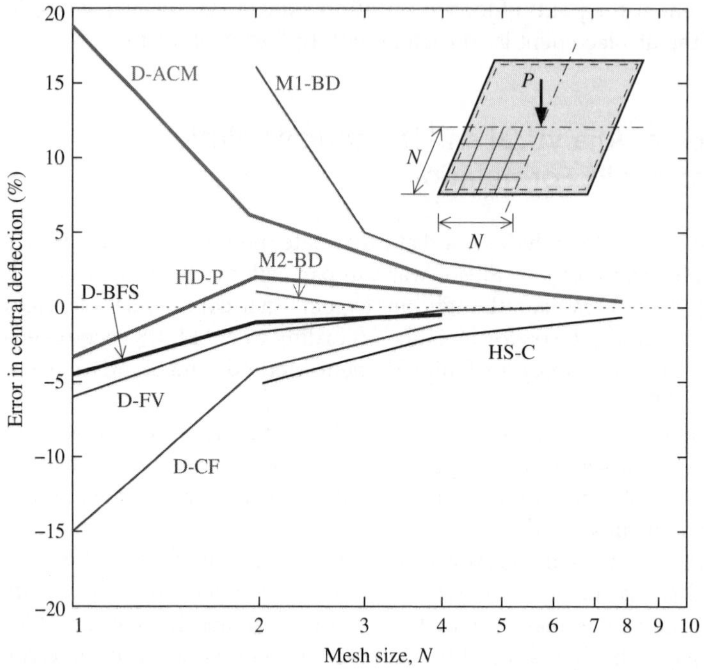

Figure 17.11 Simply supported square plate with force at the centre – comparison of numerical results using different types of rectangular FEs. Source: Yang (1986). Reproduced with permission of Yang.

Finally, the reader is referred to Chapters 5 and 18. They contain short descriptions of a few selected FE models for membranes, plates and shells, which are either simple or advanced from the viewpoint of theoretical formulation and computer implementation.

References

Adini A and Clough RW 1961 Analysis of plate bending by the finite element method. Technical Report G-7337, University of California, Berkeley.

Ahmad S, Irons BM and Zienkiewicz OC 1970 Analysis of thick and thin shell structures by curved finite element. *International Journal for Numerical Methods in Engineering* **2**(3), 419–451.

Andelfinger U and Ramm E 1993 EAS-elements for two-dimensional, three-dimensional, plate and shell structures and their equivalence to HR-elements. *International Journal for Numerical Methods in Engineering* **36**(8), 1311–1337.

Bathe KJ and Dvorkin EN 1986 A formulation of general shell elements – the use of mixed interpolation of tensorial components. *International Journal for Numerical Methods in Engineering* **22**(3), 697–722.

Batoz JL 1982 An explicit formulation for an efficient triangular plate-bending element. *International Journal for Numerical Methods in Engineering* **18**(7), 1077–1089.

Batoz JL and Dhatt G 1990 *Modélisation des Structures par Élément Finis.* Hermes, Paris.

Bićanić N and Hinton E 1979 Spurious modes in two-dimensional isoparamteric elements. *International Journal for Numerical Methods in Engineering* **14**(10), 1545–1557.

Bischoff M, Wall WA, Bletzinger KU and Ramm E 2004 Models and finite elements for thin-walled structures. In Stein E, de Borst R and Hughes TJR (ed) *Encyclopedia of Computational Mechanics: Solids and Structures*, vol. **2**. John Wiley & Sons, Chichester, UK, chapter 3, pp. 59–137.

Bogner FK, Fox RL and Schmit LA 1966 The generation of interelement-compatible stiffness and mass matrices by the use of interpolation formulas *Proceedings of the Conference on Matrix Methods in Structural Mechanics*, pp. 397–443 number AFFDL TR-66-80. Wright-Patterson Air Force Base, Ohio.

Bron J and Dhatt G 1972 Mixed quadrilateral elements for bending. *AIAA Journal* **10**(10), 1359–1361.

Chang TY, Saleeb AF and Graf W 1989 On the mixed formulation of a 9-node lagrange shell element. *Computer Methods in Applied Mechanics and Engineering* **73**(3), 259–281.

Clough RW and Felippa CA 1968 A refined quadrilateral element for analysis of plate bending *Proceedings of 2nd Conference on Matrix Methods in Structural Mechanics*, pp. 399–440 report AFFDL TR-68-150. Wright-Patterson Air Force Base.

Cook RD 1972 Two hybrid elements for analysis of thick, thin and sandwich plates. *International Journal for Numerical Methods in Engineering* **5**(2), 277–288.

Cook RD 1995 *Finite Element Modeling for Stress Analysis*. John Wiley & Sons, Inc., New York.

Cook RD, Malkus DS and Plesha ME 1989 *Concepts and Applications of Finite Element Analysis* 3rd edn. John Wiley & Sons, Inc., New York.

Felippa CA 2003 A study of optimal membrane triangles with drilling freedoms. *Computer Methods in Applied Mechanics and Engineering* **192**, 2125–2168.

Fraeijs de Veubeke B 1968 A conforming finite element for plate bending. *International Journal of Solids and Structures* **4**(1), 95–108.

Hartmann F and Katz C 2007 *Structural Analysis with Finite Elements*, 2nd edn. Springer.

Hinton E and Owen DR 1984 *Finite Element Software for Plates and Shells*. Pineridge Press.

Hughes TJR and Cohen M 1978 The 'Heterosis' finite element for plate bending. *Computers & Structures* **9**(5), 445–450.

Irons BM and Razzaque A 1972 Experience with the patch test for convergence of finite elements. In Aziz AK (ed.) *The Mathematical Foundations of the Finite Element Method with Applications to Partial Differential Equations*. Academic Press, New York. pp. 557–587.

Jang J and Pinsky PM 1987 An assumed covariant strain based 9-node shell element. *International Journal for Numerical Methods in Engineering* **24**(12), 2389–2411.

MacNeal RH and Harder RL 1985 A proposed standard set of problems to test finite element accuracy. *Finite Elements in Analysis and Design* **1**(1), 3–20.

Melosh RJ 1963 Basis for derivation of matrices for the direct stiffness method. *AIAA Journal* **1**(7), 1631–1637.

Pian THH 1966 Element stiffness matrices for boundary compatibility and for prescribed boundary stresses *Proceedings of the Conference on Matrix Methods in Structural Mechanics*, pp. 457–478, report AFFDL TR-66-80. Wright-Patterson Air Force Base, Ohio.

Radwańska M 1990 Analysis of stability and large displacements of shell structures using FEM. Technical Report Monograph 105, Cracow University of Technology, Cracow (in Polish).

Reddy JN 1999 *Theory and Analysis of Elastic Plates.* Taylor & Francis, London.

Reddy JN 2007 *Theory and Analysis of Elastic Plates and Shells* 2nd edn. CRC Press/Taylor & Francis, Boca Raton-London-New York.

Scordelis AC and Lo KS 1969 Computer analysis of cylindrical shells. *ACI Journal* **61**, 539–561.

Stolarski H and Belytschko T 1982 Membrane locking and reduced integration for curved elements. *ASME Journal of Applied Mechanics* **49**(1), 172–176.

Stolarski H and Belytschko T 1983 Shear and membrane locking in curved C^0 elements. *Computer Methods in Applied Mechanics and Engineering* **41**(3), 279–296.

Szabó BA and Sahrmann GJ 1988 Hierarchic plate and shell models based on p-extension. *International Journal for Numerical Methods in Engineering* **26**(8), 1855–1881.

Sze KY, Liu XH and Lo SH 2004 Popular benchmark problems for geometric nonlinear analysis of shells. *Finite Elements in Analysis and Design* **40**(11), 1551–1569.

Taylor RL, Beresford PJ and Wilson EL 1976 A non-conforming element for stress analysis. *International Journal for Numerical Methods in Engineering* **10**(6), 1211–1219.

Taylor RL, Simo JC, Zienkiewicz OC and Chan ACH 1986 The patch test – a condition for assessing FEM convergence. *International Journal for Numerical Methods in Engineering* **22**(1), 39–62.

Timoshenko S and Woinowsky-Krieger S 1959 *Theory of Plates and Shells.* McGraw-Hill, New York-Auckland.

Waszczyszyn Z and Radwańska M 1995 Basic equations and calculations methods for elastic shell structures. In Borkowski A, Cichoń C, Radwańska M, Sawczuk A and Waszczyszyn Z (eds) *Structural Mechanics: Computer Approach,* vol. **3**. Arkady, Warsaw. Ch. 9, pp. 11–190.

Yang TY 1986 *Finite Element Structural Analysis.* Prentice-Hall, Englewood Cliffs, New York.

Yuqiu L and Yin X 1994 Generalized conforming triangular membrane element with vertex rigid rotational freedoms. *Finite Elements in Analysis and Design* **17**(4), 259–271.

Zhu JZ and Zienkiewicz OC 1988 Adaptive techniques in the finite element method. *Communications in Applied Numerical Methods* **4**(2), 197–204.

Zienkiewicz OC, Qu S, Taylor RL and Nakazawa S 1986 The patch test for mixed formulations. *International Journal for Numerical Methods in Engineering* **23**(10), 1873–1883.

Zienkiewicz OC, Taylor RL and Too JM 1971 Reduced integration technique in general analysis of plates and shells. *International Journal for Numerical Methods in Engineering* **3**(2), 275–290.

18

Advanced FE Formulations

18.1 Introduction

This chapter is an extension of Chapter 5 and both of them deal with the description of finite elements used in plate and shell modelling. In the oldest displacement-based FE formulation functions approximating generalized displacement fields (translations and rotations) were employed, treating the strain and stress as secondary fields. Since researchers realized that this formulation involves undesirable numerical effects, described in Sections 17.3 and 17.4, new FE models have been worked out, based on alternative formulations and advanced approximation techniques.

The finite element models can be derived in different manners, frequently providing equivalent formulations. One can distinguish the procedure starting from a variational principle, which invokes the stationarity of an appropriate functional. This leads directly to finite element equations and is followed in this chapter. A broad discussion of this methodology is presented for instance in the book by J.N. Reddy (1986) on *Applied Functional Analysis and Variational Methods in Engineering*. Moreover, in the works of Pian and Tong (1969), Washizu (1975), Kleiber (1985) and Radwańska (2006) one can find the description of finite element formulations based on appropriate variational principles which involve classical or modified functionals.

It is mentioned that the approach based on the principle of virtual work belongs to this group and can be used to formulate a computational model even if a functional cannot be proposed for a mechanical problem.

An alternative general approach to deriving a finite element formulation when the BVP is posed in a strong (local) form is the weighted residual method. An overview of the concepts together with mathematical background and numerous examples is provided, for instance, in Chapter 2 of Reddy (2005) and in Chapters 4–5 of Cook et al. (1989).

In fact, in the last 40 years the motivation to figure out new formulations of the FEM has been to reach the following main goals: (i) enhancement of the approximation of all fields (displacements, strains and stresses), (ii) satisfaction of the continuity conditions for kinematic fields across interelement boundaries and (iii) assertion of the equilibrium conditions for tractions along interelement lines.

The number of FEs based on various formulations is large and the relevant literature is very broad. The reader interested in the literature survey is referred to the following papers by Yang et al. (1990) – 287 references, Gilewski and Radwańska (1991) – 329 references, and Yang et al. (2000) – 379 references.

Plate and Shell Structures: Selected Analytical and Finite Element Solutions, First Edition.
Maria Radwańska, Anna Stankiewicz, Adam Wosatko and Jerzy Pamin.
© 2017 John Wiley & Sons Ltd. Published 2017 by John Wiley & Sons Ltd.

18.2 Link between Variational Formulations and FE Models

When FEM is employed to analyse shell structures, the following two-dimensional fields are approximated depending on FE formulation (one-, two-, three-field models as well as hybrid ones are possible):

- within element domain Ω^e: generalized displacements $\mathbf{u}^e(\mathbf{x})$, $\boldsymbol{\varphi}^e(\mathbf{x})$ (translations and rotations, respectively); membrane, bending and transverse shear strains $\boldsymbol{\epsilon}^e(\mathbf{x}), \boldsymbol{\kappa}^e(\mathbf{x}), \boldsymbol{\gamma}^e(\mathbf{x})$; stress resultants $\mathbf{n}^e(\mathbf{x}), \mathbf{m}^e(\mathbf{x}), \mathbf{t}^e(\mathbf{x})$ (membrane forces, moments and transverse shear forces)
- on interelement lines $\mathrm{d}\Omega^{(ef)}$: boundary displacements $\mathbf{u}^{(ef)}(s)$ and/or tractions $\mathbf{t}^{(ef)}(s)$ (interactions between adjacent elements e and f)

The physical degrees of freedom $(\mathbf{q}_u, \mathbf{q}_\sigma, \mathbf{q}_\epsilon)$ or mathematical dofs $\boldsymbol{\alpha}$ are used as coefficients in the combinations of the approximating functions.

In particular, the following formulations can be distinguished:

- one-field model (compatible displacement, equilibrium stress)
- mixed model with two fields, that is either displacement-strain or displacement-stress
- mixed model with three fields (displacement-strain-stress)
- hybrid displacement, hybrid stress or hybrid mixed model (the concept of hybrid description is explained further on)

The following functionals are taken into account in the formulation of FE models:

- potential energy $I_\mathrm{p}[\mathbf{u}] = \Pi_\mathrm{p}[\mathbf{u}]$
- complementary energy $I_\mathrm{c}[\boldsymbol{\sigma}]$
- functionals of Hellinger–Reissner $I_{\text{H-R}}[\mathbf{u}, \boldsymbol{\epsilon}]$ or $I_{\text{H-R}}[\mathbf{u}, \boldsymbol{\sigma}]$
- functional of Hu–Washizu $I_{\text{H-W}}[\mathbf{u}, \boldsymbol{\epsilon}, \boldsymbol{\sigma}]$

In the general Hu–Washizu principle, all variables $\mathbf{u}, \boldsymbol{\epsilon}, \boldsymbol{\sigma}$ are treated as independent fields and can be varied. The three fields are understood in shell structures as generalized displacements, generalized strains and resultant forces. In fact, the Hu–Washizu principle is used not only for the derivation of the mixed three-field model, but it also gives the theoretical background for alternative concepts.

Before we briefly describe some selected FEs we next show three functionals for Kirchhoff–Love plates in bending, written in Cartesian coordinates. In particular we have:

- the potential energy functional

$$I_\mathrm{p}[w] = \int_\Omega \left\{ \frac{D^m}{2}[(\nabla^2\, w)^2 + 2(1-\nu)(w_{,xy}^2 - w_{,xx}\, w_{,yy})] - \hat{p}_z w \right\} \mathrm{d}\Omega + \int_{\partial\Omega_\sigma} [-\hat{t}_\nu\, w + \hat{m}_\nu\, w_{,\nu} + \hat{m}_{\nu s}\, w_{,s}]\, \mathrm{d}(\partial\Omega) \tag{18.1}$$

- the complementary energy functional

$$I_\mathrm{c}[m_x, m_y, m_{xy}] = -\int_\Omega \frac{6}{E\,h^3}[m_x^2 + m_y^2 - 2\nu\, m_x\, m_y + 2(1+\nu)m_{xy}^2]\, \mathrm{d}\Omega \tag{18.2}$$

- the Hellinger–Reissner functional

$$
\begin{aligned}
& I_{\text{H-R}}[w, w_{,xx}, w_{,yy}, w_{,xy}, m_x, m_y, m_{xy}] \\
& \quad = \int_{\Omega} \left\{ \frac{6}{E\,h^3}[(m_x + m_y)^2 + 2(1+\nu)(m_{xy}^2 - m_x\, m_y)] \right. \\
& \qquad \left. + [-m_x\, w_{,xx} - m_y\, w_{,yy} - 2m_{xy}\, w_{,xy}] - \hat{p}_z\, w \right\} \mathrm{d}\Omega \\
& \quad + \int_{\partial\Omega_\sigma} [-\hat{t}_\nu\, w + \hat{m}_\nu\, w_{,\nu} + \hat{m}_{\nu s}\, w_{,s}]\, \mathrm{d}(\partial\Omega) \\
& \quad + \int_{\partial\Omega_u} [-t_\nu\,(w - \hat{w}) + m_\nu\,(w_{,\nu} - \hat{w}_{,\nu}) + m_{\nu s}\,(w_{,s} - \hat{w}_{,s})]\, \mathrm{d}(\partial\Omega)
\end{aligned}
\tag{18.3}
$$

The requirements for the approximation, formulated on interelement lines, that is the continuity of generalized displacements and/or the equilibrium of tractions, are constraints to be incorporated in a variational manner, The constraint equations can be introduced into the functionals using Lagrange multipliers, forming in this way additional components in augmented functionals. The Lagrange multipliers have the mechanical interpretation of tractions $\mathbf{t}^{(ef)}$ or displacements $\mathbf{u}^{(ef)}$on interelement boundaries. The multipliers are treated as additional functions, approximated on element edges. This is the basis of the formulation of the class of hybrid FEs.

The calculation of relevant matrices and vectors for different types of finite elements requires integration: (i) within element domain Ω^e and (ii) over respective parts of domain boundary $\mathrm{d}\Omega_u^e \subset \mathrm{d}\Omega_u$ or $\mathrm{d}\Omega_\sigma^e \subset \mathrm{d}\Omega_\sigma$ as well as (iii) over common edges of adjacent FEs $\mathrm{d}\Omega^{(ef)}$. Note that the type of numerical integration (NI) quadrature affects the properties of the FE matrices and thus the quality of the final solution, see Section 17.1.

The functional employed determines the couplings of different fields approximated inside an element and on its boundary, and as a result the relationships between suitable element matrices and dof vectors. The variety of sets of FE matrix equations and their interpretations in the context of local BVP formulations are presented among others in the book by Zienkiewicz et al. (2005) and also in the paper by Radwańska (2006).

We further describe various FEs providing information about: (i) the functional on which a particular formulation is based, (ii) approximated fields and dof vectors used and (iii) FE matrix equation structure. For membrane, plate and shell FEs we introduce the notation for dof vectors with indices related to the approximated fields of generalized displacements $\mathbf{q}_w^e$, $\mathbf{q}_\varphi^e$, generalized strains $\mathbf{q}_\epsilon^e$, $\mathbf{q}_\kappa^e$, $\mathbf{q}_\gamma^e$ and generalized resultant forces $\mathbf{q}_n^e$, $\mathbf{q}_m^e$, $\mathbf{q}_t^e$.

As an introductory example we first present the four-node displacement-based flat rectangular membrane FE in Figure 18.1. Compressing the information from Subsection 5.2.1 we concisely characterize the element in three points: (i) the formulation is based on the potential energy functional $I_{\text{p}}[\mathbf{u}]$, (ii) two displacement fields in vector $\mathbf{u}_{(2\times1)} = [u_x(x,y), u_y(x,y)]^{\text{T}}$ are approximated in a plane using dof vector $\mathbf{q}_{u(8\times1)}^e$ and (iii) stiffness matrix $\mathbf{k}_{(8\times8)}^e$ is computed in this one-field displacement-based model.

In further derivations we skip the upper index e that denotes element-level quantity.

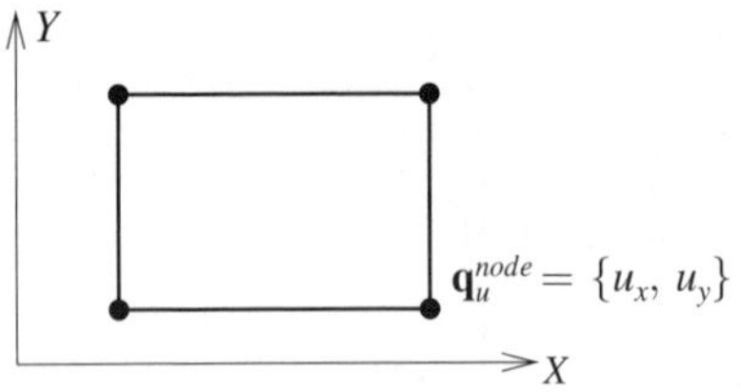

Figure 18.1 One-field four-node rectangular membrane FE

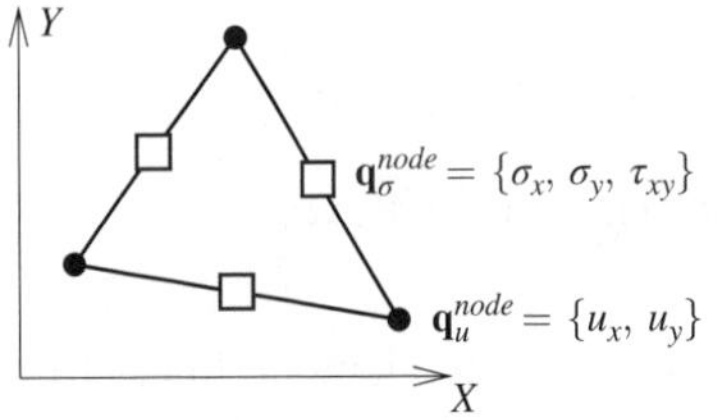

Figure 18.2 Two-field triangular membrane FE

18.2.1 Triangular Membrane FE – Mixed (Displacement-Stress) Model

The element shown in Figure 18.2 is based on the Hellinger–Reissner functional $I_{\text{H-R}}[\mathbf{u}, \boldsymbol{\sigma}]$

$$I_{\text{H-R}}[\mathbf{u}, \boldsymbol{\sigma}] = I_{\text{H-R}}(\mathbf{q}_u, \mathbf{q}_\sigma) \tag{18.4}$$

which, after the introduction of finite element interpolation, becomes a function of dofs.

The approximation of two translation fields $\mathbf{u}_{(2\times1)} = [u_x(x,y), u_y(x,y)]^{\mathrm{T}}$ employs two degrees of freedom at each of three corner nodes. The approximation of three stress fields $\boldsymbol{\sigma}_{(3\times1)} = [\sigma_x(x,y), \sigma_y(x,y), \tau_{xy}(x,y)]^{\mathrm{T}}$ is based on three dofs at each of three edge midpoints. The element has the following physical dofs: $\mathbf{q}_{u(6\times1)}$ and $\mathbf{q}_{\sigma(9\times1)}$. Along the interelement edges one obtains the continuity of the displacement fields and at the edge midpoints the continuity of stresses is imposed. The coupling of the displacement and stress fields, represented by matrix relations between $\mathbf{q}_u$ and $\mathbf{q}_\sigma$ guarantees the consistence of the strains, obtained both from the displacements and from the stresses.

In order that the matrix of this mixed displacement-stress FE (in which a zero diagonal block occurs) is not singular, the following condition must be checked, see Zienkiewicz et al. (1986)

$$\dim\ \mathbf{q}_\sigma^e > \dim\ \mathbf{q}_u^e \tag{18.5}$$

and in the considered case it is satisfied.

18.2.2 Quadrilateral Membrane FE – Mixed (Displacement-Stress) Model

The element shown in Figure 18.3 is based on modified two-field formulation Lee and Lee (1990). The main idea is to generate a linear combination of the potential energy functional $I_{\mathrm{p}}[\mathbf{u}]$ and the Hellinger–Reissner functional $I_{\text{H-R}}[\mathbf{u}, \boldsymbol{\sigma}]$, using a scalar parameter α

$$I_{\text{L-L}}[\mathbf{u}, \boldsymbol{\sigma}] = \frac{\alpha}{\alpha - 1} I_{\mathrm{p}}[\mathbf{u}] - \frac{1}{\alpha - 1} I_{\text{H-R}}[\mathbf{u}, \boldsymbol{\sigma}] \tag{18.6}$$

This way, one obtains positive semidefinite matrix of this displacement-stress FE without a zero block.

In this membrane element bilinear shape functions based on corner nodes are used for the approximation of both the displacement fields and the membrane force fields,

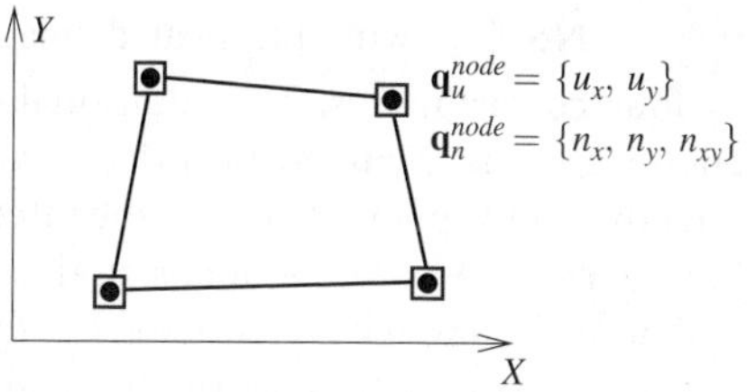

Figure 18.3 Two-field four-node membrane FE

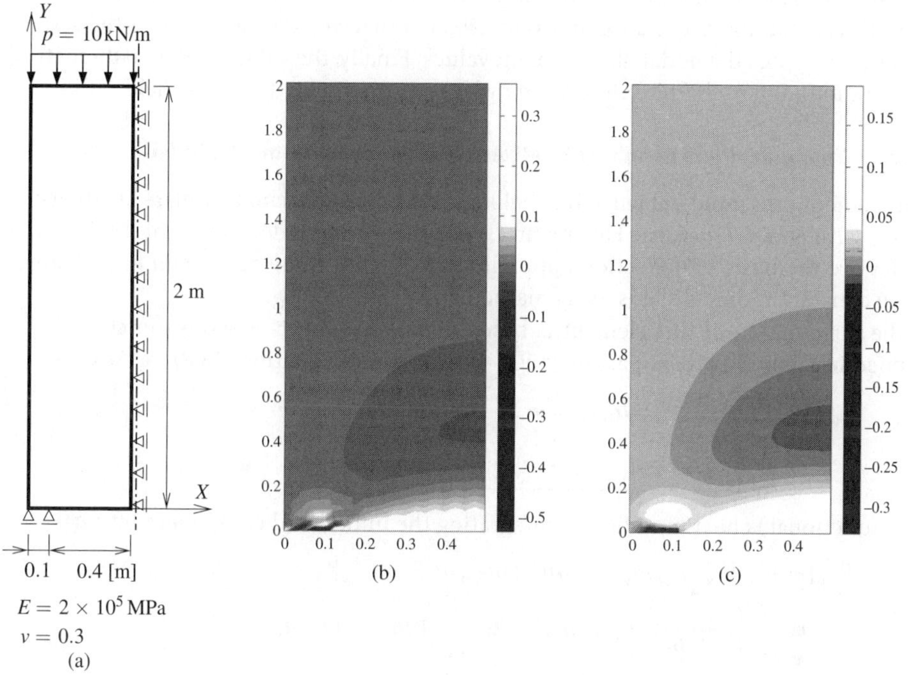

Figure 18.4 Deep beam: (a) configuration, contour maps of membrane force n_x obtained using two FEs based on the following functionals: (b) $I_{\text{H-R}}$ and (c) $I_{\text{L-L}}$ ($\alpha = 1.5$)

where the respective physical dof vectors are: $\mathbf{q}_{u(8\times1)}$ and $\mathbf{q}_{n(12\times1)}$. For a discretized membrane one obtains a superior solution in comparison with elements based on $I_{\text{H-R}}$ functional itself. This is visible in smooth contour maps of membrane forces, which for pure displacement-stress model exhibit oscillations.

In Figure 18.4 the results from Wosatko (1999) are quoted, which show the contour maps of force $n_x = h\sigma_x$ for a deep beam loaded at the top edge, obtained using the two considered approaches based on functionals $I_{\text{H-R}}$ and $I_{\text{L-L}}$, respectively.

18.2.3 Rectangular Membrane FE – Hybrid Stress Model

The rectangular hybrid stress membrane element is formulated on the basis of modified complementary energy functional $I_{\text{c,m}}[\boldsymbol{\sigma}, \mathbf{u}^{(ef)}]$, see Bathe (1982)

$$I_{\text{c,m}}[\boldsymbol{\sigma}, \mathbf{u}^{(ef)}] = I_{\text{c}}[\boldsymbol{\sigma}] + G^{(ef)}[\boldsymbol{\sigma}, \mathbf{u}^{(ef)}] = I_{\text{c,m}}(\boldsymbol{\alpha}_{\sigma}, \mathbf{q}_u^{(ef)}) \tag{18.7}$$

One approximates: (i) the stress field in the element domain using mathematical degrees of freedom $\boldsymbol{\sigma}_{(3\times1)} = \mathbf{P}\boldsymbol{\alpha}_{\sigma(7\times1)}$, (ii) the displacement field on the element boundary

$\mathbf{u}^{(ef)}_{(2\times1)} = \mathbf{N}\mathbf{q}^{(ef)}_{u(8\times1)}$ with physical displacement dofs on element boundary represented by four corner nodes. The additional component $G^{(ef)}[\boldsymbol{\sigma}, \mathbf{u}^{(ef)}]$ of the complementary energy $I_c[\boldsymbol{\sigma}]$ is included in order to guarantee the equilibrium of tractions $\mathbf{t}^{(ef)}$ using Lagrange multipliers that have the interpretation of displacements $\mathbf{u}^{(ef)}$ on interelement boundaries $\mathrm{d}\Omega^{(ef)}$. We emphasize that the approximation both in the element domain and on its boundary is characteristic for hybrid models.

The matrix equation for the finite element contains four submatrices (including one zero diagonal block). It is possible to condense out the stress mathematical dofs at the FE level, thus obtaining a pseudo-stiffness matrix. In the next step one performs the assembly and computes the nodal displacement values. Finally, the values of the mathematical dofs are found in each finite element in order to approximate the stress field.

18.2.4 Triangular Plate Bending FE – Mixed (Displacement-Moment) Model

The element presented in Figure 18.5 belongs to a group of triangular plate bending elements with $NEDOF = 6$, see Herrmann (1967). The physical dofs $\mathbf{q}_{w(3\times1)}$ at three corner nodes are the basis of deflection approximation. In the element domain one assumes three constant moment fields using mathematical dofs $\boldsymbol{\alpha}_{m(3\times1)}$.

The formulation of this element is based on the Hellinger–Reissner functional $I^{\mathrm{I}}_{\text{H-R}}$ defined in Equation (18.3), part of which is rewritten in Equation (18.8)

$$I^{\mathrm{I}}_{\text{H-R}}[w, w_{,xx}, w_{,yy}, w_{,xy}, m_x, m_y, m_{xy}] = \int_\Omega \left\{ \ldots + \underline{[-m_x\, w_{,xx} - m_y\, w_{,yy} - 2m_{xy}\, w_{,xy}]} + \ldots \right\} \mathrm{d}\Omega \tag{18.8}$$

This functional is next modified by integrating the underlined component by parts

$$\begin{aligned} I^{\mathrm{II}}_{\text{H-R}}&[w, w_{,x}, w_{,y}, m_x, m_y, m_{xy}, m_{x,x}, m_{y,y}, m_{xy,x}, m_{xy,y}] \\ &= \int_\Omega \left\{ -\frac{6}{E\,h^3}[(m_x + m_y)^2 + 2(1+\nu)(m_{xy}^2 - m_x\, m_y)] \right. \\ &\qquad \left. + \underline{(m_{x,x} + m_{xy,y})\, w_{,x} + (m_{y,y} + m_{xy,x})\, w_{,y}} - \hat{p}_z\, w \right\} \mathrm{d}\Omega \\ &+ \int_{\partial\Omega_\sigma} [-\hat{t}_\nu\, w + (\hat{m}_\nu - m_\nu)\, w_{,\nu} + (\hat{m}_{\nu s} - m_{\nu s})\, w_{,s}]\, \mathrm{d}(\partial\Omega) \\ &+ \int_{\partial\Omega_u} [-t_\nu\, (w - \hat{w}) - m_\nu\, \hat{w}_{,\nu} - m_{\nu s}\, \hat{w}_{,s}]\, \mathrm{d}(\partial\Omega) - \underline{\int_{\partial\Omega^{(ef)}} m_{\nu s}\, w_{,s}\, \mathrm{d}(\partial\Omega)} \end{aligned} \tag{18.9}$$

The use of three constant moment fields results in the cancellation of the surface integrals that contain first moment derivatives. As is visible in the second underlined part of $I^{\mathrm{II}}_{\text{H-R}}$ the component coupling the approximation of the deflection and moment is an integral over the element boundary. Moreover, due to integration by parts, the

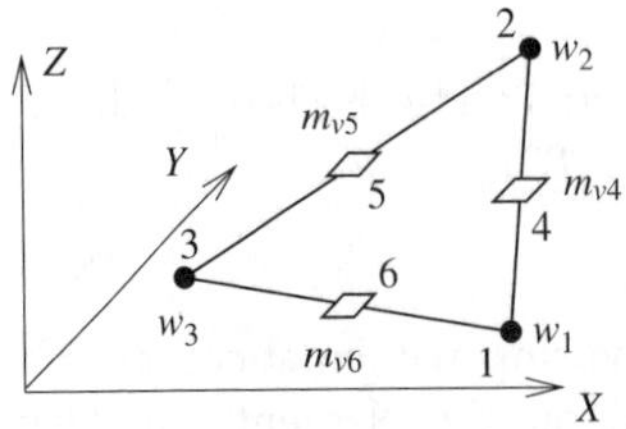

Figure 18.5 Two-field triangular plate FE

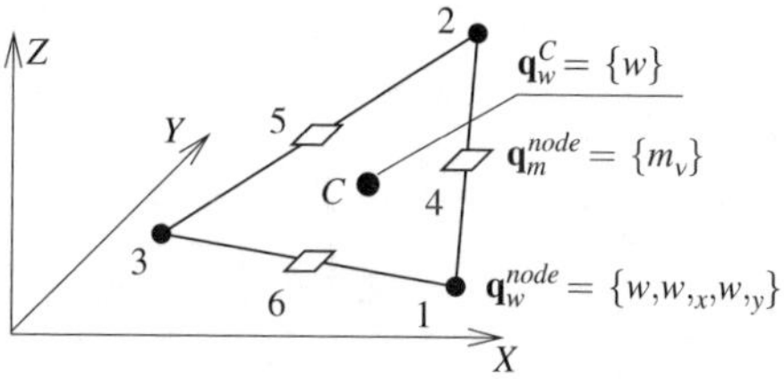

Figure 18.6 Hybrid displacement triangular plate FE

contour integrals related to static and kinematic boundary conditions changed. The transformation between the three moments in the domain and two (normal and twisting) moments on the boundary is used. Then the constant moment field in the domain can be represented by three normal moments at three midpoints of the triangular element sides. Therefore, in Figure 18.5 and in the further derivation the vector of moment dofs $\mathbf{q}_{m(3\times1)}$ occurs. The described FE satisfies the algebraic requirements for mixed formulations since $\dim \mathbf{q}_m^e = \dim \mathbf{q}_w^e = 3$.

At this stage we list other triangular plate bending elements for which $NEDOF = 6$. These are: (i) elements based on the equilibrium stress model, that is on the complementary energy functional I_c with constant moment fields, described in Fraeijs De Veubeke and Sander (1968), Morley (1968), (ii) displacement-based element, derived from the potential energy functional I_p, see Morley (1971) and (iii) element HSM6 based on the hybrid stress model and on the modified complementary energy functional $I_{c,m}$, see Batoz and Dhatt (1990).

18.2.5 Triangular Plate Bending FE – Hybrid Displacement Model

The hybrid displacement K–L plate bending element described in Harvey and Kelsey (1971), see Figure 18.6, is derived from the potential energy $I_{\mathrm{p,m}}[\mathbf{u}, \mathbf{t}^{(ef)}]$, modified by adding the component $H^{(ef)}[\mathbf{u}, \mathbf{t}^{(ef)}]$

$$I_{\mathrm{p,m}}[\mathbf{u}, \mathbf{t}^{(ef)}] = I_{\mathrm{p}}[\mathbf{u}] + H^{(ef)}[\mathbf{u}, \mathbf{t}^{(ef)}] = I_{\mathrm{p,m}}(\mathbf{q}_w, \mathbf{q}_m^{(ef)}) \tag{18.10}$$

In this displacement-based model the deflection field is approximated and the basis is composed of dofs $\mathbf{q}_{w(10\times1)}$ at three corner nodes and an additional node at the centre. The component $H^{(ef)}[\mathbf{u}, \mathbf{t}^{(ef)}]$ is introduced in order to enforce the continuity of the rotation angle $w_{,v}$ in the plane normal to the element edge. The hybrid formulation is obtained using Lagrange multipliers interpreted as interelement tractions $\mathbf{t}^{(ef)}$, which are three moments m_v at side midpoints, forming vector $\mathbf{q}_{m(3\times1)}^{(ef)}$. The element passes the patch test for hybrid formulations (Zienkiewicz and Taylor 2005). The moment dofs are condensed out at the element level and after the assembly of the pseudo-stiffness matrices the generalized displacements at the nodes are computed. Finally, the deflection field in the element domain is approximated.

18.2.6 Triangular Plate Bending FE – Hybrid Stress Model

The element in Figure 18.7, called HSM9 by Batoz et al. (1980), is also described in Batoz and Dhatt (1990). The following modified complementary energy functional $I_{\mathrm{c,m}}[\boldsymbol{\sigma}, \mathbf{u}^{(ef)}]$ is used to formulate the hybrid moment plate bending element

$$\begin{aligned} I_{\mathrm{c,m}}[\boldsymbol{\sigma}, \mathbf{u}^{(ef)}] &= I_{\mathrm{c}}[\boldsymbol{\sigma}] + G^{(ef)}[\boldsymbol{\sigma}, \mathbf{u}^{(ef)}] \\ &= I_{\mathrm{c}}[\mathbf{m}] + G^{(ef)}[t_v, m_{vs}, m_v, w, w_{,s}, w_{,v}] = I_{\mathrm{c,m}}(\boldsymbol{\alpha}_m, \mathbf{q}_w^{(ef)}) \end{aligned} \tag{18.11}$$

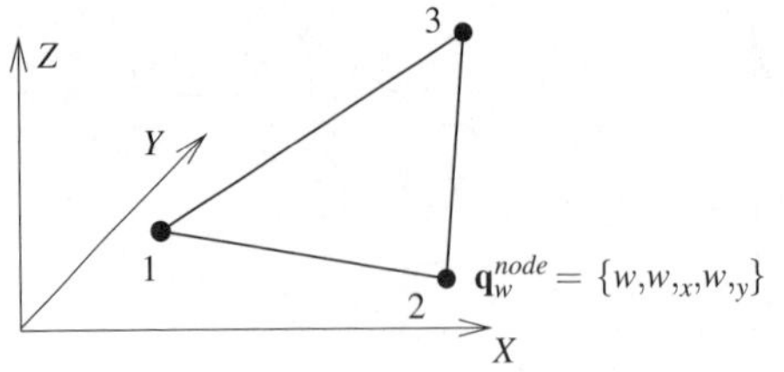

Figure 18.7 Hybrid stress three-node plate FE

In the element domain Ω^e three moment fields are approximated using mathematical dofs as $\mathbf{m}_{(3\times1)} = \mathbf{P}\boldsymbol{\alpha}_{m(9\times1)}$. In this hybrid element the modification of the complementary energy by adding the component $G^{(ef)}[\boldsymbol{\sigma}, \mathbf{u}^{(ef)}]$ serves the purpose of imposing the equilibrium conditions for the transverse shear force t_ν, the twisting moment $m_{\nu s}$ and the normal moment m_ν along the element edges. The displacement functions $w(s), w_{,s}(s), w_{,\nu}(s)$ play the role of Lagrange multipliers and are constructed using vector $\mathbf{q}^{(ef)}_{w(9\times1)}$ that contains three dofs at each corner node. The starting point of the formulation of element equations is a suitable form of the functional I_c^{III} obtained after integrating $I_c^{\text{I}} = I_c$ from Equation (18.2) by parts twice

$$
\begin{aligned}
&I_c^{\text{III}}[w, m_x, m_y, m_{xy}, m_{x,xx}, m_{y,yy}, m_{xy,xy}] \\
&= \int_\Omega \{\ldots + [-m_{x,xx} - 2m_{xy,xy} - m_{y,yy}]\, w + \ldots\}\, \mathrm{d}\Omega + \ldots
\end{aligned} \tag{18.12}
$$

In the integrand the plate equilibrium equation occurs, which is expressed using second partial derivatives of the moments, see Equation (8.18). This equation is satisfied in the case of linear moment approximation in element HSM9. In the derivation of vectors and matrices for this element the relevant formulae and transformation, known from the plate theory, are used to express: (i) the generalized edge forces $t_\nu, m_{\nu s}, m_\nu$ on the basis of moment approximation in the element domain Ω^e and (ii) the deflection derivatives $w_{,\nu}, w_{,s}$ on the edges $\partial\Omega^{(ef)}$. The element satisfies the condition $n^m_{DOMAIN} \geq n^w_{BOUNDARY} - 3$, valid for hybrid formulations (Zienkiewicz and Taylor 2005), where the number of internal parameters $n^m_{DOMAIN} = 9$ must not be smaller than the number of displacement parameters on element boundary $n^w_{BOUNDARY} = 9$ minus three rigid-body modes. Owing to this the singularity of the final matrix is avoided.

At the element level the condensation of the mathematical dofs is performed. When the pseudo-stiffness matrices are assembled and the generalized nodal displacements are computed for the structural model, one can approximate the moment field within this hybrid stress-based element.

18.2.7 Nine-Node Thin Plate and Shell FE – Mixed Two-Field Model

Plate/shell mixed finite element (displacement-strain formulation) is based on the modified Hellinger–Reissner principle (Lee et al. 1985). The description of the geometry and deformation is at first consistent with the degenerated solid approach. The displacement field is interpolated between nine nodes using five physical dofs per node $\mathbf{q}_u$. The description of bending strains is derived using the kinematic equations and the displacement approximation. However, the membrane and transverse shear strains are approximated independently using the mathematical dofs $\boldsymbol{\alpha}^n_{(14\times1)}$ and $\boldsymbol{\alpha}^t_{(10\times1)}$, respectively. To compute the element matrices and vectors for this nine-node element 3×3 Gauss NI quadrature is employed.

The matrix set of equations provides relations between the mathematical and physical dofs $\boldsymbol{\alpha}^n(\mathbf{q}_u)$ and $\boldsymbol{\alpha}^t(\mathbf{q}_u)$. Next the mathematical dofs are eliminated and the pseudo-stiffness operator is obtained. When the generalized displacements have been computed for the structural model, one returns to the element level to determine the displacement field and bending strains from the kinematic relations. Finally, the membrane and transverse shear strains are calculated using the element mathematical dofs.

In the paper cited previously (Lee et al. 1985), satisfactory results of several benchmark problems for structures of different slenderness are presented. In particular, the following cases are considered: (i) square plate with $L/h = 10^2, 10^3, 10^5$, (ii) circular plate with $2R/h = 10^2, 10^3, 10^4, 10^5$ and (iii) pinched cylindrical shell with two support variants (end diaphragm or clamped edge) with dimension ratios $R/h = 100, 300, 500$.

To sum up Section 18.2 devoted to selected two-field and hybrid finite elements, we emphasize the fact that a large number of new elements are now available that can be used to discretize thin/moderately thick plates/shells in a nonclassical (other than one-field displacement-based) manner. The new finite elements are always examined with respect to quality and robustness by selected patch tests and benchmarks, see for instance Zienkiewicz et al. (1986) and Zienkiewicz and Taylor (2005).

18.3 Advanced FEs

In this section we discuss advanced finite element formulations that make it possible to solve more and more complicated problems of linear analysis. Moreover, some of these concepts are used in the analysis of geometrical and/or material nonlinearities, for which the accuracy and efficiency play an essential role in repeated computations carried out according to an incremental-iterative algorithm. We mention here that a set of benchmarks for geometrically nonlinear analysis of shells is for instance given in Sze et al. (2004). The information provided next extends the discussion presented in Chapters 5, 16 and 17 and in the previous sections of the current chapter.

18.3.1 Enhanced Degenerated FEs

The essential features of degenerated shell finite elements have been discussed in Subsection 5.2.6, see Ahmad et al. (1970), Kanok-Nukulchai (1979) and Parisch (1979). Many advanced models use the idea of continuum degeneration for the geometry and deformation description, but the latter is further enriched.

In degenerated plate/shell FEs (four-, eight-, nine-node), employed for the discretization of both thin and moderately thick shells, transverse shearing is also considered beside the membrane and bending deformation. However, these elements, without additional numerical enhancements, are characterized by excessive stiffness when very thin plates or shells are considered, and also unsatisfactory representation of bending-shearing and membrane-bending couplings. Several methods to improve the results obtained using degenerated FEs have been developed. First, the use of uniformly or selectively reduced integration (URI, SRI) is reported in the literature since lower order of numerical quadrature usually reduces the negative effects. Moreover, two interesting papers by Belytschko et al. (1985a,b) are recalled, in which a stabilization technique is applied. The technique combines the decomposition of mechanical effects, projection of stresses to a subspace and additional use of a stabilization matrix with URI.

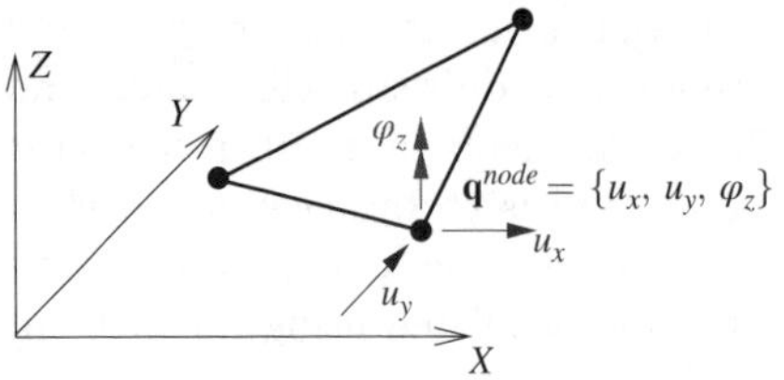

Figure 18.8 Displacement-based three-node membrane FE with three dofs per node

In Subsections 18.2.7, 18.3.8 and 18.3.9 we show that the formulation of degenerated finite elements has also been a starting point of more advanced techniques with non-standard strain interpolation, for example in the following elements SHEL9 (Lee et al. 1985), SHELM9 (Chang et al. 1989) and also EAS4-ANS and EAS7-ANS (Andelfinger and Ramm 1993).

18.3.2 Drilling Rotations in Discretization of Membranes and Shells

It is sometimes necessary to enrich the displacement field in the middle plane of a membrane or in the middle surface of a shell. It is not trivial to consider the angle of rotation around a normal to the middle surface as an independent degree of freedom of the membrane state and hence a couple of formulations of this problem have appeared. One of them, shown in Figure 18.8, is authored by Allman (1984, 1988) who uses nonstandard interpolation and 'artificial' rotations.

In the Sanders shell theory, the rotation φ_n about the normal depends on the displacements and their derivatives, see Equation (1.44). Therefore, in the derivation of equilibrium equations from the principle of virtual work the sixth equation does not appear (see Box 3.2 in Chapter 3). However, in finite elements the third rotation at a node φ_n and the associated moment M_n occur. If they are not properly described by a mechanical theory then after the assembly, which involves dof transformations from the local coordinate set to global directions, the sixth dof can result either in a singularity or in improper rotational stiffness magnitudes related to the drilling rotations, see Zienkiewicz and Taylor (2005). Different manners of incorporation of the drilling rotation field and respective dofs in the approximation of the membrane state can be found in Bergan and Felippa (1985), Frey (1989) and Hughes and Brezzi (1989). The use of drilling dofs improves the results of the analysis of membrane states (this can be observed for instance in cantilever beam or Cook's panel). It also enables the analysis of folded plate structures (e.g. Z-shaped thin-walled cantilever beam), multi-shell structures, shell connections to beams or stiffeners and also of slab-column connections (e.g. rectangular membrane or slab supported by columns).

Allman's triangular membrane element is combined with the triangular plate element in order to be able to analyse shells of arbitrary shape, discretized using flat finite elements, see the well-known benchmarks Scordelis–Loo roof and hemispherical shell with or without a hole, described in Section 17.6.

18.3.3 Triangular and Quadrilateral FEs with Discrete Kirchhoff Constraints – DKT, DKQ, SEMILOOF

In plate elements formulated according to the more general Mindlin–Reissner theory independent approximation of three fields, the deflection and two rotations, is applied. The fields are implicitly coupled by the kinematic relations for transverse shear strains.

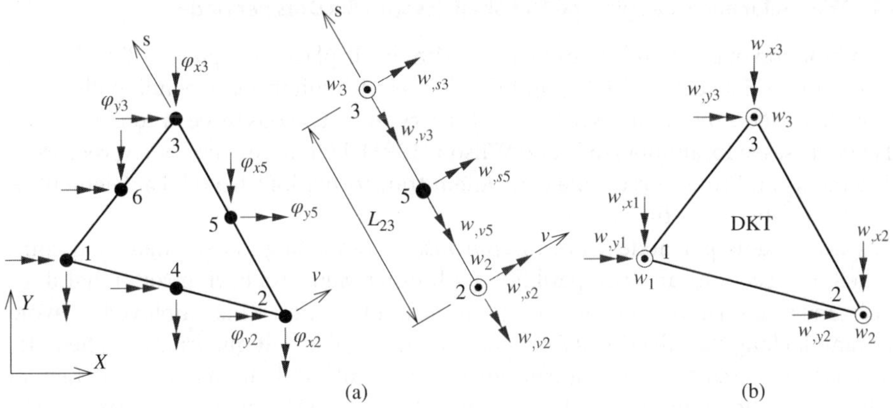

Figure 18.9 The initial (a) and final (b) DKT finite element with reduction of dofs

$$\bullet\!-\quad \mathbf{q}_i = \{u, v, w\}_i, \quad i = 1,\dots, 8$$

$$+\!-\quad \mathbf{q}_j = \{\varphi_v, \varphi_s\}_j, \quad j = 1,\dots, 8$$

$$\bullet\!-\quad \mathbf{q}_C = \{w, \varphi_x, \varphi_y\}_C$$

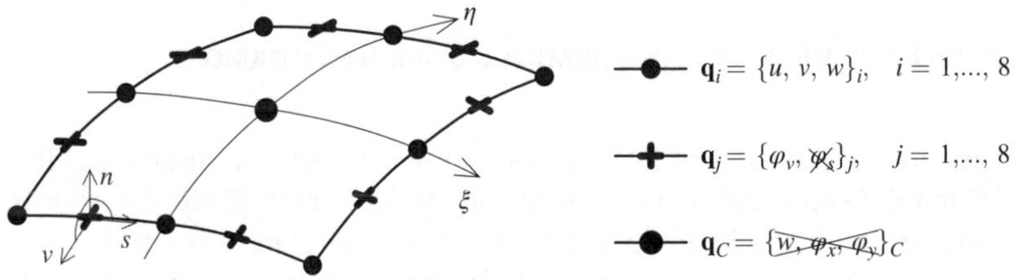

Figure 18.10 Semiloof shell FE with three types of nodes; crossed dofs are eliminated in the FE formulation

In the so-called Discrete Kirchhoff finite elements – either triangular (DKT) or quadrilateral (DKQ) – the fields are explicitly coupled by enforcing zero transverse shear strains at selected points. At these points the deflection and two rotational fields are related by the Discrete Kirchhoff (DK) constraints. Moreover, the constraints are used to reduce the number of nodal parameters. The DKT element (Figure 18.9) has at the beginning 15 dofs and finally $3 \times 3 = 9$ dofs at the three corner nodes and is regarded as a robust three-node plate element. The details of the formulation can be found in Batoz and Tahar (1982), Batoz et al. (1980) and Cook et al. (1989).

It is worthwhile to describe briefly the so-called SEMILOOF element that belongs to the second generation of degenerated plate/shell elements with DK constraints, see Irons (1976) and Figure 18.10. In this formulation different shape function families have been used for the approximation of the geometry, the translations and rotations. A specific feature of the element is constituted by three node types: four corner nodes and four side centre nodes have three displacement dofs each (u, v, w), eight so-called Loof nodes have two rotational dofs each (φ_v, φ_s), and the central node has three dofs $(w, \varphi_x, \varphi_y)$. The Semiloof element requires the condensation of degrees of freedom from 43 to 32, via a modification of the shape functions using the DK constraint equations. The constraints are a consequence of the Kirchhoff hypothesis concerning zero transverse shear and are imposed at a finite number of points. In Figure 18.10 the dofs finally used are the ones that are not crossed out.

18.3.4 Triangular and Quadrilateral Flat Shell FEs with Six Dofs per Node

The combination of the membrane elements with the displacement plus drilling degrees of freedom (see Subsection 18.3.2) and the discrete Kirchhoff plate bending elements DKT or DKQ (see Subsection 18.3.3) provides very advantageous three or four-node flat shell elements, see Ibrahimbegovic and Wilson (1991). Full three-component vectors of displacements and rotations enable convenient transformations from local (element or node) coordinate sets to the global one.

FEs with six dofs per node serve the purpose of modelling folded shells and complex shell intersections and compatibility with other elements having rotational dofs. In the numerical formulation, reduced numerical integration RI is employed to avoid membrane locking (see Section 17.4). Both the triangular and quadrilateral shell elements pass the patch test for uniform tension and uniform bending. The accuracy of this element is shown by Ibrahimbegovic and Wilson (1991) for two important tests: the cantilever beam (checking the membrane state) and a hemispherical shell with a hole (complex deformation state).

18.3.5 Discrete Kirchhoff–Mindlin Triangle DKMT and Quadrilateral DKMQ for Plates

In the plate elements with acronyms DKMT/DKMQ – Discrete Kirchhoff–Mindlin Triangle/Quadrilateral (Katili 1993) a combination of K–L and M–R theories is employed to represent both the bending and transverse shear states. The transverse shear strains, denoted by $\boldsymbol{\gamma}(w, \boldsymbol{\varphi})$ are computed from kinematic equations as in M–R theory with the approximation of deflections w and rotations $\boldsymbol{\varphi}$. However, an additional interpolation of the transverse shear strain field, denoted by $\overline{\boldsymbol{\gamma}}$ and constant along the element edges, is also used. The model is rooted in the modified potential energy functional for thick plates, taking into account both bending and transverse shear states

$$I_{\mathrm{p,m}} = U^m[\boldsymbol{\varphi}] + U^t[w, \boldsymbol{\varphi}, \overline{\boldsymbol{\gamma}}, \mathbf{T}] - W \tag{18.13}$$

In order to enforce the condition of equality of transverse shear strains $\boldsymbol{\gamma} = \overline{\boldsymbol{\gamma}}$ (defined in two ways described previously) in a variational manner Lagrange multipliers $\mathbf{T}$ (transverse shear forces) are introduced. The equations are written for the central points of element sides assuming constant transverse shear forces along element edges.

In the case of thin plates the effect of transverse shear is automatically reduced. In the opinion of Katili (1993) the new DKMT and DKMQ elements have beneficial properties: they pass the patch test for thin and thick plates (constant curvature test, constant transverse shear deformation test), give good results for typical bending plate tests, exhibit no spurious zero-energy modes, no shear locking for thin plates, are relatively insensitive to geometric distortion and have good mesh convergence characteristics. Therefore, they are considered as computationally efficient.

18.3.6 Continuum-based Resultant Shell (CBRS) FEs

The formulation of this approach starts from the Continuum-Based Theory (CBT-3D) and leads to a Continuum-Based Resultant Shell Theory (CBRST-2D) using shell kinematic and static hypotheses as well as generalized strains and resultant forces. A thorough description of the CBRS element family is contained in Stanley et al. (1986).

The advantageous properties of this element are provided by the following concepts: z-dependence is incorporated in the definitions of stress resultants, strains and constitutive equations, an improved description of strains related to natural coordinates is used. Consequently, locking and spurious mechanism are avoided and the FEs are not sensitive to irregular element shape.

Example applications are presented in the quoted work: (i) analysis of thin pinched cylinder and (ii) thin pinched hemisphere, see Subsection 17.6. In fact, they are the most meaningful validation tests for thin shell analysis. In the first test good convergence and distortion insensitivity are proven. In the second example the mesh compatible with principal spherical coordinate lines and the use of nine-node FEs whose natural coordinates are aligned with the directions of principal curvatures result in a rapid convergence. According to Stanley et al. (1986) the successful application of this formulation to a nonlinear analysis is possible, because this FE is accurate, reliable and robust.

Another broadly applied and general formulation is proposed by Simo et al. (1989). The concept is not derived from the geometrical approximation of a three-dimensional body, but starts from an exact kinematic description of so-called two-dimensional Cosserat surface (Bischoff et al. 2004). This model contains a geometrically exact shell representation, analytical integration over the thickness and appropriate stress resultant definitions. The model is equally good for linear and nonlinear applications.

18.3.7 FEs based on Advanced Formulations of Shell Models

This approach was used to formulate a couple of shell elements, known in the literature under the names MIT4, MIT6, MIT9 (Mixed Interpolation of Tensor components). We focus on the eight-node shell FE – MITC8 described in Bathe and Dvorkin (1986). The formulation employs a convected coordinate system and covariant transverse shear strain components. The displacements are interpolated as in the degenerate isoparametric elements. A mixed interpolation of strain tensor components is used, see Figure 18.11. The membrane and bending strains are calculated from the displacement interpolation. The transverse shear strains are interpolated in two steps: the strain

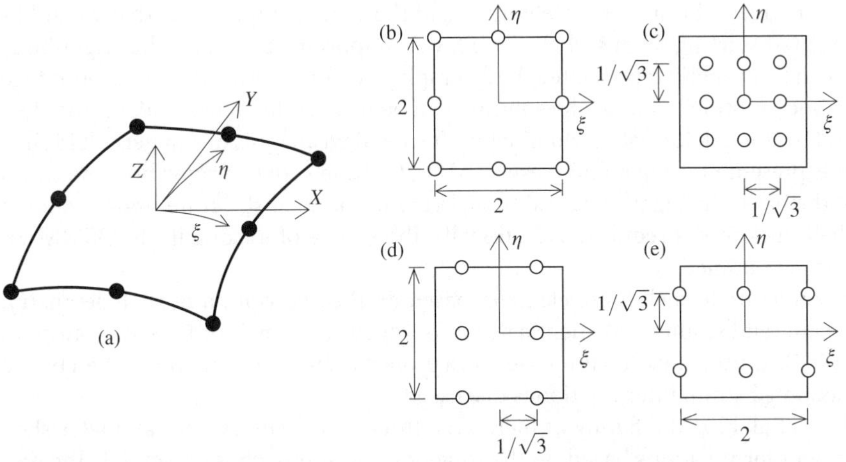

Figure 18.11 (a) FE with curvilinear internal coordinate system; positions of points used for interpolation of: (b) displacements and rotations, (c) in-surface strains and (d), (e) two transverse shear strains, respectively

components are first evaluated at certain basis points directly from the displacement interpolation, next new functions and new points are used for further approximation. Such interpolation makes it possible to avoid membrane and transverse shear locking.

The general concept of using mixed interpolation of tensorial strain components is the basis of widely applicable shell elements. In Bathe and Dvorkin (1986), the results of a set of linear analysis tests are provided: membrane with a hole, curved cantilever, square plate, pinched cylindrical shell and Scordelis–Loo roof. It has also been shown that various tests for patches of distorted elements are passed.

In order to analyse both thin and moderately thick plates another approach called Covariant Shear Strains Interpolation (CSSI) has been worked out, see Militello and Cascales (1987). It is in some sense similar to the MITC formulation. Special attention is given to the approximation of the transverse shear strains, based on the M–R theory for plate bending. The approximation of the covariant components of shear deformation is performed using lower-order polynomials than what would result from kinematic equations. Twelve selected interpolation points, shown in Figure 18.11d,e, are used.

Owing to this theoretically and numerically complex formulation, locking and spurious modes are avoided. In Militello and Cascales (1987) the results of two patch and one-element tests are presented. For the well-known rectangular configuration with five nine-node elements (see Section 17.4.7 and Figure 17.5c), applying FEs with straight or curved sides, the constant strain component fields for bending and transverse shear have been reproduced exactly. Moreover, the results for one-element test of a plate under twisting are given.

A similar formulation of shell FEs is called Basic Shell Mathematical Model (BSMM), see Lee and Bathe (2005). It is based on a shell theory combining membrane, bending and transverse shear states. When MITC9 element based on BSMM is constructed, the interpolation of the geometry and deformation is performed, whereby the fundamental forms and metric tensor of the shell middle surface and the Christoffel symbols are used in the mathematical model. Each of the strain parts (membrane, bending, transverse shear) is calculated from the kinematic shell relations expressed in terms of displacements, rotations and their derivatives. This FE can be used to discretize general shell structures, both moderately thick and thin. The authors show that when FEs are formulated starting from a shell theory, it is important to ensure that rigid-body motions can be properly represented. If the employed interpolation does not reproduce the rigid-body motions exactly, the solution can be inaccurate, in particular if the shell is thin and bending deformation dominates. The detailed analysis of a hyperboloid shell problem is presented using a fine mesh of MITC9 FEs in order to show how the strain terms of the basic shell mathematical model vary in the shell structure under bending as the shell thickness is reduced. Additionally, the source of locking in the BSMM FE simulation is examined.

Another nine-node shell finite element, based on the description of shell geometry, displacements and strains in element natural coordinate system, is proposed in Jang and Pinsky (1987). The formulation is focused on a proper strain representation, whereby the term of assumed strain interpolation is discussed.

In Echter et al. (2013) a family of hierarchic three-, five- and seven-parameter shell finite element formulations based on the isogeometric approach is presented. The discretization employs NURBS (Non-Uniform Rational B-Splines) as the shape functions, see Hughes et al. (2005). The hierarchic shell models offer the possibility to enrich the kinematics of the shell taking into account transverse shear deformations and higher order displacement approximation in the thickness direction. To avoid the transverse shear locking and so-called curvature thickness locking (not discussed in this book) a hierarchic difference vector concept is applied, which augments the K–L and M–R kinematics. This generates a group of new isogeometric FEs called 3p, 5p-hier and 7p-hier.

Two different options to avoid membrane locking are considered: the Discrete Strain Gap (DSG) concept and a Hybrid Stress (HS) formulation based on a two-field Hellinger–Reissner principle. This way, two groups of more advanced FEs are proposed with abbreviated names as follows: 3p-DSG, 5p-hier-DSG, 7p-hier-DSG and 3p-HS, 5p-hier-HS, 7p-hier-HS. To show the accuracy and robustness of these elements the following examples are analysed in Echter et al. (2013): simply supported plate, cylindrical shell strip and Scordelis–Lo roof.

18.3.8 Assumed Natural Strain (ANS) Approach

We describe here the nine-node isoparametric shell element SHELM9, formulated in Chang et al. (1989). First this FE is treated as a displacement-based degenerated element, but next, introducing nonstandard strain interpolation technique called the Assumed Natural Strain (ANS), its model can be classified as mixed displacement-strain. According to the five-parameter M–R theory three translation and two rotation fields are treated as independent. The physical dofs $\mathbf{q}_u$ are used to approximate both the generalized displacement fields and the membrane, bending and transverse shear strains derived from the kinematic relations $\boldsymbol{\varepsilon}_u = \mathbf{B}\mathbf{q}_u$. In the ANS approach an additional, independent strain approximation is employed $\boldsymbol{\varepsilon}_\alpha = \mathbf{P}\boldsymbol{\alpha}_\varepsilon$, where $\boldsymbol{\alpha}_\varepsilon$ are mathematical dofs. The generalized resultant forces are calculated from the assumed strains $\boldsymbol{\epsilon}_\alpha$ using the constitutive equations $\mathbf{s} = \mathbf{C}\boldsymbol{\epsilon}_\alpha$.

The variational formulation is based on the modified three-field Hu–Washizu functional $I_{\text{H-W}}[\mathbf{u}, \boldsymbol{\varepsilon}, \mathbf{s}]$, transformed into a modified Hellinger–Reissner functional

$$I_{\text{H-W}}[\mathbf{u}, \boldsymbol{\varepsilon}, \mathbf{s}] \quad \rightarrow \quad I_{\text{H-R,m}}[\mathbf{u}, \boldsymbol{\varepsilon}_u, \boldsymbol{\varepsilon}_\alpha] = I_{\text{H-R,m}}(\mathbf{q}_u, \boldsymbol{\alpha}_\varepsilon) \tag{18.14}$$

or, alternatively, into a modified potential energy functional

$$I_{\text{p,m}}[\mathbf{u}, \boldsymbol{\varepsilon}_u, \boldsymbol{\varepsilon}_\alpha, \mathbf{s}] = I_{\text{p}}[\mathbf{u}] + F[\boldsymbol{\varepsilon}_u, \boldsymbol{\varepsilon}_\alpha, \mathbf{s}] = I_{\text{p,m}}(\mathbf{q}_u, \boldsymbol{\alpha}_\varepsilon) \tag{18.15}$$

Imposing the strain equality $\boldsymbol{\epsilon}_u = \boldsymbol{\epsilon}_\alpha$ with Lagrange multipliers that are generalized forces $\mathbf{s}$, one can relate the mathematical dofs with the physical ones $\boldsymbol{\alpha}(\mathbf{q}_u)$. Next the mathematical dofs are eliminated and a pseudo-stiffness matrix is obtained that has the dimension equal to the number of displacement dofs.

The SHELM9 FEs have been tested analysing: Cook's panel, a rhombic thin plate, a cylindrical roof, a pinched cylinder and a spherical shell. The following conclusions have been formulated: (i) the FE is useful for both thin and moderately thick shells, (ii) no kinematic spurious deformation modes are exhibited by the FE, (iii) the FE is free from transverse shear and membrane locking, (iv) the FE is not sensitive to geometrical distortions and (v) the FE provides accurate resultant force fields.

18.3.9 Coupled Enhanced Assumed Strain (EAS) and Assumed Natural Strain (ANS) Techniques

The four-node plate/shell finite elements EAS4-ANS and EAS7-ANS, described in this subsection, combine several concepts: degeneration, enhanced assumed strain and assumed natural strain, see Andelfinger and Ramm (1993), hence they have a broader domain of application and improved properties. These plate or shell elements are based on the degenerated solid approach in which a continuous body (3D) is reformulated into a surface structure (2D) according to an approach called CBRST, using generalized strains and resultant forces (see Subsection 18.3.6).

In these elements the membrane, bending and transverse shear states are not coupled. To enhance the description of the behaviour of shell models two essential concepts are used: (i) the Enhanced Assumed Strain (EAS) approach for membrane and bending components and (ii) the Assumed Natural Strain (ANS) approach for the transverse shear components. The four- or seven-parameter approximation for the enhanced strain can be used as optimal choices for bilinear elements, that is the so-called EAS4-ANS and EAS7-ANS elements.

In the EAS concept the physical dofs for generalized displacements and strains are used, just as in the two-field (displacement-strain) mixed model. After condensation of the enhanced strain parameters membrane and bending pseudo-stiffness matrices are derived. In the transverse shear state a natural coordinate system is employed and two-stage ANS approximation of covariant strains using different basic sampling points and approximation functions is proposed. This way the transverse shear stiffness matrix is obtained.

Satisfactory numerical results are described in Andelfinger and Ramm (1993) for the following problems: (i) two-dimensional distortion test, (ii) eigenvalue analysis of different four-node plate elements and (iii) known benchmarks, that is Morley's skew plate and Scordelis–Lo roof. In second example it has been proven that for EAS7-ANS FEs, smooth and qualitatively correct distributions of membrane shear forces are obtained. The authors stress that the element is suitable for application in geometrically and physically nonlinear problems.

In the review by Yang et al. (2000), 34 papers on the enhanced strain approach are referenced. As an example of more recent developments we cite Chróścielewski and Witkowski (2006) where EAS elements are derived for the nonlinear six-field shell theory.

18.3.10 Automatic *hp*-Adaptive Methodology

As a final symptom of progress we provide a short summary of a paper by Tews and Rachowicz (2009) on 'Application of an automatic *hp* adaptive Finite Element Method for thin-walled structures', related to the book by Demkowicz et al. (2007). In the *hp*-adaptive finite element technology the following advanced concepts are used: evaluation of solution quality via error estimation and optimal improvement of discretization by (i) mesh size refinement (*h*-adaptivity) and (ii) increase of the order of approximation (*p*-adaptivity). In the process of computations sequences of meshes are generated, in which elements are subdivided on the basis of a comparison of the results obtained for the coarser and the denser mesh. The automatic selection of the size h and polynomial order p leads to a fast error reduction and hence an improved convergence to the exact solution.

The *hp*-adaptive method is a advanced optimization process, especially in the case of so-called goal-oriented adaptivity. This is particularly important when a solution of an complex engineering problem is to be found. The error minimization concerns the quantity, which is of particular importance to the analyst, called the quantity of

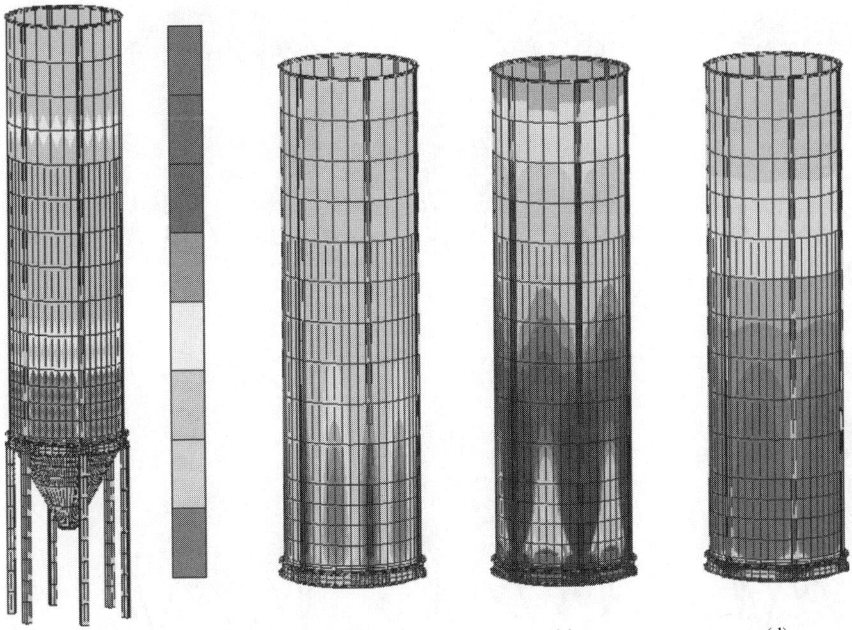

Figure 18.12 Silo structure: (a) fine *hp*-mesh; contour plots of: (b) radial displacements u_r, (c) azimuthal stresses $\sigma_{\varphi\varphi}$ and (d) vertical stresses σ_{zz}. Source: Tews and Rachowicz (2009). Reproduced with permission of Elsevier

interest by Tews and Rachowicz (2009). In the numerical analysis of engineering structures the maximum displacement or stress at a certain point are typical quantities of interest.

The described formulation provides a fully automatic computational strategy, makes use of advanced computer science technology and has a wide application potential since the approach is fully applicable to complex structures. In modelling it is possible to combine beams, plates, shells and solid blocks, consider openings and so on.

This section and the whole book is ended with a short description of finite element solutions of two engineering examples. These are complex shell-bar structures shown in Figures 18.12 and 18.13 and discussed in detail in Tews and Rachowicz (2009). Note that the original notation from that paper is used in the figure captions.

Example 1 is a silo structure. It is a combination of a cylindrical shell with a conical funnel ended with an opening at the bottom. Both shells are connected along a horizontal circular rim that is supported on six equidistant columns. The silo is loaded by a constant internal pressure. Using the goal-oriented strategy a solution satisfactory for a structural engineer (designer and constructor) and theoretical mechanics specialist has been obtained. This is because: (i) the complex joints of the component members (shells, stiffening ring, columns) have been modelled including all structural details and (ii) the large stress concentrations in certain problem subdomains have been approximated with high accuracy. The computations have been performed for both a representative segment of 1/6 of the configuration and the whole structure. In Figure 18.12 the final *hp*-refined mesh and the results in the form of contour maps for selected mechanical quantities

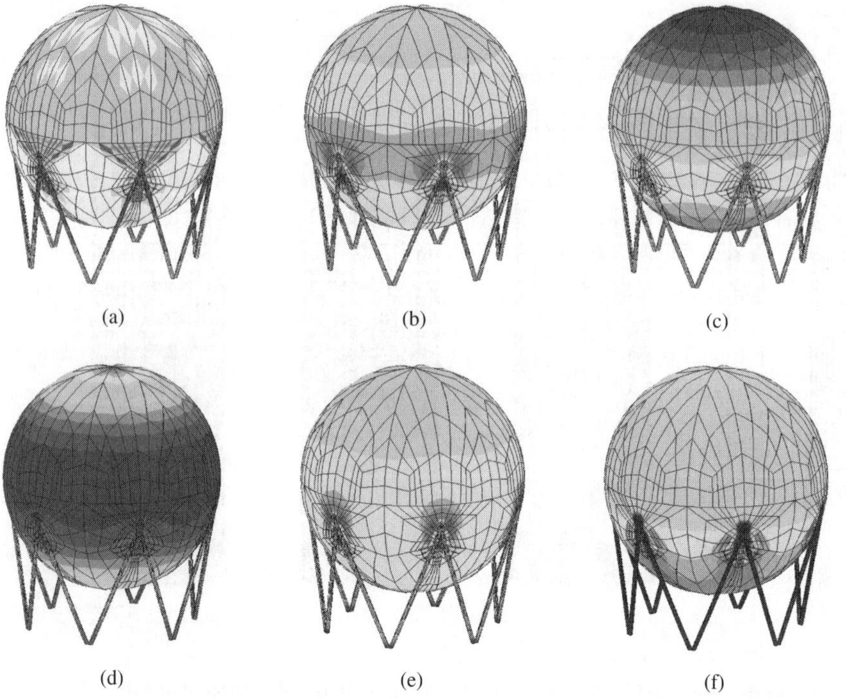

Figure 18.13 Spherical container: (a) fine *hp*-mesh; contour plots of: (b) radial displacements u_r in cylindrical coordinates, (c) and (d) radial u_ρ and meridional u_θ, (e) azimuthal stresses $\sigma_{\phi\phi}$ in spherical coordinates, (f) meridional stresses $\sigma_{\theta\theta}$. Source: Tews and Rachowicz (2009). Reproduced with permission of Elsevier

are shown, see Tews and Rachowicz (2009). The presentation is limited to a qualitative illustration of the structural behaviour.

Example 2 is a container structure, composed of a spherical shell and 12 equally distributed inclined columns whose axes are tangent to the middle surface of the shell. The container is loaded by internal hydrostatic pressure. It is stressed in Tews and Rachowicz (2009) that the geometrical modelling of intersections of the shell and the supporting columns has been performed using a special package for an implicit parametrization technique. The results of computations, namely the contour plots of selected displacement and stress fields, are presented in Figure 18.13. The final mesh obtained using *hp*-adaptivity is also shown. The spherical shell exhibits a quasi-membrane state since the columns have small flexural rigidities and the components of the resulting reaction forces, normal to the shell, are quite small. However, the results also confirm the occurrence of local bending effects at the shell-column joints.

It is emphasized that the high accuracy of the analysis is a result of the automatic adaptivity routine, in particular representing the cyclic symmetry of the adaptive mesh, consistent with the symmetry of the structure and the load, and the repeated local mesh refinement around the points of attachment of the supporting columns.

We also draw the attention of the reader to the fact that the problem of point loads and supports is complex from the theoretical, computational and practical viewpoints. Considering the analytical approach, one can refer for instance to the analysis of singularities in a plate under bending caused by a concentrated force or moment applied at an internal point, see Timoshenko and Woinowsky-Krieger (1959) and Łukasiewicz (1976).

This book has been devoted to the computational mechanics of plates and shells. We conclude this last chapter by quoting the opinion of Ramm and Wall (2004), which we share:

> 'Computational mechanics is more than just an accumulation of different building stones from disciplines like mathematics, material science, computer science, mechanics, scientific computing etc. ... Computational structural mechanics is also more than just the highest level of sophistication of models at the integration or material point level. Understanding the overall actual physical phenomena and the incorporation of this knowledge into the respective computational approaches is of vital importance for every serious undertaking in this field.'

This holds in particular for shell structures.

References

Ahmad S, Irons BM and Zienkiewicz OC 1970 Analysis of thick and thin shell structures by curved finite element. *International Journal for Numerical Methods in Engineering* **2**(3), 419–451.

Allman DJ 1984 A compatible triangular element including vertex rotations for plane elasticity analysis. *Computers & Structures* **19**(1-2), 1–8.

Allman DJ 1988 Evaluation of the constant strain triangle with drilling rotations. *International Journal for Numerical Methods in Engineering* **26**(12), 2645–2655.

Andelfinger U and Ramm E 1993 EAS-elements for two-dimensional, three-dimensional, plate and shell structures and their equivalence to HR-elements. *International Journal for Numerical Methods in Engineering* **36**(8), 1311–1337.

Bathe KJ 1982 *Finite Element Procedures in Engineering Analysis*. Prentice-Hall.

Bathe KJ and Dvorkin EN 1986 A formulation of general shell elements – the use of mixed interpolation of tensorial components. *International Journal for Numerical Methods in Engineering* **22**(3), 697–722.

Batoz JL and Dhatt G 1990 *Modélisation des Structures par Élément Finis*. Hermes, Paris.

Batoz JL and Tahar MB 1982 Evaluation of a new quadrilateral thin plate bending element. *International Journal for Numerical Methods in Engineering* **18**(11), 1655–1677.

Batoz JL, Bathe KL and Ho LW 1980 A study of three-node triangular plate bending elements. *International Journal for Numerical Methods in Engineering* **15**(12), 1771–1812.

Belytschko T, Liu WK, Ong JSJ and Lam D 1985a Implementation and application of a 9-node Lagrange shell element with spurious mode and control. *Computers & Structures* **20**(1-3), 121–128.

Belytschko T, Stolarski H, Liu WK, Carpenter N and Ong JSJ 1985b Stress projection for membrane and shear locking in shell finite elements. *Computer Methods in Applied Mechanics and Engineering* **51**(1-3), 221–258.

Bergan PG and Felippa CA 1985 A triangular membrane element with rotational degrees of freedom. *Computer Methods in Applied Mechanics and Engineering* **50**(1), 25–69.

Bischoff M, Wall WA, Bletzinger KU and Ramm E 2004 Models and finite elements for thin-walled structures. In Stein E, de Borst R and Hughes TJR (eds), *Encyclopedia of Computational Mechanics: Solids and Structures*, vol. **2**. John Wiley & Sons, Ltd, Chichester, UK, chapter 3, pp. 59–137.

CALFEM 2004 CALFEM – a finite element toolbox. Technical report, Structural Mechanics, Lund University, Lund.

Chang TY, Saleeb AF and Graf W 1989 On the mixed formulation of a 9-node lagrange shell element. *Computer Methods in Applied Mechanics and Engineering* **73**(3), 259–281.

Chróścielewski J and Witkowski W 2006 Four-node semi-EAS element in six-field nonlinear theory of shells. *International Journal for Numerical Methods in Engineering* **68**(11), 1137–1179.

Cook RD, Malkus DS and Plesha ME 1989 *Concepts and Applications of Finite Element Analysis* 3rd edn. John Wiley & Sons, New York.

Demkowicz L, Kurtz J, Pardo D, Paszyński M, Rachowicz W and Zdunek A 2007 *Computing with hp-Adaptive Finite Elements: Frontiers Three Dimensional Elliptic and Maxwell Problems with Applications* vol. 2 1st edn. Chapman & Hall/CRC.

Echter R, Oesterle B and Bischoff M 2013 A hierarchic family of isogeometric shell finite elements. *Computer Methods in Applied Mechanics and Engineering* **254**, 170–180.

Fraeijs De Veubeke B and Sander G 1968 An equilibrium model for plate bending. *International Journal of Solids and Structures* **4**(4), 447–468.

Frey F 1989 Shell finite elements with six degrees of freedom per node In *Analytical and Computational Models of Shell, ASME Winter Annual Meeting*, ed. Noor AK, Belytschko T and Simo JC (eds), vol. 3, pp. 291–315. ASME, San Francisco.

Gilewski W and Radwańska M 1991 A survey of finite element models for the analysis of moderately thick shells. *Finite Elements in Analysis and Design* **9**(1), 1–21.

Harvey JW and Kelsey S 1971 Triangular plate bending with enforced compatibility. *AIAA Journal* **9**(6), 1023–1026.

Herrmann LR 1967 Finite element bending analysis for plates. *Proc. ASCE J. Eng. Mech. Div.* **93**(EM5), 13–26.

Hinton E and Owen DR 1984 *Finite Element Software for Plates and Shells*. Pineridge Press.

Hughes TJR and Brezzi F 1989 On drilling degrees of freedom. *Computer Methods in Applied Mechanics and Engineering* **72**(1), 105–121.

Hughes TJR, Cottrell JA and Bazilevs Y 2005 Isogeometric analysis: CAD, finite elements, nurbs, exact geometry, and mesh refinement. *Computer Methods in Applied Mechanics and Engineering* **194**, 4135–4195.

Ibrahimbegovic A and Wilson EL 1991 A unified formulation for triangular and quadrilateral flat shell finite elements with six nodal degrees of freedom. *Communications in Applied Numerical Methods* 7(1), 1–9.

Irons W 1976 The semiloof shell element. In Ashwell DG and Gallagher RH (eds) *Finite Elements for Thin Shells and Curved Members*. John Wiley & Sons, Ltd, Chichester, UK. pp. 197–222.

Jang J and Pinsky PM 1987 An assumed covariant strain based 9-node shell element. *International Journal for Numerical Methods in Engineering* **24**(12), 2389–2411.

Kanok-Nukulchai W 1979 A simple and efficient finite element for general shell analysis. *International Journal for Numerical Methods in Engineering* **14**(2), 179–200.

Katili I 1993 A new discrete Kirchhoff-Mindlin element based on Mindlin–Reissner plate theory and assumed shear strain fields. Part I: An extended DKT element for thick-plate bending analysis, Part II: An extended DKQ element for thick-plate bending analysis.

International Journal for Numerical Methods in Engineering **36**(11), 1859–1883, 1885–1908.

Kleiber M 1985 *Finite Element Method in Nonlinear Continuum Mechanics.* PWN, Warsaw Poznan. (in Polish).

Lee PS and Bathe KJ 2005 Insight into finite element shell discretizations by use of the 'basic shell mathematical model'. *Computers & Structures* **83**(1), 69–90.

Lee SW, Wong SC and Rhiu JJ 1985 Study of a nine-node mixed formulation finite element for thin plates and shells. *Computers & Structures* **21**(6), 1325–1334.

Lee YJ and Lee BC 1990 A new mixed functional proposed for the analysis of linear elastic problems. *Computers & Structures* **37**(6), 1031–1035.

Łukasiewicz S 1976 *Concentrated loads in plates and shells.* PWN, Warsaw (in Polish).

MacNeal RH 1998 Perspective on finite element for shell analysis. *Finite Elements in Analysis and Design* **30**(3), 175–186.

Militello C and Cascales DH 1987 Covariant shear strains interpolation in a nine-node degenerated plate element. *Computers & Structures* **26**(5), 781–85.

Morley LSD 1968 Triangular equilibrium element in solution of plate bending problems. *Aeronautical Quarterly* **19**(2), 149–169.

Morley LSD 1971 The constant-moment plate-bending element. *Journal of Strain Analysis* **6**(1), 20–24.

Parisch H 1979 A critical survey of the 9-node degenerated shell element with special emphasis on thin shell application and reduced integration. *Computer Methods in Applied Mechanics and Engineering* **20**(3), 323–350.

Pian THH and Tong P 1969 Basis of finite element methods for solid continua. *International Journal for Numerical Methods in Engineering* **1**(1), 3–28.

Radwańska M 2006 A survey of various FE models with conceptual diagrams for linear analysis. *Computer Assisted Mechanics and Engineering Sciences* **13**(2), 209–233.

Radwańska M 2007 An overview of selected plate and shell FE models with graphic presentation of governing equations. *Computer Assisted Mechanics and Engineering Sciences* **14**(3), 431–456.

Ramm E and Wall WA 2004 Shell structures – a sensitive interrelation between physics and numerics. *International Journal for Numerical Methods in Engineering* **60**(1), 381–427.

Reddy JN 1986 *Applied Functional Analysis and Variational Methods in Engineering*. McGraw-Hill, New York.

Reddy JN 2005 *An Introduction to the Finite Element Method* 3rd edn. McGraw-Hill, New York.

Simo JC, Fox DD and Rifai MS 1989 On a stress resultant geometrically exact shell model. Part II: The linear theory; Computational aspects. *Computer Methods in Applied Mechanics and Engineering* **73**(1), 53–92.

Stanley GM, Park KC and Hughes TJR 1986 Continuum-based resultant shell elements. In Hughes TJR and Hinton E (eds) *Finite Element Methods for Plate and Shell Structures: Element Technology*. Pineridge Press. pp. 1–45.

Sze KY, Liu XH and Lo SH 2004 Popular benchmark problems for geometric nonlinear analysis of shells. *Finite Elements in Analysis and Design* **40**(11), 1551–1569.

Tews R and Rachowicz W 2009 Application of an automatic *hp* adaptive finite element method for thin-walled structures. *Computer Methods in Applied Mechanics and Engineering* **198**(21), 1967–1984.

Timoshenko S and Woinowsky-Krieger S 1959 *Theory of Plates and Shells.* McGraw-Hill, New York-Auckland.

Washizu K 1975 *Variational Methods in Elasticity and Plasticity* 2nd edn. Pergamon Press.
Wosatko A 1999 *Stress-displacement finite element models for static plane stress analysis* Master's thesis Cracow University of Technology. Supervisors: M. Radwańska M. and J. Pamin (in Polish).
Wosatko A and Radwańska M 2001 Stress-displacement finite elements for plane stress analysis. *Computer Methods in Civil Engineering* **No 5**, 21–44 (in Polish).
Yang H, Saigal S and Liaw DG 1990 Advances of thin shell finite elements and some applications – version I. *Computers & Structures* **35**(4), 481–504.
Yang HTY, Saigal S, Masud A and Kapania RK 2000 A survey of recent shell finite elements. *International Journal for Numerical Methods in Engineering* **47**(1–3), 101–127.
Zienkiewicz OC and Taylor RL 2005 *The Finite Element Method for Solid and Structural Mechanics* 6th edn. Elsevier Butterworth-Heinemann.
Zienkiewicz OC, Qu S, Taylor RL and Nakazawa S 1986 The patch test for mixed formulations. *International Journal for Numerical Methods in Engineering* **23**(10), 1873–1883.
Zienkiewicz OC, Taylor RL and Zhu JZ 2005 *The Finite Element Method: Its Basis and Fundamentals* 6th edn. Elsevier Butterworth-Heinemann.

A

List of Boxes with Equations

Plate and Shell Structures: Selected Analytical and Finite Element Solutions, First Edition.
Maria Radwańska, Anna Stankiewicz, Adam Wosatko and Jerzy Pamin.
© 2017 John Wiley & Sons Ltd. Published 2017 by John Wiley & Sons Ltd.

B

List of Boxes with Data and Results for Examples

Plate and Shell Structures: Selected Analytical and Finite Element Solutions, First Edition.
Maria Radwańska, Anna Stankiewicz, Adam Wosatko and Jerzy Pamin.
© 2017 John Wiley & Sons Ltd. Published 2017 by John Wiley & Sons Ltd.

Index

Plate and Shell Structures: Selected Analytical and Finite Element Solutions, First Edition.
Maria Radwańska, Anna Stankiewicz, Adam Wosatko and Jerzy Pamin.
© 2017 John Wiley & Sons Ltd. Published 2017 by John Wiley & Sons Ltd.

t